D1765352

Water Loss Control

About the Authors

JULIAN THORNTON is involved in numerous professional water-related activities. He specializes in water loss management and has practiced in many locations around the world, including the United States, Canada, the United Kingdom, Africa, the Middle East and Asia. He has been a member of the American Water Works Association's Water Loss Control Committee for the past 16 years and is currently chair of the subcommittee writing the third edition of the AWWA M36 publication *Water Audits and Loss Control Programs*. Mr. Thornton is also an active member of the International Water Association's Water Loss Task Force and is internationally recognized as an expert in pressure management as a water loss control technique.

REINHARD STURM has specialized in water loss management since the early stages of his professional career and is currently in charge of west coast operations for Water Systems Optimization, Inc. (WSO), a water loss management consulting firm. He has worked in many locations throughout the world, including Asia, Africa, Eastern Europe, and North America. Mr. Sturm served as coprincipal investigator for the research project "Leakage Management Technologies" sponsored by the American Water Works Association Research Foundation. During the 3 year effort that ranked this project as one of the largest water loss research projects ever conducted in the United States, Mr. Sturm was responsible for the evaluation of current leakage management technologies and their transferability to North America. Mr. Sturm is an active member of the American Water Works Association's Water Loss Control Committee and the International Water Association's Water Loss Task Force. He also serves as an advisor to the California Urban Water Conservation Council on the revision of the Council's Best Management Practice 3 (BMP3) "System Water Audits and Leak Detection."

GEORGE KUNKEL, P.E. is a registered professional engineer with the Philadelphia Water Department in Philadelphia, Pennsylvania. He has been very active in implementing water loss control measures in the City of Philadelphia and promoting the need for improved water efficiency in utility operations throughout North America. Mr. Kunkel is very active in the American Water Works Association, having served as chair of the Water Loss Control Committee and as a trustee in the Distribution & Plant Operations Division. He has participated in a number of water loss research projects and surveys and has presented workshops on water loss control methods in several states. He is a coauthor of the AWWA Free Water Audit Software and technical editor for the subcommittee writing the third edition of the AWWA M36 publication *Water Audits and Loss Control Programs*. Mr. Kunkel is also a member of the International Water Association's Water Loss Task Force.

Water Loss Control

Julian Thornton
Reinhard Sturm
George Kunkel, P.E.

Second Edition

New York Chicago San Francisco
Lisbon London Madrid Mexico City
Milan New Delhi San Juan
Seoul Singapore Sydney Toronto

The **McGraw·Hill** Companies

Library of Congress Cataloging-in-Publication Data

Thornton, Julian.
    Water loss control / Julian Thornton, Reinhard Sturm, George Kunkel—2nd ed.
       p.   cm.
    Includes bibliographical references and index.
    ISBN 978-0-07-149918-7 (alk. paper)
    1. Water leakage. 2. Water—Distribution—Management. 3. Loss control.
 I. Sturm, Reinhard. II. Kunkel, George III. Title.
    TD495.T46   2008
    628.1'4—dc22   2008009857

McGraw-Hill books are available at special quantity discounts to use as premiums and sales promotions, or for use in corporate training programs. To contact a special sales representative, please visit the Contact Us page at www.mhprofessional.com.

Water Loss Control, Second Edition

The first edition was published as *Water Loss Control Manual* (McGraw-Hill, 2002).

1 2 3 4 5 6 7 8 9 0 DOC/DOC 0 1 4 3 2 1 0 9 8

ISBN 978- 0-07-149918-7
MHID 0-07-149918-0

Sponsoring Editor Larry S. Hager	**Proofreader** Deepa Pathak
Acquisitions Coordinator Alexis Richard	**Indexer** Broccoli Information Management
Editorial Supervisor David E. Fogarty	**Production Supervisor** Pamela A. Pelton
Project Manager Aparna Shukla	**Composition** International Typesetting and Composition
Copy Editor Ragini Pandey	**Art Director, Cover** Jeff Weeks

I would like to dedicate this book to my wonderful son (and fishing buddy), Nicholas, who came into my life shortly after the release of the first edition of this manual—while of course remembering my daughter, Victoria, who since the first edition has grown into a beautiful young woman

—Julian Thornton

I would like to dedicate this book to my beloved wife, Amabel, who patiently supported me throughout the process of long hours of working on this book. You are the best—LUNUWU (she will know what that means!)

—Reinhard Sturm

This book is dedicated to my wonderful family: my wife Elisabeth, daughter Emily and son George, who patiently listen to my water war stories and tolerate my considerable time spent networking with colleagues and writing in solitary confinement.

—George Kunkel

Contents

Preface

The second edition of *Water Loss Control* has been written in the same spirit as the first edition, as a comprehensive guide to water auditing and hands-on reduction of water and revenue losses in water utility operations and management. The goal of the authors was to update the book with information on important innovations and technologies that have been developed since the first edition was released in July 2002.

Climate issues, growing populations and deteriorating water supply infrastructure are exerting unprecedented pressure on water resources throughout the world. As a result, government and regulatory bodies and water utilities are experiencing a growing awareness of the importance of accurately assessing and efficiently controlling water losses as a means to preserve water resources while facilitating growing communities. Hence raising awareness of the extent of the problem and current practices in many systems in North America and around the world is still a very important component of this book. The book covers the tools required to perform an IWA/AWWA standardized water audit both on paper and in the field. Every water utility has unique characteristics and losses and a variety of effective tools must be available in the practitioner's tool kit. This book provides valuable information for water utility managers to select the correct tools and methodology for the water and revenue losses encountered in their operations. The emphasis of the book is to promote the use of effective water loss control methods and tools as a cost-efficient means of controlling unchecked losses in water utilities. The book is suitable either as an educational tool for the inexperienced operator or as a reference manual for the more experienced operator.

A number of useful water loss publications are available to the water supply industry, however, this publication integrates ideas, techniques, methodologies and references from many international sources, making it a truly flexible and very comprehensive guide, which can be used in a variety of field situations.

Case study accounts of individual water utility experiences are an important way to communicate that a particular method or approach is feasible and has succeeded in a given setting. Referencing a case study account of a successful water loss control program is an effective way for a water utility manager to enhance his case when making a proposal for a new project or a change in rationale. It is very effective in gaining support for a proposal to provide evidence that a similar program has been carried out in an efficient and economical manner. Updates to some of the first edition case studies are included in this second edition and the authors urge interested readers to refer to the first edition accounts wherever possible. Case study accounts are included in Appendix A.

Throughout the book references are made to types of equipment, techniques and software, all of which are generally accepted in the industry. The intent of this book is not to promote one particular product, consultant, contractor or process but to promote awareness of the water loss problems encountered in the water supply industry and the innovative means to address them.

Julian Thornton
Reinhard Sturm
George Kunkel, P.E.

Disclaimer

While every effort has been made to avoid endorsements of any particular brand or model of equipment, consultant, contractor, software or process, the authors and the publisher accept no responsibility or liability for any omission or claim to loss of revenue, caused by the omission of a process type or alternative service provider. The sole intention of this book is to pass on practical field knowledge to end-users with an interest in water loss management.

Acknowledgments

Coauthors

Thanks to David Pearson and Stuart Trow who prepared Chapter 9, which is an excellent chapter on the economics of water loss control.

Acknowledgments for Case Study Material Submitted

Thanks go to the authors of the original case studies and papers presented in this book. The authors are recognized in their individual sections.

Additionally the authors would like to thank the multitude of water utility professionals who permitted their data to be published in this text. (Readers are urged to see the original case studies as some changes have been made during editing).

A special note of thanks goes to the American Water Works Association who approved the reproduction of many of their copyrighted articles, manual excerpts and case study accounts. (Readers are urged to see the original AWWA publications as some changes have been made during editing).

Special thanks also go to the American Water Works Association Research Foundation, who approved the reproduction of data, tables and charts published in their latest water loss research publications. (Readers are urged to see the original AwwaRF reports since they provide an excellent source of additional water loss control information).

Committees

The authors would like to thank all of their colleagues on the Water Loss Control Committee of the American Water Works Association, and the Water Loss Task Force of the International Water Association, many of whom are bringing forth the innovative technologies detailed in this text and serve tirelessly in these volunteer roles. Their joint efforts have resulted in great advances in standardization of methodology and greatly improved awareness of the need to better manage our limited water supplies.

General

The authors would like to also extend thanks to all of their clients, colleagues and fellow service providers who are advancing the efficiency in the water supply industry throughout the world.

CHAPTER 1

Introduction

Julian Thornton

Reinhard Sturm

George Kunkel, P.E.

1.1 Background

The world's population exploded during the twentieth century. At the close of the year 2000 approximately 6 billion inhabitants called the planet earth home, up from 4 billion in 1974.[1] That such growth could occur is a testament to man's unique ability to provide the essentials of clean air, water, food, and health care to its masses. However, during the latter half of the same century, man also recognized that the world's resources couldn't continue to sustain this rate of growth indefinitely; at least, not by using the same methods to which we have become accustomed. Our resources are finite.

> **A**s of February 2008, the world population was estimated to be approximately 6.6 billion!

The availability of safe water has been a major contributing factor in the growth of the world's population, by serving man's drinking water and sanitation needs. The ability to create large water supply systems to abstract or withdraw, treat, and transport vital water to whole communities' fingertips stands as one of history's great engineering marvels. Yet notable caveats exist to this success story. Many developing countries still do not have the water supply infrastructure to provide clean water to individual customers; or to supply it on a continuous basis. In such places, modern water systems are lacking due to the same social, political, and economic complexities that challenge all aspects of development in these lands. While these populations struggle to gain basic levels of service, many highly developed water systems, in technologically advanced countries, suffer an insidious problem that threatens the long-term sustainability of water resources for the future—water loss. Most of the world's water systems, or undertakings, have been highly successful in delivering high-quality water to large populations. However, most of these systems have done so with a notable amount of water loss occurring in their operations. In years past, the seemingly infinite supply of water in expanding "new worlds" allowed water loss to be largely overlooked. With water readily available and relatively inexpensive, losses have been ignored by water utilities, or assumed to be naturally inherent in operating a water supply system.

1

But with the demands of growing populations, realization of the limits on our natural resources and increasing costs from regulations and customer demands, it is becoming increasingly unrealistic to allow water loss to be ignored.

Upon close evaluation it appears that many of the reasons for water loss from meter error, leakage, or data mishandling are actually based on human failings and lack of maintenance. Dickinson[2] has concluded that while it is difficult to generalize, the most common reasons for water utilities not to address water loss in an appropriate manner are: "political infeasibility of admitting system leakage, falsifying water accounting records, lack of recognition that recapturing nonrevenue water with an upfront investment is a still great business case with fast payback, and inherent mistrust of anyone outside the utility examining their system."

The intention of this manual is to explain the reasons why suppliers should reduce lost water and identify how to resolve water loss problems using today's technology in an economically sound manner.

All water utilities and industrial and residential end users should practice water loss control and water conservation regardless of the size of their system or nature of their use. The level of water loss management effort that is being exercised by water suppliers worldwide varies widely. Unfortunately, most of the water industry in the United States and many parts of the world accord water loss only secondary priority since the true economic and social impact of water loss has not yet been realized by policy-makers. In this status water loss continues to suffer from a lack of good auditing practices and a failure to reduce leakage proactively; instead waiting for the next customer complaint to prompt the supplier to reactively repair the next problem leak. However, in a small but growing number of countries throughout the world, comprehensive water efficiency goals have been established. Water conservation, watershed protection, reuse and the new discipline of leakage management have been implemented as required practice by the highest level of government and supplier performance is closely monitored and sometimes regulated. This new model of water resources management is the way of the future because it must be, if mankind is to continue to sustain its growth and its environment.

1.2   The Purpose of This Manual and Its Structure

This manual discusses in great detail methodologies to assess the volume of water losses, water loss control methods and technology, and is aimed at providing the practitioner with all the necessary background and theory to apply proactive water loss management. However, this book also seeks to promote awareness, foster positive attitudes, and pull together not just the ideas of the authors, but also those of other specialists in the field. In addition to our ideas and thoughts stemming from many years of hands-on field intervention against water loss and inefficient use, this book also highlights up-to-date case studies and industry-specific papers to reinforce the concepts and methods already being successfully applied in the field.

> This book provides many useful case studies, which may be used to justify implementation of a more aggressive water loss management program in your utility.

Case studies are an excellent tool for assisting operators in preparing a master plan that takes an aggressive stance against water loss and inefficiency.

The fact that somebody else has done it before makes, in many cases, the job of selling an aggressive program and budget to an executive manager or board of directors more feasible. The steps undertaken in a water loss control program are discussed and reviewed in detail throughout this manual. The chapters are self-contained and do not need to be used in order although an operator with no experience in progressive water loss control methods is urged to read the entire book. The manual focuses heavily upon the progressive methods pioneered in the England and Wales in the 1990s and transferred widely on an international basis. It also consistently evaluates the more "traditional" conditions that exist in North America, and other nations, where water loss has not been a foremost priority. This is done to demonstrate that the need to proactively control lost water exists in even the most developed nations, and that easily transferable technology now exists to control water losses.

The manual includes sections that allow the reader to

- Understand the nature and scope of water loss occurring in public water supply systems
- Learn about the latest analytical methods and tools
- Assess water losses for any system by using a standardized water audit and component based analysis of real losses
- Follow through all steps of a successful water loss control (optimization) program
- Implement field interventions to control real losses
- Implement field interventions to control apparent losses
- Implement demand control
- Perform cost to benefit calculations
- Identify when and how to use a contractor or consultant

This manual is intended to be a hands-on tool for water system managers who are motivated to understand the nature of water loss and take meaningful action to reduce it. Its content provides a detailed road map for any water system operator to implement a program that is the appropriate response for an individual water system's needs.

References

1. Central Intelligence Agency. *The World Fact Book* [Online]. Available: *www.cia.gov/cia/publications/factbook*. [Cited: March 10, 2007].
2. Dickinson, M. A., "Redesigning Water Loss Standards in California Using the New IWA Methodology." Proc. of the Leakage 2005 Conference, Halifax, Canada: World Bank Institute, 2005.

CHAPTER 2

Water Loss Control: A Topic of the Twenty-First Century

Reinhard Sturm

Julian Thornton

George Kunkel, P. E.

2.1   How Much Water Are We Losing?

Throughout the world water losses are occurring at both the end-user's plumbing and the water supplier's distribution piping. Water losses are a universal problem and they do occur in both developed and developing countries.

Water loss is defined as occurring in two fundamental ways:

1. Water lost from the distribution system through leaking pipes, joints, and fittings; leakage from reservoirs and tanks; reservoir overflows; and improperly open drains or system blow-offs. These losses have been labeled *real losses*.

2. Water that is not physically lost but does not generate revenue because of inaccuracies related to customer metering (under recording customer meters), consumption data handling errors, or any form of theft or illegal use is referred to as *apparent losses*.

The sum of real and apparent losses plus unbilled authorized consumption is defined as nonrevenue water (NRW) according to the standard International Water Association (IWA) water balance methodology.[1]

The World Bank estimates that the worldwide NRW volume amounts to 12,839 billion gal/year (48.6 billion m³/year) (Table 2.1) and that the volume of real losses occurring in developing countries alone is sufficient to supply approximately 200 million people. The monetary value of the global annual NRW volume was estimated by the World Bank to amount to $14.6 billion U.S. per year.[2] The World Bank states in its report that a high NRW level is normally a surrogate for a poorly run water utility that lacks the governance,

Did you know that the worldwide volume of NRW is approximately 12,893 billion gal?

5

	Real Losses	Apparent Losses	NRW	Units
Developed Countries	9.8	2.4	12.2	billion m³/year
Eurasia (CIS)	6.8	2.9	9.7	billion m³/year
Developing Countries	16.1	10.6	26.7	billion m³/year
Total	**32.7**	**15.9**	**48.6**	**billion m³/year**
Developed Countries	2589	634	3223	billion gal/year
Eurasia (CIS)	1796	766	2562	billion gal/year
Developing Countries	4253	2800	7053	billion gal/year
Total	**8638**	**4200**	**12839**	**billion gal/year**

TABLE 2.1   Global Water Loss Volumes Estimated by the World Bank

Did you know that many locations in the United States suffer from periodic water shortages, or project a long-term deficit in water supply? Surprisingly, there are no federal regulations governing how much water a supplier can lose!

There are 55,000 community water systems in the United States alone, water losses are suspected to be around 6 billion gal a day!

The amount of water lost in the United States is more than enough to meet the delivery needs of the country's 10 largest cities!

the autonomy, the accountability, and the technical and managerial skills necessary to provide reliable service to their population.

Another study conducted by the U.N. Environment program estimates that by the year 2025, as much as two-thirds of the world population may be subject to moderate to high water stress. The same study estimates that water withdrawal as percentage of the total water available will rise in the United States from 10 to 20% (as of 1995) to between 20 and 40%.[3] This demonstrates the growing stress on water resources globally and in the United States and the urgent need to apply proactive water loss management.

There are more than 55,000 community water systems in the United States alone, which process nearly 34 billion gal water per day.[4] Due to the current lack of standard assessment and reporting methods for water losses, it is difficult to quantify the amount of water lost in U.S. distribution systems. The U.S. Geological Survey estimates that almost 6 billion gal/day[5] of the total 34 billion gal processed a day are approximated to occur as "public uses and losses," with the losses likely much greater than public use for most systems. Inaccuracies or inconsistencies in the reported data also contribute to the difference between the total water delivered and total consumed. The amount of water lost in the United States is more than enough to meet the delivery needs of the 10 largest cities in the United States. This massive waste of resources should be viewed as a considerable concern for the country with the third largest population in the world.

2.2   The Need for Water and Basic Facts about the Resource Water

Human body weight is approximately 50 to 65% water[6], which must be replenished on a daily basis; with a minimum of eight glasses per day recommended for each person. A human can survive without food for several weeks *but* without water we die in around 3 to 4 days! Water stands as the second most urgent body need after air. Like the human body, many of the fruits and vegetables, which we eat, are also mostly water. Obviously, water is an extremely important resource even though people in many developed countries often take its relative abundance and high quality for granted. The availability of fresh water is essential for our societies to thrive and flourish.

> **W**ater is the second most urgent body need after air.

The world's surface is made up of approximately 80% water, which is an indestructible substance. Of this water approximately 97% is salt water, 2% frozen in glaciers, and only 1% is available for drinking water supply using traditional treatment methods. Through the natural patterns of world climate conditions and the hydrologic cycle, the availability of this water varies widely over time and distance. In any point in time, some part of the world is enduring severe drought while other parts are experiencing floods. Rarely does this natural cycle coincide with the routine variation in man's use of water. The amount of water on the earth is fixed and limited. Our predecessors have probably drunk several times in the past the water we drink today! The water cycle hasn't really changed much since the beginning of time. The water cycle is essentially evaporation, cloud formation, rainfall, and passage to the sea by rivers and streams. In a 100-year period, a water molecule spends 98 years in the ocean, 20 months as ice, about 2 weeks in lakes and rivers, and less than a week in the atmosphere. People interfere with the later stages of the cycle and redirect that passage back to the sea through water piping or distribution systems, human bodies, sewer systems, and then back to the sea.

> **O**nly 1% of the earth's water is freshwater that is readily available for water supply using traditional treatment methods—we should take more care of it!

Although the water cycle hasn't changed since the passage of time, the treatment technology used to make it usable, and the distribution technologies, have changed considerably. This is particularly true with the advent of consolidation of populations into major city centers; usually with increasing industry, pollution, and demands for services. The more polluted water becomes, the more expensive it is to treat. The farther away the source from the population center, the higher the transportation cost of water. Given continuing worldwide population expansions and relocations, it is inevitable that the provision of water is becoming increasingly expensive.

Recent initiatives to better utilize water resources include water conservation, recycling, and the use of reclaimed water. Desalination, a way of tapping into the vast resources of sea water, has historically been very energy intensive and costly; however, improvements in the technology have reduced costs and pressures from supply shortages and population growth have resulted in a growing number of desalination plants around the world. Still, desalination is an option largely for coastal cities at this time. Water conservation is a proven technique for customer consumption management. It is now realized that conservation is not just a stopgap action during drought, but an efficient and cost-effective way of life for sustainable communities. Various technologies to reclaim, reuse, or recycle water for nonpotable uses are now required practice in many forward-thinking communities, as these methods satisfy multiple needs for water supply. Some communities are constructing separate, dual distribution systems to convey reclaimed water for uses such as outdoor irrigation and fire fighting. All of these innovations reflect progressive thinking on ways to supply growing populations despite static or declining resources. Still, these modified methods of supply and demand management require notable investments in infrastructure, public education, and legislation. It makes as much sense to seek to economically control losses since loss volumes represent water that has already been treated and energized for delivery to prevailing standards, only to fail to reach customer use (real losses) or generate revenue to the water utility (apparent losses).

2.3   Historic Water Supply and Milestones in Water Loss Control

Water distribution systems have been in use for thousands of years. The ancient Egyptians, Greeks, and Romans all captured, treated, and distributed water in ways not dissimilar to those we use today. The technology has changed, however, the basics remain much the same:

- Source
- Primary lift stations
- Storage
- Pumping or gravity supply
- Transmission system
- Distribution system
- Customer service connection piping, some with, and some without water meters

Even ancient people were concerned with controlling their water losses. Around 40 million gal of water per day were supplied to ancient Rome through a network of 260 mi (420 km) of pipe work and channels. The pipelines and channels were made of brick and stone with cement linings along with some lead pipes.[7] It appears that service connections were 20 mm or ¾ in with simple stopcock arrangements, not so different to what we use today! The first system was installed in 312 B.C. There were approximately 250 reservoir sites and the system was gravity fed. A commissioner and his team consisting of engineers, technicians, workers, and clerks administered this system. One of the priority jobs was to locate and repair leaks.

The durability of the workmanship of the ancient aqueducts is evidenced by the fact that one system installed between A.D. 98 and 117 is still in use in Spain. Not many water systems, or infrastructure of any kind, can boast such a history!

Innovations in water distribution system management evolved as community water systems became standard infrastructure in developing countries. Important developments included

- *1800s:* Formulas for unavoidable leakage (Kuichling)
- *1800s:* Pitot rod district measurements
- *1800s:* Simple wooden sounding rods

- *1900s:* Simple mechanical geophones
- *1900s:* First mechanical meter recording devices are used
- *Circa 1940s:* First electronic geophones and listening devices are introduced
- *Circa 1970s:* First computerized leak noise correlators come into play
- *Circa 1980s:* First battery-operated data-loggers come into play
- *Circa 2000:* Digital equipment and GIS-linked equipment is used for leak detection
- *2000:* International Water Association issues recommendations for a standardized water audit and performance indicators for water supply services, including *unavoidable annual real losses (UARL)* and the *infrastructure leakage index (ILI)*.

Innovations in accountability and loss control continue to occur and cost-effective technology is not usually the limiting factor in implementing a sound water loss control program. Often the greatest challenge in creating a water-efficient system is the need to muster the managerial and political will to launch the water loss control program into existence.

2.4 The Occurrence and Impact of Lost Water

Every water system in the world has a certain volume of real losses, and it is well known among leakage practitioners that real losses cannot be eliminated completely, and even in newly commissioned distribution networks there is a minimum volume of real losses. However, it is also well known and proven that real losses can be managed so that they stay within economic limits.

Unfortunately, it is a fact that water distribution systems have often suffered for many years from the "out-of-sight, out-of-mind" syndrome; particularly where water has been inexpensive and plentiful. The problems associated with water loss are numerous. High real losses indirectly require water suppliers to extract, treat, and transport greater volumes of water than their customer demand requires. The additional energy needed for treatment and transport taxes energy-generating capabilities, which often rely upon large quantities of water in their process. Leaks, bursts, and overflows often cause considerable damage and inflate liability for the supplier. Most leakage finds its way into community waste or storm water collection systems and may be treated at the local wastewater treatment plant—two rounds of expensive treatment without ever providing any beneficial use! Watersheds are taxed unnecessarily by inordinately high withdrawals. In this way, high losses may limit additional growth in a region due to restrictions on available source water. The full effect of leakage losses has yet to be assessed, but the economics of leakage, discussed later in this manual, show that its impact is substantial.

Apparent losses don't carry the physical impact that real losses impart. Instead, they exert a significant financial effect on suppliers and customers, and distort consumption data needed for water resource planning. Apparent losses represent service rendered without payment recovered. The economic impact of apparent losses is often relatively much greater than real losses since the apparent losses are generally valued at the retail rate charged to customers, while the baseline cost of real losses is generally the variable production cost (power, chemicals, and so on) for 1 unit of water. For water suppliers the unit retail cost to customers may be 10 to 40 times the production costs for treatment and delivery. However, for water utilities threatened by droughts and supply shortages, or those applying demand side conservation, or those in need of new water sources, it is appropriate to value real losses at the retail rate, since the water saved by leakage reduction represents a new source of water. Such "newly found" water can be sold to new customers or can help avoid demand restrictions during periods of drought or water shortage. Apparent losses occur at the "cash register" of the water utility and directly impact the water supplier's revenue stream. Yet many systems around the world have such unstructured water accounting and billing practices that they don't even comprehend that such loss is occurring. It is evident that reducing water loss would not only improve water supply operations but would also result in increased revenue. Sound water loss management, therefore, is a practice that usually generates a direct and quick payback to the water utility!

> **M**any water systems around the world can't account for their lost water. Can you imagine if banks couldn't account for all of our money?

2.5 Forces Driving Change in the Way Water Loss Is Viewed and Managed

Managing water losses to an optimum has many benefits for the public, for the water supplier, and for the environment. Some of the most beneficial reasons to reduce water losses are among the leading forces driving change in drinking water supplies, including

- Improved public health protection
- Reduced pressure on water resources and therefore the environment
- Increased level of service to customers through increased reliability of supply
- Recovered losses often stand as best source for new water resources
- Cost efficiencies for the water supplier and better control of water rates for the customer population
- Deferment of capital expenditure on water resources and supply schemes
- Improved public perception of water companies
- Reduced liability to water suppliers due to use of best leakage management practice

The technical aspects behind these driving forces will be discussed in more detail throughout the manual. The following two subsections provide insight into some of the more commonly known forces driving change in how water losses are viewed and managed.

2.5.1 Water Losses and Their Impact on Public Health

Many areas of the world have water shortages and are unable to provide a continuous supply of treated water 24-hours/day. The World Bank reports that over one billion people in the world today lack access to safe drinking water and three million people die every year from avoidable water-related diseases.[8] This situation has often been viewed as a problem faced only by developing countries, but this is not true. In the United States alone 24% of waterborne disease outbreaks reported in community water systems during the past decade were caused by contaminants

> **O**ver one billion people lack access to safe drinking water and three million die from avoidable water related diseases every year!

that entered the distribution system and not by poorly treated water. The rapidly expanding world population is requiring more treated drinking water. Much of this additional population has congregated in cities that are already experiencing water stress or in new areas that are removed from readily available water sources.

2.5.2   Climate Change and Its Potential Impacts on Water Supply

The past 200 years have seen a drastic change in emission of greenhouse gases though the ever increasing use of fossil fuels such as coal and oil.

> **I**n the United States alone 24% of waterborne disease outbreaks were caused by contaminants entering the distribution system. Leaks are an ideal place for contaminants to enter. We should pay more attention to the public health aspect of leakage management.

This trend has occurred at the same time as large-scale deforestation in many areas around the globe. In recent years, a large body of scientific evidence has been gathered showing that human activities such as these are responsible for dramatic changes in the composition of the atmosphere and that global warming is taking place as a result. Many leading scientists have predicted that global warming will increase rapidly over the next century.

In 2005, a study lead by the SCRIPPS Institution of Oceanography and published in the November 17, 2005, issue of the journal *Nature* investigated the effects of global warming on water supplies around the world. This study concluded that global warming will reduce glaciers and storage packs of snow in regions around the world, causing water shortages and other problems that will impact millions of people. Especially ice and snow-dependent regions will experience costly disruptions to water supply and water management systems. For example, it is estimated that vital water resources from the Sierra Nevada range in California may suffer a 15 to 30% reduction in the twenty-first century as a result of reduced snow pack runoff. Studies warn that even more severe problems may occur in regions depending on water from glaciers since their meltwater cannot be replaced. Vanishing glaciers will have the greatest impact on water supplies in China, India, and rest of Asia.[9]

These stark realities of climate change, combined with the occurrence of high levels of water loss around the world, make it very clear that there is an urgent need for water suppliers to reduce the volume of water losses to an optimum in order to be able to meet demand in a sustainable future.

2.6   What is Being Done Around the World to Reduce Lost Water?

The challenges for us today are the same as they were during the days of the Romans; we just have more advanced methodologies and technologies to apply to the problem. We can look back at past efforts and smile and think that we are so much better, but to be honest we just have better tools. An open mind, unwillingness to accept existing inefficiencies and a wish to improve are the basic tools a water system operator needs to have today. The rest can be purchased as work progresses. Water audits and water loss control programs will only be successful if the operator and his utility are willing to accept what they find and act on it openly. Therefore it is critical that system operators understand the extent and impact of water loss, and the control of lost water hold a priority of paramount importance throughout the entire organization.

Due to a number of dramatic late twentieth-century changes in the water supply business model worldwide, a new breed of water utility manager has entered the water supply scene. One who strives to increase the performance of the utility, increase profits, and yet be accountable for the efficient use of one of nature's most precious resources, water! The need for this new breed of water system operator has come about by pressure from a number of stakeholder groups who no longer tolerate abuse and inefficient use of natural water resources. These include the environmental community, which has been successful in raising grass-roots consciousness to the level of environmental regulation at the national and international level. Consumer advocates now carefully monitor the value of service per unit cost paid by the customer, expecting the utility to provide quality service at reasonable cost. Competitive forces have also increased, focusing utilities on improving both technical and business efficiency. The power of the internet, media, and other communication forums has helped to accelerate all of these forces, which are mandating that water loss not be tolerated or overlooked as it has been in the past.

> **W**ater system operators are now under pressure from various stakeholder groups to operate systems more efficiently, reduce losses, and improve performance.

A new model of water loss management was developed, taking root in England and quickly spreading to a number of other nations. The National Leakage Initiative was an extensive research endeavor carried-out by British and Welsh water companies in the early 1990s. Its results formed the basis for the development of a progressive leakage management structure that arguably now exists as the world's best practice model. The crux of this structure is basic applied engineering, stressing a proactive approach toward eliminating and preventing leakage, and contrasting dramatically with the largely reactive modes existing in most water systems worldwide. In less than 10 years, this structure has been successful in eliminating up to 85% of all recoverable leakage in England and Wales.[10] Proactive water loss management based on the model developed in the United Kingdom has been promoted and applied in many places around the world and it has proven to be an easily transferable technology for nations around the globe. Nowadays, more than ever, it is evident that the world's water suppliers not only have a need to reduce and proactively manage their losses, but also have the methods and technology to do so effectively.

The successful structure established in England and Wales was implemented in a relatively short period of time and was driven by a number of the forces mentioned above. British water companies were privatized and reorganized along watershed boundaries in 1989. They also fell under a heavy regulatory structure at that time; a structure that focused upon effectiveness and impact of company operations and cost to the customer. The ability for water companies to pass costs along to customers is greatly limited by this structure, which ties approvals to increase rates or tariffs to performance of the company. Consequently, innovative was accelerated as the companies sought ways to improve performance, cut costs, and increase profits. Environmental concerns and the relatively high density of the population also have elevated support for the wise use of water in the United Kingdom. A notable catalyst in the mid-1990s was the severe drought that hit the country. This event triggered the establishment of new leakage reduction requirements and targets, which the companies where able to implement, having the results of the National Leakage Initiative to guide them. While achieving great success in reducing leakage, the U.K. water industry still continues to study all aspects of water loss, as well as conservation, reuse, and other water efficiency practices. The relatively sophisticated system that is in place continues to be refined due largely to the motivation of the government, environmental, and consumer sectors, which have placed a high value on protecting water resources.

The British water loss control methodologies and technologies have had a dramatic effect on other nations as these methods have begun to take hold in perhaps several dozen countries. National or regional governments in South Africa, Malaysia, Australia, New Zealand, Brazil, and Canada during the late 1990s have adopted major new programs that emphasize leakage reduction. Strong programs in Germany and Japan are being refined. Extensive initiatives were completed in past several years in Malaysia and Brazil that extended for 10 years or more with ongoing investments of over $100 million in each project and based on the success of these projects new projects have been started since then. The projects include auditing, pressure management, improved leakage monitoring, detection and repair, and revenue enhancement.

The past has shown that the leakage management methodologies and technologies used in the United Kingdom are easily transferred to systems around the globe. Its techniques can be applied to water systems of varying characteristics and its performance indicators allow comparisons to be drawn for systems around the world. This aspect of the technology is perhaps its most compelling and is likely a primary reason why it has spread so quickly in its use in the United Kingdom and around the world. A recently completed American Water Works Association Research Foundation (AWWARF) study assessed the transferability of international (mainly from the United Kingdom) applied leakage management technologies to North America. Comprehensive field testing carried out during this project has proven that these technologies are transferable to North America where some water suppliers previously denied the transferability because of the different characteristics and requirements of the distribution system (mainly fire flow and insurance requirements).

The World Bank and its capacity development arm, the World Bank Institute, has acknowledged the serious problems arising from excessive water losses and has therefore launched an initiative promoting the IWA best practice in NRW reduction and water loss management through training courses and manuals provided to water utilities in developing countries around the world.

2.7   Program Needs and Requirements for Water Loss Control

According to American Water Works Association (AWWA) estimations, approximately $325 billion needs to be spent on upgrading distribution systems in the United States in the next 20 years.[11] Using average demand figures, the annual value of lost water and revenue, and therefore the approximate annual value of the water loss control market in the United States and worldwide, can be approximated. Interestingly, water loss control is estimated at approximately 29% of the above AWWA figure, or $94 billion. These estimations can be found in Table 2.2 and are approximations only. However, even if in error by 50%, this finding represents a huge, virtually untouched potential market that exists for water

> **A**WWA projects that $325 billion needs to be spent on water system upgrades in the United States over the next 20 years.

U.S. market potential	
U.S. population	250,000,000
Average consumption (kgal/year)	36.5
Average loss	16%
Split of real losses	60%
Average cost treated	$2.50
Average cost sold	$4.00
Recoverable %	75%
Total losses	1,460,000,000
Total real losses (kgal/year)	876,000,000
Total apparent losses	584,000,000
Value of recovered product	$ 2,190,000,000
Value of recovered revenue	$ 2,336,000,000
Recoverable %	$ 1,642,500,000
Recoverable %	$ 1,752,000,000
Market size per year	$ 3,394,500,000
Simple calculation for world market size	
Assumes like numbers as	
Other countries use less but with high loss	
U.S. water is cheap compared to others	
Loss value per capita in United States	$ 13.58
World population	6,000,000,000
World market size per year	$ 81,468,000,000

TABLE 2.2   Approximate Value of Water Loss Control Market

loss control; which can be approached by water system operators, consultants, contractors, plumbers, and facility managers.

A complete water loss control program is often referred to as a Water Loss Optimization Program. Optimizing basically means doing everything possible to improve the technical and financial performance of the water system, whether a public, private, or demand-side system. Optimization usually entails reduction of operating overheads and enhancement of revenue streams. Figure 2.1 shows a typical optimization graph. In this case it can be seen that the profitability in the beginning is low as the cost of the water loss project is being borne on a performance basis.

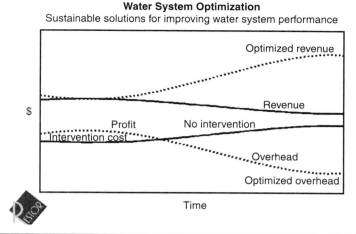

Water loss optimization programs are sometimes undertaken on a performance basis. This means that the utility enters a special partnership agreement with a contractor or consultant. The contractor or consultant is paid a portion of the money recovered from the project over a certain time frame. This is an excellent way of undertaking a project, especially for utilities that do not have a substantial initial budget to allocate for loss control, but do have an existing operating budget, which includes a fixed cost to operate the system *with* losses. The performance approach allows the utility to continue budgeting their normal allocation, however the actual cost of operation will drop and the revenue stream increase as the work continues. At a certain point the contractor drops out of the equation and the annual operating budget either reduces with an increased income, therefore profitability; or the additional funds can be redirected into other maintenance or training functions as required.

2.7.1   The General Structure of a Water Loss Control Program

In general water loss control programs are implemented in four phases:

- *Phase one:* Water audit, assessment of economic optimum volume of water losses, and performance indicators.
- *Phase two:* Pilot study to demonstrate initial recommendations of the water audit analysis in the field.
- *Phase three:* Global intervention using apparent and real loss reduction methods.
- *Phase four:* Ongoing maintenance of the loss control mechanism.

Budgets may be relatively restricted for phases one and two, until methodologies and techniques have been identified with paybacks in line with the expectations of the utility for their system.

Operators must learn to be proactive and identify realistic programs and budgets to combat loss. They must learn to identify efficient, inventive methods to reach economic levels of loss, not just apply a minimal budget to loss control and then resolve the rest by way of a "pencil" audit, writing off a major portion of loss as unavoidable. The traditional rule-of-thumb notions of the amount of water loss viewed as "unavoidable" has changed with new methods that calculate system-specific levels of technical unavoidable annual real losses. This level of loss is much smaller than the traditional ways (Kuichling equation) due to the advent of new technologies, which allow us to control losses economically to much lower levels.

Some of the tasks included in a water loss control program are

1. Overhead reduction tasks (real losses)
   a. Leakage reduction
   b. Hydraulic controls (pressure management)
   c. Pipe repair and replacement
   d. Customer service pipe replacement
   e. Condition assessment and rehabilitation
   f. Energy management
   g. Resources management

2. Revenue stream enhancement tasks (apparent losses)
   a. Baseline analysis
   b. Meter population management
   c. Meter testing and change out
   d. Meter correct sizing and change out
   e. Periodic testing
   f. Automatic meter reading (AMR)

3. Billing structure analysis and improvements
   a. Nonpayment actions
      • Turn off supply
      • Reduce supply to minimum
      • Legal action
      • Prepayment schemes
      • Reduction of fraud and illegal or unregistered connections
      • Continuous field inspections and testing
   b. Rate or tariff management
   c. Customer base management
   d. Modeling for efficient installation
   e. Modeling to assure economic efficiency

Automation is often a common component in an optimization program.

In most cases water loss management is extremely cost effective with paybacks measured in days, weeks, and months; not years as with other programs.

Water loss control and management is usually a highly cost-efficient endeavor since so many water supply systems currently suffer excessive water loss. The greatest challenge for today's progressive water manager is to change dated mindsets that view water as infinite and inexpensive. Once policy and decision-makers understand the true value of water, implementing the intervention techniques can be a relatively straightforward and reassuring undertaking.

References

1. International Water Association. *Performance Indicators for Water Supply Services.* Manual of Best Practice: London IWA, 2000.
2. Kingdom, B., R. Liemberger, and P. Marin. "The Challenge of Reducing Non-Revenue Water (NRW) in Developing Countries—How the Private Sector Can Help: A Look at Performance-Based Service Contracting." Water Supply and Sanitation Sector Board Discussion Paper Series—Paper No. 8. Washington, DC.: The World Bank, 2006.
3. United Nations Environment Program. Chapter Two: "The State of the Environment- Regional synthesis—Freshwater." Global Environment Outlook 2000. Available online: www.unep.org/geo2000/english/0046.htm. [Cited March 10, 2007.]
4. American Water Works Association, Stats on Tap. Available online: www.awwa. org/Advocacy/pressroom/STATS.cfm. [Cited March 10, 2007.]
5. U.S. Geological Survey, "Estimated Use of Water in the United States in 1995." Circular 1200. USGS, 1998.
6. MacQueen, I.A.G. (ed.). *The Family Health Medical Encyclopedia.* Book Club Associates/ William Collins Sons & Co., 1978.
7. Readers Digest. *How Was It Done? The Story of Human Ingenuity Through the Ages.* Readers Digest Publishers, 1998.
8. World Bank Reports.
9. SCRIPPS Institution Of Oceanography. Scripps-led Study Shows Climate Warming to Shrink Key Water Supplies around the World, Available online: http://scrippsnews. ucsd.edu/article_detail.cfm?article_num=703. [Cited March 13, 2007.]
10. Lambert, A.O. International Water Data Comparisons, Ltd. Personal conversation, October 2000 reinterpretation of United Kingdom Office of Water Services (Ofwat) Reported Leakage Results.
11. American Water Works Association. Stats on Tap, Available online: www.awwa. org/pressroom/statswp5.htm.revised. [Cited February 15, 2001.]

CHAPTER 3

Understanding the Types of Water Losses

Reinhard Sturm

Julian Thornton

George Kunkel, P.E.

3.1 Defining Water Supplier Losses

Understanding the types of water losses and having consistent and clear definitions for the types of water losses occurring in distribution systems is the first step to be able to manage the problem of water losses.

Simply stated, the problems of water and revenue losses are[1]

- *Technical:* Not all water supplied by a water utility reaches the customer.

- *Financial:* Not all of the water that reaches the end user is properly measured or paid-for.

- *Terminology:* Standardized definitions of water and revenue losses are essential to quantify and control the losses.

The International Water Association (IWA) defines two major categories under which all types of supplier water loss occurrences fall:

- *Real losses* are the physical escape of water from the distribution system, and include leakage from pipes, joints, and fittings; leakage from reservoirs and tanks; and water losses caused by reservoir overflows. Real losses occur prior to the point of end use.

- *Apparent losses* are caused by inaccuracies associated with customer metering, consumption and billing data handling error, assumptions of unmeasured use, and any form of unauthorized consumption (theft or illegal use).

While these two definitions are distinguished by a stark physical differentiation, in most cases a dramatic economic difference also exists. Real losses, which are most usually leakage, are typically valued at the variable production cost of the water. Apparent losses, which occur at the customer destination, penalize the water supplier at the retail cost; a rate usually much higher than the production cost. The variable production costs frequently include only the short-term costs; however, in many cases it is appropriate to include long-term costs in the valuation of real losses, the cost implications of real and apparent losses require that a careful assessment of each be undertaken to design the most appropriate and cost-effective water loss control program.

3.1.1 Real Losses

The quantity of real losses in a given water systems is a good indicator of how efficient a water supplier is in managing its assets (the distribution network) and the product it delivers to its customers. Volumes of real losses that are significantly higher than what is economically justifiable indicate that action needs to be taken if the water supplier is to be viewed as water-efficient, customer-responsive, and a responsible steward of water resources.

Real losses are made up of three components (see Fig. 3.1):[2]

- *Reported breaks and leaks:* They typically have high flow rates, are visibly evident and disruptive, and have a short run time before they are reported to the utility by customers or utility personnel since they cause nuisance to the customer (pressure drop or supply interruption).

- *Unreported breaks and leaks:* They are typically hidden from above-ground view, have moderate flow rates, and a long run time since utilities must seek out these leaks to become aware of them. They are located through active leak detection.

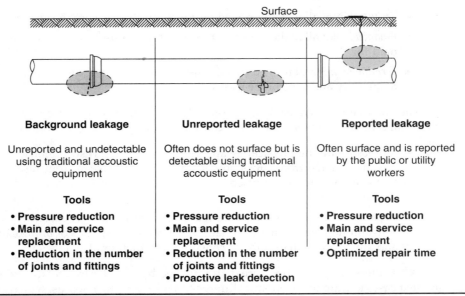

FIGURE 3.1   Components of real losses and tools for intervention. (*Source:* Ref. 2.)

- *Background leakage:* They are the collective weeps and seeps in pipe joints and connections. They have flow rates that are typically too small (1 gpm (gallons per minute) or 250 L/hr) to be detected by conventional acoustic leak-detection equipment. They run continuously until they gradually worsen to the point when they can be detected. The only ways of reducing background leakage is through pressure management or infrastructure replacement.

Why Do Real Losses (Leakage) Occur

Real Losses exist in virtually every water-distribution network. They can never be completely eliminated and even newly commissioned sections of a network can have a certain minimum volume of real losses (unavoidable volume of real losses). However, how much the volume of real losses is in excess of the unavoidable minimum depends on general characteristics of the distribution network and the leakage management policy employed by the water utility.

The most common causes of leakage are

- Poor installation and workmanship
- Poor materials
- Mishandling of materials prior to installation
- Incorrect backfill
- Pressure transients
- Pressure fluctuations
- Excess pressure
- Corrosion
- Vibration and traffic loading
- Environmental conditions such as cold weather
- Lack of proper scheduled maintenance

Where Do Leaks Occur

In general, leaks can occur on three different sections of the network: transmission mains (see Fig. 3.2), distribution mains (see Fig. 3.3), or service pipes (see Fig. 3.4). Depending on where they occur they will have different characteristics such as flow rate, tendency to cause supply interruptions, and likelihood to surface and be visible above ground.

British leakage management terminology distinguishes *reported* versus *unreported* leaks, or, more literally, *reported bursts* and *unreported leaks*. Dramatic pipe bursts are the most recognizable example of a reported leak, which, due to their damage-causing nature, are usually quickly reported, responded to and contained. However, unreported leaks, often running at a small rate of flow on underground pipes, frequently escape the attention of the water supplier and the public, but account for larger amounts of lost water since they run undetected for long periods of time. Historically in the United States, the terms reported and unreported are not employed, therefore the distinction between a "leak" and a "break" (burst) is rather subjective, and is one of a number of examples of inconsistent terminology. Efforts are underway in the United States, however to

FIGURE 3.2   Transmission main break. (*Source:* WSO—Guido Wiesenreiter.)

FIGURE 3.3   Distribution main break. (*Source:* WSO—Guido Wiesenreiter.)

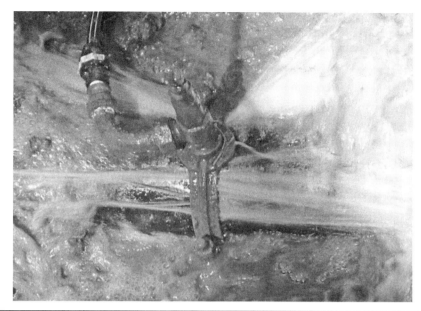

FIGURE 3.4   Service line leak. (*Source:* WSO—Guido Wiesenreiter.)

advocate for the use of this terminology. The third version of the American Water Works Association's M36 publication *Water Audits and Loss Control Programs* supports this terminology.

A recently published American Water Works Association Research Foundation (AWWARF) report on main break prediction, prevention, and control[3] estimates that water utilities in the Unites States suffer between 250,000 and 300,000 main breaks per year, causing about $3 billion of total annual damages and indirect consequences. It is unknown how many small leaks and service leaks occur, but annual leaks likely outnumber main breaks several times over in typical water supply systems; likely resulting in 500,000 to 1,500,000 leaks per year. The United States has approximately 880,000 mi of distribution mains, many of which are old unlined cast iron in need of repair, rehabilitation, or replacement. However, good leakage control practices can help prolong the life of the existing infrastructure by reducing the occurrence of leaks and breaks and forces leading to water main failures.

> **T**he United States has approximately 880,000 mi of mains!

Which Leaks Are Causing the Greatest Volume of Real Losses

It is a common misconception that major main breaks, which are surfacing quickly and causing supply disruptions, are responsible for the bulk of water lost through leaking pipes. Very often it is not understood that even though dramatic pipe failures loose huge volumes of water they do so only for a short period of time since water utility crews respond quickly to contain these disruptive events. Conversely, small hidden leaks and breaks may run for years causing significantly greater volumes of real losses before they are repaired (see Chap. 10). A significant finding of leakage research efforts during the 1990s has been the large amount of water loss occurring on the customer service piping branching from the water main and supplying water to a single or multiple user premises. For many systems, leaks on these small-diameter pipes represent the greatest number of leaks encountered in water supply operations especially in systems with a high service connection density. Often supplier policies require the customers to own their service lines and execute repairs or replacement when necessary.

> **S**ervice leaks often cause the largest volumes of real loss.

Unfortunately, many customers are often unaware of their ownership responsibilities and, when advised to repair known leaks, are neither timely nor effective in getting relatively expensive repairs executed. Consequently, customer service piping leaks can run for considerably long periods, even after being reported, and account for substantial water loss. Severe drought in England in the mid-1990s resulted in emergency regulations that required some water suppliers to implement repairs on leaking customer service lines. The resulting savings in lost water was found to be so effective and the repair methods so efficient that national regulations were soon established requiring all water companies to implement policies for company-executed customer service line leak repairs. Two other notable aspects of this: the customers still retained ownership of the lines and, once high initial backlogs of customer leaks were repaired, the rate of occurrence of new leaks was sufficiently slow that the repair policies for the water companies were found to be manageable and cost-effective. This experience demonstrates dramatically the principle that leakage losses are dependent on two primary variables: rate of flow and time permitted to run. Both parameters must be considered in developing leakage-management strategy. Too often water suppliers lose track of small volume leaks, allowing indefinite leak time to occur and losses to mount.

What Else Influences the Volumes of Water Lost through Leaks and Breaks

Another tenet employed in recent times by progressive leakage management programs around the world is the science of pressure management. In designing water infrastructure engineers have frequently specified distribution system pressure levels with the primary objective of providing service above a minimum design pressure. However, local guidelines for providing fire flows, expansion capacity, and safety factors have frequently resulted in systems supplying water pressures far above minimum requirements, without consideration for the impact of the excessive pressure. By the late 1990s, fundamental relationships between pressure and leakage rates were established and show that certain types of leaks are highly sensitive to changes in pressure. It can now be taken that, while certain minimal levels of pressure need to be provided, maximal levels for pressure should also be established and not exceeded. Excessive water pressure not only increases certain types of leakage, but also influences main break rates and the amount of needless energy costs a supplier expends. In progressively managed water systems, water pressure is now controlled within an appropriate range that meets the needs of the customer and the supplier without causing waste or harmful impact to the infrastructure.

> **P**ressure has a much greater impact on leakage than originally suspected. System design should take into account maximum pressure limits as well as minimum ones.

Considerable research work has been conducted in the past decade on the nature and impact of leakage and highly effective practices and technologies have been developed and successfully implemented around the world to reduce, control, and manage real losses. It should be in the foremost interests of all water suppliers to closely evaluate leakage occurring in their systems and take advantage of these methods which may be considered the best practice model in controlling leakage losses.

3.1.2 Apparent Losses

It is important to notice that apparent losses are not caused by leakage. They do not include any physical losses of water, since the water has reached the destination of an end user. However, this successful supply function was inaccurately metered, archived improperly in the billing system, or the use of water was unauthorized. Apparent losses are a very important component for the water supplier to keep under control as they have a direct negative impact on suppliers' revenue generation for a product that was delivered to the customer.

Accurate metering of customers provides valuable information on consumption trends needed to evaluate loss control and conservation programs. It also elevates the value of water in the mind of the consumer by linking a price with a volume. With improved metering, automatic meter reading, and data-logging technologies now widely available, customer consumption information has become a critical resource to better manage water-utility operations and the water resources of individual watersheds or regions.[4]

Before discussing the specifics of these losses, it is appropriate to review the typical metering and billing structures used by water suppliers. With the establishment of modern indoor plumbing, customer service pipes have been tapped directly into local water pipes or mains to bring water directly into the homes of the consumer. Figure 3.5 shows a typical direct-feed situation.

Many water suppliers have chosen to incorporate customer water meters at the end-user premises and gather regular meter readings for the purpose of billing per unit

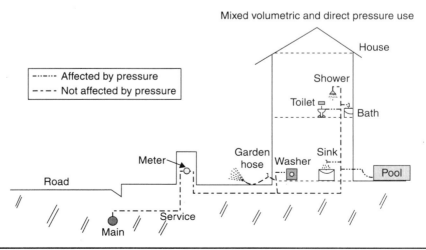

FIGURE 3.5 Typical direct pressure residential supply situation.

volume of actual water used. Customer meters also allow the user to monitor his or her own water usage and provide the customer the option to exercise restraint against excessive use and identify waste. Outwardly, this approach seems to follow the norms of typical free market commodities, payment is based upon the volume of product or service delivered. Yet, the use of customer meters and usage-based billing is far from universal in the water industry in the United States or the world at large. For a large portion of public water supply customers, service is provided without any measurement of their actual water usage and billings are based upon flat rate charges assigned by customer user type. In the United States, perhaps only one-half of all users have water meters, with sentiments regarding metering sharply divided in certain areas of the country. In England and Wales, traditionally only the industrial, commercial, and institutional (ICI) customers were metered. Environmentalists and regulators support the establishment of universal residential customer metering, and a slow transition is occurring with meters being installed in new construction and upon customer request. Approximately 25% of all residential properties were metered in England and Wales as of the close of the year 2006.

> **T**he guiding institution on water supply in the United States, the American Water Works Association (AWWA), recommends that every water utility meter all water taken into its system and all water distributed from its system at its customer's point of service. Customers reselling utility water—such as apartment complexes, wholesalers, agencies, associations, or businesses—should be guided by principles that encourage accurate metering, consumer protection, and financial equity.[5]

Why Do Apparent Losses Occur

Apparent losses occur in three primary ways:

1. Customer meter inaccuracies
2. Errors in water accounting
3. Unauthorized consumption

In comparison to real losses, apparent losses have a much greater negative effect on the utilities revenue generation since they directly impact the utility's cash register. Apparent losses should always be valued at the retail value of the water sold. Another important factor regarding apparent losses is that an understatement of the apparent loss volume results in real losses being overstated in the water audit. This can potentially misguide water loss control planning by placing inappropriate emphasis on leakage while highly potential revenue recovery goes unattended.

How Customer Meter Inaccuracies Occur

Errors in measurement can occur in several ways. First, water meters reading can be in error due to a variety of mechanical or applications reasons. Due to widely varying water consumption patterns among customer populations, a number of different meter sizes, and sometimes types, can be found in any single water utility. Standard displacement or velocity meters provide accurate flow measurement for residential users while large ICI users may experience dramatic differences in daytime and night time flows; requiring meters that are accurate through a wide range of flow rates, that is, compound type meters. Other factors place demands on the water supplier to provide accurate metering. Some of the major reasons why water meters fail to measure water flow accurately include

- Wear over time
- Water quality impact
- Chemical build up
- Poor finish and workmanship
- Environmental conditions such as extreme heat or cold
- Incorrect installation
- Incorrect sizing
- Incorrect specification of meter type for the application
- Tampering
- Lack of routine testing and maintenance
- Incorrect repair

Recommended maintenance practices for customer meters include monitoring recorded consumption patterns and rotating the meter out of use on a regular basis for testing, calibration, repair, or replacement.

Many systems use estimates of customer consumption for accounts where water meters are nonexistent, defective, or unreadable. Estimates, which are used both temporarily or permanently, can be inaccurate if they are not devised in a rational manner or kept up-to-date with changing customer consumption patterns; hence another form of inaccurate water measurement can occur here.

Meter reading is the next step in obtaining accurate water consumption data. Errors in meter reading are essentially errors in measurement. With the growing use of automatic meter reading (AMR) systems, the opportunity for meter reading error is probably being reduced relative to that occurring in traditional manual meter reading operations. However, all systems seeking to optimize should include at least a brief assessment of the accuracy of meter reading operations in transferring actual measured water consumption into the information handling (billing) system.

How Errors in Water Accounting Occur

Errors in the handling of customer accounts can occur in a number of ways, some of which include

- Customer water consumption data is modified during billing adjustments.
- Some customers who use water are inadvertently or intentionally omitted from billing records and go unmonitored.
- Certain users are accorded nonbilled (free or subsidized) status and actual consumption is not recorded.
- Human error occurs during data analysis and billing.
- Weak policies create loopholes in billing and water accounting.

- Poorly structured meter reading or billing systems.
- Poor tracking of changes in real estate ownership or other changes in customer account status.
- Lack of understanding of technical and managerial relationships in assessing, reducing, and preventing apparent loss.

Most errors in water accounting occur mainly due to a lack of structure and controls in the accounting process.

In the United States, "water accounting" is not an established practice as is "financial" accounting, which has substantial controls and accountability built into its standardized process. The fact that consistent standards for water accounting don't exist likely results in many water systems understating actual customer usage and failing to capture full billing potential.

Unauthorized Consumption

The last of the three primary occurrences of apparent water loss is unauthorized consumption. While human nature holds a high regard for the quantity-cost relationship, it is also true of human nature that a certain small segment of a population will attempt to illegally obtain service without making payment. Unauthorized consumption is likely a more common phenomenon in systems where customer meters are in use and water is billed per unit volume. Where flat rates are charged and consumption is not routinely monitored, customers can draw greater quantities of water to lower their own effective unit cost. These customers would need to evade inclusion in the billing process altogether in order to obtain water service without paying.

Unauthorized consumption can occur in a number of manners. Much unauthorized consumption occurs at the point of established end users. Some customers tamper with meters or meter-reading equipment in order to lower meter readings. Fortunately, many AMR systems have tamper detection features that help thwart such activity. Unscrupulous users with large water meters have been known to open valves on unmetered bypass piping, thereby routing their supply around the active water meter. Some users or contractors may consciously or unwittingly connect branch plumbing pipes to customer service lines upstream from the water meter, which also provides supply without passing through the meter.

Urban systems in the northeast section of the United States have encountered a frequent occurrence of customer restoration of terminated service connections. Closing and locking curb-stop valves on the customer service line is a common means of terminating service used by water utilities in the United States against delinquent customers. Illegal restoration occurs when delinquent customers reactivate their own water service after the water supplier due to nonpayment has stopped it. These situations evidence

Theft of water can be a common occurrence in the United States and is not just a third world problem.

the need for water suppliers to continue to monitor terminated accounts, after they are shutoff, for resumed, unauthorized consumption. The city of Philadelphia provides such monitoring and has achieved success in reducing illegal restorations; lowering their discovery rate from 35% of all terminated accounts to less than 20% since the installation of their AMR System in 1999. During its 2007

Fiscal Year, Philadelphia uncovered 2984 accounts that had been illegally restored, and was able to collect $341,000 in missing revenue in motivating delinquent customers to make payment. With its AMR system meter reading and consumption continue to be monitored even if an account has been shut off for nonpayment. In contrast to the U.S. experience, regulations do not allow water companies in England and Wales to terminate water service to customers under any circumstances.

Unauthorized consumption has also been known to occur when persons find ways of withdrawing water from a location in the distribution system other than the customer service line. With fire hydrants constructed as above ground appurtenances in the United States, illegal opening of these devices happens regularly in many cities. In some areas, using fire hydrants to fill street cleaning equipment, landscaper trucks, and construction vehicles has occurred so casually that upstanding businesses perceive this to be acceptable practice. Water utilities in such places have a public education challenge to instill the value of water as a commodity in the business community. Establishing bulk water dispensaries is now common for water systems that wish to allow, and even promote, water sales outside of the normal customer service line connection. Some systems allow water to be used from fire hydrants in an authorized manner with the filing of a permit. With concerns for cross connection protection and the accountability of water, such a practice is not a preferred one for most water utilities.

All water suppliers should be mindful that the potential for unauthorized consumption exists to some degree in their systems. Just as retail establishments must take safeguards against "shoplifters," water systems should have appropriate controls to monitor for unauthorized consumption and keep such occurrences in check.

3.2   Conclusion

This chapter provided a general overview on the two components of water losses, namely real and apparent losses. Both exist in every system to a certain extent, depending on the efficiency of the water utility. Both components need to be carefully assessed, monitored, and managed in order to be able to operate at an economic optimum level. Chapters 16 to 19 provide further details about real losses and a detailed insight into the available intervention tools against real losses.

Chapters 11 to 15 provide further details about apparent losses and a detailed insight into the available intervention tools against apparent losses.

References

1. International Water Association. *Performance Indicators for Water Supply Services.* Manual of Best Practice. London: IWA, 2000.
2. Tardelli, J. Chapter 10. In *Abastecimento de Agua*. São Paulo, Brazil. Tsutiya M Escola Politecnica, Universidade de São Paulo: 2005.
3. Grigg, S. N. *Main Break Prediction, Prevention, and Control*. Denver, Colo.: AWWARF, AWWA, and IWA, 2005.
4. American Water Works Association. *Water Audits and Leak Detection*, 3rd ed. Manual M36. Denver, colo: AWWA, in press.
5. American Water Works Association. *Statements of Policy on Public Water Supply Matters*, Available online: http://www.awwa.org/about/oandc/officialdocs/AWWASTAT.cfm. [Cited March 29, 2007.]

CHAPTER 4

Water Loss Management in the United States and Internationally—What is Necessary to Control the Water Loss Problem?

Reinhard Sturm

Julian Thornton

George Kunkel, P.E.

4.1   Introduction

Water loss is a chronic, and often severe, global problem; spanning from highly developed countries with extensive infrastructure to developing countries with limited resources. Climate change, drought, and water shortages, often occurring in arid or semiarid regions of expanding population, are having an increasing impact on water supplies and water is becoming a limiting factor for economic growth and environmental sustainability. Given this stark reality, it is inconceivable that most countries do not require reliable tracking of water supplies and losses. Commonly heard justifications from water utility managers for their inaction are a perceived lack of resources and the burden of many other priorities of system operation. Some utilities downplay their losses out of fear of public resentment, especially in cases where the utility is asking the customer to conserve water or pay higher rates or tariffs. In areas with limited water audit regulations, some utility managers distort their true losses on paper using "pencil," audits that are not scrutinized by outside authorities. Most of these practices, however, are merely a reflection of the lack of a regional or national agenda for water loss control for these utilities.

> **M**any utilities use "pencil" audits as a way of hiding their real volume of water losses. This practice reflects a lack of a regional or national priority for water loss control and is especially surprising in cases where the same utilities are asking their customers to conserve water or are planning to tap into new water resources.

Throughout the world, the water supply/demand balance is in jeopardy. In many developing and some developed countries, some water systems do not provide customers a continuous water supply on a 24-hour per day basis, particularly during times of drought. Other systems are faced with seemingly limited water resources to supply rapidly developing communities. Water utilities in resort communities serve a heavy holiday and tourist trade, resulting in weekend and holiday peaks many times higher than normal operating peak flows. These systems often borrow significant funds and install costly new water sources that are utilized only on a part-time basis. The rest of the time the costly investment sits unused and inefficient. For systems in these conditions, water loss management offers multiple advantages of capturing treated water volumes now lost to leakage while and recovering additional needed revenue by managing apparent losses. A successful water loss control program can defer the cost of loans for capital investments, stretch existing water resources and improve customer satisfaction; and usually provides a very fast payback.

The first step into the right direction is to assess and acknowledge the problem followed by dedicating resources and funds to efficiently control water losses. This chapter explains how water loss is managed in various countries, focusing on the contrasting structures in the United States and England and Wales; as well as a number of other countries who have taken a progressive stance on water loss. Insight is given into the regulatory structures, standards, and water loss management practices of these countries.

4.2   Water Loss Management in the United States

The United States is a country truly blessed with bountiful natural resources. Water is a primary resource that has been consistently developed to help the country grow to the level of strength and prosperity that it enjoys today. Unfortunately, the availability of plentiful water during the country's early history may have contributed to a water supply infrastructure and American psyche that now tolerates significant water loss. A general lack of awareness of this fact by the public and many water supply professionals is a large part of the problem.

Today the U.S. drinking water industry is facing growing challenges in providing water supplies necessary to sustain the country's economic and population growth. Some of the fastest growing cities in the United States, such as Phoenix and Las Vegas, are located in semiarid and arid climates. Water resources are limited in these dry areas, requiring developing and transporting water supplies from very distant sources. The Colorado River is a critical lifeline of water supply, but often runs dry at its mouth to the Gulf of California while its waters provide supply to several states which are often at odds with each other on how best to manage the river while achieving their water supply goals.[1]

The last 20 years have seen water restrictions due to multiyear droughts become routine in many areas while the development of new sources has become less attractive and costlier due to enhanced water quality and environmental protections, coupled with funding constraints. Despite these pressures, water loss policy is still not adequately addressed at the national level even though the water saved through reduction of water loss represents one of the least expensive new sources of water.

The term *water accountability* has been used casually in the United States for the last several decades to label a variety of activities that impact the delivery efficiency of water utilities. Historically in the United States, water accountability practices (unaccounted-for water percentages) have existed more as art than science, with methods often generating as much confusion as explanation in interpreting water loss conditions. Symptomatically, this confusion stemmed from inconsistent terminology, unreliable percentage measures, and a lack of procedures to rationally evaluate and compare water loss performance. On a broader level; however, outdated water accountability methods are a weak discipline due to the lack of awareness of the extent of water loss occurring in the United States. Lacking recognition is a significant concern for many water industry stakeholders, no national agenda exists for water utilities to reliably quantify or control their losses.

> **N**o consistent national methods are employed in the United States to quantify water loss accurately—however, there are strong signs of change in a number of state and regional governments!

Conversely, the field of *water conservation* has become a well-structured discipline in a number of states; achieving considerable success in limiting unnecessary water consumption; particularly in the dry regions of the country where significant population growth is occurring and water is both limited and expensive. Water conservation focuses largely on water reductions by the end user by improving usage efficiency and reducing waste. It has achieved recognition at the national level with legislation in place that sets requirements for household water appliances and other water uses. The National Alliance for Water Efficiency is launching, with the support of the United States Environmental Protection Agency (USEPA), a multitude of successful regional water conservation efforts on a national scale. USEPA has also recently launched its *WaterSense* Program and water appliances are sold with a WaterSense label, just as appliances have carried an *EnergyStar* energy efficiency rating for many years. Unfortunately, supply side losses occurring due to leakage and poor accounting by water utilities are often many times greater than the end-user savings achieved through conservation; yet are still not adequately recognized.

> **T**he success of many water conservation efforts in the United States sets the stage for improved structures to motivate water loss control; particularly since water loss management offers the ability to supplement conservation savings many times over with the often high volume savings potential of water loss recovery.

4.2.1 Cultural Attitudes

Americans are the world's consumers. As shown in Fig. 4.1,[2] their water consumption ranks them as the world's highest per capita water users, when assessing source water withdrawals for all uses: including the majority uses of power generation and agriculture, in addition to drinking water supply. The authors would like to mention that the

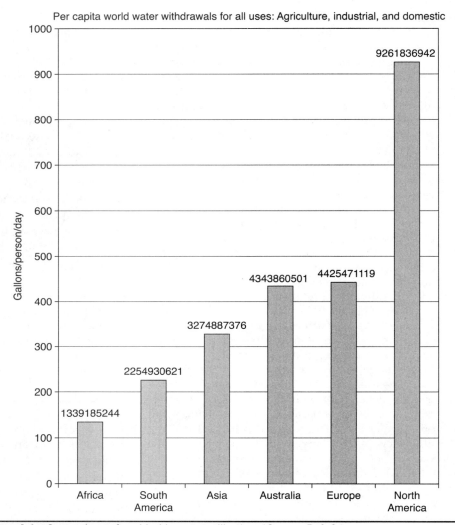

Per capita world water withdrawals for all uses: Agriculture, industrial, and domestic

Figure 4.1   Comparison of world wide water utilization. (*Source:* Ref. 2.)

consumption volumes shown in Fig. 4.1 are subject to a certain level of error. Basically, good data does not exist in many countries, so assessments like these always must be interpreted in a very general manner. However, Fig. 4.1 provides a good general picture on the significant differences in world water withdrawals for all uses: agriculture, industrial, and domestic.

"Conserving" is sometimes viewed as "doing with less," a notion that sometimes runs contrary to the American way of thinking, which is often geared toward building, development, and exploitation of resources. For many utilities water is unmetered, thus removing the "finite" sense of the resource from the thinking of both the consumer and the supplier. Like other parts of the world, water is often under-valued—literally and

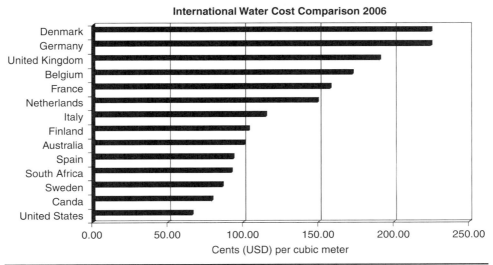

International Water Cost Comparison 2006

Figure 4.2   International comparison of water cost. (*Source*: NUS Consulting Group International Water Survey & Cost Comparison, July 2006.)

emotionally—in the United States. Costs to the consumer are often intentionally suppressed for social or political reasons (See Fig. 4.2 for a comparison of international water cost.)

4.2.2   Geography and Demographics

The fastest population growth is seen in the "sunbelt" states where water is often scarce and expensive. The critical role of water in assisting development results in a good appreciation for conservation in these areas, and generally younger infrastructure encounters less loss due to leakage. However, the frequent need to import water over vast distances requires complex planning and negotiations and the need for large, energy-intensive infrastructure (reservoirs, pipelines, and pumping stations), which makes it even more important to reduce water losses to an economically optimum level.

In contrast to the fast growing populations in the "sunbelt," population growth has slowed in the former industrial states where water has been relatively plentiful and inexpensive. Often having still-abundant resources and excess capacity, but a declining customer base and aging infrastructure, losses are often overlooked in these systems, even as they continue to grow.

4.2.3   Water Utility Organization and Structure

Most of the 55,000 water suppliers in the United States are extremely small utilities existing in rural areas; while a relatively small number of medium- and large-sized systems supply the largest share of consumers in densely populated areas. About 3700 of the largest water supply systems in the United States provide water to about 80% of the country's population. Most water utilities are municipally owned and operated. A small number of large private companies operate systems in multiple states. Some of the water suppliers are identified as irrigation districts since they were originated to provide water for agricultural irrigation. There are several large water wholesalers, providing bulk volumes of water to small suppliers. The organizational and management structure of water utilities varies widely with many systems operated by local governments; either as municipalities or authorities; and many large and small privately operated systems existing as well. System boundaries usually coincide with political boundaries rather than natural (watershed) boundaries.

Typically, water accountability practitioners are distribution system operators and water conservationists are public affairs or policy professionals. Lacking a national awareness and consensus on the overall water loss problem, these two camps historically did not interact widely or integrate their efforts under a single water conservation/efficiency mission. Fortunately, this has begun to change as stakeholders from both disciplines are now coordinating on a number of important initiatives.

Establishing standards amid this wide array of conditions is complex but as demonstrated by the implementation of complex water quality mandates under the U.S. Safe Drinking Water Act (1974, 1996 amendments), not insurmountable.

4.2.4   Environmental Perspective

The United States' environmental consciousness has grown steadily over the past several decades and is now a balancing force in planning and development decisions in the country. The establishment of the USEPA confirmed that consideration for the environment must be part of the decision-making process.

High water losses indirectly result in oversized infrastructure, excess energy usage and unneeded withdrawals or abstractions, from source water supplies; all of which exert a potentially unnecessary—and sometimes damaging—impact to the environment.

It is likely that a notable number of new source water abstractions and infrastructure expansions could be avoided if loss reduction was achieved, that is, water loss reduction could possibly represent one of the largest components of untapped water resources and potential for energy reduction currently existing in the United States.

4.2.5   The Current Regulatory Structure for Water Loss Management

The structure of the U.S. drinking water industry is highly fragmented, both in ownership and organizational oversight. The regulatory structure varies from state to state, with many water utilities falling under the auspices of two or more regulatory agencies that may include government environmental agencies, public utility commissions, river basin commissions, water management districts; as well as one or more federal agencies. Other important stakeholder organizations, such as county conservation districts, planning commissions, and watershed associations may also be party to the input and discussion about water resources management.[1]

In the late twentieth century, significant federal governmental involvement created extensive water quality legislation and rules for clean streams and drinking water. Conversely, federal requirements for auditing water delivery and customer consumption have historically existed with only minimal structure and degree of impact.

Considerable concern has grown for the need to replace aging infrastructure and identify appropriate funding mechanisms. Yet the scope of infrastructure needs is often based on projections that don't include improvements from loss reduction. A more modest estimate of national infrastructure needs might be derived if realistic loss reduction and conservation were consistently included in the analysis.

In 2001, the American Water Works Association (AWWA) conducted a comprehensive survey of state and regional water loss standards, policies, and practices entitled "Survey of State Agency Water Loss Reporting Practices."[3] The survey report concluded that even though a reasonable number of state and regional agencies hold a water loss policy, targets and standards vary widely from agency to agency. The survey confirmed that the structures in place to monitor drinking water supply efficiency are superficial in nature, of limited sophistication (in most cases "unaccounted for water" percentage is the sole performance indicator), and include scarcely any auditing or enforcement mechanism to validate the performance of drinking water utilities. The study clearly identified that in most cases the agencies do not provide incentives for achieving the required targets nor do they take action for failure of meeting targets. A very important finding of this study was that it is necessary to refine current definitions, measures and standards for evaluating water losses in the United States. The establishment of a uniform system of water accounting, with valid and reliable data, was proposed by this study.

4.2.6 Current Water Loss Management Practices

The starting point for successfully managing water losses is to accurately assess water supply and consumption volumes by conducting a standardized IWA/AWWA water audit. Many water audits are performed by utilities in the United States annually, but they lack uniformity. The audit methods used, the performance indicators and expressions of water losses calculated, and the time intervals between audits vary significantly from utility to utility. The majority of water utilities do not use the IWA water audit methodology recommended by the AWWA Water Loss Control Committee (WLCC). Therefore, it is impossible to accurately compare water losses between utilities since the assessment is not uniform. The historic indicator used to describe water losses (% volume of nonrevenue water) is highly unreliable and inappropriate. This percentage is unduly influenced by the denominator (system input volume) resulting in understated losses for water utilities with growing populations and overstated losses for utilities with contracting populations. Also, this simple percentage reveals nothing about specific loss volume quantities and costs, which are two of the most important parameters in the analysis.

The following simplified example clearly demonstrates how misleading and inappropriate percentage figures are when used as performance indicator for water loss management. In our example, we look at a standard U.S. water utility with 20,000 residents (no commercial or industrial customers) and an average per capita consumption of 400 gal/cap/d with a total metered consumption of 2920 mg/year. Assuming the utility has 325 mg of real losses per year the utility has a total system input of 3245 mg/year. The percentage loss figure for this utility is therefore 10%. If the same utility reduces the per capita consumption to 200 gal/cap/d through a successful demand side conservation program the total yearly metered consumption is reduced to 1460 mg. With no reduction in real losses the total system input is therefore reduced to 1785 mg/year, which results in a percentage loss figure of around 18%. This simple example explains why expressing water losses as a percentage of system input volume is a poor performance indicator.

North American utility with typical per capita consumption of 400 gal/cap/d:

Total system input volume:	3245 mg
Total consumption volume:	2920 mg
Total losses:	325 mg

Percentage of losses as % system input volume: 325/3245 mg = 10%
Same utility in with per capita consumption of 200 gal/cap/d:

Total system input volume:	1785 mg
Total consumption volume:	1460 mg
Total losses:	325 mg

Percentage of losses as % system input volume: 325/1785 mg = 18%

Figure 4.3 provides another example highlighting the weakness of the percentage indicator (in this case, *metered water ratio*) as it shows little variation despite a significant reduction in nonrevenue water over the 12-year period as shown by the trend line. This occurs since consumption in Philadelphia has also been in decline.

The current lack of structures, regulations, and uniform assessment methods of water losses contribute to the fact that water loss management is still a rather weak and neglected discipline in the United States. The AWWA survey "Survey of State Agency Water Loss Reporting Practices" and the AwwaRF report "Leakage Management Technologies"[4] both clearly highlight that most water utilities employ only reactive leakage management, which consists solely of repairing broken or burst water mains and leaks that have caused customer complaints and/or became visible on the surface.

Broken water mains are the most recognizable example of a reported leak, which, due to their damage-causing nature, are usually quickly reported, responded to and contained. However, unreported leaks, which frequently escape the attention of the water supplier and the public, account for larger amounts of lost water since they run undetected for long periods of time. While most water suppliers in the United States provide reasonable response to reported leaks, those that conduct regular unreported leak searches, or leakage surveys, (usually at 1- to 5-year intervals) probably represent a minority of the country's systems. Many systems conduct no surveys to detect unreported leaks. Generally, only the larger water systems employ specific "leak detection" personnel and purchase sophisticated leak correlators or other electronic equipment. Smaller systems typically rely upon leak detection consultants to provide pinpointing

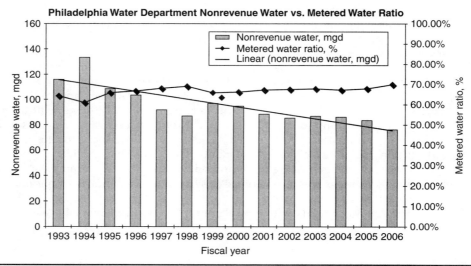

FIGURE 4.3   Nonrevenue water vs. metered water ratio. (*Source*: George Kunkel.)

services for hard-to-find leaks and to conduct periodic surveys of their systems to search for unreported leaks.

More sophisticated leakage management technologies such as district metered areas (DMA) or flow-modulated pressure control are only used by a handful of utilities in the United States.

4.2.7 Positive Developments in the United States—Regulations, Standards, and Practices

Water and energy conservation has become increasingly important for water utilities and policy makers, and utility managers are increasingly realizing that the improved accountability and loss control is important from environmental, political, and economical points of view. This trend is strengthened by factors such as ongoing droughts, increasing population in the U.S. western states, expensive water resources, and possible future regulations for distribution systems by the USEPA.

Significant progress was made over the past 5 years with several groundbreaking regulations and publications— leading the United States into active and efficient management of water losses.

Since the first edition of this manual was published in 2002, several very important and positive initiatives took place, preparing the way for successful water loss management in the United States.

The most important initiatives are listed below:

- In 2001, the American Water Works Research Foundation (AwwaRF) Research Advisory Council funded project #2811 "Evaluating Water Loss and Planning Loss Reduction Strategies"[5] to help refine water loss definitions, measures, and standards for North America. The final report of this important project was published in 2007, and is now a standard reference for water loss management in North America.

- In 2003, the AWWA-WLCC recommended both the IWA water balance and the IWA performance indicators (including the *infrastructure leakage index*) in their committee report as the current industry best practice for assessing water losses.[6]

- In 2003, the Texas State Legislature passed House Bill 3338, which includes in its language a requirement for drinking water utilities to submit a water audit every 5 years. The Texas Water Development Board (TWDB) was charged to identify the method to be used for these water audits and has established the method developed by IWA. Texas is the first state in the United States that has adopted the IWA best management practice for water audits. Texas has set a clear signal that it supports standardized and unambiguous assessment of water losses.

- Since 2003, several other water oversight agencies have set forth to improve water supply efficiency and long-term sustainability. The following organizations are reviewing state regulations, statutes, and water plans:
  - California Urban Water Conservation Council (CUWCC)
  - California Public Utilities Commission
  - Delaware River Basin Commission (DRBC)
  - States of Georgia, New Mexico, Washington, Tennessee, Maryland, and Pennsylvania

- In 2003, AwwaRF's Research Advisory Council and the USEPA-funded project #2928 "Leakage Management Technologies" to review internationally applied proactive leakage management technologies, assess the applicability of these technologies in North America and to provide guidance on how to practically and cost effectively implement these technologies in North America. A comprehensive report covering all aspects of this important research project was published in 2007.

- The AWWA-WLCC is rewriting the AWWA M36 Manual of Water Supply Practices, *Water Audits and Leak Detection*, to provide guidance on the IWA water audit method, as well as progressive apparent and real loss controls. The new AWWA M36 manual, entitled *Water Audits and Loss Control Programs*, is scheduled for publication by early 2009.

- A free, introductory software developed by the AWWA WLCC became available in early 2006. The software includes a water balance and performance indicators, based on the AWWA-approved standard IWA water audit methodology and performance indicators. The software can be downloaded from the AWWA Web site's *WaterWiser* homepage.

Significant progress was made in the United States over the past 5 years in a similar way to the initial transitions that occurred in the United Kingdom during the late 1980s and 1990s. The United Kingdom is now among the leading nations in terms of active and efficient leakage management.

4.3   International Leakage Management

Leakage management projects funded by governments, utilities, and international funding agencies are being implemented through out the world. However, only a few countries have established successful nationwide leakage management regulations and practices. This section provides the reader with a general overview of effective leakage management structures in several countries around the globe, with a special focus on England and Wales.

4.3.1   Leakage Management in the United Kingdom

This section refers to England and Wales when talking about the United Kingdom, since those are the two regions with the most structured leakage management regulations in the United Kingdom.

An interesting contrast can be drawn between the proactive system addressing water loss in England and Wales and the current conditions in the United States. A number of factors contributed to the establishment of England's progressive demand and leakage management structure in the 1990s. The reorganization, privatization, and regulation of the small number of large water companies in 1989 created an important change in the business model used for water supply. With revenue growth potential limited due to government regulation of customer rates or tariffs, leakage reduction was one of many efficiency improvements targeted by the companies to cut costs and improve their bottom line. The National Leakage Initiative of the early 1990s was a major research project underwritten by the water companies to determine the best methods to employ to reduce leakage. The severe drought of the mid-1990s prompted mandatory targets for leakage reduction from the government's economic regulator,

Office of Water Services (Ofwat); which most companies have achieved due to their ability to quickly implement the recommendations of their leakage reduction research. Enormous efforts to control water losses were undertaken in the United Kingdom since the early 1990s, with water loss reduction being a major operational task for water utilities. Today, water companies in the United Kingdom have a detailed understanding of their components of water losses and the economic optimum of their losses. The water companies now operate "trans-

Severe drought in the United Kingdom in the mid-1990s prompted mandatory leakage reduction—a scenario that could materialize in the United States given its many drought-stricken regions.

parently" in calculating and publicizing data on their water loss volumes. Most companies claim that they have reached, or will reach soon, their economic optimum level of leakage. Total leakage in England and Wales was reduced from 1350 mgd (5112 ML/d) in 1994–95 to 856 mgd (3243 ML/d) in 2000–01. This represents a reduction of 37% in leakage or a volume of 528 mgd (2000 ML/d), enough water to supply more than 12 million people.

A rise in leakage volume during the 2001–02 year (see Fig. 4.4) was caused by increasing leakage volumes at Thames Water, which have continued to increase against the general downward trend seen from all other England and Wales water companies. In 2002–03, Severn Trent Water showed a rise in its leakage volumes as well. Both companies are under strict scrutiny by Ofwat to ensure that they improve their performance according to their set targets. Thames Water and Severn Trent Water aside, leakage volumes for all of the remaining water companies have continued to fall further.

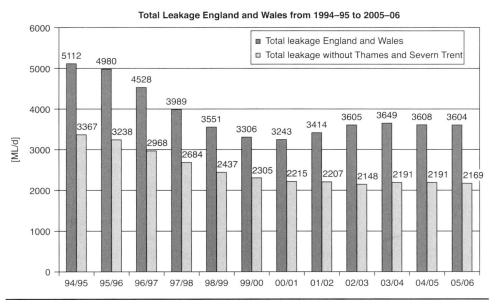

FIGURE 4.4   Leakage reduction in England and Wales between 1994–95 and 2005–06. (*Source*: Ofwat, compiled from annual leakage reports—in public domain.)

Regulations for Leakage Management

After the privatization of the water supply industry in England and Wales in 1989, 10 large regional water supply and sewerage companies were formed, which, together with 16 statutory water supply only companies, cover the entire water supply of England and Wales regulated by the Water Act of 1991. U.K. governance features two primary regulatory bodies for the water industry, Ofwat and the Environment Agency (EA). Ofwat serves as the economic regulator and the EA as the environmental regulator. Each water company is required by Ofwat to produce a detailed annual report on the volume of water supplied, consumed, and lost in each component part of the network, using a standardized water balance (similar to the standard IWA water balance). The water balance results have to be cross checked via minimum hour flow analysis of data from DMA, which are discrete zones established to distinguish leakage events from customer consumption.

The results of these reports are used by Ofwat to assess the performance of each utility, to set performance targets and for intercompany comparisons. The mandatory leakage targets set by Ofwat for each water company (in ML/d) must be met by the company in order to avoid sanctions by Ofwat.

Overall, the assessment, reporting and management of water losses are highly regulated in England and Wales. This is paired with clear definitions, measures, and standards for assessing and evaluating water losses. The two regulatory agencies monitor the performance of all water companies closely and set performance and efficiency targets driven by an economic optimum volume of leakage established for each water company.

Leakage Management Practices

Developments of the past 20 years have resulted in a detailed understanding of the interaction between the four fundamental leakage management practices:

- Infrastructure management
- Pressure management
- Active leakage control
- Speed and quality of repair

The understanding and accurate assessment of the economic optimum volume of leakage is another major development and forms an integral part of a utility's water loss management strategy. Coherent leakage management strategies and oversight by regulatory bodies rely upon a uniform way of assessing water losses and setting economically and environmentally justified loss reduction targets. The main pillars of the highly successful leakage management practices used in England and Wales are

- *Improved business focus*: Departments and teams were created with the sole purpose of managing and reducing water losses to an optimum volume.
- *Improved data quality*: It was realized that the quality of data used for the water audits and establishment of targets was fundamental for a successful leakage management strategy.
- *Routine calculation of water balances and performance indicators*: In order to define and refine intervention targets and measures, standardized water balances and

performance indicators are calculated routinely. These calculations are supported by DMA minimum hour flow analysis.

- *Network zoning and DMA establishment*: It was acknowledged that one of the most efficient ways to reduce the volume of real losses is by reducing the runtime of leaks. DMAs and the related minimum hour flow measurements allow the leakage manager to deploy the leakage reduction recourses to those areas where leakage levels have reached a volume that justifies intervention.

- *Pressure management*: It is now a well-known fact that pressure management is the most effective and efficient way of reducing leakage. The general benefits of pressure management are threefold: reduction of background leakage, reduction of break rates on mains, and service connections and reduction of flow rate from any leak.

- *Reduced response time to repair leaks*: Once it was recognized that the run time of a leak is major factor contributing to the overall real loss volume; steps were taken to ensure that the average repair times were drastically reduced.

- *Customer side leakage*: After it was understood that a significant portion of the leakage volumes can occur on customers side of the service pipe, effective management of this leakage component was included in the over all leakage reduction strategy.

- *Improved leak detection efforts*: A leakage reduction program is only as good as its field personnel finding the leaks. Therefore, comprehensive training programs were developed in order to increase the skill level of the leak detection personnel.

- *Asset management*: It was realized that leakage management is an integral part of asset management. Infrastructure replacement is the most comprehensive improvement to an asset, but this action is also the most expensive step of the four management practices. A concentrated effort was launched to develop sophisticated asset management techniques to plan infrastructure investments and replacements on a strategic basis.

These leakage management practices are discussed in further detail in Chaps. 10 to 14.

4.3.2 Additional Examples of Improved International Leakage Management in Several Countries

A brief description of progressive international water loss management activities is provided to reflect the growing recognition of water loss impacts among countries around the world; and the actions that they are taking to promote water-efficient utilities.

Germany

The German water market exists of a multitude of small- and medium-sized enterprises and municipal companies. Water utilities are operated in different legal forms with the most common form being: municipal department, municipal utility, municipal company, joint venture, operator model and management, and service contract.[7] Currently there are about 5260 water supply enterprises in Germany. Germany has very strict guidelines and ambitious performance indicators for water loss management. However, it is interesting that those guidelines are driven by hygienic, supply sufficiency, safety, and environmental reasons; unlike England and Wales which are managed largely by economic considerations. In 2003, national guidelines were published for the

German water sector entitled *W 392—Network Inspection and Water Loss—Activities, Procedures, and Assessments.* They require the water supplier to assess and analyze the water system condition, to calculate and analyze water losses and to employ efficient water loss reduction measures.

Following the W 392 guidelines, German water utilities pursue a comprehensive maintenance strategy which includes asset management as well. Distribution network maintenance in Germany comprises regular inspections of the system and its components, preventive and corrective maintenance, and repair and rehabilitation. A very interesting aspect of the German approach is that maintenance activities are taken into full account in the planning and construction phases. Establishing DMAs during construction of network extensions and installing bulk meters on transmission mains for timely leak detection are examples of this holistic approach. The German guidelines stress the need for comprehensive metering of production, distribution, and customer consumption in order to balance flows and monitor real loss levels with great accuracy. German regulations require water utilities to conduct an annual water audit using a method with precise definitions of each component of the water balance. The recommended format of the water balance is in accordance with the IWA recommendations from the IWA *Manual of Best Practice.*[8] The W 392 guidelines discourage the use of output/input percentages as a real loss indicator by stating: "expressing real losses as a percentage of the system input volume is unsuitable as a technical performance indicator since it does not reflect any of the influencing factors. Systems with higher system input volumes (e.g., urban systems) will automatically have an (apparently) lower level of water losses if expressed in percentages. Systems with low water consumption (e.g., rural systems) will show high percentage figures of real losses. Therefore, comparisons using percentages will always favor systems with high system input."[9,4]

Australia[10]

The Australian water industry consists of over 300 water utilities. Most authorities/utilities are publicly owned in Australia, with many part of national or local government. Australia entered into a multiyear period of severe drought starting in 2002; an event that is threatening the existence of its agriculture industry and has thrust water loss management into the national political limelight. Over the past 2 to 3 years water loss management activities in Australia have grown substantially in importance for the water industry, as sustainable water management has become an issue of concern for the broader Australian community. The increased focus on water loss management has been led by IWA Water Loss Task Force Deputy Chair Tim Waldron who is the CEO of a medium-sized utility in Australia, Wide Bay Water Corporation.

On a world scale, water losses in Australia are quite low (infrastructure leakage index is typically between 1 and 1.7). These low levels of water loss are the result of; relatively new infrastructure, quick response times to known bursts and high standards for assets selection and asset management throughout the Australian water industry.

Despite these relatively low levels of water loss, the recent focus and investment by the water industry in water loss management has been driven by three fundamentals:

1. Very severe droughts and water scarcity in many of Australia's largest cities and populated regions
2. Government regulation regarding water loss management
3. Increased government funding for water loss management activities

The Australian water industry has adopted the methodologies of the International Water Association's Water Loss Task Force as an organizing concept for much of this work. The adoption of this framework has institutionalized by the fact that key water industry membership organizations such as the Water Services Association of Australia (WSAA) and the various water directorate organizations have made available software tools and information packages which use the water balance and terminology promoted by the IWA thereby creating this as a de facto Australian standard.

Regulatory reform has been led by the Queensland Government which requires all water service providers in the State of Queensland to

- Prepare a system loss management plan using IWA methodologies (water losses are to be valued at the retail sale price of water)
- Implement cost effective water loss management actions (e.g., active leakage control, pressure management, etc.). Cost effective actions are defined as any activity that will achieve a payback in less than 4 years.

This regulatory regime is now being reviewed by other state governments and commonwealth government regulators and it appears likely as if it may form a model for future regulatory action by government agencies in Australia. The Australian Government (the Commonwealth) through the Australian Water Fund has provided government funding to a number of key trial water loss management projects. The recently elected federal government (December 2007) made water loss management an election issue through the announcement of a major national funding package for water loss management. This national government funding has been reinforced with a number of state governments providing significant funding to assist water authorities to implement water loss management activities (In Queensland, one of the key drought ravaged areas, this state subsidy is 40% of overall project costs).

Thus the regulatory drivers and funding drivers are pushing the water industry in Australia to implement some very large water loss management projects. Most notably in South East Queensland water service providers are currently working on a system to implement DMAs and pressure management in communities currently servicing more than 2 million consumers. The savings that are being achieved through these programs are still significant despite the relatively low levels of losses prior to project implementation.

Since 2003, Gold Coast Water has engaged Wide Bay Water Corporation to implement one of the largest water loss management projects in Australia. The savings that have been achieved by this program are as follows:

Consumption (System Input Volume)

Total system input volume declined by 22.22% from 73,750.7 to 57,361.8 ML/year.

Overall demand has reduced from 1640 to 1091 L/conn/d, reflecting an overall reduction in demand of 549 L/conn/d.

Real Losses

The unit value for current system leakage has dropped from an initial 164 to 46 L/conn/d. (It should be noted that as a result of this performance Gold Coast Water is technically exempt from the preparation of a system loss management plan as the act exempts large water service providers if their real losses are less than 60 L/conn/d.)

The gap between system input and billed consumption has closed from 9134.7 ML/year, to less than half at 3637.7 ML/year. Current system leakage has reduced by 4951 ML/year or 13.56 ML/year.

These impressive results have been achieved by

- Establishment of district metered areas (50% of service connections)
- Establishment of leakage test zones (14% of service connections)
- Pressure management in appropriate zones
- Reservoir maintenance and repairs
- Mains replacement
- Replacement of service connections and water meters
- Asset condition assessment and replacement
- Improved *burst response time*

The implementation of extensive pressure management activities has led to a significant reduction in reported bursts in pressure managed areas.

4.4   The Need for Meaningful Regulations

When looking at the success stories of leakage management on a country by country base it is evident that the countries where water loss management is succeeding are those where well-structured and balanced federal or state water loss management regulations are in place. The United States largely lacks such structure; however a lack of uniform and proactive regulations is not limited to the United States since a similar lack of recognition of water loss problems exists around the world.

Many areas of the United States have suffered significant periods of drought in the past 20 years. A severe drought in California from 1987 to 1992 triggered strict customer demand restrictions, yet very little emphasis was placed on the need for water suppliers to accurately quantify and manage their water losses.

Severe drought in parts of the United States has been a primary reason why customer water conservation programs have become well established and backed by regulations and incentives coming from federal and state levels. Many of these programs, however, would not exist had local, state, or federal regulations failed to be enacted. In the United States, it is inevitable that meaningful, industry-wide accountability and loss control improvements will come about as new federal regulations are passed requiring such. The highly fragmented water regulatory structure in the United States makes regulatory decisions and structures highly complex; however, federal and state regulatory authorities should strongly consider the need to begin to formulate a basic regulatory structure to motivate water suppliers to assess and manage their water losses in accordance with recognized best management practices. The 1996 Amendments to the Safe Drinking Water Act are a good example that federal regulations can be applied to the U.S. drinking water industry. These regulations motivated new programs and structures that have clearly increased the quality of drinking water across the United States. Similarly, a regulatory structure for water accountability and loss control is possible in the United States; but awareness of the issues must be heightened and political will has to be mustered.

As discussed, there have been many very positive changes in the U.S. water industry since the start of the new millennium, with several states and regulatory authorities adopting and/or promoting standardized water loss management. The authors believe that it is only a matter of time until efficient water loss management is required on a federal level in the United States, with many projected benefits for water consumers, water utilities, and the environment.

4.5 Summary

Water loss is truly a global problem that requires focused attention and awareness from a wide variety of stakeholders: federal, state, and local governments, water suppliers, environmental groups, and consumers. The most successful water loss management programs around the world exist in countries which have enacted regulations requiring the water supplier to apply best management practices. The causes and remedies of water and revenue loss are now well understood, and innovative technology makes loss control efficient and cost effective. As demonstrated in a number of states in recent years, it is now necessary for the insidious issue of water loss to assume a position of priority on the policy and regulatory agenda of the United States.

Table 4.1 provides a comparison of general characteristics, water loss manage-ment methods, and regulatory structures in the Unites States, England and Wales, and Germany.

Parameter	United States	England and Wales	Germany
General Characteristics			
Number of water suppliers	More than 59,000	23	More than 5000
Legal form of water suppliers	Great majority public	Private	Great majority public
Per capita consumption	100 to 200 gal/cap/d (376 to 752 L/cap/d)	38 gal/cap/d (145 L/cap/d)	34 gal/cap/d (130 L/cap/d)
Service density	70 to 100 con/mi (44 to 63 con/km)	40 to 150 con/mi (25 to 94 con/km)	40 to 150 con/mi (25 to 94 con/km)
Pressure	~71 psi (50 mH)	~71 psi (50 mH)	~ 43 psi (30 mH)
Proportion of metered residential customers	95 to 100%	5 to 60%	95 to 100%
Break rate	250 breaks/1000 mi/year (156 breaks/1000 km/year)	350 breaks/1000 mi/year (219 breaks/1000 km/year	not collected
Real losses	75 gal/con/d (282 L/con/d)	30 gal/con/d (113 L/con/d)	19 gal/con/d[1] (71 L/con/d)

(*Source:* Ref. 4.)

TABLE 4.1 Comparison of General Characteristics of Water Loss Management Methods and Regulatory Structures in the United States, England and Wales, and Germany (*Continued*)

Parameter	United States	England and Wales	Germany
Water Loss Assessment and Leakage Management Performance Indicators			
Water audit formats	AWWA Manual M36 and custom audits, IWA/AWWA recommended Audit used rarely	Standardized Water Audit comparable to IWA/AWWA recommended audit format	Standardized Water Audit in accordance with IWA/AWWA recommended audit format
Use of audits	Overall very limited. Required only by certain states	Required for all water utilities by regulator	Required for all water utilities
Implications of water audit results	Rather limited overall, varies by state	Serve as basis for setting leakage management and performance targets	Serve as basis for setting leakage management and performance targets
Leakage management performance indicators	Percent of system input volume is mostly used, although has proven to be unreliable indicator	Volumetric and financial indicators used in accordance with IWA recommendations	Volumetric performance indicators used
Water loss standards	Limited in extent, detail, and, where mandated, level of enforcement; regulations vary widely at the state, regional, and local levels	Extensive and detailed: uniformly enforced by central government regulator	Extensive and detailed standards—details about enforcement not available
Leakage Management Practices			
District metered areas	Generally not used to a wide extent	A well-established and required practice	A well-established and required practice
Pressure management	Standard pressure management is prevalent—advanced pressure management used rarely	Standard and advanced pressure management is used, a standard component of leakage management	Standard and advanced pressure management is used, a standard component of leakage management
Repair of customer service connections	Usually responsibility of customer	Company-paid or subsidized for first or subsequent leaks	NA

(*Source:* Ref. 4.)

TABLE 4.1 Comparison of General Characteristics of Water Loss Management Methods and Regulatory Structures in the United States, England and Wales, and Germany (*Continued*)

Parameter	United States	England and Wales	Germany
Reduced response time to leaks	Varies greatly from utility to utility	Main component of leakage management practice	Main component of leakage management practice
Use of leak detection equipment	Only a small number of utilities have necessary technology to effectively detect leaks	All utilities are equipped with necessary leak detection technology to meet set performance targets	Leak detection technology used as necessary to meet targets

TABLE **4.1**   (*Continued*)

References

1. Kunkel, G. "Developments in Water Loss Control Policy and Regulation in the United States." In *Proc. of the Leakage 2005 Conference*. Halifax, Canada: The World Bank Institute, 2005.
2. Pacific Institute. The World's Water 2006–2007. Available online: www.worldwater. org/index.html. [Cited January 10, 2008.]
3. Beecher Policy Research, Inc. "Survey of State Agency Water Loss Reporting Practices" *Final Report to the Technical and Educational Council of the American Water Works Association* 2002.
4. Fanner, V. P., R. Sturm, J. Thornton, R., et al. *Leakage Management Technologies*. Denver, Colo.: AwwaRF and AWWA, 2007.
5. Fanner, V. P., J. Thornton, R. Liemberger, et al. *Evaluating Water Loss and Planning Loss Reduction Strategies*. Denver, Colo.: AwwaRF and AWWA, 2007.
6. Kunkel, G. "Applying Worldwide Best Management Practices in Water Loss Control, AWWA Water Loss Control" Committee Report. *Jour. AWWA*. vol. 95. 2003.
7. BMU (Bundesministerium für Umwelt, Naturschutz und Reaktorsicherheit). *The German Water Sector Policies and Experiences*. Berlin, Germany: BMU, 2001.
8. Alegre, H., W. Hirner, J. M. Baptista, et al. *Performance Indicators for Water Supply Services—IWA Manual of Best Practice*. London: IWA Publishing, 2000.
9. DVGW (Deutsche Vereinigung fuer das Gas- und Wasserfach). *W 392 A: Rohrnetzinspektion und Wasserverluste—Massnahmen, Verfahren und Bewertung*: 2003.
10. Wiskar, D. The Information on Water Loss Control Policies in Australia. Provided by: General Manager—Business Services Wide Bay Water Corporation (personal communication).

CHAPTER 5

Steps and Components of a Water Loss Control Program

Reinhard Sturm

Julian Thornton

George Kunkle, P.E.

5.1   Introduction

There are many factors such as financial constrains, infrastructure condition, skills and technologies available, cultural and political conditions all of which are influencing a utility's ability to manage water losses. However, it should be the aim of every water utility to improve the current operational practice in order to achieve higher efficiency and to be able to provide better service to the clients. A water loss control program is without doubt an excellent tool to improve efficiency and the service provided. In order to implement a water loss control program it is first necessary to understand and assess the problem through a diagnostic approach and then design and implement actions/programs to solve the problem. This principle applies to any water company in the world.

This chapter will provide an overview on the various steps and components of a water loss control program. The content is kept brief since all components of a water loss control program will be discussed in detail in following chapters. This chapter should serve as a road map for the reader to understand the general concept and steps involved in a water loss control program. Figure 5.1 depicts a road map of a water loss control program.

5.2   Top-Down and Bottom-Up Water Loss Assessment—How Much Water Are We Loosing and Where?

On of the most important parts of a water loss control program is to assess and understand the components of water loss. However, it is equally important to understand that the accuracy of each calculated water loss volume depends on the accuracy and quality of data used for the calculations. Hence, data validation plays a key role in the assessment of water loss volumes.

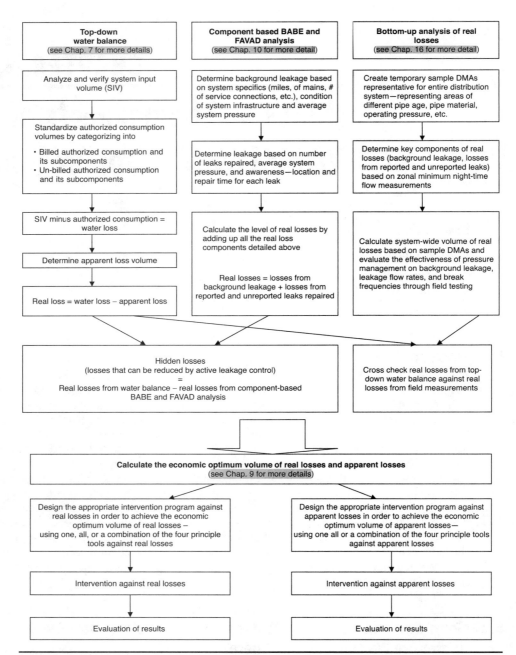

Figure 5.1   Water loss control program—road map. (*Source:* Reinhard Sturm)

5.2.1 Top-Down Water Balance

The first step in this analytical process of assessing and calculating the volume of real and apparent losses is to undertake a IWA/AWWA recommended standardized top-down water balance (see Chap. 7 for a detailed guidance on how to undertake a water audit). Good management of any resource requires that the supplier maintain accurate records of transactions and deliveries of the commodity provided to its customers. A water balance has exactly that goal, tracking and accounting for every component of water in the cycle of delivery. The water balance tracks the flow of water from the site of withdrawal or treatment, through the water distribution system up to the point of customer consumption. The water balance usually exists in the form of a worksheet or spreadsheet that details the variety of consumption and losses that exist in a water system. The water balance itself is a summary of all the components of consumption and losses in a standardized format. Every unit of water supplied into the system needs to be assessed and assigned to the appropriate component. It is certainly not best practice to have units that are unaccounted for.

It is quite common that the calculated volumes of real and apparent losses have a relatively low level of confidence the first time a water balance was established. There are many reasons for a low level of confidence in the calculated water loss volumes the first time a standardized water balance is established however the most common ones are that some of the water balance components are not metered and/or the data used has not been validated. Therefore, it might be necessary to first increase the confidence in the calculated water loss figures by validating all the volumes entered in the water balance through meter accuracy testing, improvement of record keeping, and estimation practices and if necessary installing new system input and/or export meters. The water utility will realize that the auditing process is a revealing undertaking that provides great insight to the auditor on the type and volumes of water loss (real and apparent losses) occurring in the utility.

The real loss volume calculated through the water balance includes real losses from leaks that have been repaired (through an active or reactive leakage management policy), the volume of background losses, and real losses that are due to leaks still running in the system. The losses caused by leaks which still need to be detected and repaired by the utility are called *hidden losses*. However, just by establishing the water balance it is not possible to estimate the volume of hidden losses. It is recommended as best practice by the International Water Association (IWA) and American Water Works Association Water Loss Control Committee (AWWA WLCC) that the assessment of real losses using a "top-down" water balance should be complemented by the following two methodologies:

- *Component analysis of real losses*: A technique which models leakage volumes based upon the nature of leak occurrences and durations (see Chap. 10 for more details)
- *"Bottom-up" analysis of real losses*: Using district metered area (DMA) and minimum night-time flow (MNF) analysis (see Chap. 16 for more details)

5.2.2 Component Analysis of Real Losses

As already mentioned, it is best practice that in parallel to establishing a water balance a component analysis of real losses is carried out to assess the volume of hidden losses and to get a detailed understanding of the efficiency of the current leak repair policy.

In 1994, a concept called *Burst and Background Estimates* (BABE) was published, acknowledging that the annual volume of real losses consists of numerous leakage events, where each individual loss volume is influenced by flow rate and duration of leak run time before it is repaired. A component-based leakage analysis breaks leakage down into three categories:

- *Background* leakage (*undetectable*): Small flow rate, continuously running
- *Reported breaks*: high flow rate, relatively short duration
- *Unreported breaks*: moderate flow rates, the run time depends on the intervention policy

It is not recommended that a component analysis is undertaken on its own to derive a volume of annual real losses because there is likely to be a significant level of uncertainty in much of the data used in the analysis. However, a component analysis is a very useful supplement to a top-down water balance because it provides estimates of the volumes of real losses in different elements of the distribution infrastructure. This data is so valuable because it is required to develop the most appropriate loss reduction strategy and it is essential for a robust determination of the *economic level of leakage* (ELL).

As depicted in Fig. 5.1 the water balance calculates the total volume of real losses for the audit year. However, it does not provide the information on what portion of these real losses is due to hidden losses (losses from leaks that have not been captured by the utilities current leakage management policy). By assessing the volume of real losses through component-based analysis, it is possible to determine the volume of real losses that have been captured through the current leakage control policy. Therefore, by deducting the real losses based on the component-based analysis from the real losses based on the top-down water balance, it is possible to determine the volume of hidden losses.

Hidden losses = real losses from top-down water balance
– real losses from component analysis

The results from this analysis can then be cross checked against the real loss volumes measured in DMAs (see Sec. 5.2.3).

Water balances and component analysis of real losses have to be carried out at least once a year since they are such an integral part of any water loss control program. Many utilities establish water balances on a monthly basis to keep a close eye on their water loss management performance.

5.2.3   Bottom-Up Analysis of Real Losses Using DMA and Minimum Night-Time Flow Analysis

The two ways of assessing real losses explained in the previous sections can be generalized as desktop analysis. However, an MNF analysis uses field test data to quantify the volume of real losses within the distribution network. The results can be directly compared with the volume of real losses obtained from the top-down water balance. A DMA is required in order to conduct MNF measurements. A DMA is a hydraulically discrete part of the distribution network that is isolated from the rest of the distribution system. It is normally supplied through a single metered line so that the total inflow to the area is measured (Fig. 5.2).

General DMA setup

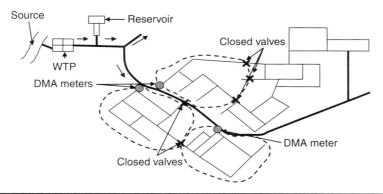

Figure 5.2 General DMA setup. (*Source:* Adapted from IWA Water Loss Task Force)

The MNF in urban situations, usually occurring between 2 a.m. and 4 a.m., is the most meaningful piece of data as far as leakage levels are concerned. During this period, authorized consumption is at a minimum and, therefore, leakage is at its maximum percentage of the total flow. The estimation of the leakage component at minimum night flow is carried out by subtracting an assessed amount of legitimate night-time consumption for each of the customers connected to the mains in the zone being studied. Typically, in European and North American urban situations, about 6% of the population will be active during the minimum night-time flow period. This activity is typically to use a toilet and the water use is almost totally related to the flushing of WC cisterns, although it can include substantial amounts of irrigation at certain times of the year. Analysis of minimum night-time flows therefore also requires the use of sophisticated techniques to determine legitimate night use. If it is known that there is significant or non-normal night use, otherwise known also as exceptional night use, within the zone, then this must also be estimated or measured by for example carrying out meter reading tests during the minimum night period.

The result obtained from subtracting the assessed night use and exceptional night use from the minimum night-time flow is known as the *net night-time flow (NNF)* and it consists predominantly of physical losses from the distribution network.

After completing these three initial components of a water loss control program it is now necessary to transfer volumes into values in order to determine the economic optimum volume of leakage.

5.3 Determine the Economic Optimum for Your Water Losses

Water loss management is an economic issue. Utilities should aim to manage losses in order to minimize overall operating costs. With any water loss reduction strategy, the lower the level of water losses achieved, the higher the cost of reducing water losses further. For this reason, it is never economic for a utility to remove all water losses. The economic optimum is the economic balance point at which the value of water lost (real or apparent losses), plus the cost to reduce the volume of real or apparent losses, is at a minimum. It is certainly best practice to determine the economic optimum point for both the real losses and the apparent losses in order to see if there is room for economically justifiable real and/or apparent loss reductions.

Models to determine the long-term economic optimum volume of water loss, like used in England and Wales, for example, can be highly complex and very labour- and data-intensive.

However, by using a short-term economic analysis, which is basically transferring water loss volumes into values and comparing them to the cost of intervention, a much less labour and data intensive approach exists to provide a utility with the economic water loss benchmarks it needs to determine its optimum intervention program. Each water system will have different types and degrees of loss and each has a potential solution and each solution has a cost. However, before the cost to benefit ratio can be defined the potential solutions have to be identified and graded technically. In addition to having a good return or cost to benefit, it is also important when considering intervention to take into account the local conditions and the sustainability of the method or solution adopted. Water losses don't go away they keep on coming back. Water loss control is not a one-time exercise it is a continuous and changing solution to an ever-changing problem.

See Chap. 9 for a detailed discussion on determining the economic optimum volume of water loss.

5.4   Design the Right Intervention Program

As we already know from previous chapters there are two basic forms of water losses—real and apparent losses—and for both types of losses the right intervention program needs to be designed. The design of the right intervention program is directly connected or interlinked with the process of determining the economic optimum volume of water losses.

This section will provide a brief overview of the common intervention methods available against real and apparent losses which form the bases for the design of every water loss control program. Chapters 11 to 15 will provide an in-depth discussion of all intervention methods available against apparent losses and Chaps. 16 to 19 will provide an in-depth discussion of all intervention methods available against real losses.

5.4.1   Real Loss Intervention Methods

The decision on which intervention methods are appropriate for the given situation will depend very much on which factors are attributing to the real losses in any particular system and the cost benefit of each intervention method. Figure 5.3 shows a component break down of intervention methods against real losses. Each of the four arrows represents an intervention method or a set of intervention methods against real losses. Depending on the local situation the final real loss intervention program may consist only of one or a combination of several or all intervention methods which will serve to bring the real losses down to the economic optimum volume.

5.4.2   Apparent Loss Intervention Methods

Just as for the real losses there is also a set of intervention methods available to reduce the volume of apparent losses down to the economic optimum point. Figure 5.4 shows a component breakdown of intervention methods against apparent losses. Each of the four arrows represents an intervention method or a set of intervention methods against

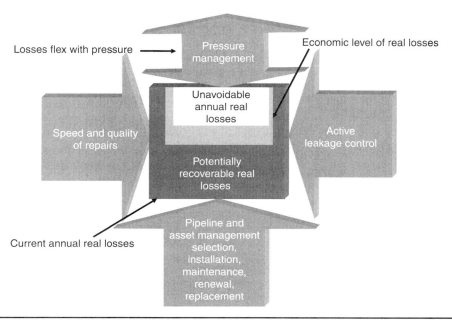

FIGURE 5.3   Four potential intervention tools of an active real loss management program. (*Source:* IWA Water Loss Task Force and AWWA Water Loss Control Committee.)

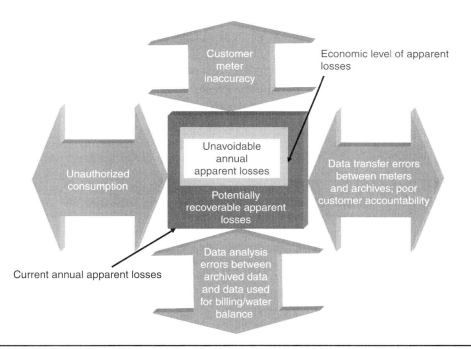

FIGURE 5.4   Four potential intervention tools of an active apparent loss management program. (*Source:* IWA Water Loss Task Force and AWWA Water Loss Control Committee.)

apparent losses. Depending on the local situation the final apparent loss intervention program may consist only of one or a combination of several or all intervention methods which will serve to bring the apparent losses down to the economic optimum volume.

5.5   Implementation Phase

Once the right intervention methods have been identified its time to implement them. The implementation is either carried out in-house or contracted out depending on the resources and the expertise of the water utility. In many cases, you can see a combination of in-house implementation and outsourcing.

5.6   Evaluate the Results

The evaluation phase at the end of the water loss control program is necessary to assess the results achieved by the program. Basically a new water balance complemented by a component analysis is undertaken and if necessary DMA measurements are carried out and the results are compared to the results before start of the water loss control program. If the intervention program took place on a DMA level then it is best to repeat the DMA measurements after completion of the intervention.

If a program extends over several years then it is advised to measure the results at least on an annual base to see if the water loss reduction efforts are moving into the right direction.

It is important to bear in mind that once the goals are achieved it is necessary to continue with the water loss control efforts in order to maintain the economic optimum volume of water losses. This is necessary because water losses increase over time if no control measures are taken. However, the efforts necessary to maintain the optimum point will be less than the efforts that were necessary to get to the optimum point.

5.7   Examples of Water Loss Control Program Costs in North America

Cost effectiveness of demand-side water conservation programs is expressed in a cost per unit of water saved. Since demand-side conservation is already widely applied in North America, especially the western parts of the Unites States, there is a wide set of demand-side water conservation cost figures available. In a paper[1] written by the authors of this manual, the cost effectiveness of several water loss control programs carried out in North America was assessed in order to compare the cost effectiveness of water loss control programs with demand-side conservation programs.

This analysis showed that water loss control program costs do vary from utility to utility. A general guideline is that water loss control programs are cheaper when the volume of real losses is high. The lower the volume of real losses the more effort is required to reduce them and therefore the overall cost for the program increases. See Fig. 5.5 for a cost comparison of several water loss control programs. It is important to note that all of these programs only reduced real losses with no intervention against apparent losses. The cost shown includes all components of a water loss control program starting at the point of assessment (water audit) and including all costs to intervene against real losses including the cost to repair the leaks.

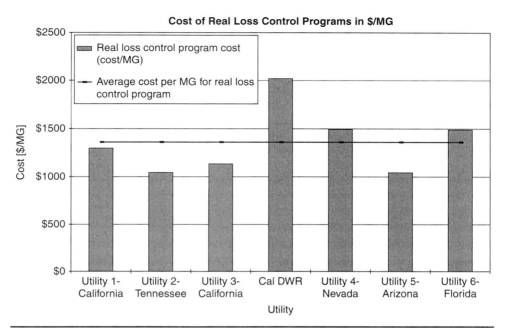

FIGURE 5.5   Cost of real loss control programs. (*Source:* Ref. 1.)

5.8   Conclusion

The reader should now have a general understanding of the steps involved in a water loss control program. Building on this general picture Chaps. 11 to 15 will provide an in-depth discussion of all intervention methods available against apparent losses and Chaps. 16 to 19 will provide an in-depth discussion of all intervention methods available against real losses.

Reference

1. Sturm, R., and J. Thornton. "Water Loss Control in North America: More Cost Effective Than Customer Side Conservation—Why Wouldn't You Do It?!" In *Proc. of the Water Loss 2007 Conference.* Bucharest, Romania: The World Bank Institute, 2007.

CHAPTER 6

Validation of Source Meter Accuracy

George Kunkel, P.E.

Julian Thornton

Reinhard Sturm

6.1 The Importance of Source Meter Accuracy to the Integrity of the Water Audit and Loss Control Program

The standard water audit methodology adopted by the International Water Association (IWA) and the American Water Works Association (AWWA) is structured to track a water supply through the water treatment and distribution process, to its arrival at the customer endpoint. In the standard IWA/AWWA water balance, the volume of water labeled as *water supplied* holds paramount importance in this assessment. This volume, which is derived from source meters (also known as production or master meters), constitutes the amount of water input to the water distribution system. The broad comparison of the water supplied volume to *billed authorized consumption* gives the amount of nonrevenue water occurring in the audit period. The validity of the water audit is greatly influenced by the accuracy of the water supplied value because this is the first major value placed into the water audit. Any error in this value is carried throughout the entire water audit and imparts its uncertainty upon the values of apparent losses and real losses. It is therefore imperative that the water utility take steps to ensure a solid level of validity in the water supplied value.

> The validity of the water audit is greatly influenced by the accuracy of the *water supplied* value because this is the first major value placed into the water audit. Any error in this value will be carried throughout the entire water audit and potentially impart its uncertainty upon the values of apparent losses and real losses.

Accurately measured source flows are critical to the efficient operations of water utilities and wise resource management as overseen by regulatory agencies. Therefore

> **A**ll water sources should include flowmeters that are technologically current, accurate, reliable, well maintained and—ideally—continuously monitored by a Supervisory Control and Data Acquisition (SCADA) System or similar monitoring system.

utility managers and regulators should give high priority to the use of accurate metering at all sources. All water sources should include flowmeters that are technologically current, accurate, reliable, well maintained and—ideally—continuously monitored by a Supervisory Control and Data Acquisition (SCADA) System or similar monitoring system.

The *water supplied* value is a summation of several registered water volumes that are routinely measured via source meters. This value is calculated as a composite value that includes as components the primary untreated and/or treated water meters, meters registering water going into and out of tanks, basins, and reservoirs, and meters measuring water across pressure zones or district metered areas (DMAs). Three requirements are necessary to ensure that the value of *water supplied* is well validated:

- Appropriate meters should be installed at the key metering locations in the supply infrastructure so that water volumes can be reliably registered.
- Source meters must be well maintained and calibrated to ensure that they produce an accurate measure of the volume registered.
- Source meter data should be reliable and accurately archived—preferably on a continuous, real-time basis—with flows into and out of all pressure zones or DMA and storage facilities properly summed and balanced to achieve an accurate volume of water entering the distribution system on a daily basis.

> **T**he water supplied value is calculated as a composite value that includes as components the primary untreated and/or treated water meters, meters registering water going into and out of tanks, basins and reservoirs, and meters measuring water across pressure zones or district metered areas.

In conducting the water audit, the auditor should assess the adequacy that these requirements are met and launch work to correct any deficiencies. Work to install, test, calibrate, repair, or replace source meters should be identified as part of the initial top-down development of the water audit. This may be particularly necessary if key metering locations lack working meters and/or metered data is believed to be in serious error.

6.2   Key Source Meter Sites for Proper Flow Balancing

Water audits are most commonly conducted to track treated drinking water in transit through retail distribution systems. Separate water audits can also be conducted on wholesale transmission systems carrying untreated (raw) water or treated water; or discrete pressure zones or DMAs inside of a retail distribution system. Table 6.1 lists system configuration locations where metering is typically employed. In this publication, the water audit process is discussed in terms of the retail distribution system and the metering sites given below are those encountered in a typical retail distribution network. Figure 6.1 illustrates a basic retail distribution system configuration for the

Location	Function
Water source (untreated water)	Measure withdrawal or abstraction of water from rivers, lakes, wells, or other raw water sources
Treatment plant or works	Process metering at water treatment plants; metering may exist at the influent, effluent, and/or locations intermediate in the process
Distribution system input volume	Water supplied at the entry point of water distribution systems; either at treatment plant, treated water reservoir, or well effluent locations.
Distribution system pressure zones	Zonal metering into portions of the distribution system being supplied different pressure. Also includes metering at major distribution facilities such as booster pumping stations, tanks, and reservoirs.
District metered areas (DMAs)	Discrete areas of several hundred to several thousand properties used to analyze the daily diurnal flow variation and infer leakage rates from minimum hour flow rates
Customers	Consumption meters at the point of end use
Bulk supply	Import/export meters to measure bulk purchases or sales
Miscellaneous	Capture use of water from fire hydrants, tank trucks, or other intermittent use

TABLE 6.1   Typical Source Meter Locations in Drinking Water Supply Systems

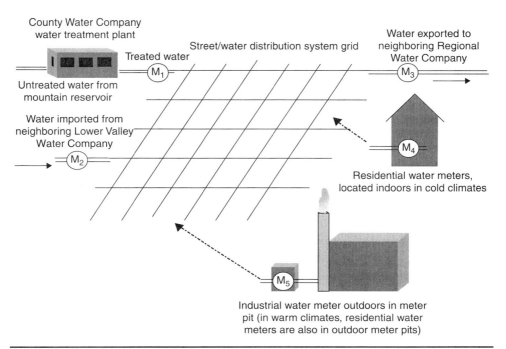

FIGURE 6.1   Typical retail water distribution system configuration. (*Source*: Ref. 1.)

fictitious County Water Company. As shown, source meters should exist at the point where the treated drinking water leaves the water treatment plant, shown as metering location (M_1). At this point the water quality has improved from untreated to potable quality and the water has been pressurized for conveyance in the distribution system; hence the monetary value of the water is greatest at this location. Source meters should also be included at any points of imported (M_2) or exported water supply (M_3). Finally, water meters should be included to measure flow entering or leaving tanks or reservoirs, and crossing pressure zones and DMAs.

Volumes of water purchased and imported from a neighboring supplier should be added to the composite metered values to obtain the water supplied value. The source meter (M_2) in Fig. 6.1 registers water purchased from a neighboring water utility by County Water Company. Interconnections between water utilities should always be metered. Such meters should be carefully maintained and monitored since the metered data provides the basis for billing large water volumes. Both the water utility supplying the water and the system purchasing the water have a strong motivation to keep this bulk measurement accurate since significant costs are at stake for each water utility.

Any water volumes sold and exported outside of the distribution system to a neighboring water utility should be monitored and adjusted with the same scrutiny given to imported water, for the same revenue implications exist. The source meter (M_3) in Fig. 6.1 registers water sold and exported out of the County Water Company grid.

Flows at storage facilities should be balanced for the water audit period. If source meters are located upstream of reservoirs and storage tanks, then stored water must be accounted for in the water audit. Generally, water flowing out of storage is replaced; as the "replacement" water flows from the source into storage, it is measured as supply into the system. If the reservoirs have more water at the end of the audit period than at the beginning, then the increased storage is measured by the source meters but not delivered to consumers. Such increases in storage should be subtracted from the metered supply. Conversely, if there is a net reduction in storage, then the decreased amount of stored water should be added to the metered supply. Table 6.2 shows how to figure the change in storage volume using data for County Water Company. Remember, *decreases* in storage are *added* to the supply; storage *increases* are *subtracted* from the supply. In this case, the net reservoir and tank storage was a drop in storage volume so the adjustment of 0.83 million gal should be added to the value of water supplied.

Reservoir	Start Volume, gal	End Volume, gal	Change in Volume, gal
Apple Hill	32,350	36,270	+3,920
Cedar Ridge	278,100	240,600	−37,500
Monument Road	978,400	318,400	−660,000
Davis	187,300	55,300	−132,000
Total change in reservoir storage			−825,580
	Volume in million gal		−0.83

TABLE 6.2   Changes in Reservoir Storage for County Water Company

	Component		Volume, million gal
1	Volume from own sources (treated water)		3480.76
2	Adjustment: source meter error	+136.89	
3	Adjustment: changes in reservoir and tank storages (±)	+0.83	
4	Other adjustments (*specify*)	0	
5	Total adjustments = lines 2+3+4	+137.72	
6	Volume from own sources (adjusted) = Lines 1 ±5		3618.48
7	Volume of water imported (adjusted)		783.68
8	System input volume = volume from own sources + water imported		4402.16
9	Volume of water exported (adjusted)		0
10	Water supplied = system input volume − water exported		4402.16

TABLE 6.3   Water Balance Calculations to Determine the Value of Water Supplied for County Water Company

Table 6.3 shows data for County Water Company and a series of tabulations that the water auditor should follow to arrive at the proper value of water supplied for a retail water distribution system. The procedure to obtain the data included in Table 6.3 is discussed below.

Compile the Volume of Water from Own Sources

Identify all water sources that are owned or managed by the water utility to supply water into the distribution system. Such sources can include raw water that is treated adjacent to sources such as wells, rivers, streams, lakes, reservoirs, or aqueduct turnouts. However, most water audits are performed on the potable water distribution system so that the "source" is often the location where *treated* water enters the distribution system; which is often the effluent of the water treatment plant. All volumes from such sources should be metered, with routine meter testing and upkeep conducted so that volumes of water taken from the sources are registered accurately. Data should be available on a daily, weekly, or monthly basis to compile into an annual volume of water supplied from each source. Meter information can be kept in a table similar to Table 6.4.

In this example, County Water Company withdraws water from three sources: an aqueduct, a well field, and an interconnection (city intertie) with a neighboring water utility. Table 6.5 is a summary of water withdrawn from these sources for the year of 2006, illustrating how source meter and flow data can be arranged and adjusted for the water audit period. The data listed is based upon uncorrected meter registrations. In this example, water withdrawn from the aqueduct and well field is presumed to be untreated water. For the simplicity's sake, it is assumed that the volumes of water for these two sources shown in Table 6.5 are the same volumes delivered to the water distribution system after the water undergoes treatment. This simplistic assumption often

| Characteristics | Water from Own Sources | | Water Imported |
	Source 1 Aqueduct Turnout 41	Source 2 Well Field	Source 3 City Intertie
Type of measuring device	Venturi	Propeller	Venturi
Identification number (may be serial number)	0000278-A	8759	OC-16
Frequency of reading	Daily	Weekly	Daily
Type of recording register	Dial	Dial	Builder type M
Units registers indicate	100,000 gal	gallons	Cubic feet
Multiplier (if any)	1.0	1.0	100.0
Date of installation	1974	1990	1978
Size of conduit	24 in	10 in	11.5 in
Frequency of testing	Annual	2 years	4 months
Date of last calibration	4/1/2006	8/21/2006	1/15/2006

TABLE **6.4**   Source Water Measuring Devices for County Water Company

2006 By Month	Source 1 Turnout 41	Source 2 Well Field	Subtotal Own Sources (unadjusted)	Source 3 City Intertie (water imported)	Total for all sources 1, 2, and 3 (unadjusted)
January	0	130.34	130.34	104.27	234.61
February	0	195.51	195.51	65.17	260.68
March	130.83	130.34	261.17	0	261.17
April	160.18	260.68	420.86	0	420.86
May	326.53	97.76	424.29	0	424.29
June	368.62	0	368.62	81.46	450.08
July	372.64	0	372.64	84.72	457.36
August	400.89	0	400.89	89.61	490.50
September	360.72	32.59	393.31	32.59	425.90
October	160.18	32.59	192.77	97.76	290.53
November	160.18	0	160.18	130.34	290.52
December	160.18	0	160.18	97.76	257.94
Annual total	2600.95	879.81	3480.76	783.68	4264.44
Daily average, million gal/day					11.68

TABLE **6.5**   Total Water Supply in Million Gallons for County Water Company (Uncorrected)

does not hold true in reality, as a portion of the water passing through a treatment process is lost due to plant infrastructure leakage and process uses such as backwashing of filters, chemical mixing, and maintenance activities such as flushing. Metering water at the source of withdrawal is essential and many regulatory agencies require this in order to track water resource utilization. However, it is recommended to also meter treated water at the location when it leaves the water treatment plant; particularly if the water treatment plant is distant from the water source.

Once a volume is established for each source for the year, the measured amounts should be reviewed and corrected for known systematic or random errors that may exist in the metering data. Figures for the total water supply, based on readings from source meters and measuring devices, are raw data. The raw data must be adjusted for a number of potential factors, including

- Meter inaccuracies (see Table 6.6)

- Changes in reservoir and storage levels (see Table 6.2)

- Any other adjustments such as losses that occur before water reaches the distribution system. One example would be losses incurred during the treatment process (filter backwashing, etc.) if the source meter is located upstream from the water treatment plant. None are included in the example data shown in Table 6.3 so a volume of zero is entered on line four of this table.

The tabulations shown in Table 6.3 arrive at a corrected value of water supplied of 4,402.16 million gal for the water audit period. This data takes into account the registered yearly volumes for three source meters (two of own sources and one imported supply), a correction for meter error on "Source 1" and the storage adjustment. This is a simplified example that includes only a few sources. It is recognized that many water utilities withdraw water from many sources, purchase/sell water at multiple interconnection points, and have many tanks and pressure zones. It is incumbent upon utility managers operating supply systems with such complex configurations to be meticulous in identifying the key source metering locations, establishing and maintaining source meters,

Source	Yearly Total: Uncorrected Metered Volume (UMV)*	Meter Accuracy (MA) percent	Meter Error Calculation UMV/MA† – UMV	Meter Error	Adjusted Metered Volume‡
1 Turnout 41	2600.95	95	(2600.95/0.95) – 2600.95	+136.89	2737.84
2 Well Field	879.81	100	(879.81/1.00) – 879.81	+0.0	879.81
				+136.89	

* From Table 6.5

† A percentage, written as a decimal (95 percent = 0.95) taken from meter testing performed regularly.

‡ The corrected meter volume for sources 1 and 2 is 2,737.84 + 879.81 = 3617.65 million gal; note that this is 136.89 million gal greater than the raw total supply given for these sources in Table 6.5.

TABLE 6.6   Volume of Water from Own Sources in Million Gallons for County Water Company—Adjusted for Source Meter Error and creating spreadsheets or databases that properly balance flows such that an accurate value of water supplied is attained and made available to the water audit each year.

6.3   Types of Source Meters

Source meters come in a wide variety of types, sizes, and flow-registering mechanisms. Some of the more common types include

> Differential pressure meters
>
> Venturi meters
>
> Dall tube meters
>
> Orifice plate meters
>
> Proportional flowmeters
>
> Magnetic meters
>
> Insertion meters
>
> Ultrasonic meters
>
> Turbine meters
>
> Propeller meters
>
> Vortex shedding meters

All types have advantages and disadvantages in any given application and each metering site must be evaluated independently to determine the optimum meter. It is important that an established meter function according to its specification and the data being recorded are compatible with the other source meter data being collected throughout the water utility. Detailed guidance on source meter types, function and management is provided in the AWWA M33 publication *Flowmeters in Water Supply*[2]. Figures 6.2 through 6.5 show a sequence of photos from the replacement of a large magnetic flowmeter on a raw (untreated) water supply main in Philadelphia.

Figure 6.2   Source meter replacement at raw water pumping station: Existing meter removed and placed next to new 48-in diameter magnetic meter prior to its installation. (*Source: Philadelphia Water Department.*)

FIGURE 6.3   Source meter replacement at raw water pumping station: New 48-in diameter magnetic meter being prepared for installation. (*Source*: *Philadelphia Water Department*.)

FIGURE 6.4   Source meter replacement at raw water pumping station: Ferrule being drilled into solid sleeve piping adjacent to new 48-in diameter magnetic meter. The ferrule will be used as a location for future meter verification testing by use of an insertion pitot rod. (*Source*: *Philadelphia Water Department*.)

Figure 6.5   Source meter replacement at raw water pumping station: New 48-in diameter magnetic meter has been lowered into its chamber and work is underway to make connections to the raw water transmission piping. (*Source*: *Philadelphia Water Department.*)

Traditional source meters are the full pipe bore type; or meter designs that consumed the entire pipe diameter. Some of these meters, such as propeller meters, induce a head loss through the pipe since part of the meter apparatus exists within the flow stream of the pipe. Full bore meters, particularly in larger sizes are costly and require considerable space. Water pipelines must be shutdown and dewatered to replace these meters. Still, many full bore meter types and brands have proven histories of reliable, accurate service for periods of many decades. In recent years, insertion-type metering devices have witnessed considerable evolution. These meters offer advantages of lower cost, less arduous space requirements and no need to shutdown/dewater the pipeline to install the meter. Insertion meters can be installed in ferrules that can be tapped while the water pipeline remains in service. The insertable electromagnetic averaging flow meter is one type of reliable insertion meter available today.

Many reliable types and brands of flowmeters exist in the commercial marketplace. Water utilities have many options at their disposal in selecting meters. The challenge becomes making the best reasonable meter decision to match the desired application in the field.

6.4   Source Meter Accuracy and Testing Program Steps

Treated drinking water is commonly measured by meters, but untreated source water from lakes, reservoirs, streams may be measured by other devices, such as Parshall flumes or weirs. Any unreasonable degree of error in a measuring device must be discovered and corrected; incorrect supply data compromises the water audit since any error in the source meters carries throughout the audit.

To be sure that meters are accurate, compare the results of meter tests to applicable AWWA standards and guidance manuals. If a meter measures incorrectly and the error exceeds the standard for its category, repair and recalibrate the meter to function within standard limits. If the meter has not been tested within the past 12 months, test the meter.

If source meters are inaccurate, inspect each one in the field. Normal wear is not the only cause of inaccurate meter readings. Check to be sure that the meter is the right type and size for the application and that it is installed correctly. See the AWWA M33 publication[2] for guidance on typical source meter types and applications. Check the size against manufacturers' recommended ranges. Be sure that the meter is level; most meters are not designed for sloped or vertical operation. Inspect the meter to see if hard-water encrustation is interfering with the measurement. Also check to verify that the proper registers were selected and installed correctly. Finally, be sure that the register is read correctly or the signal from the meter is properly transmitted through the SCADA System. Have an employee familiar with metering instrumentation perform calibration of the instrument and make a special reading of the source meter, or have a second employee accompany the meter reader to verify sample readings. Check to be sure that the meter is read and recorded correctly, and the correct conversion factor is used.

Check venturi meters for blockages in the throats of the meters or in the sensing lines. Test the primary device by comparing it with a measurement taken from a pitot rod or other insertion-type meter installed in series with the meter. Testing the meter with a pitot rod shows whether or not the installation is adequate for nonturbulent flows. The meter's primary device should be tested at different flow ranges. If pressure deflection for appropriate flows is adjusted without checking the venturi itself, the meter may still record flows erroneously.

Testing Meters

There are four ways meters may be tested. Meter testing methods are listed here in order of decreasing effectiveness.

- Test the meters in place. Some pipes may need to be replaced to make this possible. Use of an insertion pitot rod will provide a measurement to compare against that recorded by the meter.

- Compare meter readings with readings of a calibrated meter installed in series with the original meter (Sec. 12.5).

- Record meter readings for a given flow over a specified time period. Remove the meter and replace it with a calibrated meter. Record readings from the calibrated meter using the same flow rate for the same duration; compare the readings.

- Test the meter at a meter-testing facility. This is usually not feasible or cost-effective for very large meters.

Meters can be tested with portable equipment. Pump efficiency flow testing can be used to check meters; it is sometimes provided free of charge by electric utilities. Some utilities use an averaging rod meter or anubar to test meters, but results may be off by as much as 10 percent. A standard single-point pitot rod gives more accurate results, generally ± 2%. Meter testing may be done by an outside agency. Consultants, meter manufacturers, and special testing laboratories offer testing services.

In order to calculate an adjustment to account for meter inaccuracy (see Table 6.6) divide the uncorrected metered volume (UMV) by the measured accuracy of the meter (a percentage expressed as a decimal) and subtract the UMV as follows:

$$\frac{\text{Uncorrected metered volume}}{\text{Percent accuracy}} - \text{uncorrected metered volume}$$

$$= \text{meter error}$$

Then calculate the adjusted metered volume (AMV) as:

$$\text{Adjusted metered volume} = \text{uncorrected metered volume} \pm \text{meter error}$$

A checklist of activities is given in Table 6.7. Some source meters (8, 10, 12 in diameter) fall within the same size range as large customer meters. Because of this, operators with source meters in this size range can also refer to Chap. 12 for further information on accuracy testing of these meters. While many reliable meter types and brands provide excellent service for many years, water utilities should make a particular effort to

Pre-test Activities	
1	Identify and locate all meters
2	Reference all available manufacturer's specifications for those meters
3	Confirm the meter installation according to the manufacturer's specification
4	Determine whether onsite testing can be undertaken
5	Identify the type of onsite testing to be undertaken and its realistic limitations
6	Define an allowable band of error between the test volume and metered volume
7	Locate any records of prior testing or repair work and factor this information into the planning of the test
8	Identify a local supplier or contractor who can calibrate the meters which fall outside the allowable limits
9	Research manufacturers and suppliers for replacement meters for those meters which cannot be calibrated
10	Set a realistic budget for the work
11	Establish a realistic time frame for the work to be carried out
12	Identify an accountable tracking mechanism to clearly show both a baseline measurement before calibration and a calibrated measurement after testing
Post-test Activities	
	Clearly identify and record any major changes in calibration, both for span and zero
	Identify the impact on the annual water balance
	Store both raw and adjusted data for future reference
	Identify local extraction limitations and how they are affected by the new results
	Put in place a periodic testing program to ensure that the meters stay calibrated and the new results accountable

TABLE 6.7 Source Accuracy Testing Checklist of Activities

maintain the function and accuracy of those meters serving as source meters. In many ways, the reliability of the entire water audit is only as good as the source meter management is in a water utility.

6.5   What to Do if Meters Do Not Exist at Key Metering Sites

Perhaps one or more water sources are unmetered, or have meters that are not routinely monitored. In such cases the following applies:

> *If no meters exist at a water source:* Use a portable meter or estimate the flow. Portable meters can be insertion types or strap-on types and can be installed on source piping just downstream of the treatment plant effluent or other source. A minimum of 24 hours of continuous metering should be obtained. If portable metering is not feasible, one way to infer an estimate is to utilize treated effluent water-pumping records. If the water pump performance characteristics are known, a volume estimate can be derived by multiplying the number of hours that the pump was operated during the year by the average pumping rate. If water is taken from a large reservoir, an estimate of the withdrawal can be formulated by accounting for the amount of drawdown of the reservoir level, adjusted by the amount of inflow from streams and rainfall. Such methods give an approximate volume measurement, and unmetered sources should ultimately be designated for metering when possible.

> *If source water meters have not been routinely monitored:* Conduct an inspection of the source structures and meter. Note the type of metering device that exists (e.g., venturi flowmeter, magnetic flowmeter, ultrasonic flowmeter). Note basic information about the measuring device: type, identification number, frequency of reading, type of recording register, unit of measure (and conversion factor, if necessary), multiplier, date of installation, size of pipe or conduit, frequency of testing, and date of last calibration. Document this information as in Table 6.4.

Attempt to obtain a record of how much water was produced by each source during the period of the audit. Most meters have some type of register, or totaling device. Registers may be round reading or direct reading. Round-reading registers have a series of small dials with pointers, registering cubic feet, or gallons, in tens, hundreds, thousands, and ten thousands. Direct-reading registers have a large sweep hand for testing and a direct-reading dial that shows total units of volume. If the meter has not been routinely read, tested, or calibrated, efforts should be initiated to calibrate the meter and institute routine reading or polling of the meter. Many drinking water utilities now link source meters with SCADA systems that convey data in real time to centralized computers, where the flow data is totaled and archived for easy retrieval. Again, a portable meter can be utilized to obtain measurements to compare during any source meter calibration or verification activities.

6.6   Summary: Source Meter Accuracy

Source meters register the bulk water resource supply to the water utility, as well as interconnection transfers between water utilities, major treated water transmission flows into and out of tanks and other storages, and flows across pressure zones and district metered areas. These meters provide the input that goes into the value of water supplied; which is the first primary component in the water audit. Appreciable error in source meters and the water supplied value can carry throughout the other components of the water audit, corrupting the validity of the audit and, therefore, its usefulness.

> **V**erifying the working condition of source meters is the recommended first field activity to take when launching the water audit process.

Verifying the working condition of source meters is the recommended first field activity to take when launching the water audit process. This can require some investment if meters must be installed or replaced; however the credibility and effectiveness of the water audit—and the water utility's water accountability—relies heavily upon validated source meter data.

References

1. American Water Works Association. *Water Audits and Loss Control Programs.* Manual of Water Supply Practices M36 3d ed. Denver, Colo. : AWWA, 2008.
2. American Water Works Association. *Flowmeters in Water Supply.* Manual of Water Supply Practices M33. AWWA, 2006.

CHAPTER 7

Evaluating Water Losses— Using a Standardized Water Audit and Performance Indicators

Reinhard Sturm

George Kunkel, P.E.

Julian Thornton

7.1   Introduction

A standardized International Water Association's (IWA)/American Water Works Association's (AWWA) water balance provides the water utility with the necessary results and understanding of the nature and extent of its water losses. Subsequently, the water utility will be able to select the appropriate tools for intervention against real and apparent losses.

Just as businesses routinely prepare statements of debits and credits for their customers, and banks provide statements of monies flowing into and out of accounts, a water audit displays how quantities of water flow into and out of the distribution system and to the customer. Yet, as essential and commonplace as financial audits are to the world of commerce, water audits have been surprisingly uncommon in the world of public water supply throughout most of the world. In places where the intrinsic value of water has not been recognized, little motivation has existed to prompt requirements for auditing and sound assessments of water loss performance. As water is becoming a more valued commodity, however, this picture is beginning to change and Chap. 4 has

> **C**ompiling a reliable water audit or water balance is the critical first step in managing water losses in public water supplies.

provided examples of countries where standardized and regularly compiled water audits build the bases for a successful reduction and management of water losses.

Terms *water audit* and *water balance* are often interchanged. However, when talking about a water audit we mean the work related to tracking, assessing, and validating all components of flow of water from the site of withdrawal or treatment, through the water distribution system and into customer properties. The water audit usually exists in the form of a worksheet or spreadsheet that details the variety of consumption and losses that exist in a community water system. The water balance summarizes the results of the water audit in a standardized format (see Fig. 7.1).

Throughout the 1990s efforts materialized to develop a rational, standardized water audit methodology and water loss performance indicators (PI). Part of the motivation spurring this work was the focus on demand management and the wise use of water in England and Wales, which was driven by competition, drought-related water shortages, and other factors. In the late 1990s, IWA initiated a large-scale effort to assess water supply operations, which resulted in the publication of *Performance Indicators for Water Supply Services* in 2000[1] (a second edition of this publication was published in 2006 by IWA publishing[2]). While this initiative included various groups assessing all aspects of water supply operations, the Task Force on Water Loss worked specifically to devise an acceptable water audit format and performance indicators that can be used to make effective comparisons of water loss performance of systems anywhere in the world.

The methods put forth by the IWA Task Force on Water Loss, represent the current "best practice" model for water auditing and performance measurement. This is not just because of the multination process used in assembling the results, but primarily because the work was groundbreaking in providing a clear structure for a need that was void of knowledge throughout most of the world. Additionally, the work has been tested thoroughly using data from dozens of countries and since its publication numerous utilities around the globe have successfully adopted these methods as their best practice for assessing water losses. Several countries, including South Africa, Australia, Germany, Malta, and New Zealand have adopted the IWA best practice model for water auditing and performance indicators as best practice for their national water loss management

System input volume (allow for known errors)	Authorized consumption	Billed authorized consumption	Billed metered consumption	Revenue water
			Billed unmetered consumption	
		Unbilled authorized consumption	Unbilled metered consumption	Non-revenue water (NRW)
			Unbilled unmetered consumption	
	Water losses	Apparent losses	Unauthorized consumption	
			Customer metering inaccuracies and data handling errors	
		Real losses	Leakage on transmission and/or distribution mains	
			Losses at utility's storage tanks	
			Leakage on service connections up to point of customer use	

Figure 7.1 Standard IWA/AWWA water balance. (*Source:* Ref. 6.)

regulations. The American Water Works Association (AWWA) Water Loss Control Committee (WLCC) has adopted the IWA water audit methodology and performance indicators as best practice in its committee report "Applying Worldwide Best Management Practices in Water Loss Control" published in the August 2003 edition of the AWWA Journal.[3] The AWWA WLCC is currently in the process of rewriting the AWWA M36 manual of "Water Supply Practices, Water Audits, and Leak Detection" to incorporate the current best practice for water audits and in general water loss management. In addition the World Bank, the Asian Development Bank, and the European Investment Bank have also adopted the IWA methodology as best practice to assess water losses and determine performance indicators.

The water audit discussed in this chapter relates to the treated water distribution network and does not include the raw water transmission systems or the treatment process. The reason is that in the majority of systems the losses stemming from the distribution system represent an order of magnitude that eclipse the losses stemming from the raw water transmission systems or the treatment process. However, water losses from the raw water transmission system or the treatment process can be evaluated in a separate balance if necessary.

7.2 A Rosetta Stone for Water Loss Measurement

In 1799, Napoleon's soldiers found an ancient carved piece of black basalt at Rosetta, near the mouth of the river Nile. It contained a decree of the Egyptian priests of Ptolemy V. Epihanes (205–181 BC) written in Egyptian hieroglyphics, demotic characters, and Greek, permitting a simultaneous translation of these three written texts. The Rosetta stone enabled the hieroglyphics to be correctly translated for the first time by archaeologists.

This could have more to do with water loss accounting in North America than the reader may at first imagine. Remarkably, in North America, there is no single standard terminology, or commonly accepted definitions or methodology for undertaking an annual water audit of the components of a water balance. The water balance calculation seeks to identify the destinations of all water entering a distribution system, so that the water losses occurring within the distribution system can be assessed. Each state, government organization, professional institution, consultant, or contractor can (and usually does!) define the terminology and undertake the calculations in any way they please. This is perhaps because few states request or require water utilities to report such data on an annual basis. However, water is an important natural resource, and in an increasing number of developed countries similar absences of accountability for demonstrating responsible stewardship of natural resources is being actively addressed.

For example, in England and Wales, since 1992 the privatized water companies have had to produce annual independently audited calculations of water losses in a standard format, for national publication by their economic regulator. Publication of standardized data raised questions regarding performance and economic levels of water losses, which in turn (spurred by the 1995–96 drought and political impetus) resulted first in voluntary, and then mandatory, leakage targets. Some 5 years later, leakage from public water supply systems in England and Wales has been reduced overall[4] by 40%, or some 480 mgd, and U.K. expertise in modern leakage management is now internationally recognized. Would any of this happened had the English and Welsh water utilities been permitted to choose for themselves:

- Whether to undertake annual calculations of water losses, or not.
- How the calculation should be carried out and which performance indicators should be used?
- Whether the results should be published, or not.

The extent of the problem in North America can be illustrated from the results of an American Water Works Research Foundation (AwwaRF) project titled "Leakage Management Technologies," completed in 2007. The report concludes that although performing water audits is common in North America, the methods, the expressions of water loss, or time intervals are not consistent. Most utilities do not use the IWA standard method for water audits. Many audit methods currently used leave ample room for inaccurate accounting and measures of performance. In North America, water loss cannot be accurately compared between utilities because water loss is still commonly expressed as a percentage of system input volume, a practice that allows the denominator (the high per capita consumption of North America) to minimize water loss.[5] The IWA/AWWA standard water audit methodology and performance indicators are the Rosetta Stone for water auditing in North America. However, as mentioned in Chap. 4 there have been very important and encouraging developments in the United States over the past 5 years, with several state agencies and national organizations adopting and promoting the IWA/AWWA standardized water audit methodology as best practice.

7.3   The Benefits of the IWA/AWWA Standard Water Audit and Performance Indicators

Advantages of the IWA/AWWA methodology can be summarized as

- The IWA/AWWA methods are structured to serve as a standard international best practice methodology and terminology for such calculations, based on the conclusions of IWA Task Forces on water losses and performance indicators.
- The IWA/AWWA methods question the desirability of the common North American practice of counting unavoidable water losses and discovered leaks and overflows as part of authorized consumption.
- A system-specific method for calculating unavoidable real losses is included.
- The IWA/AWWA method counters the deficiencies in the performance indicators most commonly used in North America—percentage of system input volume and losses per mile of mains.
- The IWA/AWWA has dropped the term *unaccounted for water* (UFW) in favour of *nonrevenue water* (NRW), because there is no internationally accepted definition of UFW, and all components of the water audit can be accounted for using the IWA/AWWA methodology.
- The IWA/AWWA methodology does not leave room for ambiguity. Every type of water use and loss has an appropriate component in the water balance it is assigned to, which assures that the results are meaningful and comparable.
- The IWA/AWWA methodology has been successfully applied in numerous countries and utilitised around the globe.

- A meaningful comparison of water audit results and performance indicators can be undertaken, independent form location, size, and operational characteristics of the water supply system.

7.4   The IWA/AWWA Recommended Standard Water Audit

The top-down water audit is basically assembled in two steps:

1. Quantification of all individual water consumption and water loss components, via measurement or component based estimation

2. Undertaking the standardized water balance calculation

This section explains the recommended water audit approach and each component of the water audit. The effort required to conduct a top-down water audit is relatively modest depending on the availability and quality of data. The top-down audit also helps to identify components that require further validation.

The components of the water balance as shown in Fig. 7.1[6] can be measured, estimated, and calculated using a variety of techniques. Ideally, all components of the water balance (excluding those components that are calculated by adding or subtracting other components) should be based on measurements. However, in reality estimates will need to be made especially the first time a water balance is established. Once the components needing estimation are identified it is best practice to put actions in place that allow to meter the component or to improve the estimation process. Validation of water balance components is an important and integral part of conducting a water balance. Sensitivity analysis and the use of 95% confidence limits are best practices to assess the impact which individual water balance components have on the overall accuracy of the calculated volume of nonrevenue water and real and apparent losses.

A water balance should be established annually and before establishing the water balance it is important to determine the audit period (e.g., fiscal year or calendar year) and the system boundaries. The units of the water balance components must also be chosen and standardized so that the same units are used for each component of the water balance.

The calculation procedure for the water balance is as follows:

- Obtain system input volume and correct for known errors.
- Obtain components of revenue water, calculate revenue water which equals billed authorized consumption.
- Calculate nonrevenue water (= system input – revenue water).
- Assess unbilled authorized consumption.
- Calculate authorized consumption [= (billed + unbilled) authorized consumption].
- Calculate water losses (= system input – authorized consumption).
- Assess components of apparent losses, calculate apparent losses.
- Calculate real losses (= water losses – apparent losses).

The following subsections explain the various steps of a water audit based on the AwwaRF report "Evaluating Water Loss and Planning Loss Reduction Strategies."[6]

7.4.1   Determining the System Input Volume

Definition: The annual volume input to the water supply system.

In case the entire system input is metered, the calculation of the annual system input should be a straightforward task. The regular meter records have to be collected and the annual quantities of the individual system inputs calculated. This includes own sources as well as imported water from bulk suppliers.

The accuracy of the input meters should be verified on an annual basis, using portable flow-measuring devices, or if possible by conducting volumetric comparisons via reservoir drop test, for example. If any inaccuracies of the system input meters are revealed it is necessary to further investigate the problem, and if necessary, the recorded volume of water has to be adjusted to account for the inaccuracy of the system input meter. It is recommended that as well as verifying the accuracy of the meters, the entire data recording chain from the raw 4 to 20 mA signal produced by the meter to the Supervisory Control and Data Acquisition (SCADA) archive is checked when testing the input meters.

If there are unmetered sources then the annual flow has to be estimated by using any (or a combination) of the following:

- Temporary flow measurements using portable devices
- Reservoir drop tests
- Analysis of pump curves, pressures, and average pumping hours

It is important to realize that the accuracy and reliability of the water balance results are directly linked to the accuracy of the figures used for the system input volumes. It is recommended that system input meters are tested for their accuracy at least once a year so they can be recalibrated if necessary.

7.4.2   Determining Authorized Consumption

Definition: The annual volume of metered and/or unmetered water taken by registered customers, the water supplier, and others who are authorized to do so.

Billed Metered Consumption
The calculation of the annual billed metered consumption goes hand in hand with the detection of possible billing and data-handling errors, information which is required at a later stage of the water audit process for the estimation of apparent losses. Consumption of the different consumer categories (e.g., domestic, commercial, or industrial) have to be extracted from utility's billing system analyzed and validated. Special attention should be given to the group of very large consumers.

The annual billed metered consumption information taken from the billing system has to be processed for meter reading time lag to ensure that the billed metered consumption period used in the audit is consistent with the audit period.

Billed Unmetered Consumption
Billed unmetered consumption can be obtained from the utility's billing system. In order to analyze the accuracy of the estimates, unmetered domestic customers should be identified and monitored for a certain period, either by the installation of meters on

those non-metered connections or by monitoring a small area with a number of unmetered customers. The latter has the advantage that the customers are not aware that they are metered and so they will not change their consumption habits. In the unlikely case that nondomestic customers are unmetered, detailed surveys have to be carried out to check the accuracy of the estimated billed consumption figures.

Unbilled Metered Consumption

The volume of unbilled metered consumption has to be established similar to that of billed metered consumption.

Unbilled Unmetered Consumption

Each type of unbilled unmetered consumption shall be identified and individually estimated by building up from individual usage events using a component-based approach to develop a realistic estimate of use, for example:

- *Street cleaning/sewer flushing:* Components to be assessed are what is the number of street cleaning trucks in operation? What is the volume of water a street cleaning truck transports? How many times is a street cleaning truck filled per month? The street cleaning and sewer flushing departments should be able to provide the necessary data.

- *Mains flushing:* How many times per month? For how long? How much water? The operations and construction departments should be able to provide the necessary data.

- *Fire fighting:* Number of fires during year? Average volume per fire? Has there been a big fire? How much water was used? The fire department should be able to provide this data.

- *Fire flow tests:* How many tests in year? Average duration of test? Flow rate? Again the fire department should be able to provide this data.

In some circumstances, it may be appropriate to meter a small sample of these use events to obtain a better estimate of use per event.

7.4.3 Calculation of Water Losses

Definition: The difference between system input volume and authorized consumption, consisting of apparent losses plus real losses.

Water Losses are calculated by subtracting the total authorized consumption volume from the system input volume. In the subsequent process of the water audit, the volume of water losses is further broken down into real and apparent losses.

7.4.4 Assessment of Apparent Losses

Definition: This component includes unauthorized consumption, all types of customer metering inaccuracies and data-handling errors.

Unauthorized Consumption

It is difficult to provide general guidelines of how to estimate unauthorized consumption. There is a wide variation of situations and knowledge of the local circumstances will be most important to estimate this component. Unauthorized consumption can include

- Illegal connections
- Misuse of fire hydrants and fire-fighting systems, for example, unauthorized construction use of hydrant water
- Vandalized or bypassed consumption meters
- Corrupt practices of meter readers
- Open boundary valves to external distribution systems (unknown export of water)

The estimation of unauthorized consumption is always a difficult task and should at least be done in a transparent, component-based way so that the assumptions can later easily be checked and/or modified if necessary.

Customer Metering Inaccuracies and Data-Handling Errors

The extent of customer meters inaccuracies, namely, under- or overregistration, has to be established based on tests of a randomly selected representative sample of meters, (AWWA manuals M6 and M22 provide the relevant guidance). The composition of the sample shall reflect the various brands and age groups of domestic meters. Tests are done either at the utility's own test bench, or by specialized contractors. Large customer meters are usually tested on site with a test rig. Based on the results of the accuracy tests, average meter inaccuracy values (as % of metered consumption) will be established for different user groups.

In applying the accuracy test results to the whole population of different user groups of meters, it is also important to consider the issue of how quickly the utility is able to identify meters which are totally stopped by considering the utilities processes for identifying stopped meters. The average time taken to identify and replace stopped meters can have a significant impact on the overall accuracy of the meter population as a whole.

Other issues which are important to consider as part of assessing the level of meter inaccuracies are

- *Meter size in relation to actual use patterns:* Are the meters sized correctly to maximize revenue?
- *Meter type:* Is it the best type of meter for the operating range?
- *Service line size:* Is it appropriate for the operating range?

Data-handling errors are sometimes a very substantial component of apparent losses. Many billing systems are not up to the expectations of the utilities but problems often remain unrecognized for years. It is possible to detect data-handling errors and problems within the billing system by exporting billing data (of at last 12 months) and analyzing it using standard database software. Types of data-handling errors that may be encountered and should be checked for include

- Changes to consumption volume data when bills are adjusted for any reason other than an incorrect reading
- Inappropriate use of estimated consumptions
- Inappropriate determination of estimated consumptions
- Accounts incorrectly flagged as inactive
- Accounts missing from the database
- Inaccurate meter data

The detected problems have to be quantified and a best estimate of the annual volume of this component has to be calculated.

7.4.5   Calculation of Real Losses

Definition: The annual volumes lost through all types of leaks, breaks, and overflows on mains, service reservoirs, and service connections, up to the point of customer metering.

The volume of real losses is calculated by subtracting the volume of authorized consumption and the volume of apparent losses from the total system input volume.

7.4.6   Calculation of Nonrevenue Water

Definition: The difference between system input volume and billed authorized consumption.

Nonrevenue water is the portion of the water that a utility places into the distribution system that is not billed and, therefore, recovers no revenue for the utility. Nonrevenue water consists of the sum of unbilled authorized consumption (metered and unmetered), apparent losses, and real losses.

It is recommended as best practice by the IWA and AWWA WLCC that the assessment of real losses using a "top-down" water balance should be complemented by at least one of the following two methodologies:

- Component analysis of real losses, a technique which models leakage volumes based upon the nature of leak occurrences and durations (see Chap. 10)
- "Bottom-up" analysis of real losses using district metered area (DMA) and minimum night-time flow (MNF) analysis (see Chap. 16)

Both methodologies add increased refinement and confidence in the calculated volume of real losses and are described separately in this manual.

7.5   Unavoidable Annual Real Losses—Unavoidable Water Losses and Discovered Leaks and Overflows

Nowadays it is well understood among water loss practitioners that every system has a certain volume of real losses occurring, that is unavoidable. Even newly commissioned sections of the distribution network will have some volume of real losses.

Since the last century, it has been common practice in North America to estimate, using various formulae, the *unavoidable* leakage from pressurized pipework systems—those small leaks which are believed to be undetectable, or which are considered uneconomic to repair. The original intention of this was presumably to try to define a baseline or lower limit for leakage management, below which it is uneconomic to attempt further leakage control. An outline of the various methods previously used in North America can be found in Ref. 7.[7] The system-specific predictions based on an auditable component-based equation proposed by the IWA Task Force on Water Losses,[8] described later in this chapter, can be regarded as a natural progression of previous North American efforts to predict unavoidable losses.

Because of the simplified nature of some of the formulae previously used in North America, or the very generous allowances given for old pipework (particularly cast-iron pipes), the effect of the *unavoidable leakage* calculation has in practice often resulted in a considerable amount of leakage being written off as beyond control. In fact, there are infrastructure and pressure management options that now exist to reduce it.

A similar situation applies regarding *discovered* leaks from pressurized pipework and overflows from service reservoirs. The most common practice in countries outside North America is to calculate the annual volume of water losses from the water balance without making any deductions for unavoidable leakage or discovered leaks and overflows, and then to calculate the performance indicators. Accordingly, superficial comparisons of North American water losses with water losses from other countries often present a more favourable picture than is actually the case.

The IWA and AWWA recommended standard methodology for water audit calculations and performance indicators allows *unavoidable losses* and *discovered leaks and overflows* to be considered, but only as partial explanations of the total volume of water losses, which should always be explicitly stated before attempting to explain or justify the total volume. The IWA system–specific approach to unavoidable annual real losses is described in the next section.

7.5.1 The IWA Approach to Calculating Unavoidable Annual Real Losses

The IWA approach is described in detail in the December 1999 issue of the IWA *AQUA* Magazine,[8] and can be seen as a natural development of previous North American attempts to take key local factors into account. The component-based approach is based on auditable assumptions for break frequencies, flow rates, durations; background and breaks estimates concepts[9] to calculate the components of unavoidable real losses for a system with well-maintained infrastructure; speedy good-quality repairs of all detectable leaks and breaks; and efficient active leakage control to locate unreported leaks and breaks.

Parameters used in the calculation, taken from "Water Loss Management in North America"[7] and converted to North American units, are shown in Table 7.1. Table 7.2 shows these parameters in a more user-friendly format for calculation purposes.

Infrastructure Component	Background (Undetectable) Losses	Reported Breaks	Unreported Breaks
Mains	8.5 gal/mi/hr	0.20 breaks/mi/year at 50 gpm for 3 days duration	0.01 breaks/mi/year at 25 gpm for 50 days duration
Service lines, main to curb stop	0.33 gals/service line/hr	2.25/1000 service line/year at 7 gpm for 8 days duration	0.75/1000 service line/year at 7 gpm for 100 days duration
Underground pipes, curb stop to meter (for 50 ft ave. length)	0.13 gal/service line/hr	1.5/1000 service line/year at 7 gpm for 9 days duration	0.50/100 service line/year at 7 gpm for 101 days duration

gal = U.S. gallon; all flow rates are at a reference pressure of 70 psi
Source: Ref. 7.

TABLE 7.1 Parameters Values Used for Calculation of Unavoidable Annual Real Losses (UARL)

Infrastructure Component	Background Losses	Reported Bursts	Unreported Bursts	UARL Total	Units
Mains	2.87	1.75	0.77	5.4	gal/mi mains/d/psi of pressure
Service lines, mains to curb stop	0.112	0.007	0.030	0.15	gal/mi/d/psi of pressure
Underground pipes between curb stop and customer meters	4.78	0.57	2.12	7.5	gal/mi u.g. pipe/d/psi of pressure

Source: Ref. 7.

TABLE 7.2   Components of Unavoidable Annual Real Losses

"UARL Total" values, in the units shown in Table 7.2, provide a rational yet flexible basis for predicting UARL values for a wide range of distribution systems. The calculation takes into account length of mains, number of service lines, location of customer meters relative to property line (curb stop), and average operating pressure (leakage rate varies approximately linearly with pressure for most large systems). An important aspect of Table 7.2 is the value assigned to unavoidable "Background (undetectable real) Losses," shown in Col. 2. These figures are based on international data, from analysis of night flows in sectors just after all detectable leaks and breaks have been located and repaired. This component of unavoidable real losses does not appear to have been quantified previously in North American practice, yet it accounts for at least 50 percent of the unavoidable real losses components in Table 7.2. Estimates of background (undetectable) leakage following intensive leak-detection surveys in small U.S. systems have been compared with IWA unavoidable background loss predictions based on the Col. 2 of Table 7.2. Initial comparisons are encouraging, and more comparisons are being actively sought.

There are many different ways to present the UARL equation. Figure 7.2 shows UARL in gal/mi/d/psi of pressure (Y axis) plotted against density of service lines. The large variation of unavoidable losses per mile of mains for different densities of service lines shows why it is not recommended to use "per mile" for comparisons of real losses. However, Fig. 7.2 can be used to estimate unavoidable annual real losses for any system, as the following example shows.

Example   A water supply system has 60,000 service connections and 600 mi of mains (a connection density of 100 service lines per mile of mains), and the average operating pressure is 70 psi. Calculate the unavoidable annual real losses from Fig. 7.2 if the average distance of customer meters from the curb stop is (a) 100 ft or (b) 20 ft.

Answer   At a connection density of 100 per mile of mains (X axis), from Fig. 7.2 the UARL is
(a) 34 gal/mi/d/psi of pressure × 70 psi = 2380 gal/mi/d × 600 mi = 1.43 mgd (for customer meters 100 ft from the curb stop); or
(b) 23 gal/mi/d/psi of pressure × 70 psi = 1610 gal/mi/d × 600 mi = 0.97 mgd (for customer meters 20 ft from the curb stop).

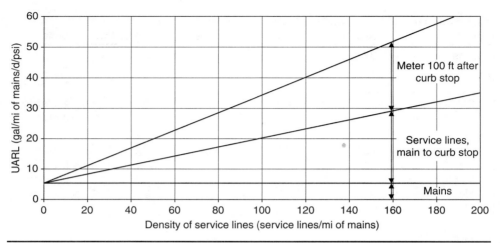

FIGURE 7.2   Unavoidable annual real losses (Gal/mile of mains/day/psi) vs. density of service connections. (*Source:* Ref. 7.)

Comparison of IWA system–specific values of unavoidable annual real losses in gallons per mile of mains per day compare well with the range of 1000 to 3000 gal/mi/d usually quoted for North American systems. However, the IWA prediction method has the considerable advantage that it allows estimates to be made on a system-specific basis, taking account of density of connections, average operating pressure, and locations of customer meters (relative to the curb stop). The last of these factors is particularly important in a region of diverse climates such as North America, where some customer meters are close to the curb stop and others are in buildings more distant from the curb stop.

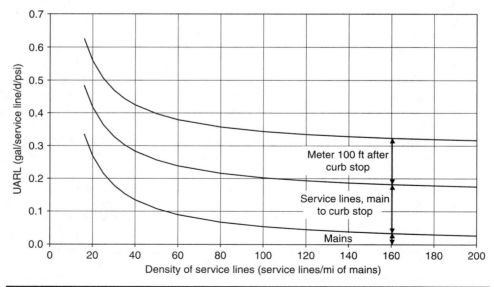

FIGURE 7.3   Unavoidable annual real losses (gal/Service Line/d/psi) vs. density of service lines. (*Source:* Ref. 7.)

The UARL values in Table 7.2 can just as easily be plotted as a graph of gallons per service line per day per psi of pressure versus density of service lines, as shown in Fig. 7.3.

In well-run systems worldwide, the greatest annual volume of real losses occurs from long-running, small- to medium-sized leaks on service connections, except at low densities of service connections. This is why the IWA Task Forces recommend using "per service connection" instead of "per mile of mains" as the basic performance indicator for real losses, for connection densities exceeding 32 per mile. Using the previous calculation example, for the system with 60,000 service connections and 600 mi of mains, the UARL derived from Fig. 7.3 would be

(a) 0.34 gal/service/d/psi of pressure × 70 psi = 23.8 gal/service/d × 60,000 services = 1.43 mgd (for customer meters 100 ft from the curb stop); or

(b) 0.23 gal/service/d/psi of pressure × 70 psi = 16.1 gal/service/d × 60,000 services = 0.97 mgd (for customer meters 20 ft from the curb stop).

The curved lines in Fig. 7.3 are relatively flat for a wide range of connection densities. In calculating unavoidable annual real losses, for example, systems with customer meters 50 ft from the curb stop, and connection densities in the range 80 to 200 per mile, an acceptable simplification from Fig. 7.3 would be to say that the UARL is 0.25 gal/mi/d/psi of pressure (=±10%).

7.6   Which Performance Indicator? What's Wrong with Percentages?

Because water utilities are of different sizes, with different characteristics, comparisons of performance in water loss management need to be made in terms other than volume per year. Traditionally, several different performance indicators are used by North American utilities to compare water losses—percent of system input volume or the metered water ratio, and "per mile of mains per day" appear to be the most common. But are these reliable indicators for comparing performance?

Why do some countries use "per property per day," or "per service connection per day," or "per kilometer of systems (mains + services length) per day?" The IWA Task Force on Water Losses, with nominated representation from the AWWA, has been considering best practice internationally, and their conclusions[8] strongly suggest that there are more reliable and meaningful performance indicators than "percent of system input" and "per mile of mains."

In emphasizing the importance of the correct choice of measuring units, another example from history is useful. Two thousand years ago, in the first century A.D., Julius Frontinius Sextus, then water commissioner for Rome, was spending the whole of his professional career trying (and failing) to achieve a meaningful balance between the quantities of water entering and leaving the aqueducts, which served the city. Failure was not due to lack of diligence on his part—he was simply using the wrong measures. The accepted Roman method was to compare only areas of flow; because they did not take velocity of flow into account also, their calculations could never be reliable for management purposes.

> **E**xpressing losses as a percentage is not the best way to compare loss-management performance, as systems with lower demands or successful customer side conservation programs will never be able to compete with those with larger demands. Instead the volume of loss per service connection per day should be used.

Because per capita consumption in North America is so high compared to most other countries, the common practice of expressing water losses as a percent of system input volume tends to produce lower figures than would be the case in the other countries. This gives a false impression of true performance when comparisons of performance are made with other countries with lower per capita consumption.

The same problem occurs when comparisons are made between North American utilities with a high consumption base and North American utilities with a low consumption base. Data of 1996 showed that 51 water supply systems in California had density of connections varying from 24 to 155 per mile, with an average of 75 per mile. The average metered consumption per connection varied from 136 to 2200 gal/service conn/d, with an average of some 600 gal/service conn/d. Suppose that each of these water utilities was achieving real losses of 60 gal/service conn/d, which is around three times the unavoidable annual real losses (21 gal/service conn/d) for a system with 75 conn/mi, pressure of 70 psi and customer meters 50 ft from the curb stop. Table 7.3 and Fig. 7.4 show that the percent real losses for various systems in California would vary from less than 3% to almost 30%, a tenfold range, depending upon their average consumption per connection, even if all of them had exactly the same actual leakage management performance of 60 gal/service conn/d.

Based on the average consumption of 600 gal/service conn/d, a target of 10% real losses or less might seem reasonable. However, from the above figures it can be shown that

- For utilities with low consumption per service connection it would be a quite unrealistic target, being almost equal to the unavoidable annual real losses.

- For utilities with high consumption it would represent real losses of around 11 times the unavoidable annual real losses.

If Table 7.3 and Fig. 7.4 were not in themselves sufficient to demonstrate the problem of using percentages for comparisons of performance in managing real losses, there would be further serious disadvantages.

- Where a utility exports water, the percentage real losses will be lower if the exported volumes are included in the calculation, and higher if they are excluded.

- The problem of expressing water losses in percentage terms is compounded when demand management measures (customer side conservation) to reduce per capita

System Consumption in gals/service line/d	Real Losses in gal/service line/d	System Input in gal/service line/d	Real Losses as % of System Input Volume
150	60	210	28.6%
300	60	360	16.7%
600	60	660	9.1%
1200	60	1260	4.8%
1800	60	1860	3.2%
2400	60	2460	2.4%

TABLE 7.3 How Percent Real Losses Vary with Consumption, for Real Losses of 60 gal/service conn/d

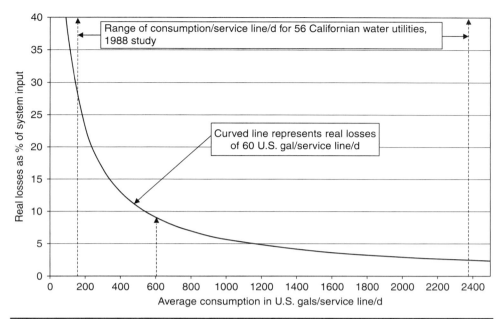

FIGURE 7.4 How percent real losses vary with consumption, for real losses of 60 gals/service line/d.

consumption (pcc) are applied—as the pcc goes down, the percent water losses goes up. Not a great incentive to demand management in its widest sense, simply because of the choice of an inappropriate performance indicator!

Technical committees worldwide (Germany, United Kingdom, South Africa) have recognized these paradoxes of using percentages, but perhaps most significantly the England and Wales Economic Regulator [Office of Water Services(OFWAT)] also recognized it and stopped publishing water losses statistics in percentage terms in 1998. Water system managers who unquestioningly accept percentages as a valid measure of technical performance in management of water losses should consider if they are falling into the same trap as Julius Frontinius Sextus, 2000 years ago—using a simple, but inappropriate, measure to draw inappropriate conclusions.

7.7 IWA/AWWA Recommended Performance Indicators for Nonrevenue Water and Real Losses

During the period 1996 to 2000, various IWA Task Forces undertook a detailed study to determine the most appropriate performance indicators for different water supply purposes. Table 7.4 below shows the PIs for nonrevenue water and real losses recommended by IWA[1,2,8] converted to North American units.

The PIs are categorized by function and by level, defined as follows:

- *Level 1 (basic)*: A first layer of indicators that provide a general management overview of the efficiency and effectiveness of the water undertaking.

Function	Ref.	Level	Performance Indicator	Comments
Financial: nonrevenue water by volume	Fi36	1 (basic)	Volume of nonrevenue water as % of system input volume	Can be calculated from simple water balance
Financial: nonrevenue water by cost	Fi37	3 (detailed)	Value of nonrevenue water as % of annual cost of running system	Allows different unit costs for nonrevenue water components
Inefficiency of use of water resources	WR1	1 (basic)	Real losses as a % of system input volume	Unsuitable for assessing efficiency of management of distribution systems
Operational: real losses	Op24	1 (basic)	gal/service line/d, when system is pressurized	Best "traditional" basic performance indicator
Operational: real losses	Op25	3 (detailed)	Infrastructure leakage index	Ratio of current annual real losses to unavoidable annual real losses

Source: Ref. 7.

TABLE 7.4   IWA Recommended Performance Indicators for Nonrevenue Water and Water Losses

- *Level 2* (*intermediate*): Additional indicators, which provide a better insight than the Level 1 indicators for users who need to go further in depth.
- *Level 3* (*detailed*): Indicators that provide the greatest amount of specific detail, but are still relevant at the top management level.

Particular points to note from the Table 7.4 are as follows:

- *Fi36*: Percentage of nonrevenue water is the basic *financial PI.*
- *Fi37*: This detailed financial PI is a development of a 1996 recommendation of the AWWA Leak Detection and Water Accountability Committee.
- *WR1*: Real losses as percentage are unsuitable for assessing efficiency of management of distribution systems for control of real losses, because of the influence of consumption.
- *Op24:* Gallons/service line/d is the most reliable of the traditional PIs for real losses, for all systems with service line densities of > 32/mile.
- *To improve on Op24, take account of three key system-specific factors:* Density of service connections, location of customer meter relative to curbstop, average operating pressure.

NOTE: *By expressing Op 24 as "Gallons/service line/d/psi of pressure," the influence of pressure is included.*

- *Op25*: The infrastructure leakage index (ILI) is a measure of how well the system is being managed for the control of real losses, at the current operating pressure.

- *ILI*: It is a dimensionless ratio between the *current annual real losses* (*CARL*) based on the results of the water balance and the *unavoidable annual real losses* (*UARL*) for a given system.

$$\text{ILI} = \text{CARL} / \text{UARL}$$

- *UARL:* They are calculated as previously described in this chapter, using the IWA methodology which takes into account average operating pressure, length of mains, number of service lines, and location of customer meters relative to the curb stop.

The infrastructure leakage index is a relatively new, and potentially very useful, performance indicator. Being a ratio, it has no units, so facilitates comparisons between countries that use different measurement units (metric, U.S. Customary). The ILI can perhaps be better envisaged from Fig. 7.5, which shows the four components of leakage management.

The large square represents the current annual volume of leakage, which is always tending to increase, as infrastructure systems grow older. This increase, however, can be constrained by an appropriate combination of the four components of a successful leakage management policy.

The small square represents UARL—the lowest technically achievable value for real losses at the current operating pressure. The ratio of the current annual real losses (the large square) to the unavoidable annual real losses (the small square) is a measure of how well the three infrastructure management functions—repairs, pipe materials management, and active leakage control—are being controlled. We will be seeing more of this diagram in future chapters where we will be discussing some of the hands-on techniques associated with in the field loss-reduction programs.

> **T**he ILI ratio is a great way of demonstrating loss management performance, as each system effectively compares the ratio of their individual best possible performance against how they are actually performing.

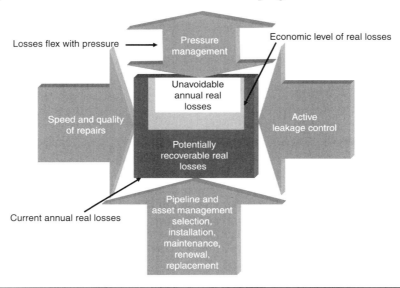

FIGURE 7.5   The four components of a successful leakage management policy.
(*Source:* IWA Water Loss Task Force and AWWA Water Loss Control Committee)

An infrastructure leakage index close to 1.0 demonstrates that all aspects of a successful leakage management policy are being implemented by a water utility. However, typically it will only be economic to achieve an ILI close to 1.0 if water is very expensive, scarce, or both. Economic values of ILI depend upon the system-specific marginal cost of real losses, and typically lie in the range 1.5 to 2.5 for most systems.

7.8   The Use of 95% Confidence Limits and Variance Analysis for Water Audits

The use of 95% confidence limits to validate the degree of uncertainty in individual components of the water balance is nowadays best practice among qualified water loss management professionals.

In order to understand the concept of 95% confidence limits, it is first necessary to understand normal distributions which are an important class of statistical distributions. All normal distributions are symmetric and have bell-shaped density curves with a single peak. To speak specifically of any normal distribution, two quantities have to be specified: the mean μ where the peak of the density occurs, and the standard deviation σ, which indicates the spread or girth of the bell curve. Different values of μ and σ yield different normal density curves and hence different normal distributions.

The normal density can be actually specified by means of an equation. The height of the density at any value x is given by

$$\frac{1}{\sigma\sqrt{2\pi}} e^{-\frac{1}{2}(x-\mu/\sigma)}$$

Although there are many normal curves, they all share an important property which is often referred to as the empirical rule:

- 68% of the observations fall within one standard deviation of the mean, that is, between $\mu - \sigma$ and $\mu + \sigma$.

- 95% of the observations fall within two standard deviations of the mean, that is, between $\mu - 2\sigma$ and $\mu + 2\sigma$.

- 99.7% of the observations fall within three standard deviations of the mean, that is, between $\mu - 3\sigma$ and $\mu + 3\sigma$.

Thus, for a normal distribution, almost all values lie within three standard deviations of the mean as can be seen in Fig. 7.6.

Using 95% confidence intervals allows generating a lower and upper limit for the water balance component. The interval estimate or lower and upper limit gives an indication of how much uncertainty there is in the volume used for each water balance component. The narrower the interval, the more precise is the value used.

The 95% confidence limits also allow for the calculation of the variance related to each water balance component. Variance is a measure of dispersion around the mean. Components with a large variance will have the biggest impact on 95% confidence limit related to the final result of the water balance. The final derived result of the water balance is the volume of real losses. This component will have a 95% confidence limit that is an accumulated value based on the variance related to each component of the water balance. The variance analysis is based on standard statistical principles of normal distribution and uses the root-mean-square (RMS) method for accumulation of error on derived values (see Table 7.5).

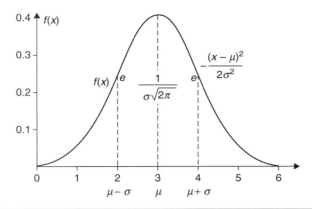

FIGURE 7.6   Normal distribution curve. (*Source:* WSO).

Component	Annual Volume (Million gal)	95% Confidence Limits	Variance (gal² × 10¹²)
Source #1	7,512.80	2.6%	9,553.7
Source #2	10,519.84	2.6%	18,732.1
Source#3	6,580.71	2.6%	7,330.2
Source#4	4,411.61	2.6%	3,294.3
Source#4	7.60	2.6%	0.0
Total system input volume (*a*)	**29,032.56**	**1.3%**	**38,910.3**
Billed metered authorized consumption	24,778.64	1.1%	20,237.7
Billed un-metered authorized consumption	0.0	NA	NA
Total billed authorized consumption (*b*)	**24,778.64**	**1.1%**	**20,237.7**
Nonrevenue water [(*a*) − (*b*)]	**4,253.92**	**11.2%**	**59,148.0**

Source: SFPUC

TABLE 7.5   Calculation of Confidence Limits for Nonrevenue Water

The standard approach to calculate the variance related to a certain volume of the water balance, based on its 95% confidence limit, is as follows:

$$\text{Variance} = (\text{volume in million gal} \times 95\% \text{ confidence limit}/1.96)^2$$

The aggregated confidence limit related to a calculated volume of the water balance is based on accumulation of error on derived values. Following these principles the 95% confidence limit related to the calculated volume of nonrevenue water is calculated as follows:

95% confidence limit for nonrevenue water

$$= 1.96 - \sqrt{(\text{variance } a + \text{variance } b)} / (\text{annual volume nonrevenue water})$$

The above equation explains the standard approach for calculating an aggregated confidence limit based on accumulation of error on derived values. Table 7.5 provides an example of the calculation of the 95% confidence limits and variances relevant for the calculation of 95% confidence limits for nonrevenue water.

Since the real losses have a confidence limit that is an accumulated value based on the variance related to each component of the water balance it is very important to accurately assign 95% confidence limits to all components of the water balance in order to see which of the components has the biggest impact (which components have the highest variance) on the confidence related to the calculated real loss volume. Once this information is available, it is best practice to take actions (e.g., improving the accuracy of metering devices or installing new metering devices where no meter was in place) in order to improve the confidence related to the real loss volume by improving the confidence related to those components that showed the highest variance.

7.9   Conclusion

The IWA/AWWA standard terminology and water balance methodology and the use of 95% confidence limits, together with the equation for unavoidable annual real losses and the recommended performance indictors such as the infrastructure leakage index (ILI), are the basis for a rational assessment of water loss volumes allowing meaningful comparisons of water loss management.

Examples of free and commercial water audit software and related water balance results are shown in Chap. 10.

References

1. Alegre, H., W. Hirner, J. Baptista, et al. "Performance Indicators for Water Supply Services." *Manuals of Best Practice*: IWA Publishing, 2000. ISBN 1 900222 272.
2. Alegre H, JM. Baptista, E. Cabrera Jr, et al. "Performance Indicators for Water Supply Services," 2nd ed. *Manuals of Best Practice Series*. IWA Publishing, 2006. ISBN 1843390515.
3. Kunkel, G. et al. "Applying Worldwide Best Management Practices in Water Loss Control." Water Loss Control Committee Report. *Journal AWWA*. 95(8):65, 2003.
4. Office of Water Services. UK. "Leakage and the Efficient Use of Water." 1999–2000 Report. ISBN 1 874234 69 8.
5. Fanner, V. P., R. Sturm, J. Thornton, et al. *Leakage Management Technologies*. Denver, Colo.: AwwaRF and AWWA, 2007.
6. Fanner, V. P., J. Thornton, R. Liemberger, et al. *Evaluating Water Loss and Planning Loss Reduction Strategies*. Denver, Colo.: AwwaRF and AWWA, 2007.
7. Lambert, A., D. Huntington, and T. G. Brown. "Water Loss Management in North America: Just How Good Is It?" AWWA Distribution Systems Symposium, New Orleans, September 2000.
8. Lambert, A., T.G. Brown, M. Takizawa, et al. "A Review of Performance Indicators for Real Losses from Water Supply Systems." *AQUA*. Decemeber 1999.
9. Lambert A. O., S. Myers, and S. Trow. *Managing Water Leakage: Economic and Technical Issues*. Financial Times Energy Publications, 1998.

CHAPTER 8

Data Collection, Formatting, and Management

Julian Thornton

Reinhard Sturm

George Kunkel, P.E.

8.1 Introduction

To undertake any water system audit and properly identify where volumes of losses are occurring and the magnitude of the loss it is necessary to collect data, which is

- Accurate
- Standardized
- Organized
- Accountable

In Appendix A, we discuss various equipment and methodologies for accurately capturing data for flows and pressures using both portable and permanent field equipment. However once we have captured the data it is important to properly organize and store the data in a meaningful manner, so that we can be accountable for the subsequent decisions which will be made.

When collecting and validating data for top down water balances it is common to collect large volumes of data from the customer information system. Usually, at least 14 months of data are collected. In large water systems, this could amount to many gigabytes of data. It is important that the operator carefully considers the environment in which the data will be stored for analysis in order not to loose data in the transfer process. Many operators use industry standard products

Not all water systems will have all of the data they need for a full audit, however it is better to make estimations and perform an audit than not to do one at all. Lower confidence can be assigned to estimates and higher confidence to measured values.

such as Excel work sheets or Access databases, however these products have size limitations, Excel (Pre 2007) can hold only 58,000 rows of information and Access can store only 2 GB of data.

In many cases, good accurate data may not be available and the operator will have to make a decision as to whether to use the questionable data or estimations or not. In many cases it is better to do something rather than stop and do nothing. In this case, we should be sure to note that the data was questionable or estimated and the operator's assessment of what should be done to improve this in future audits and how the data should be used this time round.

The following section discusses good data management techniques.

8.2   Data Collection Worksheet

One of the first things we must do before starting to download field loggers and recorders is decide on the key factors, which we will be analyzing and assign relevant measurement units and decimal places to each of the parameters. For example, in most audits we will be measuring flow, measuring pressure, analyzing volumes, measuring levels, and accounting for time periods.

Some of the units, which we might assign, are in the following section.

8.2.1   Flow

Metric

- Cubic meters per second
- Cubic meters per hour
- Liters per second
- Mega liters per day

U.S. Customary

- U.S. gallons per minute
- Imperial gallons per minute
- U.S. gallons per hour
- Imperial gallons per hour
- U.S. Kgallons per day
- Imperial Kgallons per day
- U.S. millions of gallons per day (mgd)
- Imperial millions of gallons per day (mgd)
- Cubic feet per second
- Cubic feet per hour
- Cubic feet per day
- Acre feet per day

8.2.2   Pressure

Metric

- Meters head of water
- Bar
- Kilopascals

U.S. Customary

- Pounds per square inch (PSI)
- Feet head of water

8.2.3   Volumes

Metric

- Cubic meters
- Liters
- Mega liters

U.S. Customary

- Gallons
- Kgallons
- Million gallons
- Cubic feet
- Acre feet

8.2.4   Levels

Metric

- Millimeters
- Meters
- Millibar
- Bar

U.S. Customary

- Inches column of water
- Feet column of water

8.2.5   Time Periods

- Milliseconds (used for surge analysis and leak noise correlation)
- Seconds
- Minutes
- Hours
- Days
- Months
- Years

So, as we can see there are many options for recording all of our varying parameters. It is important to use parameters and units, which are both meaningful to the country or area in which we are working, and also units, which are easily interchangeable. So, for example, we wouldn't want to mix cubic meters per hour of flow with pounds per square inch of pressure. We might however use either pounds per square inch of pressure with gallons per minute of flow or cubic meters per hour of flow with meters head of water pressure.

8.2.6   Balancing Flows

When undertaking audits, which involve dynamic flows and not just volumes, it is important to balance our flow inputs. To do this, we usually select a unit of flow, for example, cubic meters per hour.

> **T**op down annual audits use volumes; bottom up audits often use night flows.

We will then identify key points within a 24-hour profile, usually minimum night flows if we are trying to identify leakage. The balance is a simple matter of adding and/or subtracting individual zone flows, (these might be metered areas or pressure zones) and comparing them with supply meter or production metered flows to ensure that we have all of the inflows and outflows for the system in question accounted for. (Take care if storage is located inside of the areas we are trying to balance as filling volumes will confuse the issue).

In situations where the system is not zoned in any way at all and is not intended to be for the future, the key points within the flow balance would be

- Production meters
- Import meters or bulk supply meters
- Outlets from storage (tanks, reservoirs, and towers)
- Outlet from pumps or wells

This may seem like a relatively simple procedure but can take many hours of careful analysis especially in large systems.

It is particularly important to properly define one unit of measure before attempting this exercise, otherwise the difference in one working unit and another could be confused for a missing inlet or outlet and create a lot of unnecessary work load, which in turn would create unnecessary cost.

8.2.7 Balancing Pressures

It is equally as important to balance pressures in a water system when attempting to identify losses, as the system pressure plays a large part in water loss especially leakage as discussed later in this manual.

Usually, when we want to balance pressures we work out hydraulic grade lines (HGLs). Hydraulic grade is a sum of the ground level plus the static pressure at that particular point and the lines are the chosen points connected up.

Most often in water loss control situations we will need to know

- The supply or inlet pressure
- Average zone pressure
- Critical point pressure
- Minimum service pressure required
- The number of hours the system is pressurized in cases where there is intermittent supply

8.2.8 Balancing Levels

In systems with large storage capacity, it is also important to include the various tank or reservoir levels in the water balance, as the change in volume over time may represent significant flow and could be mistaken for loss.

In systems with a small amount of storage capacity, this is not so important, however, this should not be overlooked. It is always better to overanalyze than underanalyze!

8.2.9 Putting Data into a Common Format

Putting data into a common format is extremely important. Metric and U.S. Customary units should never be mixed and even when using one or the other it is still a good idea to think about the method of data recording, which has taken place in the field and the required reporting units.

Don't mix incompatible units.

If working in metric, for example, it is much easier to work in cubic meters per hour flow if you measure your velocity in meters per second and calculate your pipe effective area in square meters. The result will always be automatically in meters and then it is just a simple case of deciding the time units.

For example, we measure a velocity of 2 m/s in a pipe, which has a diameter of 400 mm (400 mm is actually 0.400 m). To calculate the area we would use our formula $\text{Pi} \times R^2$, which in this case would be $3.142 \times 0.2 \times 0.2$. The answer would be 0.12568 m². We have a velocity of 2 m/s so we would multiply this figure by two, which would give us 0.25136 m³/s. Now we must decide on a unit of time. Usually when working in the field with cubic meters we would use cubic meters per hour of flow. There are 60 seconds in a minute and 60 minutes in an hour so we would multiply our flow of 0.25136 m³/s by 3600, which would give us 904.896 m³/hr. We know that there are a 1000 L in 1 m³ so we could also say that we have a flow of 904,896 L/hr. This number is quite large and if added to other large numbers could lead to mistakes. If we wanted to express our flow in liters we would most likely use liters per second. If that were our desired final number we would have taken our figure above of 0.25136 and multiplied by 1000 to take our flow units from cubic meters to liters. We would not need to multiply anything else as our original number was in seconds. Our flow would then be 251.36 L/s. If we are in a situation where we need to translate data from liters per second to cubic meters per hour then our common figure is 3.6. When altering liters per second to cubic meters per hour we just need to multiply our original number by 3.6 to have cubic meters per hour and vice versa.

Alternatively if we were working in U.S. Customary we might have the following:

We measure a velocity of 2 ft/s in a pipe, which has a diameter of 36 in (36 in is actually 3 ft). To calculate the area we would use our formula $Pi \times R^2$ which in this case would be $3.142 \times 1.5 \times 1.5$. The answer would be 7.0695 ft^2. We have a velocity of 2 ft/s so we would multiply this figure by two, which would give us 14.139 ft^3/s. Now we must decide on a unit of time. Often when working in the field with cubic feet we would use cubic feet per hour of flow. There are 60 seconds in a minute and 60 minutes in an hour so we would multiply our flow of 14.139 ft^3/s by 3600, which would give us 50,900 ft^3/hr. We know that there are 7.48 gal in 1 ft^3 so we could also say that we have a flow of 380,734 gal/hr. This number is quite large and if added to other large numbers could lead to mistakes. If we wanted to express our flow in gallons we would most likely use gallons per minute. If that were our desired final number format we would have taken our figure above of 380,735 gal/hr and divided by 60, we would not need to divide anything else as our figure was already in gallons. Our flow would then be 6,345 gpm.

8.3   Data Calibration Form

Often when measuring devices are tested there is a small margin of error. It is not always possible to recalibrate the flow meter before measuring in the field; although that option is preferable. If the flow-measuring device cannot be recalibrated mechanically or electronically then it is still possible to use the data; however the data must be calibrated theoretically using a spreadsheet.

The spreadsheet will be constructed using the calibration curves from the meter tests prior to data collection and will show errors for brackets of flow. The data will then be imported into the form or spreadsheet, and automatically be changed by the error attributed to that flow range. The resultant data is closer to the truth than the original. Obviously there are some cases where error will still occur, especially in the case of a particularly sensitive or unstable measurement device.

8.3.1   Equipment Calibration Form Pressure and Level

As with flow measurement devices pressure and level sensors can also have errors, which cannot be recalibrated before testing is undertaken. The same process can be undertaken to ensure that pressures and levels are closer to the true value.

8.4   Summary

Good data management will ensure that the whole project has accountable, baselines from which to judge performance and allocate new budgets.

So as we can see it is vitally important that the data is managed properly from the start of the program. Accountability is a word, which we are using more often in the water industry now. Accountability doesn't mean that we guarantee that all of our data is accurate. What is important is that where we have doubts as to the accuracy of the data we leave an audit trail explaining what was done estimated or calculated. If we perform accountable audits with a data trail, we can always improve data accuracy over the years to come until eventually all data is top class.

The following checklist covers many of the aspects necessary for good data management.

8.4.1   Data Management Checklist

- Data should be accurate.
- Data should be organized.
- Data should be accountable.
- Bad data should be clearly highlighted.
- Estimations can be made but should be clearly marked as such.
- Raw data should be kept as well as calibrated data.
- Constant measurement values should be used.
- Constant units should be used.
- A column alongside the audit sheet with relevant comments will help future auditors figure out what you did when you made your audit.

CHAPTER 9

Identifying Economic Interventions against Water Losses

David Pearson

Stuart Trow

9.1   Introduction

The level of losses from water systems is often considered by observers from outside the industry to be unacceptable. In many countries, environmentalists and regulators have expressed concerns at the level of losses, and believe that lower levels should be achievable. However, any water company has to work within current operating budgets and seek additional finance if these are not sufficient. Leakage control can be expensive, and water companies will seek to achieve an economic balance between the costs of leakage control and the benefits that accrue. This balance between costs and benefits is common in many fields, and the idea of the economic level of operation is commonplace in many industries. The concept of an economic level of leakage (ELL) dates back several decades, and there have been many previous attempts to determine a practical definition and methodology. Previous methodologies tended to confuse the impact of the various leakage management options available. It is only over the past 15 years that we now have a better understanding of all the issues.

9.2   Definition

Looking at economic theory, there are two levels at which the economic level can be considered. Taking manufacturing as an example, production can increase by taking on more labour. Increased costs would be incurred in terms of labour costs, raw material costs, and costs of production—typically power, which are a function of the level of production. As levels of production are increased, for example, by increasing the number of shifts, production will rise until the capacity of the production plant itself becomes a limiting factor. At some point it may be more economic to extend the plant. However, in this case, major capital expenditure will be involved and long term payback of this capital expenditure has to be taken into account. These two levels of economic optimum (firstly, by varying revenue items alone, and then looking at capital expenditure) are known, respectively, as the short- and long-run economic levels.[1] The formal economists definitions[2] of these are: "The *short run* is a period of time in which the quantity of at least one input is fixed and the quantities of the other inputs can be varied. The *long run* is a period of time in which the quantities of all inputs can be varied, and other new inputs can be introduced."

Examples that are generally quoted, using manufacturing industry, refer to labour, materials, and power as variables that can be changed in the short run, whilst plant capacity can only be changed in the long run.

The current thinking on the economic level of leakage (ELL) is based on the knowledge that each and every activity aimed at reducing leakage follows a law of diminishing returns; the greater the level of resources employed, the lower the additional marginal benefit which results. This understanding forms the basis of a new methodology in which every activity is analysed in a similar way to compare its marginal cost with that of other interrelated activities, and with the marginal cost of water in that supply zone.

This approach can be applied to the four primary activities that impact on leakage control, that is, pressure management, active leakage control (ALC), quality and speed of repairs, and infrastructure improvements, which are often illustrated as shown in Fig. 9.1. To further the comparison with the examples used in manufacturing industry, the elements such as active leakage control and repair activity can be considered to be revenue items and would therefore be considered in the evaluation of the short-run ELL, whereas pressure management and mains rehabilitation would require an investment decision,

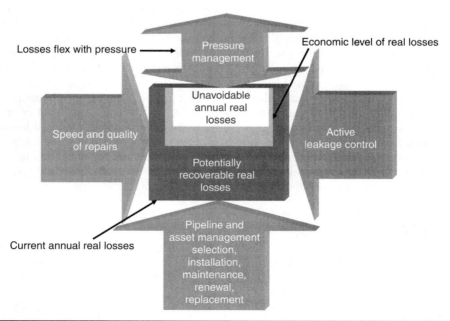

Figure 9.1 The four primary methods of controlling water losses. (*Source:* IWA Water Loss Task Force and AWWA Water Loss Control Committee.)

and would therefore be considered in the evaluation of the long-run ELL. There are other activities that can impact on leakage such as sectorization, customer meter reading policy, customer side repair policy, extent of customer metering, and so on.

9.3   Short-Run ELL

9.3.1   Active Leakage Control

The purpose of active leakage control (ALC) is to find leaks that do not surface or otherwise come to the attention of the operating company through customer contact, for example, poor supply, loss of water, and so on. These leaks are often referred to as *reported* leaks. The process of active leakage control involves teams of leakage detection staff sweeping an area to find leaks generally using sounding techniques or similar. This may be in response to an increase in a nightline if the area is sectorized, an increase in the output from a treatment works or service reservoir/tank or simply as a result of a regular sounding programme at an agreed interval.

This ALC activity will locate *unreported* leaks, which will then be repaired, and leakage levels will be maintained. If sweeping is carried out at more frequent intervals then leakage will be maintained at a lower level. Thus, there is a relationship between average leakage level and the time between surveys. This is shown as curve A-A in Fig. 9.2 and is referred to as the *active leakage control curve*. The vertical axis is usually expressed in cost terms and is simply the annual cost of the leakage detection resources. The horizontal axis is the average leakage level, over the same period (usually a year). On the assumption that some leaks would never come to the attention of the operating company if they did not come to the surface (e.g., if they break through to a sewer) and would therefore accumulate on the system, then the curve will asymptote to the horizontal axis. The curve will also asymptote to a line parallel to the vertical axis. This line B-B, will be equivalent to the level of leakage that would result if infinite resources were deployed on leakage control activity. This minimum level of leakage would equate to

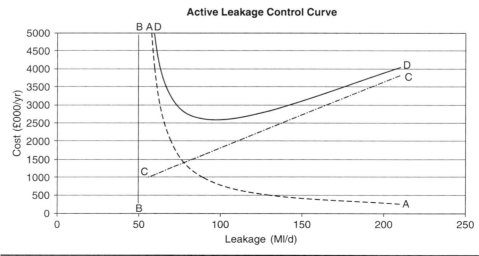

FIGURE 9.2   Active leakage control (ALC) cost curve. (*Source:* Dave Pearson.)

background leakage, that is, leakage below the level of detection, plus the leakage from reported leaks plus the leakage from unreported leaks during the period they run between detection and repair, resulting from any given leakage control policy. This is sometimes referred to as *the policy minimum level of leakage.*

There has been much debate about the shape of the curve between these asymptotes. In the most simplistic model of regular sounding, the curve will be hyperbolic. This is based on the fact that the curve will be defined by the leakage during the period which unreported leaks run until they are detected. This will be directly related to the length of time they run before being detected and hence the intervention interval. As the intervention interval will be inversely related to the resources (doubling the resources will half the intervention interval) then leakage will be inversely proportional (i.e., a hyperbole) to the level of resources and hence the ALC cost. If the area is sectorized, or if other forms of flow measurement are used to direct resources more efficiently compared to simple regular sounding, the curve will be flatter than a pure hyperbole.

If the cost of the water lost at different levels of leakage is plotted on the same graph this would be represented by the line C-C. The cost will be the simple difference in cost in producing one more or less unit of water in terms of power, chemicals, and possibly labour. The slope of this line is referred to as the marginal cost of water. If the marginal cost of water is constant, line C-C will be a straight line. If the marginal cost of water production is not constant, then line C-C will be made up of a number of straight lines; usually increasing in slope with higher leakage as more expensive water is used. Curve D-D is the total cost of operation, that is, cost of leakage control plus cost of water production. As can be seen, the curve will be high initially due to the high cost of leakage detection required to achieve very low levels of leakage. The total cost then reduces before increasing again as the cost of water production increases with increasing levels of leakage. The point at which the total cost is lowest will be the short-run economic level of leakage. At this point, the marginal cost of leakage detection activity will be equal to the marginal cost of water. This point will also define the economic level of resources to be deployed on leakage detection and the economic period between interventions.

It can be shown that the minimum total cost of lost water and intervention costs occur when the accumulated value of lost water since the last intervention equals the cost of intervention. This simple relationship has been used by a number of people to develop methodologies to calculate the economic intervention period for a system.

The solution to the calculation of the economic intervention period in the case of regular sounding, that is, where all parts of the system are swept with the same frequency, is reasonably straightforward[3] and this has been developed[4] into methodologies that can be readily applied to distribution systems.

Where the system has been sectorized and information therefore exists for the rate at which leakage accumulates on different parts of the network then a more specific approach can be taken.[5,6] In this approach, the actual volume of leakage is accumulated using night-line information since the last intervention and proactive detection is initiated when the value of this is equal to the cost of intervention on that sector. The advantage of this approach is that it can take into account sector-specific cost of water (say due to local boosting of water) and also sector-specific survey costs (say due to urbanisation or pipe materials).

An alternative approach has been to try and define the ALC curve itself. This can be carried out in a number of ways, which can be classified as either empirical or theoretical.

The former relies on the establishment of a number of points along the curve by analysing the results from actual ALC operations. When a number of points have been derived then a curve is fitted. This may assume a given shape to the curve.[7] The difficulty with this approach is that the current position on the curve represents a static situation of the balance between average leakage over a number of years at a constant resource level. It may take a number of years to reach stability when detection resources are changed. It is therefore a long process to develop accurate estimates of a number of points on the curve.

Alternatively, a theoretical approach using component loss modelling methodologies[8] can be used to define the ALC curve, but this will require a number of assumptions, such as burst flow rates, although attempts can be made to calibrate these from actual data. A compromise is to establish the ALC curve by building a component loss model of the system and then to calibrate this such that it passes through the current operating position established by analysing the actual cost of operations. The economic intervention period can then be found by direct differentiation of this curve or by numerical methods.

9.3.2   Background Leakage and Backlog Removal

Background leakage is generally defined as the leakage below the level of detection (with current technology). The level of background leakage can be assessed using a number of methodologies.[7] However, the level of background leakage is a function of the extent and method of leakage detection employed, which itself will have different operating costs associated with different levels of leakage. Therefore, a matrix of leakage detection costs versus level of background leakage can be derived, from which a view can be taken on the appropriate economic method of detection, and the associated level of background leakage.

Background levels of leakage have been related to system characteristics.[9,10] Such as pressure, length of mains, and number of connections. From these, unit background losses at standardized pressure have been estimated. These can then be related to asset type, material, age, and condition. From this work it is possible to provide an estimate of the background level of leakage that might be expected in an area. By comparing this to the actual minimum achieved on that area, a view can be taken as to whether background levels have been achieved or whether it is likely that there are leaks on the area that would be possible to find.

These leaks will have gradually accumulated on the system over a number of years, and are essentially hidden in accepted minimum historic night flows. The leaks may be on parts of the network that are not normally checked for leaks, for example, large industrial complexes, mains which are believed to have been abandoned, private supply pipes, complex road junctions. The number of backlog leaks and hence the associated repair bill can be substantial, but they are one-off costs and the cost benefit can be readily assessed. However, it may be appropriate to take other action, for example, pressure reduction (described later) to reduce the frequency at which the system is breaking in order to allow for this backlog to be reduced over a period of time within the current repair budget. Alternatively, it is possible that these could be considered as a one-off capital cost depending on local accountancy rules.

9.3.3   Transition Costs

Once an economic level of leakage has been established, then a company should move toward this ELL. However, as this is likely to be at a lower level of leakage than the current level, moving to this point will involve one-off costs. As each point on the ALC

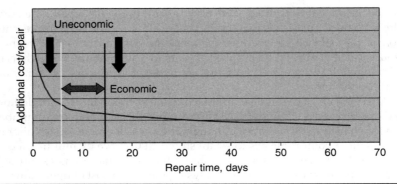

FIGURE 9.3   Economic repair time. (*Source:* Stuart Trow.)

curve is a static situation, then there are less leaks running at any one time at lower levels of leakage. Thus, moving from one point to a lower point will mean that additional leaks are brought in for repair before the situation reaches equilibrium again. Transitional costs should generally be fairly low and they can be added, with appropriate discounting as they are a one-off investment, into the calculation of the economic level of leakage to obtain a slightly revised economic level of leakage.

9.3.4   Leak Repair Activity

A similar methodology to that for ALC can be applied to developing the economic level of speed of repair. Very short repair times can be achieved but at the cost of possible overtime for weekend and evening working for the repair teams. This may or may not be economic. There will be a relationship between cost and repair time as in Fig. 9.3. Leakage level will be related to the average repair time, and so a similar curve to the ALC curve can be produced. The benefit from reducing repair times can be estimated using a component loss model. The economic repair time can therefore be determined in the same way as described above for ALC. At this point the marginal additional cost of repair will equal the marginal cost of water production.

9.4   Long-Run ELL

Some leakage control activities will involve an investment decision, and hence a payback longer than the short-run period. This will typically apply to options such as pressure management and mains rehabilitation. In these cases, it will be economic to make an investment on pressure management or rehabilitation to reduce leakage if the cost of water saved over the investment period would pay for the cost of carrying out the works. Once the investment has been made, there will be a new (lower) economic level of leakage, which has to be recalculated using the method above.

9.4.1   Pressure Management

Leakage will reduce as a result of pressure reduction due to two factors, namely,

- Both background and leak flow rates will reduce, as leakage flow is directly related to pressure by a factor called the $N1$ relationship.[11]

- Burst frequency rates will reduce, due to reduced stress on the pipe network, the so called *N*2 relationship.[12]

Bursts and leaks can be caused by surges on the network. These surges can be caused by defective operator or customer equipment or the lack of surge suppression equipment on pumped systems. Short-period logging should be used to investigate whether a system is experiencing surges before any pressure reduction is investigated.

In the case of pressure reduction, the investment costs will include the one-off cost of construction of the chambers, the cost of purchasing the pressure reducing valves (PRVs) and their replacement as well as ongoing maintenance costs. As pressure management is deployed in an area, the average pressure will reduce. Schemes will be deployed on the basis of those which give most benefit first and therefore as more and more schemes are installed, the marginal benefit of each scheme on the average pressure for the system as a whole will reduce. Figure 9.4 shows a typical curve relating the benefit from scheme deployment on average zone night pressure (AZNP). As leakage is proportional to pressure, there will be a break-even point at which the additional cost of scheme deployment equals the marginal cost of water production.

The process involved in calculating this breakpoint is as follows:

- The potential for pressure reduction from the installation of pressure management valves, and other schemes, is estimated using hydraulic modelling and/or logging of areas.

- The cost of construction is estimated, and the cost discounted into an equivalent annual cost using financial accounting methods (usually agreed with the finance department of the operating company) such as discounted cash flow analysis (DCF).

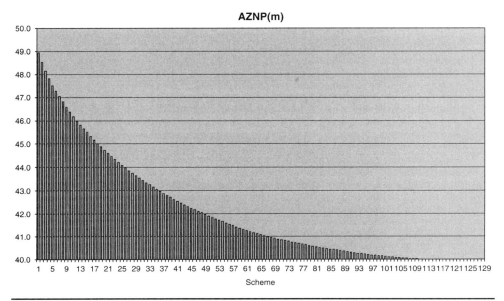

FIGURE 9.4 Ranking pressure management schemes by the benefit of reducing AZNP.
(*Source:* Dave Pearson—Northumbrian Water.)

- The cost of the valve and its replacement (as recommended by the vendor) is discounted in a similar way.

- The annual cost of maintenance of the PRV (as recommended by the vendor) is estimated.

- The benefit in terms of leakage reduction is estimated using a component loss model or similar approach.

- The reduction in operating costs due to the lower burst frequency is assessed in terms of
  - Reduced repair bill
  - Reduced customer contact costs for reported leaks
  - Reduced visit/inspection costs for reported leaks
  - Lower active leakage control costs for unreported leaks

- The marginal cost/benefit is calculated as the net cost divided by the leakage saving

All the schemes with a cost/benefit lower than the cost of water would be deployed. This will establish the economic level of pressure reduction and the associated leakage level. Examples of this approach have been published recently.[13]

There will also be less tangible benefits such as

- Reduced risk of discoloured water events

- Reduced interruptions of supply

These benefits will lead to improved levels of service and customer satisfaction and a reduction in the risk of any regulatory action. A notional monetary value can be placed on these less tangible benefits in order to allow for these in the calculation.

9.4.2   Network Rehabilitation

Network rehabilitation (both mains and service pipes) will reduce the rate at which leaks break out on the network. This will reduce leakage, as well as reducing costs associated with inspections and active leakage control activity highlighted above. Figure 9.5 shows a typical burst frequency distribution curve. This shows that there is a distribution of the frequency at which pipes burst on the network. A small proportion will burst

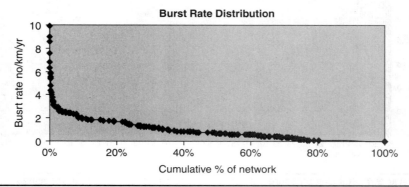

FIGURE 9.5   Burst frequency distribution. (*Source:* Dave Pearson—Sofia Water.)

at a high frequency, whilst other parts of the network will burst at a much lower frequency. In order to have the greatest impact on leakage one would try to identify those pipes with a high frequency of failure and replace these first. The benefit of replacing further sections of pipe will then be less. Again the law of diminishing return applies, and a point will be reached when it is not economic to replace pipes. A similar curve will exist for the distribution of service pipe bursts across the network.

It has been suggested that there will also be a distribution of background leakage, which will not necessarily be the same as that for burst frequency. Those mains with high burst frequencies may have a low background leakage level and vice versa. This is because background leakage is primarily driven by leakage at joints on service pipes rather than mains themselves. Therefore, network rehabilitation should be targeted at burst and background leakage separately.

To find the economic point, the following calculations are performed:

- The benefit of replacing a section or group of essentially similar pipes in the same locality in terms of reduction in burst frequency and/or background leakage is assessed.
- The cost of replacing these pipes is estimated.
- The reduction in leakage is estimated using component loss modelling.
- The savings in costs in inspections, repairs, and active leakage control are assessed.
- The marginal cost/benefit is assessed as the cost less the sum of the savings divided by the leakage saving.

All the schemes with a cost benefit lower than the cost of water would be deployed. This will establish the economic level of network rehabilitation and the associated leakage level.

9.4.3   Sectorization

It is common practice in some parts of the world to split the water network into sectors and monitor flows into and out of these sectors at night. Data about the flows into sectors provides information to be able to locate leaks faster and therefore improve leakage detection efficiency. However the introduction of sectorization involves costs in the following areas:

- One-off cost of construction of meter chambers
- Cost of meter and replacements and/or refurbishment
- Cost of data logging equipment
- Ongoing cost of data retrieval (either manual or by telemetry)

The benefit of introducing sectorization in terms of leakage will be a function of the natural rate of rise of leakage in the sector. Not all sectors will have the same rate of rise, and so again there will be a curve showing diminishing returns. Other factors affecting costs will be the environment, the complexity of the network, and the degree to which sectorization has already been established. The calculations are similar to the ones described above for pressure management and rehabilitation. They can be carried out to establish an economic breakpoint that would give the economic level of sectorization and the optimum size of sectors.

9.4.4   Combination of Activities

The methodologies described above all require the assessment of the benefit in leakage terms from the proposed activity. Each case has been considered independently, that is, the assessment of the economic level of pressure management or rehabilitation. However, the implementation of one option will affect the economics of the implementation of the other, that means, the benefits from rehabilitation will be reduced if average pressures have already been reduced due to pressure management. In practice, an operating company will want to develop a strategy that looks to establish the economic balance between all activities, that is, active leakage control, leakage repair, mains rehabilitation, service pipe replacement, sectorization, and pressure management.

The normal approach to solving this problem is to choose a small increment of activity in each area and work out the cost/benefit. These are ranked and the one with the best benefit is implemented. The leakage benefit for the other schemes are then reassessed due to the change that this scheme imposes and compared again. The next scheme is then chosen and the leakage benefits reassessed and so on. This process is continued until the marginal cost of any activity is equal to or greater than the marginal cost of water. This then establishes the economic level of leakage and the list of schemes that will be implemented and their associated costs to achieve this level.

By following this procedure of "squeezing the box" (i.e., the box containing the level of losses shown in Fig. 9.1) using each of the primary activities of a well-developed program of leakage management in turn based on best value, a point will be reached where any further activity is uneconomic, that is, its marginal cost will be greater than the marginal cost of the water saved. At this point the marginal cost of further leakage control activity will be the same for all activities.[14]

9.5   Deficiency in Water Supply Reliability

9.5.1   The Supply-Demand Balance

The calculations described above establish the economic level of leakage against the marginal cost of water production. In effect, this could be called the *unconstrained ELL*. In practice, this level of leakage, when combined with consumption, may be insufficient to provide the necessary reliability of supply for the operator. The excess of water available for supply compared to the demand is often referred to as *headroom*. Some countries have standards for determining the appropriate level of headroom[15] in order to provide the required security of supply against factors such as climate change, and the like. If, after working out the unconstrained ELL there is insufficient headroom, then an operating company needs to decide whether it is more economic to carry out further leakage control or whether to develop a new water resource, or to implement measures to reduce customer demand.

In order to evaluate the least cost solution to meet the supply-demand balance, the cost of leakage control activity described above should be compared to the marginal cost of the optional water resource development. This marginal cost is calculated as follows:

- The one-off capital cost of construction is estimated and discounted using an agreed discount rate.
- The ongoing maintenance cost of the resource once constructed is estimated.
- A "sensible" yield of the scheme is assessed.

- The cost of water production is estimated.

- The marginal cost is assessed as the sum of the discounted cost plus the maintenance cost divided by the yield plus the production cost.

- Environmental and social costs associated with the resource development can be assessed and added to the cost of the option.

Leakage activity schemes, developed using the methodologies described earlier, would be implemented if these were cheaper than this marginal cost. As the marginal cost of the new scheme will be significantly higher than the production cost from existing sources, as it includes the discounted cost of the construction of the works, then it will be economic to carry out further leakage control measures consisting probably of more pressure control, a higher level of active leakage control, and possibly more rehabilitation and sectorization. Schemes should be implemented until the necessary level of headroom is attained. This level of leakage could be referred to as the *constrained ELL.* The marginal cost of leakage management at this new level of leakage could be referred to as the *marginal value* of water. The marginal cost of carrying out additional activity in any area of leakage or demand management, or resource development will be equal to or greater than this value.

9.5.2 External Drivers

In practice, there will be many external influences on the various aspects of the supply demand balance. Figure 9.6 illustrates this.

Figure 9.6 shows that it will be necessary to look at apparent loss management strategies as well as real loss strategies. Although apparent loss management strategies do not in themselves reduce water production they will generally increase the recorded

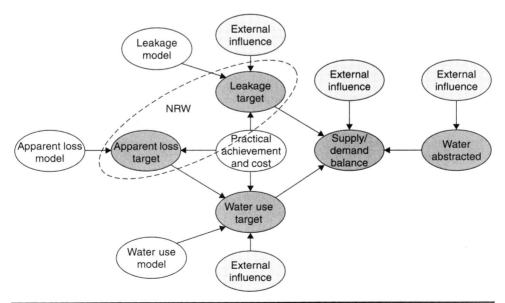

FIGURE 9.6 Supply/demand drivers. (*Source:* Dave Pearson/Stuart Trow.)

water used making this more transparent. This in turn may then make demand management strategies more cost effective. A well-developed strategy for apparent losses will also reduce wasted expenditure looking for real losses which do not actually exist.

However, it is strictly the supply/demand balance itself that drives the final solution. In this case, demand is the sum of real losses and consumption. The evaluation has to be carried out at water resource zone level, that is, where all the customers have the same level of security of supply taking into account all possible internal and external drivers.[16]

External drivers on water abstracted may include

- Environmental concern over low flows
- Environmental damage from over abstraction
- Environmental drivers, for example, European directives such as the Water Framework, Birds and Habitats directives, or equivalent
- Carbon footprint of water production

External drivers on water use may include

- Regulatory water efficiency targets[17]
- Sustainable water use targets[18]

External drivers on the supply/demand balance may include

- Security of supply requirements
- Risk of supply restrictions in drought conditions
- Impact on social and economic progress
- Risk of additional environmental damage in drought conditions

External drivers on leakage performance may include

- Regulatory minimum comparative performance
- Social and economic cost of disruption
- Possible political target
- Carbon footprint of repairs
- Carbon footprint of detection activity

The least-cost solution to meeting the supply-demand balance can be found using a standard optimisation method, for example, genetic algorithm or unconstrained mixed integer optimiser, using a formulation such as

Minimise the total cost of operating the system including

- Repairs
- Pressure management
- Proactive leakage detection
- Reactive leakage detection
- Rehabilitation

- Water production
- Demand management options
- Apparent loss strategies
- Resource development
- Abstraction mitigation

Subject to achieving (at least)

- Security of supply target
- Leakage target
- Water use target
- Apparent loss target
- Carbon footprint target
- All environmental constraints (low flows, habitats, etc.)

9.6   History and Experience

9.6.1   England and Wales

England and Wales (E&W) have a well-developed supply system with over 99% of properties connected to public water supply networks. Continuous supply is available 24 hours a day with less than 0.02% of premises receiving low pressure (usually taken to be less than 15 m) at any time during the year.[19] Only 34% of properties are metered,[20] the rest pay for water based on the value of the house. However the network is of mixed age with some parts of the network well over 100 years old. There is a small number of operating companies (less than 25 covering over 20 million properties), which were privatised in 1989, and there is a strong environmental and economic regulatory regime. Figures on leakage are reported to the regulators each year and audited by independent assessors. Every 5 years the companies have to develop business plans for the following 20 years, which include a full engineering assessment of their assets and a financial model of forecast income and expenditure. This is used to establish the price limits for the next 5 years. Part of the engineering submission involves the assessment of the economic level of leakage and whether this is constrained by headroom or not. Following the severe drought in 1995–96, leakage levels have been reduced by over a third and leakage targets are set by the regulator each year based on companies' assessment of their ELL. Most companies are operating at or close to their assessed ELL. Several companies are operating at a level that is constrained by headroom.

The assessment of ELL within England and Wales has a long history. Although there were many papers on ELL, the first national study and report on the topic was published in 1980.[21] This set down a methodology for the assessment of ELL, and it identified the benefits of pressure control and sectorization in managing leakage. This led to the implementation of sectors (DMAs) in most companies in England and Wales. The findings of this report were updated by a major national research programme that reported in 1994.[9] This and subsequent reports have led to greater understanding of the relationship between pressure and leakage and other activities which allow the construction of models to forecast the effect of changes in operating regime on leakage. There is a very high level of monitoring, and hence data availability, within England and Wales, for example, 15 minutes flow and pressure data on each sector. Most companies now have fully calibrated all mains hydraulic models of their networks. As a result of the drought in 1995–96 a number of companies initiated major leakage management programmes based on economic assessment outlined in this paper. One of these involved the construction and implementation of over 2000 pressure management schemes within a 3-year period. As a result of this, a company supplying over 3.2 million properties reduced their average night time pressure from over 50 m to less than 40 m.[3] All companies implemented a free or heavily subsidised programme for the repair or replacement of customer supply pipes in order to speed the repair of leaks that previously required the serving of statutory notices.

9.6.2 International Experience

The situation in other parts of the world is quite different from England and Wales. Water supply is often still in the hands of local municipal authorities each covering a relatively small number of properties. Most connections are metered, but it is common for supplies to be intermittent due to resource shortages. Sectorization is very rare and proactive leakage control is limited. The benefits of pressure management are not widely appreciated and there is generally no assessment of the economic level of leakage. Only limited data is available and there are generally very few hydraulic models. There is therefore the need for advice on the application of ELL in a staged manner in the situation of limited data.

9.7   Practical Application

Application of the ELL analysis in many situations has shown that pressure management is by far the most cost-beneficial activity. Its benefit in reducing burst frequency[12] is such that pressure-reducing schemes will often have payback periods significantly less than 12 months. In fact, the initial schemes can have such a quick and direct influence on the repair budget that they will free up sufficient money to pay for further pressure management schemes, and also some leakage detection resources to start proactive leakage detection. If this resource can be effectively targeted to identify backlog leaks, then it will be found that leakage can be reduced significantly within the existing budget.

The priority in terms of the identification of pressure management schemes should be

- Identify any occurrence of surges or instability in pressure on the network using very short-time interval logging and identify solutions to the problem.
- Identify and, where possible, move from fixed to variable speed pumps.
- Look for areas of high pressure (greater than 40 m) that can be controlled by pressure management.
- Look for areas with high diurnal flow and pressure variation and look to control these using flow-modulated pressure control valves.

As the benefits of pressure management start to be achieved, the economic level of regular sounding can be calculated[4] and appropriate targets can be implemented. If the area is sectorized, then economic leakage detection can be applied practically at sector level.[5,6]

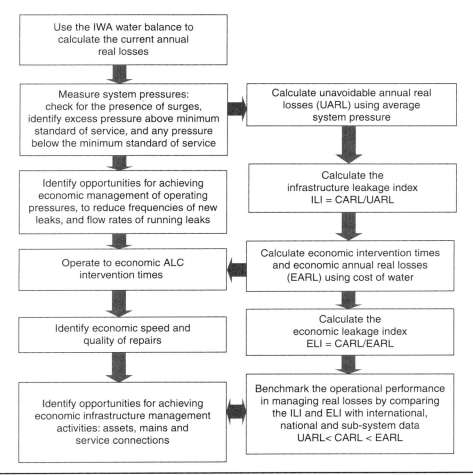

FIGURE 9.7   Practical application—flowchart. (*Source:* Allan Lambert/Dave Pearson.)

Throughout the leakage reduction plan, the performance of the network should be assessed using the IWA ILI approach[10] and information systems should be set up to collect data on the topography, pressure regime, burst frequencies, and so on so that more detailed analysis of ELL can be carried out as reductions in leakage are made. Whereas initial estimates of ELL will rely on default values and assumptions, the calculations can be refined using actual data from the specific operations which are implemented.

This approach can be described by a flowchart (Fig. 9.7).

9.8   Summary

For any system, the economic level of leakage is that which results from a combination of a range of leakage management activities that comprises (in priority)

- An optimised overall pressure management policy in which
  - The presence of surges are identified and steps are taken to minimise their adverse effects
  - Projects are implemented to adopt basic simple reductions of excess pressures
  - Further projects are implemented in order of cost/benefit
- An optimised repair time policy for all bursts
- An economic intervention policy for awareness, location, and repair of unreported (hidden) bursts which is
  - Influenced by the level of investment in leakage management infrastructure, that is, telemetry/SCADA, DMAs, advanced pressure management
  - Influenced by the exit level (background and other leaks remaining after interventions)
- An economic level of investment in mains and services renewals which takes account of all regulatory factors

If each of these activities is pursued to a logical conclusion in terms of cost and benefit, then the definition of the economic level of water loss can be summarised as:

"That level of water losses which results from a policy under which the marginal cost of each individual activity for managing losses can be shown to be equal to the marginal *value* of water in the supply zone."

References

1. Wikipedia. Production theory basics. [Online]. Available: http://en.wikipedia.org/wiki/Production_theory_basics. [Cited: 10 April 2008]
2. Parkin, M. *Economics.* 5th ed. Addison Wesley, 2005. ISBN 0201537621.
3. Lambert, A., S. Myers, and S. Trow. *Managing Water Leakage: Economic and Technical Issues.* Financial Times Energy, 1998. ISBN 1 84083 011 5.
4. Lambert, A. and A. Lalonde. "Using Practical Predictions of Economic Intervention Frequency to Calculate Short-Run Economic Leakage Level, with or without Pressure Management." *IWA Leakage2005 Conference*, Halifax, Canada: The world Book, IWA, 2005.
5. Rizzo, A. "Tactical planning for effective leakage control. Leakage Management." *A Practical Approach Conference:* IWA, Lemesos, Cyprus: IWA, 2002.
6. Dellow, D. and S. Trow. "Implementing a New Operational ELL Model." *Water U.K. 8th Annual Leakage Conference.* WaterUK, Oct 2007.
7. Department for Environment, Food, and Rural affairs, Environment Agency, and Office of Water Services. Future Approaches to Leakage Target Setting for Water Companies in England. Ofwat, 2003. http://www.ofwat.gov.uk/aptrix/ofwat/publish.nsf/Content/tripartitestudycontents
8. Lambert, A. and J. Morrison. "Recent Developments in Application of 'Bursts and Background Estimates' Concepts of Leakage Management." *Journal of Chartered Institute of Water and Environmental Management.* 10 April 1996:100–104.
9. Water Research Centre. *Managing Leakage.* WRC, 1994. ISBN 1 898920 0 87.
10. Lambert A., T.G. Brown, M. Takizawa, et al. "A Review of Performance Indicators for Real Losses from Water Supply Systems." *AQUA.* 48(6), 1999:ISSN 0003–7214.

11. Lambert, A. and J. Thornton. "Progress in Practical Prediction of Pressure: Leakage, Pressure: Burst Frequency and Pressure: Consumption Relationships." *IWA Leakage2005 Conference*. Halifax, Canada: The World Bank, IWA, 2005.

12. Pearson, D., M. Fantozzi, D.Soares, et al. "Searching for N2: How Does Pressure Reduction Reduce Burst Frequency?" *IWA Leakage2005 Conference:IWA*. Halifax, Canada: The World Bank, 2005.

13. Fantozzi, M. and A. Lambert. "Including the Effects of Pressure Management in Calculations of Short-Run Economic Leakage Levels." *IWA Water Loss* 2007. Conference, Bucharest, Romania: Romanian Water Association, IWA, Oct 2007. ISBN 978-973-7681-24-98.

14. Lasdon. *Optimisation Theory for Large Systems*. London:Macmillan 1970, 2002 (reprinted). ISBN 0486419991

15. UKWater Industry Research and EA "A Practical Method for Converting Uncertainty into Headroom." *Report 98/WR/13/1*: UKWIR, 1998.

16. Office of Water Services. "Providing Best Practice Guidance on the inclusion of Externalities in the ELL Calculation." *Main Report:* Ofwat, Nov 2007.

17. Office of Water Services. "Water efficiency targets." *RD 15/07:* Ofwat, Aug 2007. Available on: http://www.ofwat.gov.uk/aptrix/ofwat/publish.nsf/Content/rd1507

18. Department for Communities and Local Government. Code for Sustainable Homes: DCLG, February2008 http://www.communities.gov.uk/documents/planningand-building/pdf/codesustainhomesstandard.

19. Office of Water Services. "Levels of service for the water industry in England and Wales 2006–07.": Ofwat, 2007. http://www.ofwat.gov.uk/aptrix/ofwat/publish.nsf/Content/levelsofservice_0607.

20. Office of Water Services. "Security of Supply for the water industry in England and Wales 2006–07.": Ofwat, 2007. http://www.ofwat.gov.uk/aptrix/ofwat/publish.nsf/Content/SecuritySupply_06-07.

21. National Water Council "Leakage Control Policy and Practice." *NWC Standing Technical Report 26:* NWC, 1980, 1985 (reprinted).

CHAPTER 10

Modelling Water Losses

Julian Thornton

Reinhard Sturm

George Kunkel, P.E.

10.1   Introduction

Quantities of the various water loss volumes occurring in a water utility can be approximated by employing a mathematical representation, or model, of the loss values. Depending upon the type and nature of the apparent or real losses being modeled, a model can be a simple spreadsheet of estimates of loss volumes attributed to a specific type of loss occurrence, or it can be a complex set of calculations that rely upon a number of data inputs to calculate a reliable quantity of loss. Models are an excellent tool to assist the operator with the preparation of a water audit and water loss management planning; however they should be used with care and due diligence. Models are not magic nor do they give us hind sight or act as a crystal ball; they are only as good as the concepts they employ, the data that is put into them, and the skill and experience of the user; training in their use is essential. So care should be taken to ensure that field data captured and coefficients and variables used represent real conditions as closely as may be necessary for a result of required accuracy. If accountable data is not available estimated data may be used, however, the model should be notated with comments reflecting the estimated inaccuracy for each component and calculating the final weighted potential inaccuracy. Many industry standard water loss control models now incorporate the use of 95% confidence limits, which are applied to each component of data input and calculated for each component of data output. Further information on the use of 95% confidence limits is covered in Chap. 7. This chapter presents examples of some basic water loss models.

Good data in means good data out!

95% confidence limits is used in order to assign confidence to each input component and to calculate aggregated confidence in the final result.

Modeling flows in pipe networks and components of consumption has been an integral part of *hydraulic network analysis modeling* (hydraulic models) for over 30 years, but in these models nonrevenue water has generally been treated very simplistically as a fixed residual. Accordingly, a separate series of concepts for modeling water loss has been developed since the early 1990s for the following components of nonrevenue water:

- Apparent loss (customer meter inaccuracy, systematic data-handling error in billing systems, and unauthorized consumption)
- Real loss (leakage and overflows)
- Pressure/leakage, pressure/consumption, and pressure/break (frequency relationships)

The reliability and effectiveness of water loss modeling makes it a standard part of the loss management practitioner's tool kit.

It is important to emphasize that water loss management models are not the same tools as hydraulic models. Many water utility personnel, consultants, and contractors have used or seen a hydraulic model, which mathematically calculates values of water flow and pressure in a distribution network, subject to specific inputs and consumption patterns. Hydraulic models are an extremely powerful tool for distribution system analysis, allowing the operator to simulate varying operating scenarios within the system. However, the concepts used for simulating water loss management in most hydraulic models are often oversimplified, to the point where the estimated current leakage is nominally distributed globally around the nodes of the model; and assumed to be fixed over time and pressure-invariant. While such simplified assumptions may be valid for modeling flows and pressures in water distribution piping systems, they are not valid for models which seek to quantify key water loss components.

The water loss modeling approaches discussed include

- Top-down water audit spreadsheet models
- Component analysis of apparent (nonphysical) losses
- Component analysis of real (physical) losses, such as the breaks and background estimates (BABE) model
  - The fixed and variable area discharge (FAVAD) concept for modeling pressure/leakage rate relationships and pressure/consumption relationships and making predictions
  - Pressure/break frequency analysis concepts for making predictions of the reduction in break frequency on mains and services with reduction in operating pressure
  - Application of component analysis and FAVAD concepts for night-flow analysis in discrete zones or district metered areas
  - Consumption analysis models
  - Short run economic leakage levels

10.2   Top-down Water Audit Spreadsheet Models

The water audit methodology recommended for use in this publication was jointly developed by the International Water Association (IWA) and the American Water Works

Association (AWWA) and published in 2000. By compiling a water audit using the standardized IWA/AWWA methodology, water utility auditors gain an understanding of the nature and extent of their system water loss volumes and, via the validation process, allows the utility to calculate the mathematical confidence in those annual volumes. Good management of any resource requires that the supplier maintains accurate records of transactions and deliveries of the commodity provided to its customers. A water audit has exactly that goal, tracking and accounting for every component of water in the cycle of delivery. The water audit typically tracks and validates the volumes of water from the site of withdrawal or treatment, through the water distribution system up to the first point of customer consumption. The water audit usually exists in the form of a worksheet or spreadsheet that details the variety of consumption and losses that exist in a water system. The water balance itself is a summary of all the components of consumption and losses in a standardized format. Every unit of water supplied into the system needs to be assessed and assigned to the appropriate component. Once volumes of valid authorized consumption and losses (apparent and real) have been assigned, the cost impact of these components can be calculated. Subsequently, the water utility will be able to select the appropriate tools for intervention against real and apparent losses as discussed further in Chaps. 11 to 19 of this manual.

Several effective top-down water audit spreadsheet models are available for free download. In 2006, the AWWA Water Loss Control Committee launched its free water audit software, which can be downloaded from www.awwa.org. Instructions for use of this top-down model are provided with the software; however, instructions to conduct a detailed, bottom-up water audit using the same methodology are provided in the third edition of the AWWA M36 publication *Water Audits and Loss Control Programs* (proposed 2008). This publication is also compiled by the AWWA Water Loss Control Committee. Table 10.1 provides an example using the AWWA free software, showing the input and output from the top-down water audit for the Philadelphia Water Department (PWD) for its fiscal year ending June 30, 2006.

> **I**t is easy to program a spreadsheet to automatically perform audit calculations. However, the operator must fully understand the concepts being modeled—very useful and user-friendly spreadsheets are available for free from several sources, including AWWA the World Bank, and various consultants, see References.

The AWWA's free water audit software is an excellent tool that water utilities can utilize to start the auditing process in a top-down manner. However, as the utility progresses with more detailed, bottom-up auditing, it becomes advantageous to incorporate 95% confidence limits as discussed in Chap. 7. Table 10.2 illustrates an example of a water balance compiled using free water audit software that is available from the World Bank. The example shows the resultant statistical confidence value in each key component of the water balance.

A number of other software packages are available which offer additional features including variance analysis (Aqua Solve, LEAKS/PIFastCalcs, and the like). The Aqua Solve package is shown in Tables 10.3 and 10.4. Table 10.3 shows a balance from San Francisco Public Utilities Commission (SFPUC) and Table 10.4 shows how variance analysis can be used to identify the components which have the most impact on the aggregated uncertainty of the water loss components.

AWWA WLCC Water Audit Software: Reporting Worksheet
Copyright © 2006, American Water Works Association. All Rights Reserved.   WASv3.0

`Back to Instructions`

`?` Click to access definition

Water Audit Report for: **Philadelphia Water Department**
Reporting Year: **2006**

Please enter data in the white cells below. Where possible, metered values should be used; if metered values are unavailable please estimate a value. Indicate this by selecting a choice from the gray box to the left, where M = measured (or accurately known value) and E = estimated.

All volumes to be entered as: MILLION GALLONS (US) PER YEAR

WATER SUPPLIED

Volume from own sources :	`?`	M	92,931.500	million gallons (US)/yr (MG/Yr)
Master meter error adjustment :	`?`	E	294.200	over-registered MG/Yr
Water imported :	`?`			MG/Yr
Water exported :	`?`	M	6,971.500	MG/Yr
WATER SUPPLIED :			85,665.800	MG/Yr

AUTHORIZED CONSUMPTION

Click here: `?`
for help using option
buttons below

						Pcnt:		Value:
Billed metered :	`?`	M	57,633.500	MG/Yr				
Billed unmetered :	`?`		0.000	MG/Yr				
Unbilled metered :	`?`	M	0.300	MG/Yr				
Unbilled unmetered :	`?`		892.500	MG/Yr		1.25%	○ ●	892.500
AUTHORIZED CONSUMPTION :			58,526.300	MG/Yr				

↑ Use buttons to select
percentage
OR
value

WATER LOSSES (Water Supplied - Authorized Consumption)   27,139.500   MG/Yr

Apparent Losses

					Pcnt:		Value:
Unauthorized consumption :	`?`		1,579.000	MG/Yr	0.25%	○ ●	1579.000
Customer metering inaccuracies :	`?`	E	114.600	MG/Yr		○ ●	114.600
Systematic data handling errors :	`?`	E	3,826.400	MG/Yr			
Apparent Losses :			5,520.000	MG/Yr			

Real Losses

Real Losses = (Water Losses − Apparent Losses) :		21,619.500	MG/Yr
WATER LOSSES :		27,139.500	MG/Yr

NON-REVENUE WATER

NON-REVENUE WATER :	28,032.300	MG/Yr

SYSTEM DATA

Length of mains :	`?`	M	3,084.0	miles
Number of active AND inactive service connections :	`?`	M	551,959	
Connection density :			179	conn./mile main
Average length of customer service line :	`?`	E	12.0	ft
Average operating pressure :	`?`	M	55.0	psi

(pipe length between curbstop and customer meter or property

COST DATA

Total annual cost of operating water system :	`?`	M	$190,162,000	$/Year
Customer retail unit cost (applied to Apparent Losses) :	`?`	M	$4.50	$/1000 gallons (US)
Variable production cost (applied to Real Losses) :	`?`	M	$160.48	$/million gallons (US)

DATA REVIEW - Please review the following information and make changes above if necessary:

- Input values should be indicated as either measured or estimated. You have entered:

10 as measured values
 6 as estimated values
 0 as default values
 2 without specifying measured, estimated or default

- Water Supplied Data: No problems identified

- Unbilled unmetered consumption: No problems identified

- Unauthorized consumption: No problems identified

- It is important to accurately measure the master meter - you have entered the measurement type as: measured

- Cost Data: No problems identified

PERFORMANCE INDICATORS

Financial Indicators

Non-revenue water as percent by volume :	32.7%
Non-revenue water as percent by cost :	15.0%
Annual cost of Apparent losses :	$24,840,000
Annual cost of Real Losses :	$3,469,497

Operational Efficiency Indicators

Apparent Losses per service connection per day :	27.40	gallons/connection/day
Real Losses per service connection per day* :	107.31	gallons/connection/day
Real Losses per length of main per day* :	N/A	
Real Losses per service connection per day per psi pressure :	1.95	gallons/connection/day/psi
`?` Unavoidable Annual Real Losses (UARL) :	2,185.90	million gallons/year
`?` Infrastructure Leakage Index (ILI) [Real Losses/UARL] :	9.89	

* only the most applicable of these two indicators will be calculated

Source: AWWA WLCC

TABLE 10.1   Water Audit Example from PWD

Home

Water Balance for XYZ Water Company, Year 2006			
System input volume 2,465,753 m³/d Error margin [±]: 5.0%	Authorized consumption 1,674,658 m³/d Error margin [±]: 0.2%	Billed authorized consumption 1,643,836 m³/d	Billed metered consumption 1,643,836 m³/d	Revenue water 1,643,836 m³/d
			Billed unmetered consumption 0 m³/d	
		Unbilled authorized consumption 30,822 m³/d Error margin [±]: 10.0%	Unbilled metered consumption 0 m³/d	Nonrevenue water 821,918 m³/d Error margin [±]: 15.0%
			Unbilled unmetered consumption 30,822 m³/d Error margin [±]: 10.0%	
	Water losses 791,096 m³/d Error margin [±]: 15.6%	Commercial losses 256,945 m³/d Error margin [±]: 24.0%	Unauthorized consumption 10,370 m³/d Error margin [±]: 0.0%	
			Customer meter inaccuracies and data-handling errors 246,575 m³/d Error margin [±]: 25.0%	
		Physical losses 534,151 m³/d Error margin [±]: 25.8%		

Source: WB Easy Calc software

TABLE **10.2** Water Balance Example with Confidence Limits for Each Component

125

Water Audit Results

System Input Volume	Authorized Consumption / Water Losses	Billed / Unbilled / Losses	Detailed breakdown	Revenue / Nonrevenue Water
System Input Volume 29,033 Million gal (100%)	**Authorized Consumption** 25,990 Million gal (90%)	**Billed Authorized** 24,779 Million gal (85%)	**Billed Metered Water Exported** — Million gal (0%)	**Revenue Water** 24,779 Million gal (85%)
			Billed Metered Authorized 24,779 Million gal (85%)	
			Billed Unmetered Authorized — Million gal (0%)	
		Unbilled Authorized 1211 Million gal (4%)	**Unbilled Metered Authorized** 1148 Million gal (4%)	**Nonrevenue Water** 4254 Million gal (15%)
			Unbilled Unmetered Authorized 63 Million gal (0%)	
	Water Losses 3043 Million gal (10%)	**Apparent Losses** 163 Million gal (1%)	**Unauthorized Consumption** — Million gal (0%)	
			Meter Error 163 Million gal (1%)	
		Real Losses 2880 Million gal (10%)		

Source: SFPUC Water Audit 04/05

TABLE 10.3 Water Balance

	Annual Volume (Million gal)	95% Confidence Limits	Variance
NONREVENUE WATER	4253.92	11.2%	59,148
WATER LOSSES	3042.60	15.7%	59,148
APPARENT LOSSES	162.25	0.7%	0
REAL LOSSES	2879.65	16.6%	59,148.3

# A – Z	Water Audit Component	Item	Annual Volume (Million gal)	95% Confidence Limits	Variance	Rank A – Z
2	System Input Volume	San Andreas #2 country line meter	10,519.84	2.6%	37,542	1
1	System Input Volume	Crystal Springs #2 county line meter	9965.75	2.0%	19,147	2
3	System Input Volume	Lake Merced Pump Station to Sunse	7512.80	2.6%	14,691	3
31	Billed Metered Authorized Consumption	CPRM (City Paying Multi Family)	6580.7	2.6%	10,341	4
5	System Input Volume	Lake Merced Pump Station to Sutro	7134.27	2.0%	6,602	5
27	Billed Metered Authorized Consumption	CPCM (City Paying Commercial)	6611.357	2.0%	5,300	6
32	Billed Metered Authorized Consumption	CPRS (City Paying Single Family)	4411.61	2.6%	4,551	7
30	Billed Metered Authorized Consumption	CPMU (City Paying Municipal)	472.04	2.0%	23	8
29	Billed Metered Authorized Consumption	CPIN (City Paying Industrial)	94.82	2.0%	0.94	9
51	Meter Error	3" without Affidavit	2.16	50.0%	0.30	10
46	Meter Error	5/8" without Affidavit	9.32	0.7%	0.00	14

Source: SFPUC Water Audit 04/05

TABLE 10.4   Variance Analysis (*Continued*)

# A – Z|	Water Audit Component	Item	Annual Volume (Million gal)	95% Confidence Limits	Variance	Rank A – Z|
26	Billed Metered Authorized Consumption	CPBC (City Paying B&C)	26.79	2.0%	0.07	11
28	Billed Metered Authorized Consumption	CPDS (City Paying Docks and Ships)	19.28	2.0%	0.04	12
50	Meter Error	2" without Affidavit	2.78	0.2%	0.00	15
49	Meter Error	$1^1/_2$" without Affidavit	0.29	0.1%	0.00	18
6	System Input Volume	Lake Merced Pump Station to Lake	7.60	2.6%	0.02	13
47	Meter Error	3/4" without Affidavit	0.17	0.9%	0.00	17
48	Meter Error	1" without Affidavit	0.73	0.4%	0.00	16
52	Meter Error	4" without Affidavit	0.01	0.0%	0.00	19

Source: SFPUC Water Audit 04/05

TABLE 10.4   Variance Analysis (*Continued*)

By ranking the water balance input components which have the greatest impact, the auditor can quickly identify those components that should be field validated. Obviously field validation is the best means to confirm that the output from a model truly represents field conditions. However, field measurements require time and resources (staffing, equipment) and it is often desirable to limit the extent of field validation in order to contain activities within a reasonable water audit budget. In this way, the key variables are field validated and the auditor works down the list until the desired aggregated confidence limit is reached. It is important to note at this stage that the operator should strive to model ranges of volume for each key component of water loss. Water loss volumes are not absolute volumes.

A detailed procedure for preparing the standard top down water balance can be found in the third edition of the AWWA M36 publication.

10.3   Component Analysis and Modeling of Apparent Loss

Modeling components of apparent losses has been done in many forms for many years. One example of apparent loss modeling is the attempt to quantify the volume of water not registered due to customer meter underregistration. However, in recent years component analysis of apparent losses has been approached in a similar manner as the methods of real losses modeling; where components of apparent loss are shown as multiples of an unavoidable annual volume.

In Table 10.5, first attempts at a component analysis model for apparent losses can be seen.

The IWA Water Loss Task Force Apparent Loss Team is currently working to develop an unavoidable annual apparent loss (UAAL) formula that calculates the minimum

Apparent Loss Components

System	XYZ			Date	12/09/07
Mains length	5555	km		Connections	800,000
		Component			Volume m³
System volumes	System input volume				444,555.00
	Authorized metered consumption	Small meters			145,555.00
		Large meters			138,768.00
Current annual apparent loss	Unauthorized consumption			1.00%	4445.55
	Meter under registration	Small meters		12.00%	19,848.41
		Large meters		5.00%	7303.58
		—			—
		—			—
		—			—
	Total current annual apparent loss volume				31,597.54
Performance indicator	L/conn/d				108.21

Source: Thornton International Ltd.

TABLE 10.5 Example Component Analysis of Apparent Losses (*Continued*)

Apparent Loss Components

Unavoidable annual apparent loss	Unauthorized consumption		0.25%	1111.39
	Meter under registration	Small meters	2.00%	2970.51
		Large meters	2.00%	2832.00
		—		—
		—		—
		—		—
	Total unavoidable apparent loss volume			6913.90
Performance indicator	L/conn/d			23.68
ALI (Aparent loss index)				4.57

Source: Thornton International Ltd.

TABLE 10.5 Example Component Analysis of Apparent Losses (*Continued*) amount of apparent loss that a water utility would suffer in spite of enacting all reasonable and effective apparent loss control activities. Similar to the infrastructure leakage index (ILI) that assesses real loss standing as the ratio of current real loss volume over the unavoidable annual real losses (UARL) (See Chap. 7) an apparent loss index (ALI) would be the ratio of the current apparent loss volume in the water audit over the UAAL.

However, in lieu of a reliable UAAL measure, the team's interim recommendation is that 5% of the metered consumption be assumed as a reference value for apparent losses. The authors feel that this may be high for water utilities in developed countries, which typically have good customer meter management and where buildings do not have roof tanks which present the opportunity for very low flows that pass unregistered through many water meters. (See Sec. 12.4.) These utilities typically also have reasonable policies and safeguards that prevent exorbitant unauthorized consumption. Therefore the 5% assumption may be high in developed countries, but reasonable for developing countries.

The Water Loss Task Force is actively engaged in work to develop a set of apparent loss performance indicators and further information will be made available by the IWA team as research progresses.

10.3.1 Modeling Customer Meter Accuracy

Customer meters have been called the "cash register" of the utility and are responsible for ensuring an equitable distribution of water volume and income throughout various different customer types within a utility. It is therefore extremely important to assess the accuracy of the meters on a regular basis and make repairs or replace groups of meters to keep the customer meter population at an overall high level of accuracy. Accurate metered consumption data is also necessary for engineering functions such as hydraulic models, evaluation of water conservation programs, and sizing of infrastructure for water resources development. The reader should also refer to Chap. 16 which provides detailed information on meter performance, as well as procedures for meter accuracy testing.

It is necessary to model average weighted meter accuracy for the entire customer meter population and include it in the water audit. The water balance calculations are used to deduct the volumes of apparent losses from the total volume of losses in order to arrive at a top-down approximation of the annual volume of real losses.

In attempting to quantify the volume of apparent loss due to customer meter inaccuracy in the water audit, it is important to recognize that three primary occurrences cause a meter population to become inaccurate, namely

- Eventual decline of the inherent (mechanical) accuracy of a meter population through wear.
- The meter or the meter reading device may fail or "stop" altogether.
- Meters may not be of the proper size or type to accurately register the full range of water flows encountered in a given customer supply.

It is necessary to disaggregate, or separate, the activities of the water utility's meter management in these three occurrences in order to properly construct a representative picture of the annual volume of apparent loss attributed to customer meter inaccuracy and the reasons for each disaggregated volume. In this way, planning can be carried out to remedy the specific causes of meter inaccuracy in the most economic manner.

Loss of Accuracy Due to Mechanical Wear

Well-manufactured water meters can lose appreciable mechanical accuracy due to

- Aggressive water quality
- High rates of flow being measured
- Chemical or residual buildup
- Abrasive materials such as sand in suspension carried by the water
- Air running through the meter after a system outage

As the cumulative volume passed through the meter increases toward meter life cycle levels then the mechanical failures are compounded. Chapter 16 provides detailed information on the assessment of life cycle accuracy of customer meters and means to control losses that occur in this subcomponent of apparent losses.

Zero Consumption Billings from Stopped Meters or Vacant Properties

Meters or meter-reading devices can fail to register for various reasons. However, meters that show no registration might also reflect a customer property with no use, such as that which may occur at a vacant property. Large numbers of customer meters that mechanically fail to register any flow from billing cycle to billing cycle can account for large volumes of apparent losses and uncaptured revenue.

Many water utilities employ the use of an estimated consumption volume if they encounter periodic low or zero consumption volumes generated from meter reading. This practice can be effective if the zero reads are only periodic. However, when estimation is undertaken for many consecutive months, estimated volumes will likely deviate from the actual consumption volumes. If all values of consumption for a given account are based upon estimates for an entire audit year, then the volume assigned to that particular account for the water audit can be seriously in error. Water utilities should routinely review billing data and assess the occurrence of zero consumption bills, particularly those that register zero consumption for several consecutive months. It is worthwhile for the utility to dedicate personnel to physically inspect the meter site of a representative sample of customer accounts to determine the reason for the continuing zero consumption registrations. The findings of such inspections provide data that can be used to model the occurrence of apparent loss in the zero consumption population throughout the entire system.

By applying the above analysis, it is possible to model best case and worst case scenarios for customer meter losses occurring due to meters registering zero consumption. The best case reflects the overall accuracy of the entire meter population without including zero consumption meters, a scenario that would occur only in the ideal case of the water utility responding quickly to accounts registering zero consumption and correcting meter or meter-reading problems just after they occur. The worst case reflects customer meter population accuracy including the greatest potential extent of zero consumption meters, reflecting a water utility policy that ignores zero consumption registrations, allowing them to mount throughout the audit year. Calculate the apparent losses in both the best and worst case, then the average meter accuracy can be calculated for water balance purposes, representing the average inherent accuracy of the meter population, including the average response time to correct accounts that chronically register zero consumption.

For this type of analysis to be accurate, it is necessary that there is a large enough test sample of data from field inspections of zero consumption accounts in order to properly represent the total customer account population.

Improper Size or Type of Meter

Many brands of customer meters are known to become appreciably inaccurate when very high or very low flows (relative to the design range of the meter) are registered. If the size or type of meter in a given application results in the majority of flow occurring in these extreme ranges, then the meter will fail to register a large portion of the customer flow. Section 12.4 provides a detailed discussion on meter sizing impacts and the best practices to employ to ensure that losses due to poor sizing or typing are minimized. With direct-feed pressure systems as are typical in North American water utilities, customer meters need to be selected and sized to record a wide range of flow rates. Any underregistration of metered consumption is considered an apparent loss in the water audit, as the lost water is reaching the customer, but a portion of the consumption is not being registered or billed. A number of software models have been developed for this type of loss analysis.

Similar modeling techniques can be applied for the apparent loss components of data transfer error, systematic data handling error in customer billing systems, and unauthorized consumption. Detailed spreadsheet models for these components are not as common as those modeling customer meter inaccuracy; however, it is up to the water auditor to assess the occurrences of these losses and attempt to model their extent in their utility operations.

10.4 Modeling Components of Real Losses Using Breaks and Background Estimates Concepts[1]

In the early 1990s, during the U.K. National Leakage Control Initiative, a systematic approach to modeling components of real losses (leakage and overflows) was developed by Allan Lambert.

Recognizing that the annual volume of real losses is the result of numerous leakage events, each individual volume loss being influenced by flow rate and duration, Lambert considered leakage events in three categories:

- *Background* (*undetectable*) *leakage*: Small flow rate, runs continuously
- *Reported breaks*: High flow rate, relatively short duration
- *Unreported breaks*: Moderate flow rates, duration depends on intervention policy

For each separate component of the distribution system—mains, service reservoirs, service connections (main to curb stop), service connections (curb stop to meter) —the value for each component of annual losses can be calculated using the parameters in Table 10.6 below for some given standard pressure. The effect of operating at different pressures can then be modeled by applying FAVAD principles to each of the individual components of real losses, using appropriate specific $N1$ values. FAVAD is discussed in more detail in Sec. 10.6.3 of this chapter.

Component of Infrastructure	Background (undetectable) Losses	Reported Breaks	Unreported Breaks
Mains	Length Pressure Min loss rate/km*	Number/year Pressure Average flow rate* Average duration	Number/year Pressure Average flow rate* Average duration
Service Reservoirs	Leakage through Structure	Reported overflows: Flow rates, duration	Unreported overflows: Flow rates, duration
Service connections, main to edge of street	Number Pressure Min loss rate/conn*	Number/year Pressure Average flow rate* Average duration	Number/year Pressure Average flow rate* Average duration
Service Connections after edge of street	Length Pressure Min loss rate/km*	Number/year Pressure Average flow rate* Average duration	Number/year Pressure Average flow rate* Average duration

* At some standard pressure.
Source: Water Loss Control Manual, 1st ed.

TABLE 10.6   Parameters Required for Calculation of Components of Annual Real Losses

The BABE annual component analysis model was first calibrated and successfully tested using British data in 1993. It was rapidly extended to cover economic analysis to assess the economic frequency of active leakage control interventions, and since then has been used in many countries.

The BABE annual model can be considered as a statistical model, in that it does not seek to identify every individual leakage event and calculate an annual loss volume; rather, it groups together similar events, and does simplified calculations. The larger the number of events, the better the accuracy of the calculated values, so BABE annual models work more reliably with large systems. The BABE model used for calculation of unavoidable annual real losses (UARL) is limited to systems with more than 3000 service connections (based on detailed sensitivity analysis this value was revised down from 5000 connections in 2005).

The powerful combination of BABE and FAVAD concepts meant that, in the late 1990s, a range of simple spreadsheet models could be developed to approach a number of leakage management problems for individual systems, on a rational and systematic basis. Figure 10.1 shows the range of problems which has been successfully modeled.

BABE modeling or component analysis can also be undertaken at district or zone level breaking down night flows into the key consumption and real loss components.

10.5   Using BABE Modeling Concepts to Prioritize Activities

It is not recommended that component analysis is undertaken on its own to derive a volume of annual real losses because there is likely to be a significant level of uncertainty

Problem-Solving Using BABE and FAVAD Concepts

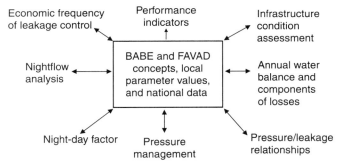

FIGURE **10.1**   Range of problems which have been successfully modelled using BABE concepts. (*Source*: Water Loss Control Manual, 1st ed.)

in much of the data used in the analysis. However, a component analysis or BABE model is a very useful supplement to a top-down water balance because it provides estimates of the volumes of real losses in different elements of the distribution infrastructure. This is valuable data because it is required to develop the most appropriate loss reduction strategy and it is essential for a robust determination of the economic level of leakage (ELL) as discussed in detail in Chap. 9.

There are several commercial versions of these models (and many homemade ones) available in the market today, most of them being extremely user friendly and flexible. If using a commercial model, the operator must first fully understand what it is they wish to perform and ensure that the commercial model has been suitably customized to the local situation. If constructing a model in a spreadsheet it is vitally important that the operator fully understands the concepts being used and their limitations. And users of the models must be adequately trained if reliable results are to be obtained.

In order to arrive at an estimation of the loss situation, most statistical component analysis models require

- Infrastructure and system data
- Coefficients and default values

10.5.1   Infrastructure and System Data

In most cases, the field data required for an annual component analysis model are leak information by category over the audit period supplemented by flow data and pressure data, which can also be used for the district component analysis. More information can be found about the collection of field data in App. B. The BABE and FAVAD approaches to modeling ensure that only a limited amount of specific data needs to be obtained, and it is obviously important to collect the data as accurately as possible to ensure that the estimation of loss is as close to the real situation as possible.

Typical infrastructure and system data needed for BABE and FAVAD models are

- Length, material, and diameter of mains
- Volume of service reservoirs/storage tanks

- Number of service connections
- Location of customer meters relative to the curb stop
- Number of households, population, and consumption
- Number of nonhouseholds and consumption
- Average zone pressure (at night, and 24-hour average)
- Numbers or frequencies of different categories of leaks and breaks
- Average duration of each category of leaks and breaks (depending on utility policies for leak detection and repair)

Although it might seem on face value that most of this data would be readily available many utilities do not have good pressure data available.

Since pressure has a significant impact on the leakage flow rates and subsequently on the annual volume of real loses it is necessary to accurately assess the average system wide pressure.

The *average zone pressure* (*AZP*) is a surrogate value for the average pressure that the average leak within a distribution zone will experience. AZP can be used to determine the average flow rate for a given type of leak within a distribution zone. AZP is therefore a key parameter in real loss component analysis. Unfortunately, some leakage practitioners and researchers try to interpret leakage data without measuring or assessing an AZP pressure, and use inlet pressures or critical point pressures instead; the modeling results then become unreliable to a greater or lesser extent.

Calculating AZP and Identifying an AZP Measurement Point

There are several ways to calculate AZP and identify an AZP measurement point. Where network analysis models are available, this calculation can be based on node point data, weighted by number of service connections. Alternatively, if hydrant pressures are recorded, the average pressure can be estimated by taking an average of the hydrant pressures recorded. Another method is to allocate numbers of service connections (or properties, or hydrants) or mains lengths within contour bands, and obtain a weighed average ground level for the selected type of infrastructure.

Once the weighted average ground level, or weighted average pressure has been calculated, a hydrant that experiences that pressure in the center of the zone can be selected as the AZP point for measurements, when field tests are being undertaken. It may be necessary to consider seasonal variations in average pressure throughout the year, if there are significant seasonal variations in demand causing seasonal pressure changes.

10.5.2   Coefficients and Default Values

Most statistical models use coefficients and default values developed from series of field testing. It is important; however, that the operator understands the nature of the coefficients and default values, how and why they were applied to the calculation so that they make any necessary changes for local conditions.

Coefficients and default values often used may include

- Typical flow rates of each category of leaks and breaks at some standard pressure (normally 70 psi or 50 m).
- Typical background leakage for mains if in good condition (per mi/hr, at some standard pressure, this can be measured in an area where all locatable leaks have just been repaired—see ICF calculations in Sec. 10.6).

- Typical background leakage for service connections if in good condition (at some standard pressure can also be measured as above—see ICF calculations in Sec. 10.6).
- Typical numbers of residents using toilets at 3 to 4 a.m. each morning (or other relevant minimum night-flow period).
- Typical toilet flush volume (toilet use is one of the largest residential individual uses and the most common use of water at night other than in areas where irrigation is being undertaken).
- Typical toilet leakage.
- FAVAD *N*1 values for different types of leaks and pipe materials.
- FAVAD *N*3 values for pressure dependant and nondependant consumption.
- ICF values for estimating background leakage volumes and separating them from reported and unreported breaks volumes.

A simple example showing the need for care when applying coefficients and defaults values is shown below:

A night flow analysis model is used to estimate the amount of leakage present in a zone. The zone consists of residential properties and no commerce or industry (*infrastructure and system data*).

One of the key factors in this model is to identify estimated legitimate night consumption and subtract it from the night flow. To do this the model makes some assumptions based on preprogrammed coefficients and default values. In our example, the model was built in the United Kingdom and is being applied in the United States.

The model assumes that most of the use at night in a residential zone is from toilet flushes. In our example, the toilet flush volume was 1.5 gal per flush (*default value*). However in the zone in which the model is being applied the toilets have not been retrofitted and the flush is really 4 gal (*default value*).

So the model will ask for the population in the zone and multiply this by the estimated number of people active at night. Let's say 6% (*coefficient*) during our analysis window of 3 to 4 a.m. in the morning.

If the population in our zone were 6000 (*infrastructure and system data*) then the model would assume that 6% were active at some time during that period, which would be 360 active flushes.

The model then identifies the flush volume from the default value and multiplies this by the number of active flushes. In our example this would be 360 flushes multiplied by 1.5 gal per flush, which would equal 540 gal used between 3 and 4 a.m. in the morning, which is 540 gal/hr or 9 gal/min.

However a closer estimation using the correct flush volume would be 360 flushes multiplied by 4 gal per flush, which would equal 1440 gal, used between 3 and 4 a.m. in the morning, which is 1440 gal/hr or 24 gal/min.

If the measured night flow was 50 gal/min (*field data*) the model would then subtract the estimated legitimate usage and identify the rest as leakage. If the coefficients and default values were incorrectly applied as shown above the model would identify the example zone as having 41 gal/min of leakage, where as really it would only have 26 gal/min of leakage.

Then there are the allowances for leaking toilets; what percent of households have leaking toilets, what is the typical leak flow rate. Leaking toilets are a significant component on night consumption in North America and many other countries.

10.6   Modeling Background Losses

Background losses are individual events (small leaks, weeping joints, and the like) with flow rates too low to be detected by visual inspection or traditional acoustic

> **I**f you are using a model from another region or country always ensure that the concepts and coefficients applied in the model are applicable to your system.

leak detection techniques. They will continue to flow unless either detected by chance or until they gradually worsen to a point where they can be detected. The level of background leakage tends to increase with increasing age of the network and is higher for systems operated at higher pressure. The type of pipe materials and jointing techniques are also factors contributing to the level of background losses. It is important when modeling components of real loss to separate out background loss from other components as the tools used to reduce background losses are limited. Managing and reducing pressure is an effective option for reducing background losses in well-maintained systems. In most cases, it is also a lower cost option than the alternative of infrastructure replacement, however, often the latter is a good long-term investment.

Table 10.7 provides flow rates for unavoidable background leakage (UBL) at a standard pressure of 70 psi, or 50 m; UBL corresponds to an infrastructure condition factor (ICF) of one.

Another common error in modeling background leakage is to assume that UBL varies linearly with pressure; this misassumption arose because of the way the data were presented at standard pressure, in a table in the original paper.[2] In fact, the standard modeling assumption, based on available reliable data from various sources, is that UBL varies with pressure to the power 1.5 (FAVAD $N1$ = 1.5).

Once the UBL values in Table 10.7 have been corrected for pressure, using a FAVAD $N1$ of 1.5, they must be multiplied by ICF. The ICF is an unknown factor to most utilities and without carrying out tests, it is difficult to estimate the ICF. Field tests used to estimate the ICF can only be undertaken in small zones temporarily or permanently established for the purpose of measuring minimum night-time flows and pressures. Methods available to estimate the ICF are

Infrastructure Component	Background Leakage at ICF = 1.0	Units
Mains	2.87	gallons per mile of mains per day per psi of pressure
Service connection—main to curb-stop	0.11	gallons per service connection per day per psi of pressure
Service connection—curb-stop to meter	4.78	gallons per mile of service connection per day per psi of pressure

Source: Adapted from Water Loss Control Manual 1st ed.

TABLE 10.7   Unavoidable Background Leakage Rates

1. *ICF based on system-wide ILI*: The ILI is a performance indicator calculated in relation with the top-down water balance. It is a dimensionless indicator describing the ratio between the unavoidable annual real loss volume and the current annual real loss volume calculated by the water balance. A quick first estimate of ICF can be taken from the ILI of the entire system. The system-wide ICF can be assumed to have a similar value to the ILI.

2. *ICF based on initial sensitivity analysis*: Undertake a sensitivity analysis which averages the two extreme possibilities of the ICF. A minimum ICF equals one, where real losses volumes are composed of the unavoidable background losses and recoverable losses. The maximum ICF happens when all leakage is due to background leakage except for a ratio of 1 for the components of reported and unreported leakage. For example, if the maximum ICF is 6 with the other two components at 1 and the minimum ICF is 1 with the other components higher, then the average ICF would be 3.5 and initial estimations could be made for the other components of leakage and potential solutions. It is recommended however that field testing is undertaken to validate this simple estimation process.

3. *ICF based on N1 step test*: If the system is predominantly a rigid or metallic system, an *N*1 step test is a valuable tool to estimate the ICF value in DMAs. Based on a representative sample of *N*1 step test results a system-wide ICF can be calculated. If the system is not predominantly rigid or metallic the principles behind the *N*1 step test and its calculations do not fully apply, and may result in an overestimation of the background leakage component as the breaks themselves may have a variable leakage path.

4. *ICF based on removal of all detectable leaks*: Once a DMA has been installed, even on a temporary basis, and all recoverable leakage has been identified and repaired, then the remaining background leakage level can be measured. In an ideal situation, night time consumers are temporarily turned off so that there is little doubt that the measure flows represent background leakage. Where this is not possible then it is necessary to use a process similar to that described in the previous example to build up a picture of night consumption (including toilet leakage) and subtract that from the measured night flows. However, confidence in such a result would not be as good as the first option of turning off consumers for the period of the test. Results from representative DMAs can be used to estimate a system-wide ICF.

10.6.1   Calculating Losses from Reported and Unreported Breaks

After collecting the annual numbers of reported breaks on mains and service connections (and other system components such as valves and hydrants if so desired), flow rates and durations have to be established. Unless the utility has investigated average leak flow rates and has detailed data available the figures from Table 10.8 can be used as a starting point.

The break/leak duration can be split in three elements—time needed for

Awareness duration: The length of time taken from a leak first occurring—whether it is reported or unreported—to the time when the utility first becomes aware that a leak exists, although not necessarily aware of its exact location. For reported leaks and breaks, this duration is usually very short, while for unreported leaks and breaks, it is a function of the active leakage control policy interventions.

Location of Break	Flow Rate for Reported Breaks [gal/hr/psi pressure]	Flow Rate for Unreported Breaks [gal/hr/psi pressure]
Mains	44	22
Service connection	6	6

Source: Julian Thornton, Reinhard Sturm, George Kunkel, P.E

TABLE 10.8   Example Reported and Unreported Leakage Flow Rates

Location duration: For reported leaks and breaks, this is the time it takes for the water utility to investigate the report of a leak or break and to correctly locate its position so that a repair can be effected; for unreported leaks and breaks, the location duration is zero since the leak or break is detected during the leak detection survey and awareness and location occur simultaneously.

Repair duration: The time it takes to make the repair or shut off once a leak has been located.

The overall volume of water lost through each running break and leak is determined from the overall time of these three components and the flow rate of the leak at the current system pressure. This is shown graphically in Chap. 17.

The water balance calculates the total volume of real losses for the audit year. However, it does not provide the information on what portion of these real losses is due to background losses, reported losses, and unreported losses.

By assessing the volume of real losses through component-based analysis, it is possible to model the volume of real losses that are due to each component and identify suitable tools for their reduction.

A more in-depth analysis of components of real loss may include an analysis of the frequency of breaks on different system components against the baseline UARL frequencies, which in conjunction with measured ICF values might help to dictate the longer term need for infrastructure replacement.

10.6.2   Analyzing the Effects of Changing System Pressure—FAVAD and BFF Concepts

Pressure management can be used to mitigate the adverse effects of excess pressure in a distribution system. Later in Chap. 12, we will be addressing pressure management as a means of controlling leak volumes, reducing leak frequency, and reducing wasteful consumption, as part of a water conservation strategy. However, prior to installing pressure-management systems it is important to understand the effects of our control.

10.6.3   Modeling the Effects of Changing System Pressure on Leakage Flow Rates and Volumes Using FAVAD

Predicting the Reduction in Break Flow Rates

Theoretical hydraulics[3] tells us that the equation for fully turbulent flow Q_f through a fixed orifice of area A_f at static head h follows the square root principle, whereby Q_f is proportional to the orifice area A_f and the real fluid exit speed V_f (which varies with the square root of the static pressure h, and a discharge coefficient C_d):

$$Q_f = C_d A_f \sqrt{2gh} \tag{10.1}$$

However, if the area of the orifice, and/or the coefficient of discharge C_d, also changes with pressure, then the flow through the orifice will be more sensitive to pressure than the "square root" relationship predicts. So Eq. (10.1) can be expressed as

$$Q_f = k_f p^x \tag{10.2}$$

where x is the leakage exponent
$\quad p$ is the static pressure
$\quad k_f$ is the leakage coefficient

As there is no international convention for the exponent, the IWA Water Losses Task Force uses the alphanumeric characters $N1$ for the exponent in Eq. (10.2); obtaining the following expressions:

$$Q_f \cong P^{N1} \tag{10.3}$$

$$\frac{Q_{f_1}}{Q_{f_0}} = \left(\frac{P_1}{P_0}\right)^{N1} \tag{10.4}$$

where Q_{f_1} is the leak flow rate after the change in pressure
$\quad Q_{f_0}$ is the leak flow rate before the change in pressure
$\quad P_1$ is the pressure after implementing the change
$\quad P_0$ is the pressure before implementing the change

This general form of equation [Eq. (10.4)] between leak flow rate L and pressure P has been used since 1981 in Japan, where a weighted average exponent of 1.15 is used.[4] A different relationship (the leakage index curve) was used in the United Kingdom from 1979, but after May (1994) the *fixed and variable area* concept, now known as FAVAD [Eqs. (10.3) and (10.4)], are now recommended as best practice in the United Kingdom and by the IWA Water Losses Pressure Management Team.[5]

Measuring *N1* in the Field

Values of the $N1$ exponent can be obtained from tests in distribution system zones, by reducing inlet pressures in several steps at night, during the period of minimum consumption. Leakage rates (L_0, L_1, and L_2), obtained by deducting an appropriate allowance for night consumption from the inflow rates, can be compared with pressures (P_0, P_1, P_2) measured at the average zone pressure , to obtain estimates of the $N1$ exponent. Analyses of more than 150 field tests in distribution zones in various countries (Table 10.9) have confirmed that the exponent $N1$ is generally between 0.5 and 1.5, but may occasionally reach values of 2.5 or more. A limited number of tests carried out to date in North America have produced $N1$ exponents within the range 0.5 to 1.5.

Tests in systems after all the detectable losses have been repaired or put out of service, have generally produced higher values of $N1$, close to 1.5, for background leakage.

Table 10.9 clearly shows that leak flow rates in distribution systems are usually much more sensitive to pressure than the traditional $N1$ value of 0.5. A physical explanation for this apparent paradox was proposed by May[6] in 1994, using the FAVAD

Country	Number of Zones Tested	Range of N1 Exponents	Average N1 Exponents
United Kingdom (1970s)	17	0.70–1.68	1.13
Japan (1979)	20	0.63–2.12	1.15
Brazil (1998)	13	0.52–2.79	1.15
United Kingdom (2003)	75	0.36–2.95	1.01
Cyprus (2005)	15	0.64–2.83	1.47
Brazil (2006)	17	0.73–2.42	1.40
Totals	**157**	**0.36–2.95**	**1.14**

Source: Ref. 2

TABLE 10.9   Range of N1 Values

concept. May considered what would happen if the area of some types of leakage paths changed with pressure, while the velocity changed with the square root of the pressure. This would mean that different types of leaks can have different relationships for pressure: flow rate (velocity × area), for example,

- Fixed areas leaks (for example, orifices in thick-walled rigid pipes) would have an exponent of 0.5.
- Variable area leaks (for example, cracks where the length changes with pressure) would have an exponent of 1.5.
- Variable area leaks (for example, cracks where the length and width change with pressure) would have an exponent equal to 2.5.

An interesting finding in zones where all detectable leaks had been repaired prior to reliable $N1$ tests is that the remaining background leakage (small undetectable leaks) consistently showed $N1$ values close to 1.5.

How Significant are N1 Exponents in Practical Terms?   Using Eq. (10.4), consider how flow rates of existing leaks in a distribution zone would change if management of excess pressures produced a 20% reduction in average pressure ($P_1/P_0 = 0.8$).

- If $N1 = 0.5$, then $L_1/L_0 = (0.8)^{0.5} = 0.89$, or an 11% reduction in leak flow rates.
- If $N1 = 1.0$, then $L_1/L_0 = (0.8)^{1.0} = 0.80$, or a 20% reduction in leak flow rates.
- If $N1 = 1.5$, then $L_1/L_0 = (0.8)^{1.5} = 0.72$, or a 28% reduction in leak flow rates.
- If $N1 = 2.0$, then $L_1/L_0 = (0.8)^{2.0} = 0.64$, or a 36% reduction in leak flow rates.
- If $N1 = 2.5$, then $L_1/L_0 = (0.8)^{2.0} = 0.58$, or a 42% reduction in leak flow rates.

N3 pressure consumption exponent $N3$ is used as a coefficient for changes in consumption flow or volume due to changes in pressure. In most cases, the change in consumption for the direct pressure use components will correspond to the traditional square root relationship of $N3 = 0.5$ and the volume use components will correspond to an $N3 = 0$ (invariant). If volumetric and direct pressure consumption was evenly distributed a first estimate for a compound $N3$ value could be $N3 = 0.25$. In this case, the impact of reduction in leakage will mostly be proportionally greater than the impact of reduced demand. In some cases, reduction of demand might actually be desired as in the case of water conservation projects.

10.6.4   Modeling Break Frequency Factor BFF[7]

The need for a better understanding of the influence of maximum system pressure on breaks has recently been addressed by the IWA Water Loss Task Force's pressure management team. An extended data set of 112 systems from 10 countries as reported by Thornton and Lambert in IWA Water 21[8] is summarized in Table 10.10. The following can be noted:

- " Before" pressure (meters) ranges from 23 to 199, median is 57 and average 71.
- Percent pressure reduction ranges from 10 to 75%, median 33%, average 37%.
- Percent reduction in breaks ranges from 23 to 94%, median 50%, average 53%.
- The data shows no significant difference between average % break reductions on mains and service connections.

The data from Table 10.10 are also shown in Fig. 10.2 as a plot of % reduction in pressure versus % reduction in new break frequency, for mains and services together.

A simple interpretation, likely to give generally conservative predictions, is to assume that the % reduction in new breaks = BFF × % reduction in maximum pressure, where BFF is a break frequency factor, this can be checked against the data in Fig. 10.2.

The average value of BFF for mains and services together from Fig. 10.2 is 52.5%/ 38% = 1.4, so a line drawn through the data in Fig. 10.2 with a slope of 1.4 gives an average prediction.

An Upper line, with a BFF of 2.8 (twice the average) encompasses all but two of the data points which give larger reduction in new break frequencies

A 'Lower' line, with a BFF of 0.7 (half the average) encompasses all the data points which give smaller reductions in new break frequencies

When applying this simplified prediction approach, it is important to ensure that in cases where both the BFF and the % reduction in maximum pressure are both large, the prediction does not reduce the break frequency below the values used in the UARL formula.

10.6.5   The Latest Conceptual Approach

The latest conceptual approach currently being used by the Pressure Management Team of the IWA WLTF, in attempting to develop an improved practical understanding of pressure/break frequency relationships, is shown in the following series of figures.

In Fig. 10.3 the X-axis represents system pressure and the Y-axis represents failure rates. When a new system is created, mains and services are normally designed to withstand maximum pressures far greater than the range of daily and seasonal operating pressures for a system supplied by gravity. The system operates with a substantial factor of safety, and failure rates are low. Even if there are pressure transients in the system as shown in Fig. 10.4, the maximum pressures do not exceed the pressure at which increased failure rates would occur.

As the years pass, adverse factors based on age (including corrosion) gradually reduce the pressure at which the pipes will fail as shown in Fig. 10.5. Then, depending upon local factors such as traffic loading, ground movement, and low temperatures (which will vary

Country	Water Utility or System	Number of Pressure Managed Sectors in Study	Assessed Initial Maximum Pressure (m)	Average % Reduction in Maximum Pressure	Average % Reduction in New Breaks	Mains (M) or Services (S)	
Australia	Brisbane	1	100	35%	28%	M,S	
	Gold Coast	10	60–90	50%	60%	M	
					70%		S
	Yarra Valley	4	100	30%	28%	M	
Bahamas	New Providence	7	39	34%	40%	M,S	
Bosnia Herzegovin	Gracanica	3	50	20%	59%	M	
					72%		S
Brazil	Caesb	2	70	33%	58%	M	
					24%		S
	Sabesp ROP	1	40	30%	38%	M	
	Sabesp MO	1	58	65%	80%	M	
					29%		S
	Sabesp MS	1	23	30%	64%	M	
					64%		S
	SANASA	1	50	70%	50%	M	
					50%		S
	Sanepar	7	45	30%	30%	M	
					70%		S
Canada	Halifax	1	56	18%	23%	M	
					23%		S
Colombia	Armenia	25	100	33%	50%	M	
					50%		S
	Palmira	5	80	75%	94%	M,S	
	Bogotá	2	55	30%	31%		S
Cyprus	Lemesos	7	52.5	32%	45%	M	
					40%		S

TABLE 10.10   The Influence of Pressure Management on New Break Frequency from 112 Systems in 10 Countries

Country	Water Utility or System	Number of Pressure Managed Sectors in Study	Assessed Initial Maximum Pressure (m)	Average % Reduction in Maximum Pressure	Average % Reduction in New Breaks	Mains (M) or Services (S)	
England	Bristol Water	21	62	39%	25%	M	
					45%		S
	United Utilities	10	47.6	32%	72%	M	
					75%		S
Italy	Torino	1	69	10%	45%	M,S	
	Umbra	1	130	39%	71%	M,S	
USA	American Water	1	199	36%	50%	M	
Total number of systems		112					
	Maximum	199	75%	94%	All data		
	Minimum	23	10%	23%	All data		
	Median	57	33.0%	50.0%	All data		
	Average	71	38.0%	52.5%	M&S together		
	Average		36.5%	48.8%	Mains only		
	Average		37.1%	49.5%	Services only		

Source: Ref. 7

TABLE 10.10   *(Continued)*

from country to country, and from system to system), at some point in time the maximum operating pressure in the pipes will interact with the adverse factors, and break frequencies will start to increase. This effect can be expected to occur earlier in systems with pressure transients or with pumping, than in systems supplied by gravity.

If the system is subject to surges or large variations in pressure due to changing head loss conditions, and has a relatively high break frequency, then introduction of surge control or flow or remote node pressure modulation may be expected to show a rapid significant reduction in the new break frequency. The average pressure in the system may be unchanged, but the reduction of surges and large variations means that maximum pressures do not interact to the same extent with the adverse factors as shown in Fig. 10.6.

If there is excess pressure in the system at the critical point, over and above the minimum standard of service for customers, then permanent reduction of the pressure by installation of pressure management (PRV, subdivision of large zones, and the like) will move the range of operating pressures even further away from the pressure at which combinations of adverse factors would cause increased frequency of failure as shown in Fig. 10.7.

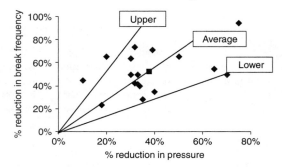

FIGURE 10.2 Plot of % reduction in pressure vs. % reduction in new break frequency (*Source:* Ref. 6.).

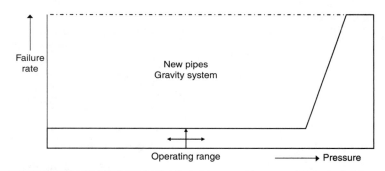

FIGURE 10.3 New system supplied by gravity operates well within design maximum pressure (*Source:* Ref. 6.).

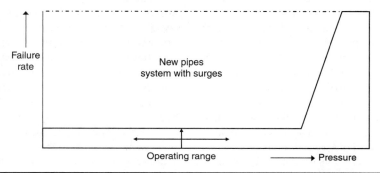

FIGURE 10.4 New system with surges also operates well within design maximum pressure (*Source:* Ref. 6.).

Pressure and Pipe Failure

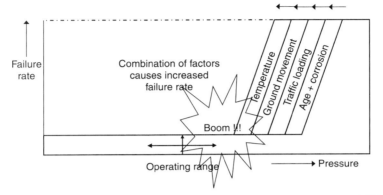

As the pipes deteriorate through age (and possibly corrosion), and other local and seasonal factors, the failure pressure gradually reduces until at some point in time, burst frequency starts to increase significantly

FIGURE 10.5 Combination of adverse factors (including surges) cause increased failure rates (*Source:* Ref. 6.).

A reduction in pressure variation and if possible a reduction in zonal pressure will increase the factor of safety for the zone. A hypothesis as to why mains and/or service connections in some systems show large % reductions in new break frequency with pressure management, but in others the % reduction is only small, can be proposed using this concept.

If, before pressure management, there is already a relatively high break frequency (point 3 in Fig. 10.8), then a relatively small % reduction in pressure may cause a large % reduction in new break frequency (toward point 2).

But if there is already a relatively low break frequency before pressure management (point 2 in Fig. 10.8), then any % reduction in pressure (from point 2 to point 1) should have little effect on new break frequency, but will create a greater factor of safety and extend the working life of the infrastructure.

Reduce Surges and Variations

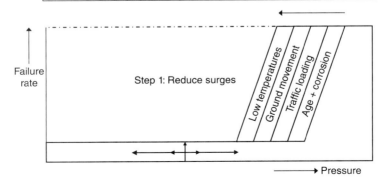

The first step in pressure management is to check for the presence of surges or variations; if they exist, reduce the range and frequency of both

FIGURE 10.6 Reduction of surges and variations limits interaction with adverse factors and increases factor of safety (*Source:* Ref. 6.).

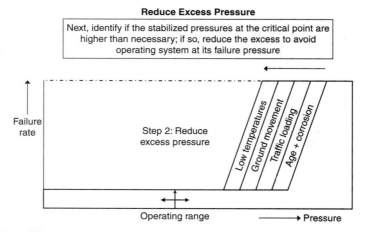

FIGURE 10.7   Reduction of average system pressure limits interaction with adverse factors and increases factor of safety (*Source:* Ref. 6.).

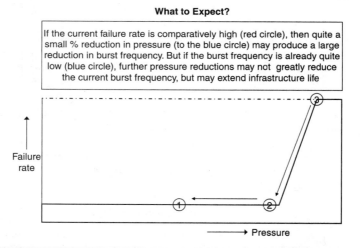

FIGURE 10.8   Percentage reductions in break frequency influenced by initial break frequency (*Source:* Ref. 6.).

10.6.6   Consumption Analysis Model

Analysis of components of consumption forms an important part of any loss reduction or conservation program. The following model is a simple model made in a spreadsheet to predict industrial restroom usage for industries with many employees. This type of model may be used to see the potential benefit of toilet changeout for water conservation or it may be used to predict the amount of water used for sanitary purposes so that it may be broken out of a measured flow profile.

Table 10.11 shows an input table with estimations for volumetric use and frequency of use for men and women within various different buildings within a fictitious industry.

Figure 10.9 shows a resultant modeled estimation of sanitary use per shift.

Parameters / Key

Men flush	2
Women flush	4
Volume flush	3.5
Volume wash	1
Urinal flush	2
Urinal volume	1
(washes)	2 / 1

Key
Blue user enter
Red calculated
Black description

Building Total Population Distribution	No Flushes	Volume/flush	Vol/person	No flushes	Vol/person	Vol/person	No washes	Vol/person	Total volume per person
Bldg. 1 — 414									
Men — 207	2	3.5	7	2	1	2	4	1	13
Women — 207	4	3.5	14			0	4	1	18
50/50									
Bldg. 2 — 65									
Men — 36	2	3.5	7	2	1	2	4	1	13
Women — 29	4	3.5	14			0	4	1	18
55/45									
Bldg. 3 — 40									
Men — 20	2	3.5	7	2	1	2	4	1	13
Women — 20	4	3.5	14			0	4	1	18
50/50									
Bldg. 4 — 200									
Men — 120	2	3.5	7	2	1	2	4	1	13
Women — 80	4	3.5	14			0	4	1	18
60/40									

Source: Water Loss Control Manual, 1st ed.

TABLE 10.11 Input Table with Estimations for Volumetric Use

Building Total Population Distribution		Men flush / Women flush	Volume flush	Volume wash	Urinal flush	Urinal volume				Key	Total volume per person
		2 / 4	3.5	1	2	1				Blue user enter	
										Red calculated	
										Black description	
Bldg. 5	270										
Men	162	2	3.5	7	2	1	2	4	1	4	13
Women	108	4	3.5	14	0		4	4	1	4	18
60/40											
Bld. 6	33										
Men	20	2	3.5	7	2	1	2	4	1	4	13
Women	13	4	3.5	14	0		4	4	1	4	18
60/40											

Source: Water Loss Control Manual, 1st ed.

TABLE 10.11 Input Table with Estimations for Volumetric Use (*Continued*)

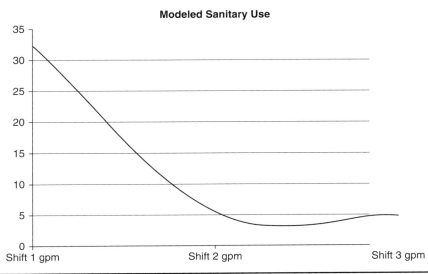

FIGURE 10.9   Resultant modeled estimation of sanitary use per shift (*Source*: Water Loss Control Manual, 1st Edition.)

Table 10.12 shows our input table again with the volume per flush changed to reflect change out to a lower volume flush toilet, in this case the volume reduced from 3.5 gal per flush to 1.6 gal per flush. (Excellent base information on usage can be found in the AWWA end-user survey and in the U.K. managing leakage series).

Figure 10.10 shows the resultant modeled reduction in use per shift.

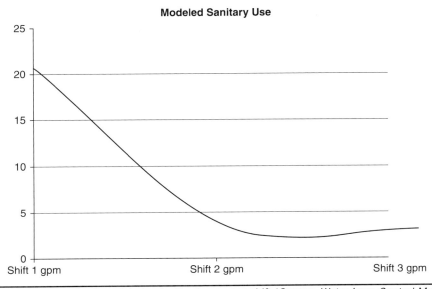

FIGURE 10.10   Resultant modeled reduction in use per shift (*Source*: *Water Loss Control Manual, 1st ed.*)

Men flush	2
Women flush	4
Volume flush	1.6
Volume wash	1
Urinal flush	2
Urinal volume	1
	2
	1

Building Total Population Distribution		No flushes	Volume/flush	Vol/person	No flushes	Vol/person	Vol/person	No washes	Volume/wash	Vol/person	Total volume per person
Bldg. 1	414										
Men	207	2	1.6	3.2	2	1	2	4	1	4	9.2
Women	207	4	1.6	6.4			0	4	1	4	10.4
50/50											
Bldg. 2	65										
Men	36	2	1.6	3.2	2	1	2	4	1	4	9.2
Women	29	4	1.6	6.4			0	4	1	4	10.4
55/45											
Bldg. 3	40										
Men	20	2	1.6	3.2	2	1	2	4	1	4	9.2
Women	20	4	1.6	6.4			0	4	1	4	10.4
50/50											
Bldg. 4	200										
Men	120	2	1.6	3.2	2	1	2	4	1	4	9.2
Women	80	4	1.6	6.4			0	4	1	4	10.4
60/40											

Bldg. 5	270	2	1.6	3.2	2	1	2	4	1	4	9.2
Men	162	4	1.6	6.4			0	4	1	4	18
Women	108										
60/40											
Bld. 6	33	2	1.6	3.2	2	1	2	4	1	4	13
Men	20	4	1.6	6.4			0	4	1	4	18
Women	13										
60/40											

Source: Water Loss Control Manual, 1st Edition

TABLE 10.12 Flush Volume Changed to Reflect Conservation.

10.7   Summary

In this chapter, we have shown models and theory covering a variety of different tasks, which make up a water loss control program.

- Top-down water audit
- Component analysis of apparent losses
- Meter accuracy
- Meter sizing
- Component analysis of real losses
- Pressure management FAVAD and BFF
- Consumption analysis

In all cases, the important factor is the validity of the data being used and the knowledge of the person operating the models. It is necessary that the operator understands the limitations of the models and the data that they are using and what impact that may have on the final decision for intervention, budget allocation, and team resource.

If good data is not available, then estimations can be used when modeling and 95% confidence limits can be applied to input components and calculated for output results; however, it is important to carefully note any estimation so that others may properly interpret the results.

> **W**hen modeling it is important to be accountable. Always clearly mark all assumptions and estimations along with the model goals and outputs.

References

1. Lambert, A.O., S. Myers, and S. Trow. *Managing Water leakage—economic and technical issues.: Financial Times Energy.* 1998. ISBN 1 94083 011 5
2. Lambert, A., T. G. Brown, M. Takizawa, et al. A review of performance indicators for real losses from water supply systems. Aqua. 48(6), December 1999.
3. Thornton, J., Garzon, F., and Lambert, A. "Pressure-Leakage Relationships in Urban Water Distribution Systems." *International Conference on Water Loss Management.* Skopje, Macedonia: ADKOM USAID GTZ, September, 2006
4. Hiki, S. Relationship between Leakage and Pressure. *Journal of Japan Water Works Association,* 51(5):50–54, 1981.
5. Thornton, J. *Managing Leakage by Managing Pressure.* IWA Publishing Water 21 ISSN 1561 9508, October, 2003.
6. May, J. "Leakage, Pressure and Control." *BICS International Conference on Leakage Control Investigation in Underground*: Assets London, 1994.
7. Thornton, J., and Lambert, A. *Pressure Management Extends Infrastructure Life and Reduces Unnecessary Energy Costs.* Bucharest, Romania: IWA Water Loss, 2007.
8. Thornton, J., and Lambert, A. "Managing Pressures to Reduce New Breaks." *IWA Publishing Water 21 Magazine.* ISSN 1561-9508, December 2006.

CHAPTER **11**

Controlling Apparent Losses—Capturing Missing Revenue and Improving Consumption Data Integrity

George Kunkel, P.E.

Julian Thornton

Reinhard Sturm

11.1   Introduction

Water losses in drinking water utilities occur as two distinct types: *real losses* are the physical losses from distribution systems, mostly leakage but also water lost from tank overflows. *Apparent losses* are the nonphysical losses that occur when water is successfully delivered to the customer but, for various reasons, is not measured or recorded accurately, thereby inducing a degree of error in the amount of customer consumption. When such errors occur systematically in an appreciable number of customer accounts, the aggregate measure of water consumption can be greatly distorted and significant billings can be missed.

This chapter explains the causes of apparent losses and describes the significant impacts that they exert on consumption data integrity and revenue capture potential in systems with metered customers. Chapters. 12 to 15 explain the major categories of apparent loss and the means to control these losses to economic levels.

Apparent losses are defined as nonphysical losses, since no water is physically lost from the water supply infrastructure. However, these inefficiencies in the accounting and information-handling practices of the water utility can exert significant impacts. They are caused by faulty, improperly sized or badly read meters, corruption of water consumption data in billing systems, and water which is taken from the distribution system without authorization. Apparent losses consist of three primary components:

- Customer metering inaccuracies
- Systematic data-handling errors, particularly in customer billing systems
- Unauthorized consumption

Certain occurrences of apparent losses are easily identified; and assumptions can be made to initially approximate the more complex components of apparent losses. Ultimately, detailed components should be verified as bottom-up work (field investigations) is conducted and the water loss control strategy develops.

11.2   How Apparent Losses Occur

Apparent losses occur due to inefficiencies in the measurement, recording, archiving, and accounting operations used to track water volumes in a water utility. These inefficiencies result from inaccurate or oversized customer meters, poor meter-reading, billing and accounting practices, weak policies, or ineffective management. Apparent losses also occur from unauthorized consumption, which is caused by individual customers or others tampering with their metering or meter-reading devices or otherwise maliciously obtaining water without appropriately paying for the service. For any type of apparent loss, it is incumbent upon utility mangers and operators to realistically assess metering and billing operations for inconsistencies, and then develop internal policies and procedures to economically minimize these inefficiencies. It is also important to clearly communicate to customers, utility executives, elected officials, financing agencies, and the media the problems of apparent losses and the need to control them.

The specific ways in which apparent losses occur are many and varied and, particularly with unauthorized consumption, always changing. Those taking water in unauthorized fashion do so for varied reasons. Some sincerely believe that water should be free and it is their right to obtain water without paying for it. Others feel that they do not have the financial resources to pay for the service. More often, however, such users take water maliciously, always thinking of new ways to "beat the system."

The water utility must therefore be vigilant in its effort to manage its product (water) via effective meter management and rational billing, auditing, collection, and enforcement policies in order to realize projected levels of revenue and maintain accurate measures of the water that it supplies.

A note regarding collections: As water utility financial managers know, not all of their customers pay their water bill as required, or pay their bill on time. The *collection rate* is a financial performance indicator that reflects the rate at which customers pay their water bills. The collected payments are measured as a percentage of the money billed each month for the utility's services. Collection rates at the 30-day, 60-day and 90-day milestones are typically tracked in order to provide a representative picture of the customer population's payment record. While the collection rate is a highly important measure that represents the pace at which revenue is gained by the water utility, collections are not included in the water audit methodology detailed in this publication because the collection rate measures payments based upon billed consumption, whether or not all water has passed through customer meters, or was accurately measured. The water audit methodology has as its terminal boundary the customer meter which generates the consumption data that is the basis for the customer billing. This publication provides utilities guidance in maximizing the efficiency of their water billing process, while collections focus on payment efficiency, which is beyond the scope of this textbook. The reader should consult publications on water rates and finance to obtain guidance on tracking their collection rate and instituting policies that maximize collections.

11.3   Customer Meter Inaccuracy

Customer meters that inaccurately measure the volumes passing through them can be a major source of apparent loss in drinking water systems. While most North American drinking water utilities meter their customer consumption, a notable number do not. For example, only 56% of all residences in Canada were metered as of 1999, therefore many customers are unmetered and typically pay a flat-rate fee for water service.[1] In unmetered water utilities, meter accuracy cannot be evaluated as an apparent loss; although these utilities are behooved to use other methods to quantify the amount of customer consumption and separate it from components of authorized consumption and water losses.

Figure 11.1 gives the American Water Works Association's (AWWA) policy statement on metering and accountability. This publication supports AWWA's recommendation to meter water supplied to distribution systems as well as all customer consumption, therefore this discussion exists in the context of water utilities having fully metered customer populations. Water utilities that do not meter their customers can obtain an approximation of customer consumption by metering and data-logging representative samples of customer accounts and statistically evaluating the results to infer general customer consumption trends.

Customer meters provide valuable information on consumption trends for long-term planning, and data needed to evaluate loss control and conservation programs. Metering also elevates the value of water in the mind of the consumer by linking a price with a volume. With highly capable metering, automatic meter-reading systems, and data-logging technologies now widely available, customer consumption information has become a critical element to better manage water utility operations and the water resources of individual watersheds or regions.

The American Water Works Association (AWWA) recommends that every water utility meter all water taken into its system and all water distributed from its system at its customer's point of service. AWWA also recommends that utilities conduct regular water audits to ensure accountability. Customers reselling utility water – such as apartment complexes, wholesalers, agencies, associations, or businesses – should be guided by principles that encourage accurate metering, consumer protection, and financial equity.

Metering and water auditing provide an effective means of managing water system operations and essential data for system performance studies, facility planning, and the evaluation of conservation measures. Water audits evaluate the effectiveness of metering and meter reading systems, as well as billing, accounting, and loss control programs. Metering consumption of all water services provides a basis for assessing users equitably and encourages the efficient use of water.

An effective metering program relies upon periodic performance testing, repair, and maintenance of all meters. Accurate metering and water auditing ensure an equitable recovery of revenue based on level of service and wise use of available water resources.

Figure 11.1   Policy statement: metering and accountability. (*Source:* American Water Works Association)

A thorough discussion of customer meters is beyond the scope of this publication. AWWA provides excellent guidance in several manuals that cover all aspects of sound meter management. The M6 publication, *Water Meters—Selection, Installation, Testing, and Maintenance,* provides comprehensive information on the basics of customer meter management.[2] The M22 publication, *Sizing Water Service Lines and Meters,* provides outstanding guidance on customer demand profiling and sizing criteria, which are critical for meter accuracy.[3]

A word of caution about data handling: Meter accuracy is only the first step in obtaining customer consumption data. While the meter must provide an accurate measure, the subsequent processes—including meter readings (gathered manually or automatically), data transfer to billing systems, and archival operations—must also be handled accurately, or the actual customer consumption will be distorted, with the data from some customer accounts lost entirely. In many water utilities, it is not uncommon to find accurate meter data transposed erroneously, adjusted improperly, or incorrectly archived. If any part of the data path lacks integrity, it is easy to misinterpret apparent losses solely as meter inaccuracy, with potentially costly consequences if loss control decisions (such as replacing large numbers of accurate meters) are based upon this faulty assumption.

11.4   Data Transfer and Systematic Data-Handling Errors

The customer water meter is only the beginning of a sometimes complicated trail that ultimately generates a large amount of customer consumption data. Since most water utilities manage data for many thousands of customers, systematic data-handling inaccuracies can easily be masked by the shear volume of the bulk data. Figure 11.2 gives an overview of the typical steps existing in the data trail from meter to historical archive.

In any of the above steps errors can be introduced into the output data that is ultimately documented as customer consumption. Some of the ways in which the integrity of customer consumption data may be compromised are

- Data transfer errors
  - Manual meter-reading errors
  - Automatic meter-reading equipment failure

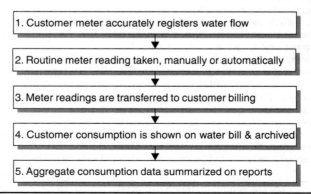

1. Customer meter accurately registers water flow
2. Routine meter reading taken, manually or automatically
3. Meter readings are transferred to customer billing
4. Customer consumption is shown on water bill & archived
5. Aggregate consumption data summarized on reports

Figure 11.2   Metered consumption data archival path. (*Source:* Ref. 6.)

- Data analysis errors
  - Use of poorly estimated volumes in lieu of meter readings
  - Customer billing adjustments granted by manipulating actual metered consumption data
  - Poor customer account management: accounts not activated, lost, or transferred erroneously
- Policy and procedure shortcomings
  - Despite policies for universal customer metering, certain customers are intentionally left unmeasured or unread. This is common for municipally owned buildings in water utilities run by local governments.
  - Provisions allowing customer accounts to enter "nonbilled" status, a potential loophole often exploited by fraud, or unmonitored due to poor management.
  - Adjustment policies that do not take into account preservation of actual customer consumption.
  - Bureaucratic regulations or performance lapses that cause delays in permitting, metering, or billing operations.
  - Organizational divisions or tensions within the utility that do not recognize the importance or "big picture" of water loss control.

The above list provides but a few of the data-handling problems that might be encountered in a drinking water utility. It is not exhaustive, however, and almost any utility might identify an apparent loss situation that is unique to their organization. Any action that unduly modifies the actual amount of customer consumption can be considered an apparent loss. The IWA Water Loss Task Force did not specifically identify data-handling error as a source of apparent loss during the initial work published by Alegre et al.[4]; however subsequent articles published by IWA and AWWA clearly define this category. The AWWA Water Loss Control Committee considers such manipulations of data as apparent losses.[5]

11.5   Unauthorized Consumption

Unauthorized consumption occurs in virtually all drinking water utilities. It typically occurs through the deliberate actions of customers or other persons who take water from the system without paying for it. The nature and extent of unauthorized consumption in a system depends on the economic health of the community and the emphasis that the water utility places on policy and enforcement.

Unauthorized consumption occurs in many ways, including tampering of customer meters or meter-reading equipment, illegal openings of fire hydrants, illicit connections, and sundry other means. Establishing the key features of a good accountability and loss control program—water auditing being foremost—will inevitably uncover situations where unauthorized consumption is occurring.

The water audit should quantify the component of unauthorized consumption occurring in the utility. For first-time water audits, or where unauthorized consumption is not believed to be excessive, the auditor should use the default value of 0.25% of the *water supplied* value in the water audit. This percentage has been found to be representative of this component of loss in water audits compiled worldwide. For water utilities with well-established water audits, or those believing that unauthorized consumption is excessive, the extent and nature of unauthorized consumption should be specifically identified, as well as policy or procedural gaps that allow water to be taken without payment. The opportunities for water to be stolen from the water utility are functions of individual customers who either cannot or will not pay for the services they are rendered. All utility systems are susceptible to the occurrence of unauthorized consumption, and this occurrence is substantial for some.

A portion of the customers in any community may live with real economic hardship, and the water utility should seek to strike a balance between service provision to this group of customers and enforcement actions against those who can afford water service but choose not to pay. A careful evaluation of utility policy is therefore necessary to operate rationally to stem unauthorized consumption.

11.6   The Impacts of Apparent Losses

Because apparent losses under-record the volume of customer consumption, they generate two major impacts on water resources management:

- Apparent losses induce a degree of error into the quantification of customer water demand, thereby impacting the decision-making processes used to determine needed source water withdrawals, calculate the appropriate capacities of water supply infrastructure, and evaluate conservation and water loss control practices.

- Apparent losses cause water utilities to underbill a portion of the water consumed by customers, thus a portion of the potential revenue is not recovered.

Both of these impacts can be significant. If a high level of apparent loss exists in a water utility, its recorded volume of customer consumption could be subject to a significant degree of error. Consider a water utility that documents customer consumption of 3.65 billion gal of water in a year [10 million gals per day (mgd)]. If routine water auditing found apparent losses equal to 1 mil gal/d (10% of consumption) then actual customer consumption during the year being audited was 4.015 billion gal, an additional 365 mil gal. Such a loss creates a distortion of the true customer consumption volume; in this case under-stating it by 365 mil gal. Activities that rely on accurate customer data are compromised by this degree of error. These can include efforts to evaluate the success of water conservation programs, using consumption data to assign demands in hydraulic models and evaluation of community drinking water requirements needed for regional water resource plans. Apparent losses therefore represent a degree of error that is interjected into a wide range of analytical and decision-making processes regarding water resource management. Given that the water industry in the United States is highly fragmented, with many different sized water utilities existing in any given region, the degree of error from apparent losses can be compounded by the varying errors existing in many disparate water utilities. Gauging true customer needs on a regional basis can be difficult without a reasonable assessment of the apparent losses existing in the region's water utilities.

From a financial perspective, apparent losses can exert a tremendous impact on the water utility's bottom line. Apparent losses cost utilities revenue, and can account for over 5% of a utility's annual billing for water and wastewater service rendered to individual customers. Many water utilities are confronted with increasing financial pressures from a variety of forces and stand to gain from the revenue recovery potential of apparent losses. Since apparent losses are quantified by the amount of water improperly recorded at the customer's delivery point, this water is valued at the retail cost that is charged to the customer. Water rates frequently also include a waste water charge that is also based upon the volume of consumption. The cost impact of apparent losses is frequently higher than the impact of real losses, which are typically valued at the variable production costs to treat and deliver the water. When water resources are greatly limited, real losses can also be valued at the retail rate based upon the theory that any water saved by real loss reduction can be sold to customers. Since the retail rates usually include fixed and administrative costs, infrastructure improvement, and debt repayment, this cost is typically much greater than the variable production costs that water utilities incur to treat and deliver water. Therefore apparent losses can have a dramatic financial impact to the water utility's revenue stream.

Apparent losses also create a problem of payment inequity for the community. Apparent losses occur when the actual amount of water delivered is understated. Hence, a portion of the customer population obtains discounted or free water service. This means that the paying customer population effectively subsidizes those customers who are underpaying or not paying. This situation is particularly troubling as water utilities encounter pressure to raise water rates, with the paying customers shouldering an even greater financial burden for the entire water-using community. Reducing apparent losses and recovering missed revenue can reduce the frequency of, or defer the need for, water rate increases by identifying underpaying and nonpaying customers and adding them to the active billing roles.

Apparent loss recovery can create a direct financial improvement to the water utility, and many apparent loss occurrences can be recovered with relatively little cost. This is important in terms of seeking early success and payback to the water loss control program. Funds recovered early in the program in this manner can serve to seed further activities in the long-term water loss control effort.

In summary, water utility managers can obtain a more realistic quantification of the actual customer demand by identifying apparent losses. Controlling apparent losses can result in the capture of significant missing revenue for the water utility. Hence, the assessment of apparent losses has bearing on all quantitative aspects of accountability and the water loss control program.

11.7   The Economic Approach to Apparent Loss Control

Figure 11.3 is a graph that represents a conceptual approach to water loss control, in this case applied to apparent losses.[6] The center boxes represent three levels of apparent losses, as defined below:

- The outer box represents the current volume of apparent losses that a water utility can quantify using the water audit process.

- The middle box represents the utility-specific target level for apparent losses. Conceptually, this is the economic level of apparent losses (ELAL), or the level at which the cost of the apparent loss control efforts equal the savings garnered from the apparent loss recovery.

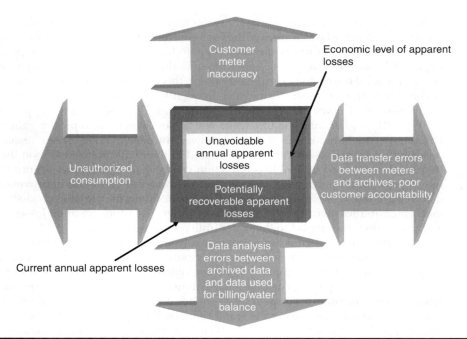

FIGURE 11.3   The four pillar approach to the control of apparent losses. (*Source:* Ref. 6.)

- The inner box is the level of unavoidable annual apparent losses (UAAL). This is a conceptual level of apparent losses representing the lowest level that could be attained if all possible apparent loss controls could be exerted. Unlike the unavoidable annual real losses (UARL) which has an established calculation, an established formula or reference value for the UAAL does not currently exist. Discussion on the means to develop a calculation for the UAAL continues among the IWA Water Loss Task Force.

- The four arrows represent means to address the four significant causes of apparent losses. The arrows indicate that, as targeted actions exert control over certain components of apparent loss, the total annual volume of apparent losses (outer box) can be reduced. The dual directional structure of the arrows reflect that lack of control of these component areas results in the total volume of apparent loss increasing.

Controlling losses in almost any field of endeavor is an effort of diminishing returns, as many losses can never be completely eliminated. When losses are rampant, relatively large reductions can often be gained early in a loss-control program; this is the "low hanging fruit." However, further loss reduction requires ever-greater cost and effort to recoup ever-diminishing returns. Figure 11.4[7] provides an example cost curve for customer meter replacement, with points plotted at replacement frequency (years) and average cumulative consumption passed through the meters (million gal). It can be seen that replacing meters at a high frequency results in less apparent loss due to meter inaccuracy. However, a high replacement frequency means higher replacement costs. So, when is the optimum time to replace meters?

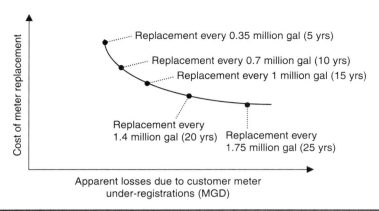

FIGURE 11.4   Cost curve for meter replacement programs. (*Source:* Ref. 7.)

When setting an apparent loss reduction target there exists a breakeven point, beyond which the effort to control the apparent losses costs more than the likely recoveries. In this case, further apparent loss control effort is not economic to pursue. This is the ELAL, or the optimum target of apparent losses to seek. The ELAL for customer meter inaccuracy is shown graphically in Fig. 11.5.[7] In this figure, the meter replacement cost curve is matched against the cost recovery line, which reflects the savings generated by apparent loss recovery. A third curve is generated by adding the two values and plotting, thus a curve of total annual apparent loss cost is derived. The ELAL for apparent loss due to meter inaccuracy is found by taking the level of loss at the minimum point of this curve, as shown in Fig. 11.5. The optimum level of apparent loss reduction at the ELAL is determined by reading back off the apparent loss reduction cost curve. For apparent losses due to customer meter inaccuracy, the optimum frequency of meter replacement can be determined by selecting the point on the meter replacement cost curve that matches the minimum point of the total cost curve.

In setting out to generate a particular curve, the economic analysis should start by determining the volume and cost value of the most significant sources of apparent loss. For each apparent loss component, it is necessary to analyze the problem and determine

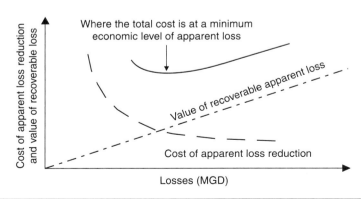

FIGURE 11.5   The economic balance for an apparent loss reduction solution. (*Source:* Ref. 7.)

why these errors are occurring. It is then possible to consider various solutions to reduce these losses. Possible solutions might range from improved auditing, new reports to identify these errors, or better training as low-cost endeavors to replacement of the entire customer meter population or a new customer billing system at the opposite end of the cost spectrum. Solutions to reducing apparent losses due to meter reading errors may range from better training for meter readers, improved auditing of meter readings, and improved software on handheld meter-reading equipment to the implementation of a complete AMR system as a long-term solution. The cost of each of these alternative solutions should be compared to the projected revenue recovery from the reduction in apparent loss, and the solutions ranked in terms of cost/benefit. Only those solutions with a sufficiently attractive cost/benefit ratio or payback period should be included in the apparent loss control plan. Clearly the scale and the shape of the cost curve for solutions to the various components of apparent loss could be very different and will vary from utility to utility. Also, comprehensive solutions, such as an AMR system, offer many additional benefits in addition to apparent loss control. Until further research has been undertaken, it is up to each water utility to develop appropriate utility-specific costs and cost curves for the various apparent loss components that they perceive to be significant.

The above approach illustrates two limitations in the current status of apparent loss target setting. Firstly, in applying the method using cost curves, considerable data must be generated. This can be a complex and time-consuming undertaking. Secondly, separate cost curves must be developed for each of the components (and subcomponents) of apparent losses that are deemed significant; one for customer meter inaccuracy, one for meter tampering, one for unauthorized use of fire hydrants, and so on. Unfortunately, there is no single, composite ELAL for a water utility. There will be an ELAL for each apparent loss control solution considered and the overall ELAL for the utility will be the sum of each solution to the different components of apparent losses selected. Therefore the present means of rigorously developing the ELAL is a demanding task that cannot be executed without considerable data. At this time, work is underway by the IWA Water Loss Task Force to develop a simpler and more straightforward method of obtaining the ELAL.

Clearly, the current approach to identify the overall ELAL is resource intensive and time consuming. While work is undertaken to develop a simpler method to calculate the ELAL, water utilities can still undertake a cursory analysis of their apparent losses and identify approximate levels of desired apparent loss reduction. If a water utility is only beginning to audit their water supply then it is very likely that considerable apparent (and real) losses exist and it will be economic to recover a cost-effective volume of both real and apparent losses. In lieu of a complex apparent loss analysis, the following recommendations are offered as standard starting points for water utilities in apparent loss control.

- Flowchart the customer meter reading and billing process—understanding this process and identifying any lapses or loopholes that allow apparent losses to occur are fundamental to the management of all apparent loss components. Additionally, this exercise can be conducted largely in a table-top manner with limited resources and costs, and may identify a number of loss components that are quickly and inexpensively corrected by policy, procedural, or computer programming changes. See Chap. 14 for details on flowcharting.
- Unless the customer meter population is very young and well documented, perform annual meter accuracy tests on a sample of customer meters. This can be

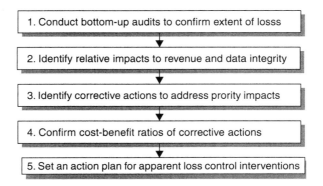

FIGURE 11.6   Establishing an apparent loss control strategy. (*Source:* Ref. 6.)

as few as 50 meter tests per year, with 25 randomly selected meters and 25 meters that have registered high cumulative consumption. Data from this testing will give a preliminary representation of the accuracy status of the current meter population, and the yearly trend will ultimately reveal the points at which meters lose accuracy significantly due to cumulative volumes passed through the meter.

The above first steps are manageable in terms of effort and expense and can provide good data and possible recoveries that can get apparent loss control efforts started productively. Once water auditing has been performed for several years, additional bottom-up data from field investigations should become available and a more robust assessment of existing apparent losses can be undertaken.

Figure 11.6 identifies a sequence of steps to take to develop and implement the apparent loss control strategy after the initial top-down water audit has been compiled. These steps, starting with the bottom-up auditing activity, should be followed in sequence in order to assure that intervention actions are economically justified and well planned. Bottom-up activities for apparent loss control include detailed investigations of metering, accounting, and billing functions. Flowcharting the billing system process is an important bottom-up activity described in Chap. 14 as the recommended first-step action. Meter accuracy testing also falls under the heading of bottom-up activities. These activities also include field investigations of customer properties to inspect connections for possible meter tampering, illegal connections, or other forms of unauthorized consumption. Many other similar activities could be conducted to track down apparent losses. Any activity that delves into the specific conditions of suspected apparent losses can be considered bottom-up activities. Bottom-up activities require more work to conduct than top-down activities but they specifically identify individual losses, allowing intervention actions to be strategically targeted to known losses and more reliable data to be generated for the water audit.

11.8   Developing a Revenue Protection Program to Control Apparent Losses

The most significant impact of apparent losses for water utility managers is uncaptured revenue. The label *revenue protection program* is used to identify the collective activities used to protect the utility's revenue base by controlling apparent losses. As noted above, a number of distinct components, and subcomponents, of apparent losses occur in water utilities; therefore a revenue protection program must be tailored to the individual needs of the water utility. Figure 11.7[6] shows an example revenue protection plan for the fictitious water utility County Water Company (CWC). Revenue protection plans should be developed by considering each of the major components of apparent losses: customer meter inaccuracy, data transfer error, systematic data-handling error,

---

SAMPLE REVENUE PROTECTION PLAN

Name of Utility: _County Water Company_	Date: _7/10/2007_

I. Revenue Protection Plan Approach

After completing County Water Company's (CWC) first annual water audit the CWC manager determines to create an ongoing revenue protection program that identifies causes of the most significant apparent loss components, and launches efforts to begin to reduce these losses to economic levels. After initial gains are evaluated, additional, less significant occurrences of apparent loss will be evaluated for reduction.

The CWC water audit quantifies apparent losses as:

Residential meter under-registration	134.33 Mil Gal @ $556,395
Industrial/commercial/agricultural meter under-registration	29.97 Mil Gal @ $108,701
Systematic data transfer error	12.57 Mil Gal @ $49,589
Systematic data analysis error	8.72 Mil Gal @ $34,400
Data policy/procedure impacts	11.63 Mil Gal @ $45,880
Unauthorized consumption (default 0.25% of water supplied)	11.0 Mil Gal @ $43,395
	--
Total apparent losses	208.22 Mil Gal @ $838,360

(Composite customer retail cost is $3,945/Mil Gal; total cost to operate the water system is $9,600,000)

From this summary the cost impact of customer meter inaccuracy is $556,395 + $108,701 = $665,096. This is equal to 6.9% of the total cost of running the system ($665,096/$9,600,000). The three sub-components of systematic data handling error add to a total cost impact of $129,869 or 1.3% of the total cost of running the water system. Unauthorized consumption is believed to be a very minor occurrence in the CWC system and is estimated using the default value of 0.25% of water supplied. From the results of the water audit, the Revenue Protection Plan should focus primarily on customer meter inaccuracy, with a secondary focus on systematic data handling error. In following the recommended first step in addressing apparent losses, the Manager of CWC plans to flowchart the workings of the customer billing system in order to ascertain the integrity of the customer consumption data and identify occurrences of systematic data handling error.

II. Customer Billing Process Analysis

II-a. The Manager determines to assign one CWC billing analyst to work part time over a period of 2 months, in conjunction with a billing system consultant, to analyze the customer meter reading and billing process. From these findings, any apparent loss that is deemed to be readily correctable will be implemented. Such corrections are recognized as relatively minor procedural or programming changes; an example of which might be a programming lapse that inadvertently left a two-year old housing development of 50 homes off of the meter reading/billing roles. The cost of this effort is basically the human resources to implement it.

II-b. Staffing costs, including wages and benefits for CWC personnel

Number of CWC staff	1	Cost, $/hour	33.50	$/day	268.00
Number of consultant staff	1	Cost, $/hour	75.00	$/day	600.00
		Total,	$/hour	108.50	$/day	868.00

II-c. Duration

Days, per project task	Flowcharting/Analysis	Corrections	Total days	Total project costs, $
CWC staff	14.00	4.00	18.00	4,824.00
Consultant	25.00	7.00	32.00	19,200.00
Total	39.00	11.00	50.00	24,024.00

Figure 11.7   Sample revenue protection plan. (*Source:* Ref. 6.)

III. Customer Meter Accuracy Testing

III-a. The water audit for CWC estimates that customer meter inaccuracy caused under-registered consumption worth $665,096 of revenue during the audit year. This amount represents the majority of the revenue recovery potential in CWC. During the water audit process CWC undertook customer meter testing on a sample of meters—50 random residential meters and 5 random large (industrial, commercial, and agricultural) meters. The findings of this meter testing were extrapolated to the entire meter population to determine an estimate of the entire apparent losses attributed to customer meter inaccuracy. Based upon the value of this testing, the CWC Manager determines to continue such testing on an annual basis; both to continually gauge meter accuracy, and to also observe the rate of long-term degradation in accuracy with increasing cumulative consumption. CWC does not have its own meter testing facility, therefore they utilize contracted testing services. The metering supervisor and one staff person participate by identifying meters for testing, rotating meters from customer properties, and performing the administrative and analysis work.

III-b. Staffing & testing service costs, including wages and benefits for CWC personnel
  Number of CWC Staff   __2__
  Supervisor   cost, $/hour __35.00__ $/day __280.00__ # of days __3__ Cost, $ __840.00__
  Service worker cost, $/hour __27.50__ $/day __220.00__ # of days __15__ Cost, $ __3,300.00__
                  CWC Staff Cost, $ __4,140.00__

III-c. Estimated Costs of Meter Testing Program-55 annual meter tests
  Meter Testing Services cost, $/small meter   __35.00__   Cost for 50 meter tests, $ __1,750.00__
  Meter Testing Services cost, $/large meter __250.00__ Cost for 5 large meter tests, $ __1,250.00__
                  Meter Testing Service Cost, $ __3,000.00__

III-d. Total cost for annual meter testing program, $ __7,140.00__

IV. Revenue Protection Program Summary

IV-a. The total cost of the two components of the initial revenue protection program are given below:

  Customer Billing Process Analysis,$ __24,024.00__

  Annual Meter Testing Program, $ __7,140.00__

  Total Revenue Protection Program Cost, $ __31,164.00__

IV-b. Economic level of revenue recovery

During its first year of its new revenue protection program, CWC anticipates spending $31,164 to launch the program. In order to recover the cost of this program, CWC would need to recover revenue equal to this amount. By applying the composite customer retail billing rate of $3,945/Mil Gal of customer consumption, an equivalent volume of consumption can be determined, as shown below:

$$\text{Breakeven Recovery Volume} = \frac{\$31,164.00}{\$3,945/\text{Mil Gal}} = 7.90 \text{ Mil Gal}$$

If CWC's initial revenue protection efforts recover merely 7.90 mil gal of consumption, then the revenue protection program will have paid for itself in its first year of operation. This level is only 3.8% of the total apparent losses of 208.22 mil gal quantified in the water audit. Since apparent losses are valued at the customer retail rate, recovering these losses can be highly cost-effective. CWC has strong potential to more than recoup its first year revenue protection program costs in its first year. If this level of revenue recovery is met or exceeded, then CWC will be well on its way to creating a very cost-effective apparent loss control and revenue enhancement program.

Figure 11.7   (*Continued*)

and unauthorized consumption. Data from the water audit should be evaluated to assess the relative impact that each component exerts on the water utility. In the CWC example in Fig. 11.7, CWC estimates that very little unauthorized consumption occurs in its system, so this component is not included in its initial revenue protection program.

As shown in Figure 11.7, the cost impact in lost revenue to CWC due to apparent losses is $838,360, which is 8.7% of the total annual operating cost of $9,600,000.

In following with the above recommendations, the CWC manager determines to launch a revenue protection program that will analyze the customer billing process and institute annual customer meter accuracy testing.

The billing process analysis (flowcharting) is envisioned as a 2-month project costing $24,024. This cost includes the analysis and any low-cost apparent loss corrections that can be immediately incorporated into the process. CWC conducted accuracy testing of a sample of customer meters during the compilation of its initial water audit and determines to continue testing a sample on an annual basis in order to track the accuracy of the customer meter population and monitor degradation of accuracy over time. The projected cost of this effort is $7,140 to test 50 residential meters and 5 large meters. The total first-year cost of the two component revenue protection program is estimated at $24,024 + $7,140 = $31,164. By applying its composite customer retail billing rate of $3,945/million gal, CWC need only recoup 7.90 million gal of apparent loss to break even during the first year of program operation. This is only 3.8% of the total apparent loss volume of 208.22 million gal quantified in the water audit. If each residential customer consumes 800 cubic feet/month of water (71,808 gal/year), then the equivalent of recovering 110 missing accounts from the billing roles would meet the cost-effective breakpoint of 7.90 million gal recovered. This is less than 1% of the total of 12,196 accounts in the customer billing system. It is evident that recovering losses valued at the customer retail rate offers a swift and high payback.

During the early phases of a revenue protection program, significant recoveries may be recouped with less costly programming and procedural refinements. However, as the program matures, the water utility will ultimately consider more extensive and costly improvements to control apparent losses. Such efforts can include wholesale meter change-out, installation of automatic meter reading (AMR) systems, or implementation of a new computerized billing system. The economics of such long-term improvements should be carefully considered, but with a mature program, sufficient data will exist to provide a basis for rational decision making.

11.9   Apparent Loss Control: A Summary

Apparent losses distort the measure of the volume of customer water consumption and cause water utilities a loss of revenue. Controlling apparent losses, however, can be very cost-effective since initial corrections may require relatively little work with potentially high payback. It is often advantageous to target apparent loss control early in the water loss control program in order to quickly generate recoveries that can seed further water loss reduction activities, particularly real loss reduction. Loss control in almost any endeavor is an effort of diminishing returns, but it is likely that many water utilities have significant apparent losses which can be cost-effectively recovered to enhance the utility's revenue stream and further promote the water loss control program.

References

1. Environment Canada. "Metering.," The Management of Water. [Online]. Available: www.ec.gc.ca/water/en/manage/effic/e_meter.htm.
2. American Water Works Association. "Water Meters—Selection, Installation, Testing, and Maintenance." *Manual of Water Supply Practices M6*: AWWA, 1999. ISBN 0-58321-017-2

3. American Water Works Association. "Sizing Water Service Lines and Meters." *Manual of Water Supply Practices M22. Denver, Colo.*: AWWA, 2004. ISBN 1-58321-279-5

4. Alegre, H., W. Hirner, J. Baptista, et al. "Performance Indicators for Water Supply Services." *Manual of Best Practice Series*: London: IWA Publishing, 2000. ISBN 1 900222 272

5. Kunkel, G., J. Thornton, D. Kirkland, et al., "Water Loss Control Committee Report: Applying Worldwide Best Management Practices in Water Loss Control." *Journal AWWA*, 2003.95(8):65.

6. American Water Works Association. "Manual of Water Supply Practices M36." *Water Audits and Loss Control Programs*, 3rd ed., Denver, Colo.: AWWA 2008.

7. Fanner, V. P., Thornton, J., Liemberger, R., et al., *Evaluating Water Loss and Planning Loss Reduction Strategies*. Denver, Colo.: AwwaRF and AWWA, 2007.

CHAPTER 12

Controlling Apparent Losses—Customer Meter Inaccuracy

George Kunkel, P.E.

Julian Thornton

Reinhard Sturm

12.1   Customer Meter Function and Accuracy

Metering production flows and customer consumption is standard practice in many water utilities throughout the world. Even in countries where metering is not universal, such as the United Kingdom, there is a strengthening movement to make customer metering standard practice. The role of metered data is also increasing due to improved technology to record, communicate, and archive the data. While customer meters continuously register water flowing through them, meter readings are traditionally gathered on a periodic basis to determine water consumption over a 30- or 90-day period for billing purposes. Rapidly developing technologies are now being used in many systems to gather customer-metered data more frequently, or continuously, via datalogging systems or fixed network automatic meter reading (AMR) systems. In fixed network AMR systems, customer consumption can be recorded every few minutes, giving the water utility a detailed profile of the consumption variation throughout the day. Such granular data can be used to indicate leakage in customer premises, to develop water consumption profiles to assist hydraulic modeling calibration and a number of other operational purposes. Given these multiple uses of customer-metered data, in addition to its fundamental purpose of generating accurate water bills, it is critical that the meter population be maintained at a high level of functionality and accuracy.

Managing a large population of customer meters requires knowledge of meter and meter reading equipment as well as billing policies and customer relations. Policy and procedures regarding the sizing and installation of customer meters also play a role in water supply efficiency and these should be reviewed to ensure that inappropriate meters are not installed inadvertently due to policy shortcomings. The benefits of accurate customer metering, however, continue to evolve as consumption data is recognized as critical to evaluate conservation programs, loss control efforts, and economic efficiency.

Many highly accurate brands of meters are available to the drinking water industry. Installation and upkeep of meters should be included as part of the ongoing functions of the water utility, therefore funds should be budgeted to accommodate regular testing and rotation of customer meters. Implementing a program that routinely tests groups of customer meters is an efficient and economical way to keep a meter population current, and provides essential data to develop a rational long-term meter change-out plan for the customer meter population.

12.2   Customer Meter Demographics and Consumption Record

Water utilities that employ best management practices for meter management usually have a thorough understanding of their customer meter demographics and the accuracy of the different meter types in their system. Many water utilities, however, are not current with the status of their meter population. It is not uncommon for an incoming water utility manager to inherit a meter population that was installed 15, 20, or 25 years ago but hasn't experienced ongoing meter testing, rotation, or right-sizing. In many such cases, the size, type, make, and performance of the meter population are poorly documented. The important first step in this case is to compile existing customer account and meter data to establish the basic demographics and accuracy levels of the meter population.

Meter demographics: If the customer meter population characteristics are not well known, the auditor can conduct research using purchase and installation records, billing records, customer complaint histories, and meter accuracy test results to compile information on the sizes, types, brands, ages, and cumulative consumption levels of customer meters. Additionally, new procedures can be instituted to require customer service and/or meter service workers to gather specific meter and account information at customer sites as they conduct their work assignments; this information can be input into the data archival system. Table 12.1 is a summary table displaying the customer meter demographics for the fictitious County Water Company (CWC).[1] Reports can be generated in a manner similar to this table to display the characteristic of the meter population.

Since meter technology is always improving, new types and models of meters are frequently introduced to the water market. Many water utilities purchase meters in lots in a competitive bidding process and, over long periods of time, gradually install a variety of makes and models in their system, particularly in the large customer meter classes. It is important that the auditor have a reasonable sense of the meter population demographics in order to formulate a sound meter testing, right-sizing, and rotation strategy.

In addition to the meter demographics shown in Table 12.1, consumption summaries are a useful management tool to track metering trends and note any unusual patterns. Table 12.2 gives the summary of consumption for County Water Company for calendar year 2006.[1] The total consumption in each customer class is tallied and shown in a monthly breakdown. It is important that water utility managers monitor consumption

Meter Size, (in)	Number of Meters	Percent of Total Meters	Type (No.)	Manufacture (No.)	Ave Age (yrs)	Percent of Metered Consumption
5/8	11,480	94.1	PD* (11,480)	Badger (11,480)	13	71.2
3/4	10	0.08	PD (10)	Rockwell (10)	26	0.1
1	338	4.4	PD (338)	Badger (250)	18	2.8
				Neptune (88)	11	
1 1/2	124	1.0	PD (124)	Badger (18)	18	2.8
				Neptune (106)	9	
2	216	1.8	PD (216)	Rockwell (54)	28	11.7
				Badger (146)	22	
				Neptune (16)	20	
3	15	0.12	Turbine (15)	Sensus (15)	15	6.6
4	7	0.05	PD (2)	Sparling (2)	26	2.2
			Turbine (5)	Sensus (5)	15	
6	6	0.05	Turbine (2)	Sensus (2)	15	2.6
			Compound (2)	Sparling (2)	29	
			Propeller (2)	Hersey (2)	40	
Total	12,196	100.00				100.0

* PD—Positive displacement.
Source: Ref. 1

TABLE 12.1 Customer Meter Population Demographics and Metered Consumption for County Water Company: January 1 to December 31, 2006

2006 by Month	Residential (million gal)	Industrial (million gal)	Commercial (million gal)	Metered Agriculture (million gal)	Total for all meters (million gal)
January	146.6	35.8	8.1	0	190.5
February	162.9	35.8	8.1	0	206.8
March	162.9	35.8	8.1	0	206.8
April	179.2	39.1	8.1	24.4	250.8
May	211.8	42.4	8.1	57.0	319.3
June	228.1	48.9	8.1	74.9	360.0
July	260.3	48.9	8.1	57.0	374.3
August	266.5	48.9	8.1	74.9	398.4
September	228.1	45.6	8.1	65.2	347.0
October	162.9	35.8	8.1	0	206.8
November	162.9	35.8	8.1	0	206.8
December	146.6	35.8	8.1	0	190.5
Annual total	2,318.8	488.6	97.2	353.4	3,258.0
Daily average, MGD	6.35	1.34	0.27	0.97	8.93

Source: Ref. 1

TABLE 12.2 Metered Water Consumption by User Category for County Water Company patterns using tables such as Table 12.2. Consumption data should be carefully tracked on a monthly and annual basis in order to detect data anomalies as they become evident. A table similar to Table 12.2 can be constructed showing monthly consumption totals broken down by meter size.

12.3 Flow Measurement Capabilities of Customer Water Meters

In general, meter accuracy is influenced by two principal factors: the physical performance of the flow sensing mechanism of the meter, and the appropriate sizing of the meter to fit the customer's consumption profile.

Water utilities provide service to a wide variety of customers, from residential service (5/8-in meters typically in the United States) to large industrial sites (up to 12-in meters). Many accurate and reliable meter types exist to measure flows in this variety of settings; each with distinctive features or advantages in performance. Displacement type meters, as shown in Fig. 12.1, are most common for smaller, residential service. Compound, turbine, or propeller meters are employed to serve large commercial or industrial connections larger than 1 in. Turbine meters are designed to accurately record flows that occur steadily in a moderate to high rate of flow. Compound meters are designed with two registers to record flows that alternate between high and low

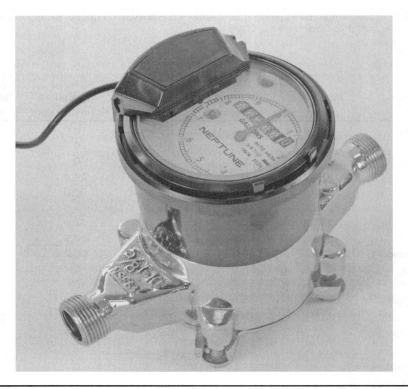

Figure 12.1 Displacement meter for residential service. (*Source:* Neptune Technology Group.)

FIGURE 12.2 Compound meter with dual registers used to meter consumption that varies between high and low rates of flow. (*Source:* Neptune Technology Group.)

levels. A compound meter is shown in Fig. 12.2. Fire connections should be metered separately with appropriate fire meters that do not restrict flow. A fire service meter with bypass line is shown in Fig. 12.3. Technology is always advancing with single jet meters, an example of a more recent innovation. Most meters available on the commercial marketplace provide good accuracy for a given application. However, any type or

FIGURE 12.3 Fire service meter with bypass piping. (*Source:* Neptune Technology Group.)

brand of meter can suffer a loss of accuracy due to a variety of reasons. Some of the common causes of loss of meter accuracy include:

- Incorrect installation, particularly meters installed vertically or askew
- Build-up of scale or deposits due to aggressive water quality
- Debris in the water
- Air entrained in the piping/meter
- Excessively high or low velocity of the flow through the meter
- Manufacturing defects
- Extreme environment: high or low temperature, humidity, vibration, and the like
- Vandalism or destruction

Properly installing appropriate meters and maintaining them by testing and rotation should ensure a high level of accuracy of the customer meter population.

Even under the best of conditions meters wear from long-term flow registration and eventually reach a threshold beyond which they will appreciably lose accuracy, some meter types deteriorating more quickly than others. Therefore meters must be tested, repaired, or replaced with new or refurbished meters (meter rotation) on a structured basis.

Historically, AWWA guidelines recommended that water meters be rotated on a set time schedule based upon meter size with small 5/8-in meters every 20 years and the largest of meters rotated as often as every 4 years. This approach has merits in terms of planning for mass deployment of meter rotation personnel and commensurate budgeting, planning, and so on. However, water meters experience different consumption patterns and, after 20 years of service, some may have lost appreciable accuracy, while others can offer many more years of reliable service. Rotating customer meters based upon fixed time intervals may have significant economic drawbacks, particularly in the large meter classes since these meters are expensive and require much more effort to rotate than small meters.

The current thought on meter rotation strategy bases meter rotation scheduling on the cumulative water volume that has passed through the meter, rather than a fixed time interval. Cumulative flow registered by a meter is the most important factor in long-term accuracy of the meter. Targeting meter rotations based upon cumulative measured volume is similar to automobile maintenance, where the 3,000 mi oil and filter change occurs not at any set time, but only when the 3,000 mi odometer reading is reached. This approach can be more efficient since heavily used meters will see a timely rotation that will ensure accuracy is maintained, while lightly used meters will not waste resources by rotating the meters too soon. Decisions regarding meter rotation based upon cumulative consumption should be formulated in conjunction with crew deployment scheduling realities, since it may be advantageous to have crews rotate multiple meters in a given area all at the same time, even if some of the meters have not yet reached their cumulative volume target. Small meter rotation scheduling may be best guided by a combination of cumulative volume target and geographic proximity, while large meter rotations are perhaps better formulated around cumulative volume targets and the characteristics of the individual meters and consumption profiles.

Davis describes an assessment conducted for the Metropolitan Domestic Water Improvement District, a small water supplier serving communities northwest of Tucson, AZ.[2] The methodology included meter testing on randomly selected and high-cumulative volume residential water meters. Meter accuracy was plotted versus cumulative volume for individual low-, medium-, and high-flow rates. The best linear fit of the data was determined and the weighted meter accuracy was plotted versus the cumulative volume. Calculated lost revenue from meter inaccuracy per year was plotted versus the cumulative volume and economic analysis was used to determine the optimum cumulative volume for meter replacement. For the District, the optimum cumulative volume was determined to be 1,420,000 gal per residential meter. Prior to the assessment, the district was replacing customer meters at the relatively frequent interval of every 10 years. Many of the district's customer meters do not achieve 1,420,000 gal of cumulative volume in 10 years, therefore the district was able to implement a meter rotation strategy that greatly improved the cost-effectiveness of its customer meter management.

> **T**argeting meter rotations based upon cumulative measured volume is similar to automobile maintenance, where the 3,000 mi oil and filter change occurs not at any set time, but only when the 3,000 mi odometer reading is reached.

12.4 Customer Meter Sizing

Water meters must be properly sized in accordance with the actual customer consumption patterns in order to accurately register the flows at all levels of consumption. Historically, water utilities sized customer service connections and meters based upon the peak flow rates that the meter was expected to encounter. Since peak flows occur only on rare occasions, most of the time meters sized in this manner registered flows in the low end of their design range. Many meter types are less accurate in the low end of their flow range with very low flows not captured at all. Current wisdom focuses on sizing the meter to accurately capture the flow range most usually encountered, not seldom-occurring peak flows. Many water utilities have recovered considerable water and revenue by *right-sizing* oversized customer meters. Between 1990 and 1992, for example, the Boston Water and Sewer Commission's meter downsizing program recovered over 100,000 cubic feet of additional water per day in apparent water loss, which translated into millions of dollars in subsequent additional billings and revenue.[3]

Data-logging technology and fixed network AMR technology (discussed in Chap. 13) provide the means to obtain detailed customer consumption profiles in increments of minutes or hours for periods of days, weeks, or months. By using this detailed data, meters can be sized to fit the individual consumption profiles of customers. Applying this user-specific approach can promote superior meter accuracy, particularly in large water utilities with widely varying user classes. As described in the AWWA M22 publication *Sizing Water Service Lines and Meters,* accurate data-logging for meter sizing is dependent on the resolution of the data.[4] Data resolution is a function of the water volume per pulse logged and the data storage interval. Both should be as small as possible so that actual flow rates are recorded, as opposed to just a collection of average flow rates, which may not accurately reflect the consumption profile. Examples of

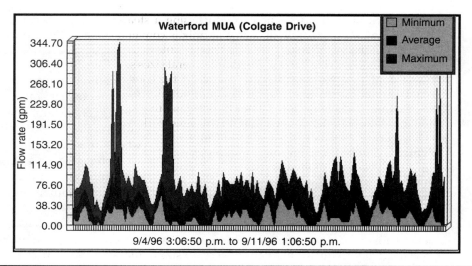

FIGURE 12.4   Graph produced from customer consumption meter data-logging showing minimum/average/maximum flow rates. (*Source:* F. S. Brainard & Co.)

customer consumption profile graphs derived from data-logging are given in Figs. 12.4 and 12.5.

If large meters have been in service for many years, current customer flows may not match the water demand variation occurring just after the meter was installed. Low flows may not be registered by some large, old meters and data-logging may prove the need to downsize the existing meter to an appropriate size. In regions with changing demographics and economies, customer consumption patterns can change significantly and this can affect water meter accuracy. For example, a 6-in turbine meter that reliably

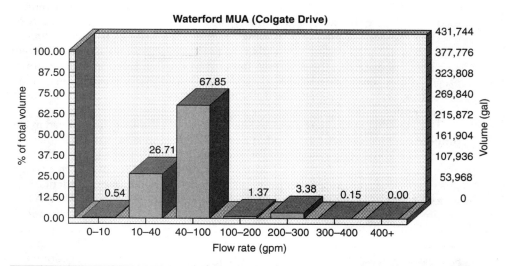

FIGURE 12.5   Graph produced from customer consumption meter data-logging showing minimum/average/maximum flow rates. (*Source:* F. S. Brainard & Co.)

measured consumption in a small factory using a steady volume of water becomes much less accurate when the factory building is converted to office space with much lower water consumption. The consumption profile in the office setting will likely motivate a switch to a smaller meter—perhaps several sizes smaller—in order to ensure that flows throughout the high and low ranges of the consumption profile are measured accurately. In order to determine whether or not meters are properly sized for existing customers, consumption profiles for a representative sample of large meter accounts should be obtained via data-logging or fixed network AMR. Data-logging devices can be attached to the customer meter and record individual meter pulses in order to develop a detailed customer consumption profile showing consumption variation at short time intervals. Meters with consumption consistently occurring in the low range of the meter suggest that the existing meter is oversized and downsizing would be beneficial to more accurately register the total flow. Figure 12.6 presents graphically customer meter test data gathered under a wide range of flows.[5] As shown, meter error increases rapidly at very low flow rates. At very high flow rates the meter can underperform due to excessive wear. The shaded area on the graphic represents flow rates that should be avoided in selecting the proper size of the meter.

When obtaining customer consumption data to develop a usage profile, recognize that it is very important not to base the decision only on 24 hours of data. A customer's consumption can vary greatly on a daily, weekly, or seasonal basis. Care should be taken to locate seasonal use information and also to understand the type of consumption for each specific case. Data should be gathered for at least several consecutive days, preferably 1 week. Separate weekly data collection periods may need to be scheduled in order to obtain consumption data from high- , medium- , and low-demand seasons.

Residential properties in warm climates often incur a significant seasonal increase in water consumption that reflects the hot weather and irrigation needs of residential landscapes. It is not unusual for more than 50% of warm climate residential consumption in industrialized nations to occur from outdoor irrigational use. Yet the high outdoor irrigation demand may only occur for 4 to 6 months out of the year. Similar swings in consumption might also occur in vacation properties that are unoccupied in the off

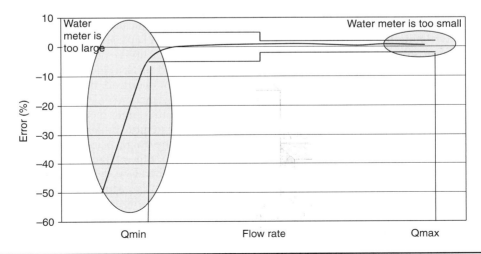

FIGURE 12.6 Range of appropriate sizing of a customer consumption meter to ensure necessary accuracy. (*Source:* Ref. 5)

season and heavily utilized during the peak season. Care needs to be taken when gathering consumption data so that consumption profile(s) are obtained to reflect the variations in demand that the customer property will incur. In many communities water consumption is notably higher during the warm or hot months of the year. Much of this increased water consumption goes to outdoor irrigation, but additional showers and bathing also occur during this time of year. Peak period consumption patterns can have a big impact on any potential meter sizing decision. When looking at consumption profiles in vacation or resort areas, obvious care has to be taken with the season. When considering the consumption profile of a large apartment block during winter and summer, the occupancy rate could change from 10 to 100%. Again the volume used will change dramatically, however most of the use will be at peak times, as people prepare for the day or evening ahead.

Large meters (1-in diameter and larger) are typically installed in multiunit residential buildings as well as commercial, industrial, or agricultural settings. Water demand profiles can vary widely among the different types of building uses and/or manufacturing processes that occur in some of these properties. Normally the largest variations are seen in commercial or industrial properties between weekday use and weekend use with minimal weekend consumption since business is closed during this period. Seasonal consumption variations depend upon the type of manufacturing or business process. Certain manufacturing processes may incur steady water consumption throughout the day, and maintain this pattern continuously. Other processes may utilize large quantities of water in batches, with high volume flows alternating with periods of minimal use. Some factories shut down processes during nights and weekends or may close for several weeks during holiday periods. The water utility manager should inquire about the water usage patterns of a particular facility before determining which periods of time to analyze using customer water consumption profiles.

> **S**easonal use variations of customer properties should be carefully checked when sizing meters.

The economics of meter right-sizing must also be taken into account. The water rate or tariff structure of most North American water utilities includes several component charges. A water charge is typically based upon consumption, with variations for class of customer as well as for volumes of consumption; typically increasing block charges, declining block charges or other billing structures. A separate waste water charge, or even storm water charge, may be included for water utilities that provide these additional services. Most water utilities also assess a fixed service charge to cover the administration expenses of metering, billing, and other overhead functions. Many water utilities base the service charge on the size of the customer meter, with the charge increasing dramatically with meter size. In downsizing from a larger meter of poor accuracy to a smaller meter of high accuracy, the water utility can more reliably capture the volume of water consumption and increase revenue from the usage charge. However, in reducing the size of the meter, the water utility could lose some revenue due to a smaller service charge. The net change in revenue to the utility, therefore, depends upon the amount of recovered revenue due to improved accuracy, offset by reduced service charges. Each customer account being considered for downsizing should, therefore, be carefully reviewed to determine the exact economic impact to the water utility. The Greater Cincinnati Water Works reported on the success of a structured large meter downsizing effort, but noted the dilemma of downsizing certain large meters when anticipating a net loss of revenue due to a significantly lower service charge.[6]

Meter downsizing decisions should be approached with sensitivity when it appears that a reduced service charge makes the downsizing decision uneconomic for the water utility. Keeping a customer account with an oversized meter means that the recorded flow remains understated and the apparent loss is not reduced. Also, it becomes evident that the customer is paying a higher service charge than necessary since they could function (more accurately) with a smaller water meter. Water utilities that specifically avoid downsizing in this manner risk customer dissatisfaction should this information reach the customer. If many customers perceive that they are being overcharged by the water utility, a public relations backlash could result in negative media attention or fines if such actions violate any regulations. If the water utility manager maintains a "big picture" perspective of the value of meter accuracy and apparent loss reduction, he or she can tolerate the uneconomic downsizings of some large meters in order to provide equity to its customers and strive to optimize its apparent loss reduction from meter sizing improvements.

Meter right-sizing initiatives typically address large meters in settings where customer consumption patterns have changed due to building occupancy changes, or where an inappropriately selected or sized meter was originally installed. However, accurate and reliable small meters also incur low flow limitations in which a portion of flow is not registered. No meter is 100% accurate. While most meters have limitations only at very low flow rates, such unregistered flows can occur in hundreds or thousands of customer meters in a water utility, therefore the cumulative volume of unmeasured water can be significant. A common occurrence in North America is of flows *below detectable limits* (BDL) occuring from toilet leaks. Slight leaks in toilet flapper valves allow a continuous trickle of water to pass into the toilet and drain to waste. It is very common that these flows are so slight as not to be registered by many reliable brands of water meters. A similar low-flow condition has been documented in Europe in communities where it is common for individual buildings to have small roof tanks. The slow closing of ball valves included in the roof tanks results in flows that are lower than the starting flow of the customer water meter. One device that has been created to address metering low flow limitations is the unmeasured-flow reducer (UFR) which changes the flow regime passing through the water meter to batches that the water meter can measure.[7] In this way, only flow rates that are sufficient to be registered by the meter are passed through the meter.[8] Innovations in meter technology, data-logging, AMR, and devices such as the UFR continue to offer water utilities the means to measure water consumption with ever-improving accuracy. It is incumbent on the water utility manager, however, to assess the overall accuracy and reliability of their customer meter population and seek to improve where needed.

12.5   Developing the Customer Meter Accuracy Testing Program

In order to assess and maintain good physical accuracy of the customer meter population, many water utilities operate their own meter test facility and equipment, and perform ongoing accuracy testing of meters that have been rotated out of service. For these operations, testing of targeted groups of meters can be readily accommodated. Water utilities that do not have their own facilities can outsource their testing to specialty companies.

Total customer consumption meter error includes meter errors from all meter sizes, including residential, industrial, commercial, agriculture, and others. In general, meter error can be assessed for small meters (5/8 in and 3/4 in), which are typically employed for residential use, and all other (larger) meters which include industrial, commercial, agricultural, and meters for other applications. Testing can serve both the general purpose of providing information to the water audit on the system-wide level of apparent loss due to customer meter inaccuracy, and to identify the accuracy of individual meters, thereby allowing meter improvements to be implemented where needed.

AWWA's guidance manuals on meters give excellent instruction on meter accuracy testing. These include the M22 publication and the M6 publication, *Water Meters— Selection, Installation, Testing, and Maintenance*, the latter of which provides comprehensive information on the basics of customer meter management.[9] Generally, accuracy tests should be conducted at low, medium, and high flow rates. For small, residential meters sample groups of meters can be tested. A randomly selected sample of several dozen to several hundred meters (depending on the size of the meter population) can be selected and tested. A separate sample of meters with high cumulative consumption should also be tested. Results of the latter testing can help to develop a long-term meter change-out strategy based upon the level of cumulative consumption at which accuracy begins to decline.

Because there are hundreds or thousands of customer meters in a drinking water utility, it is impractical to inspect and test every one each year. Instead, the water utility manager can identify sample numbers of customer meters of various sizes and types for inspection and testing. The results of such sample tests give a reasonable indication of the status of the entire customer meter population.

Residential (small) meter testing: Many utilities operate meter testing and rotation programs. Particularly for small meters, it has become more cost-effective to replace meters than to repair them. Random or specific testing to determine the accuracy of installed customer meters can be conducted to monitor the wear of meters. A representative sample of newly purchased residential meters should also be tested to confirm the acceptability of the newly delivered meters. All of this test data represents a good source of information to infer the overall degree of inaccuracy existing in the customer meter population. In this way the level of apparent loss in the system can be quantified for the water audit. Test a random sample of residential meters, 50 to 100 is a sufficient number, but the optimal number to be tested depends upon the size of the customer meter population, the degree of confidence required in the test results, and the variance in the actual test results observed. Residential meters may be tested on a test bench or sent to the factory or a testing service contractor for testing.

Tables 12.3 to 12.5 give an example of calculations using small meter accuracy test data to determine the level of apparent loss from small meter inaccuracy included in the water audit for County Water Company (AWWA 2008).[1] Weighting factors for small meter flow rates are given in Table 12.3. The weighting factors reflect common percentages of time that flows are found in the low, medium, and high flow ranges, respectively, with flows existing most often in the medium range for most properly sized meters.[10] In the example 50 randomly selected residential meters are tested for low, medium, and high flows with summary test results shown in Table 12.4. These results, shown as a percentage of accuracy, are used to calculate the total meter error at average flow rates. Table 12.5 demonstrates how to use existing meter test data to calculate total residential meter error. The resulting residential (small) meter error for County Water Company is given at the bottom of Table 12.5 as a value of 134.33 million gal for calendar year 2006.

Industrial/commercial (large) meter testing: Large industrial, commercial, and agricultural meters register a much greater portion of consumption and produce a much larger share of revenue per account than do residential meters. For many water utilities over

Percent of Time	Range, gpm	Average, gpm	Percent Volume*
15	Low 0.50–1.0	0.75	2.0
70	Medium 1–10	5.00	63.8
15	High 10–15	12.50	34.2

* Percent volume refers to the proportion of water consumed at the specified flow rate, as compared to the total volume consumed at all rates. In this example, only 2.0 % of the total water consumed occurs at the low-flow range of approximately 0.5–1.0 gpm.

    Instead of using the percentage of volumes shown here, you may compute your own percentage volume data. Using special dual-meter yokes and recording meters, you can determine the actual flow rates for your water meters.

Source: Ref.1

TABLE 12.3   Weighting Factors for Flow Rates Related to Volume Percentages for 5/8- and 3/4-in Water Meters

Test Flow Rates	Mean Registration, percent
Low flow (0.25 gpm)	88.8
Medium flow (2.0 gpm)	95.0
High flow (15.0 gpm)	94.0

Source: Ref. 1

TABLE 12.4   Mean Meter Testing Data from a Random Sample of 50 Meters for County Water Company

Percent Volume* (%V)	Total Sales Volume† (Vt) milion gal	Volume at Flow Rate (Vf) (%V × Vt) milion gal	Meter Registration (R)‡ percent	Meter Error (ME) ME = Vf/(0.01R) – Vf milion gal	Meter Error (ME) milion gal
2.0	2,318.8	46.38	88.8	[(46.38/0.888) – 46.38]	5.85
63.8	2,318.8	1,479.39	95.0	[(1,479.39/0.95) – 1,479.39]	77.86
34.2	2,318.8	793.03	94.0	[(793.03/0.94) – 793.03]	50.62
Total Residential meter error (line 8)..					134.33

* From Table 12.3.
† Based on residential water sales data in Table 12.2.
‡ From Table 12.4.
Source: Ref. 1

TABLE 12.5   Calculation of Residential Water Meter Error

50% of revenue is received from less than 20% of customers accounts with large meters. Therefore it is critical that these accounts are systematically reviewed to ensure that they are being metered and billed correctly. Large meters should be inspected for proper selection and sizing before installation. Additionally, large meters should be tested for accuracy before they are used, since not all new meters are sufficiently accurate. In the United States, meters sized 1 in and larger are typically considered to be large meters, although the specific size convention can vary from one utility to another.

All water utilities, regardless of their number of customer accounts, should strive to regularly inspect, test, and confirm appropriate sizing for the relatively small number of meters serving the largest of water consumers. These meters provide the basis for the largest billings in the water utility and every effort must be made to keep them accurate. Inspecting and testing the top 10 largest users in the system on an annual basis will help ensure that optimal customer billings are occurring. Ideally, a representative segment of the large meter population should be tested each year, including 1-, 1½-, and 2-in meters, a mid-range that sometimes is overlooked by utilities.

Tables 12.6 to 12.8 illustrate the use of meter test data to calculate total large meter error.[1] The mean registration data in Table 12.6 are used to calculate the meter error for large meters. The actual test results are shown in Table 12.7 and the resulting large meter error for County Water Company is shown in Table 12.8 as 29.97 million gal. The results of the individual large meter tests can be used to estimate the amount of revenue to be gained by improving the function of large meters by applying the appropriate cost factor.

12.5.1   Customer Meter Accuracy Testing: Methods and Procedures

Most water meters are mechanical devices. As such they wear and lose accuracy after an extended period of operation. Unfortunately, many water utilities do not carefully track the overall accuracy of their customer meters, resulting in unchecked, growing apparent losses and their negative impacts. Small meters of size less than 1 in are usually applied in residential applications and have distinct advantages in testing since one worker can easily remove and replace (rotate) the old meter and test it away from the customer location: at the water utility test bench or that of a meter testing contractor. By using this approach, water utilities ensure speedy service to their customers at their premises and accurate testing of meters at a controlled testing site. Many water utilities have moved away from testing small water meters at the installation site, just as they have moved away from repairing these meters. Old meters rotated out of customer properties and tested at the utility test facility provide meter accuracy data that allows

Flow Rates	Percent of Volume Delivered
Low	10
Medium	65
High	25

* For this example, assume flow recordings were made for 24 hours in July and February to derive the percent of volume registered by large meters at low, medium, and high flow rates.
Source: Ref. 1

TABLE 12.6   Volume Percentages for Large Meters for County Water Company*

Meter ID Number	Size (in)	Meter Type	Date of Installation	Manufacturer	Test Date	Mean Registration at Various Flow Rates: (Designated as Percent of Registration)		
						Low	Medium	High
XYZ001	3	Turbine	June 1991	Sensus	Oct 2004	89	93	100
XOOZAA	3	Turbine	June 1993	Sensus	Oct 2004	70	95.2	98
NB123	4	Displacement	July 1980	Sparling	Oct 2004	95	99	102
NB456	6	Compound	Sept 1977	Sparling	Oct 2004	98	96.5	102
AA002	6	Propeller	May 1966	Hersey	Oct 2004	98	99	103
Sum of mean registrations...........						450	482.7	505
Mean registration for five meters tested.........						90	96.54	101

Source: Ref. 1

TABLE 12.7 Meter Test Data for Large Meters for County Water Company

185

Percent Volume* (%V)	Total Sales Volume† (Vt) million gal	Volume at Flow Rate (Vf) (%V × Vt) million gal	Meter Registration (R)‡ percent	Meter Error (ME) ME = Vf/(0.01R) − Vf million gal	Meter Error (ME) million gal
10	939.2	93.92	90.0	[(93.92/0.90) − 93.92]	10.43
65	939.2	610.48	96.54	[(610.48/0.9654) − 610.48]	21.86
25	939.2	234.80	101.0	[(234.80/1.01) − 234.80]	−2.32
Total Meter error for large meters (line 19)...					29.97

* From Table 12.6.
† From Table 12.2 sum of industrial, commercial, and agricultural metered consumption.
‡ From Table 12.7.
Source: Ref. 1

TABLE 12.8   Calculation of Large Water Meter Error

the utility to keep statistics on accuracy levels versus the cumulative volume of water registered for various brands and sizes of meters. Conversely, complicated logistics usually require large meter accuracy to be carried out at the customer location. Due to the significant portion of water consumption billings that are generated by large customer meters, a formal large meter-testing program is recommended for the water utility maintenance program. Many water utilities have published accounts documenting that increased revenue and water accountability gains have substantially offset the initial investment and continuing costs of such testing programs.

It is important to keep detailed records of meter account histories and accuracy test results obtained at the various flow rates. For an on-site test, remember to record the meter's registration before and after the testing so the customer is not charged for the water used during the test.

As discussed in Sec. 12.3, meter accuracy test results and the water rates charged to customers are needed to determine the target meter replacement rate based upon eventual drop in accuracy from high cumulative flows passed through the meter. Each utility should attempt to establish the level of inaccuracy—and commensurate cumulative volume—that prescribes when meters should be repaired or replaced. In order to obtain sufficient data to determine the economic target, a reasonable number of randomly selected and high cumulative volume meters should be selected for testing each year.

The Customer Meter Accuracy Testing Methodology
Meter accuracy testing can be performed on-site at the customer premise or at a testing facility. When testing meters on-site, the methodology is to compare the accuracy of the meter being tested with a calibrated meter tester used in the process. The calibrated meter has its own performance characteristics and is not 100% accurate across its entire flow range and should have an available compensation curve describing this. Meter accuracy tests conducted at a test facility usually offer better validated results since the volume of water passed through the meter(s) being tested flows into a tank of known volume. Therefore the test process is well calibrated, since the volume passed through the meter is known precisely. Photos of the large and small meter test benches of a typical water utility are given in Figs. 12.7 and 12.8.

FIGURE 12.7   Water meter test bench for accuracy testing of water meters of size 3 in and larger.

On-site testing is usually necessary for meters of size 2 in and larger and is recommended for all sizes of current (magnetic flowmeters) and compound type meters. Few meter shops are equipped with sufficiently large tanks to handle the quantities of water needed to test the larger meters. Furthermore, the accuracy of some current and compound meters may be affected by the configuration of pipe and fittings directly ahead of the meter, therefore it is appropriate to test these meters where they exist in service.

FIGURE 12.8   Water meter test bench for accuracy testing of water meters smaller than 3-in size.

Prior to testing it is necessary to know what the typical accuracy curve is for each specific brand, model, and size of meter being tested. This information may be obtained from the meter manufacturer's literature. A local chart can be made up which lists the flow rates at which each type of meter should be tested in order to properly assess its operating condition. Techniques for performing the tests, selecting the appropriate test flow rates, determining the accuracies, and reaching conclusions must be known and carefully followed to obtain valid test results. For positive displacement meters, which are typically the small residential meters, the AWWA M6 publication provides the three flow rates (low, mid, and high), which apply to all meter brands. For turbine and propeller meters, which are used in large meter applications, either the M6 publication or the manufacturer's meter literature should be consulted; an example of the latter is shown in Fig. 12.9. Compound meters are used in large meter applications where the consumption varies from high flows to low flows. These meters have two registers: a

DISPLACEMENT METERS (AWWA C700)

Size in.	Maximum Rate (All Meters)				Intermediate Rate (All Meters)				Minimum Rate (New and Rebuilt) ①				Maximum (Repaired) ①
	Flow Rate gpm	Test Quantity gal.	Test Quantity ft.³	Accuracy Limits percent	Flow Rate gpm	Test Quantity gal.	Test Quantity ft.³	Accuracy Limits percent	Flow Rate gpm	Test Quantity gal.	Test Quantity ft.³	Accuracy Limits percent	Accuracy Limits percent (min.)
5/8	15	100	10	98.5–101.5	2	10	1	98.5–101.5	1/4	10	1	95–101	90
5/8 × 3/4	15	100	10	98.5–101.5	2	10	1	98.5–101.5	1/4	10	1	95–101	90
3/4	25	100	10	98.5–101.5	3	10	1	98.5–101.5	1/2	10	1	95–101	90
1	40	100	10	98.5–101.5	4	10	1	98.5–101.5	3/4	10	1	95–101	90
1-1/2	50	100	10	98.5–101.5	8	100	10	98.5–101.5	1-1/2	100	10	95–101	90
2	100	100	10	98.5–101.5	15	100	10	98.5–101.5	2	100	10	95–101	90

CLASS I TURBINE METERS (AWWA C701)

Size in.	Maximum Rate				Intermediate Rate				Minimum Rate			
	Flow Rate gpm	gal.	ft.³	Accuracy Limits percent	Flow Rate gpm	gal.	ft.³	Accuracy Limits percent	Flow Rate gpm	gal.	ft.³	Accuracy Limits percent
1-1/2	80	200	20	98–102	35	100	10	98–102	12	100	10	98–102
2	120	300	30	98–102	50	200	20	98–102	16	100	10	98–102
3	250	500	50	98–102	75	300	30	98–102	24	100	10	98–102
4	400	1000	100	98–102	125	500	50	98–102	40	100	10	98–102
6	1000	2000	200	98–102	200	500	50	98–102	80	1000	100	98–102
8	1500	3000	300	98–102	300	1000	100	98–102	140	1000	100	98–102
10	2200	5000	500	98–102	500	1000	100	98–102	225	1000	100	98–102
12	3300	7000	700	98–102	700	2000	200	98–102	400	1000	100	98–102

CLASS II TURBINE METERS (AWWA C701)

Size in.	Maximum Rate Flow Rate gpm	gal.	ft.³	Accuracy Limits percent	Intermediate Flow Rate gpm	gal.	ft.³	Accuracy Limits percent	Minimum Flow Rate gpm	gal.	ft.³	Accuracy Limits percent
1-1/2	90	300	30	98.5–101.5	10	100	10	98.5–101.5	4	100	10	98.5–101.5
2	120	300	30	98.5–101.5	10	100	10	98.5–101.5	4	100	10	98.5–101.5
3	275	600	60	98.5–101.5	20	100	10	98.5–101.5	8	100	10	98.5–101.5
4	500	1000	100	98.5–101.5	20	1000	100	98.5–101.5	15	100	10	98.5–101.5
6	1100	2500	250	98.5–101.5	40	1000	100	98.5–101.5	30	1000	100	98.5–101.5
8	1800	4000	400	98.5–101.5	50	1000	100	98.5–101.5	50	1000	100	98.5–101.5
10	3000	6000	600	98.5–101.5	75	1000	100	98.5–101.5	75	1000	100	98.5–101.5
12	4000	8000	800	98.5–101.5	100	1000	100	98.5–101.5	120	1000	100	98.5–101.5

COMPOUND METERS (AWWA C702) (Test at intermediate rate not necessary.)

Size	Maximum Rate Flow Rate gpm	gal.	ft.³	Accuracy Limits percent	Intermediate Rate ② Flow Rate gpm	gal.	ft.³	Accuracy Limits percent	Minimum Rate Flow Rate gpm	gal.	ft.³	Accuracy Limits percent
2	100	100	10	97–103	10–15	100	10	90–103	1/4	10	1	95–101
3	150	500	50	97–103	10–15	100	10	90–103	1/2	10	1	95–101
4	200	500	50	97–103	20–25	100	10	90–103	3/4	10	1	95–101
6	500	1000	100	97–103	25–35	100	10	90–103	1-1/2	100	10	95–101
8	600	2000	200	97–103	35–45	100	10	90–103	2	100	10	95–101
10	900	2000	200	97–103	—	—	—	90–103	4	100	10	95–101

FIRE-SERVICE TYPE (AWWA C703) ③

TURBINE MAIN LINE TYPE WITH BY-PASS ③

Meter Size in.	Minimum Rate (95 percent min. accuracy limit) Flow Rate gpm	gal.	ft.³	Cross-Over Rate (90–103 percent accuracy limit) Flow Rate gpm	gal.	ft.³	Maximum Rate (98.5–101.5 percent accuracy limit) Flow Rate gpm	gal.	ft.³
4	④	100	10	25–35	1000	100	750	2000	200
6	④	100	10	50–60	1000	100	1500	5000	500
8	3	100	10	50–60	1000	100	2500	5000	500
10	3	100	10	55–65	1000	100	4000	8000	800

TURBINE MAIN LINE TYPE WITH BY-PASS ③

Meter Size in.	Minimum Rate (95 percent min. accuracy limit) Flow Rate gpm	gal.	ft.³	Intermediate Rate (98.5–101.5 percent accuracy limit) Flow Rate gpm	gal.	ft.³	Maximum Rate (98.5–101.5 percent accuracy limit) Flow Rate gpm	gal.	ft.³
4	10	1000	100	20	1000	100	750	2000	200
6	20	1000	100	40	1000	100	1500	5000	500
8	30	1000	100	50	1000	100	2500	5000	500
10	35	1000	100	75	1000	100	4000	8000	800

① A rebuilt meter is one that has had the measuring element replaced with a factory-made new unit. A repaired meter is one that has had the old measuring element cleaned and refurbished in a utility repair shop.

② Cross-over flow rates vary depending on meter model and brand. These values are for Sensus (Rockwell) Compound Meters. Consult manufacturers for other brands.

③ The values listed are for Sensus meters only.

④ Flow rate for FireLine 1-1/2 – 3" gpm depending on bypass meter. Flow rate for UL/FM Compact at 3 gpm.

FIGURE 12.9   Test flow rates. (*Source: Water Loss Control Manual*, 1st ed.)

high side and a low side to capture the high and low flows, respectively. For compound meters, it is important to know the level of the "crossover" flow, or the level where flow switches from the high to low register, or vice versa. If the customer consumption rate occurs frequently at flows in the crossover range, poor meter accuracy at this level could result in a great loss of flow registration. Therefore, the crossover flow should be determined and the meter specifically tested at this rate in addition to the high and low flows. The manufacturer's accuracy curve is a proper information source as different brands of compound meters offer variant crossover flow rates. This information is not currently available in the AWWA M6 publication.

The test equipment and methods for determining the accuracy of small meters are not applicable to accuracy tests on larger meters. The larger meters require specialized test equipment which can handle a wide range of flow rates and provide accurate, valid data. These devices may either be purchased as a manufactured assembly or fabricated by the water utility.

The equipment for large meter testing is available as a portable test package, installed on trailers, or mounted in a van or pickup truck. Regardless of the style, these testers all contain certain basic elements, which are required to properly test turbine, compound, and propeller meters. Because of the wide flow ranges involved, a tester includes at least two, and sometimes three, calibrated test meters of varying capacities. A shut-off valve is typically located downstream of each meter to control the flow rate during the various tests. A pressure gage is required to check both the line pressure and the residual pressure at the tester. Sometimes resettable registers and/or flow raters are included to reduce the time required to conduct a complete test.

Flexible hoses are required to connect the test equipment to the test connection of the meter being tested. Due to static pressures and hydraulic forces present, all hoses must be in good condition and positioned as straight as possible between the two meters. For the larger testers, it is important that the tester itself be anchored by means of a vehicle, or similar restraining method, since significant hydraulic forces will impact the meter tester during the test. The master meters used on the testers should be protected and handled with care. They should also be tested and recalibrated periodically to ensure accurate measurement is being maintained.

Unfortunately large meters are often ignored as long as they continue to record consumption. While large meters are usually relatively few in number in water utilities, they account for a significant amount of revenue. If large meters mean so much to a water system's financial health, why are they not maintained to provide peak performance? The explanation is multifold. Large meters are difficult to repair, spare parts are expensive, assemblies are sometimes complex, and a relatively high skill level is necessary for the service personnel to maintain them. The largest sized meters are very heavy and difficult to handle and transport. Maintenance work is hindered by meter installations in crowded or cramped spaces and/or piping compromises have to be made. Many times there is no bypass piping to continue supplying water to the customer during a meter accuracy test, or it is difficult to dispose of the water discharged during the test. Also, work space around the meter may be restricted and unsafe. Liability, safety issues, and span of control of the testing personnel can sometimes be a concern to the system's management. Because many large meters are very important to the overall billings, their operating condition must be monitored on a systematic and timely basis. One common approach is testing large meters on-site by qualified test personnel.

Large customer meters may be tested on-site or at the water utility's facilities. There are certain advantages to testing large customer meters at the water utility test bench;

however, in the majority of situations, on-site testing is more economical in terms of time and resources. From a technical standpoint, the piping configuration surrounding the meter can have an appreciable impact on the meter's accuracy, and such impacts can be detected and evaluated when the testing is conducted on-site.

On-site testing of large meters is often the preferred method as the customer site is tested for suitability as well as the meter for accuracy.

Both on-site and test bench testing of large meters rely on a large volume of water being passed through the meter being tested. When tested on a test bench in the shop, water is passed into a tank of known volume. In conducting on-site testing, the flow registered by the meter being tested is compared to a meter that has been previously calibrated and known to be accurate. The two meters are connected in series, and the test water is discharged to waste. Since the calibrated meter is not 100% accurate on all flows, it may be necessary to adjust for its accuracy variance at different rates of flow, in order to ensure proper test results. One very important point to remember in field-testing is that both meters must be full of water and under positive pressure with all air removed. The control valve for regulating flow, therefore, should always be on the discharge side of the calibrated meter. A valve on the inlet side of the meter being tested or one located between two meters for controlling rates of flow should not be used, as inaccurate results may occur.

One acceptable method of maintaining proper performance for certain types of larger meters is to replace the operating components and assemblies while leaving the meter body in place. For such meters it is also recommended that an on-site meter accuracy test be conducted at the time of installation to confirm that the composite metering unit is functioning as designed. If the measurement and registration functions are within one integral assembly, no accuracy tests are required at the time of installation, and the entire unit must be tested at the regular maintenance intervals.

Some larger meters have built-in test plugs while others do not. For installations requiring test outlets, these can be fabricated in a number of ways. Service saddles and reducing tees are the most frequently used approaches. These need to be installed according to the recommendations of the meter manufacturer and located so that the connecting hose to the on-site tester is correctly located downstream to the meter. To facilitate periodic testing of the meter, it is suggested that, as part of the original installation process, a short length of pipe be permanently attached to the test outlet, along with a shut-off valve, which can be locked into position. These features will allow for quick, efficient testing at regular intervals throughout the life of the large meter.

The piping configuration around the meter must include valves to positively isolate the meter, while still maintaining an adequate flow to the end user through temporary or permanent bypass piping. If either of the isolation valves fails to seal tightly, an inaccurate test result may occur. Similarly, if leakage occurs at either of the valves or at the meter connections, the integrity of the accuracy test may be compromised. The lower the test flow rate, the higher the significance of any such leaks.

Large meter settings are relatively expensive and require considerable preliminary planning. These meters are heavy and removal of the meters for servicing or testing is costly and time-consuming. Therefore on-site testing of large meters is the preferred method in many instances. When small meters are rotated out of service, the water supply to the customer property is halted for the typically brief period of time that is needed to remove the old meter and install the new meter. Such outages are usually easily tolerated by the residential customers supplied by the small meter. This is not the case for large meter customers. Buildings serviced by large meters include factories, hospitals, military installations, shopping centers, and many important facilities that cannot easily tolerate the lengthier water supply interruption that accompanies the replacement of a large meter. Similarly many such buildings must be provided fire service capability with minimal interruptions. Many large meter installations are designed with a bypass line and valve that are used to allow continuous supply water to the customer while the meter is being serviced or tested. For nonfire line meter applications, the bypass should be sized one nominal size smaller than the meter being tested down to the 2-in size. For fire line metering applications the bypass line size should be the same nominal size. The bypass line provides the capability for a customer with critical water supply needs to receive continuous service. Without bypass lines the meters serving these customers cannot be tested or repaired which may result in the loss of significant revenue. Typical bypass line configurations for large meters are shown in Figs. 12.10, 12.11, and 12.12.

Preassembled meter packages are designed to provide the necessary equipment for the complete meter installation and help provide for fast, easy installation. These packages are supplied by most meter manufacturers and are especially valuable to many utilities that may not have the tools or equipment required to handle the installation of large meters.

Most meter manufacturers recommend that a spool piece of piping be installed downstream of the meter. The length of the spool piece should be at least twice the diameter of the pipe. This feature is used to eliminate any turbulent flows on the exit

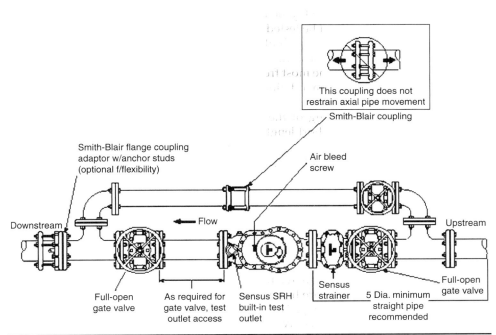

FIGURE 12.10 Installation recommendations for compound meters. (*Source: Water Loss Control Manual*, 1st ed.)

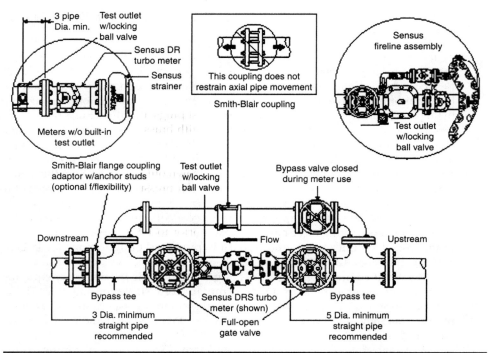

FIGURE 12.11 Installation recommendations for turbo meters. (*Source: Water Loss Control Manual*, 1st ed.)

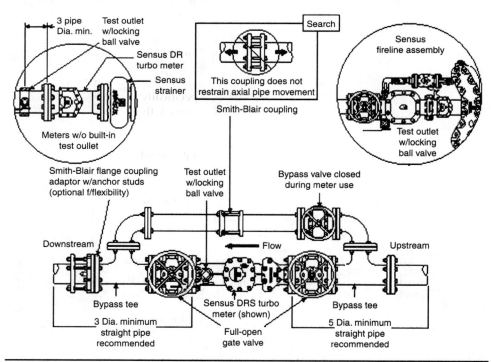

FIGURE 12.12 Installation recommendations for fire line meters. (*Source: Water Loss Control Manual*, 1st ed.)

side of the meter's measuring element. A tagging saddle, brass nipple, and ball or gate valve should be installed on top of the spool piece. These devices are used during the field-testing of the meter. Most compound and fire service meters have test plugs built into the meter casings. Many 4- and 6-in sized meters typically have 2-in test plugs, while 8-in and larger meters often have 3-in test plugs. Prior to installing the meters the test plugs should be removed and replaced with brass nipples and either ball or gate valves to facilitate field-testing of the meters. The ball or gate valve is needed to safely relieve pressure from the meter before opening the main casing. There have been numerous accidents where the test plug has blown out during removal when the main gate valves were leaking and the meter was under pressure. Use great caution as meters can encounter working pressures of well over 100 psi, which can impart destructive forces if weak or corroded fittings fail and are expelled.

Nearly all turbine meters manufactured prior to 1992 offered no test outlet in the meter body and required a separate spool piece and test nipple installation downstream of the meter. When test plugs are fitted in the meter bodies, a separate test tap is not needed. Test outlets typically range in size from 1 to 2 in, depending on various meter sizes. Additionally, some commercially available fire line meter assemblies and compound meters can be provided with a test riser outlet assembly with a locking ball valve and fire hose coupling for proper testing. See Figs 12.10, 12.11, and 12.12 for installation recommendations for compound, Turbo (turbine), and fire line meters.

The Customer Meter Accuracy Testing Process

In conducting meter accuracy testing, it is of critical importance that personnel assigned to perform the tests are properly trained and have the appropriate test equipment. Meter testers are often designated as skilled field specialists or technicians in work specifications, and training should be of sufficient caliber to reflect this skilled trade. Appropriate techniques and procedures should be followed when using test equipment. The consequences of discharging large volumes of water at high flow rates must be understood, appreciated, and considered specifically for each test. Improper use of the equipment may be harmful to testing personnel, the meter, the surrounding area, and the general public. The meter pit must have adequate space in which personnel can operate safely. In the United States, safety requirements published by the Occupational Safety and Health Administration (OSHA) should be followed.

Prior to running any test, determine the make, model, and manufacturer of the meter in question and document this data on the meter test sheet as shown in Fig 12.13.

In planning on-site testing of large meters, the technician must assess a number of factors that are critical in conducting a safe and accurate meter test. A checklist of the steps of the large meter testing process is given in Table 12.9. The technician must carefully identify the impacts of the large volume of water that must be passed through the meter tester to run the test. One thousand gpm is not an uncommon rate to test the largest of meters. Suddenly extracting such a high rate of flow from the water distribution system could reduce the supply pressure in the local water supply grid and/or release debris in the service line to the customer or adjoining water mains. The technician must also assess where to safely discharge the large volume of flow that is passed through the meter during the testing process. An uncontrolled discharge can cause considerable damage to landscapes or private property, or create a safety hazard to vehicular or pedestrian traffic. Discharge water must be safely disposed of in a manner than does not cause any damage or violate any environmental regulations.

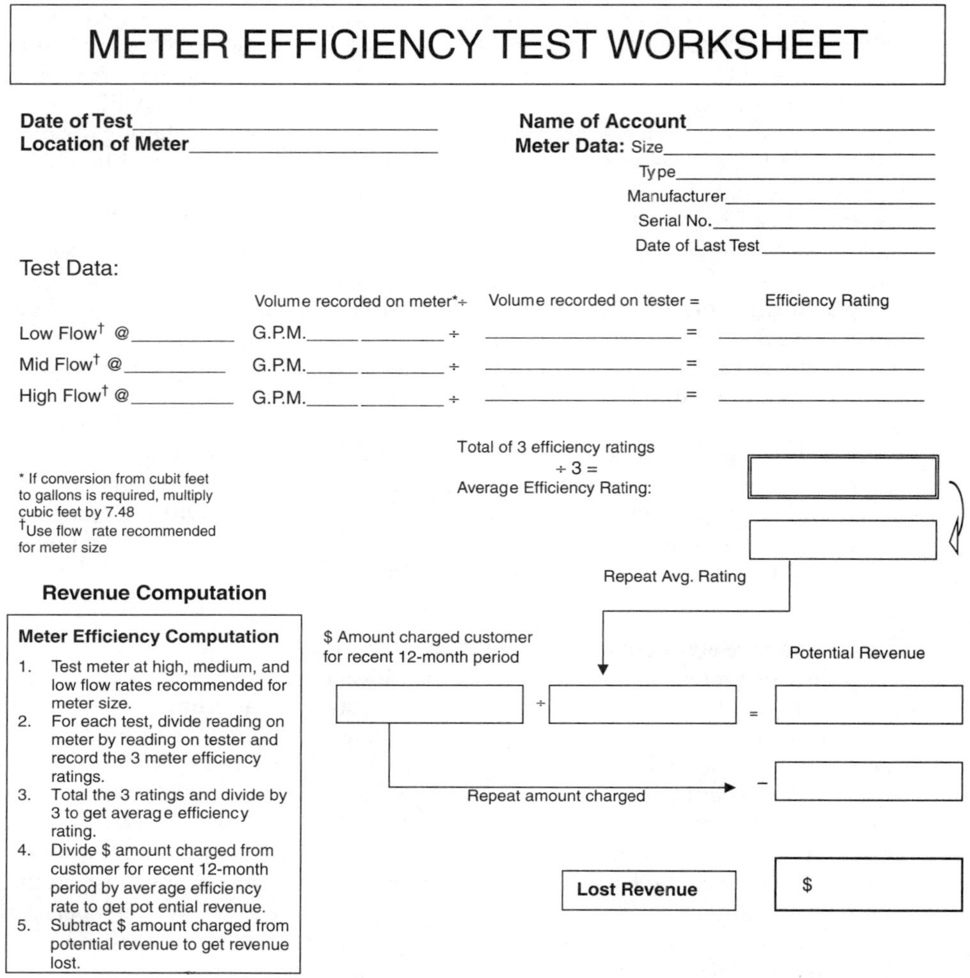

METER EFFICIENCY TEST WORKSHEET

Date of Test_____   **Name of Account**_____
Location of Meter_____   **Meter Data:** Size_____
                                                     Type_____
                                                     Manufacturer_____
                                                     Serial No._____
                                                     Date of Last Test_____

Test Data:

	Volume recorded on meter*÷	Volume recorded on tester =	Efficiency Rating
Low Flow[†] @_____ G.P.M.____ _____	÷	_____ =	_____
Mid Flow[†] @_____ G.P.M.____ _____	÷	_____ =	_____
High Flow[†] @_____ G.P.M.____ _____	÷	_____ =	_____

* If conversion from cubit feet to gallons is required, multiply cubic feet by 7.48
[†]Use flow rate recommended for meter size

Total of 3 efficiency ratings
÷ 3 =
Average Efficiency Rating:

Repeat Avg. Rating

Revenue Computation

Meter Efficiency Computation

1. Test meter at high, medium, and low flow rates recommended for meter size.
2. For each test, divide reading on meter by reading on tester and record the 3 meter efficiency ratings.
3. Total the 3 ratings and divide by 3 to get average efficiency rating.
4. Divide $ amount charged from customer for recent 12-month period by average efficiency rate to get potential revenue.
5. Subtract $ amount charged from potential revenue to get revenue lost.

$ Amount charged customer for recent 12-month period

Potential Revenue

÷   =

Repeat amount charged

−

Lost Revenue $

FIGURE 12.13   Meter efficiency test worksheet. (*Source: Water Loss Control Manual*, 1st ed.)

Care should be taken to select a test meter with sufficient capacity to deliver the high rates of flow required for the maximum flow test rate. It is often necessary to use a lower rate than that set forth for the maximum flow test of the larger meters. In many instances the maximum flow rate may be limited to 500 gpm. This rate is usually sufficient to evaluate accuracy in the high flows of all but the largest meters. Lesser flow rates should be used only as an expedient, and the established test rates should be used wherever possible. It is safe to assume that the test curve will *flatten out* after reaching peak registration, which is approximately 10% of the meter's rated capacity. Stay within the required limits for registration.

Safety is a foremost consideration in conducting large meter accuracy tests as large volumes of water under high pressure are discharged during this process.

Large Meter Testing Checklist	
1.	Adhere to all instructions on the warning tag. Never deviate from the instructions on the warning tag. The warning tag that is affixed to the tester has been developed from extensive product testing in field situations.
2.	Conduct a pretest inspection of the meter, meter pit or chamber, and adjacent area. Are there test plugs? Is there a bypass around the meter? Are there isolating valves on both sides of the meter? If any of these features don't exist, a means must be determined to safely perform the test in their absence, or they must be installed before testing. Identify an adequate area for the safe discharge and run-off of water to be passed through the meter tester. It is not unusual to discharge in excess of 10,000 gal of water in a large meter test segment. Make sure that water will not run back into the meter pit while testing. Be aware of sidewalks and streets where pedestrian traffic may occur. Even a moderate 300-gpm stream of water from a tester can be dangerous to vehicular and pedestrian traffic.
3.	Close the meter's isolation valves. Close both upstream and downstream valves to isolate the meter from line pressure. This must be done prior to removing any test plugs. Take note as to whether the valves operate smoothly or with difficulty.
4.	Bleed pressure from the meter. Bleed all residual line pressure from the meter assembly prior to removing any test plugs. Generally, this is accomplished by loosening a bleed screw found on the meter cover. If a bleed screw is not present, a main flange or drain plug may be loosened to relieve any pressure.
5.	Connect the meter tester. After ensuring that water pressure has been relieved to a safe level, remove the test plug and connect the test pipe, hoses, and meter tester. Be sure that all equipment is laid out across the ground in a straight manner with both hoses (inlet and outlet) having no sharp and/or irregular bends.
6.	Secure the meter tester. If high water pressure or flow rates are expected, chain the tester to a fixed object and/or drive large stakes through the holes in the tester to the ground to secure the equipment, thus preventing unsafe movement.
7.	Inspect hoses and connections and purge air. Inspect the meter tester hoses and connections for tightness and purge air from the equipment by opening the small valve on the tester and slowly opening the meter supply valve. Continue until the equipment is under full pressure and all air is bled from the assembly.
8.	Begin the test from zero flow conditions. Slowly flush all air until maximum flow is reached (maximum flow is achieved when either the valve is wide open or the tester pressure gage drops to 20 psi).
9.	Read and reset the registers and run the maximum flow as previously run for a quantity of at least one sweep of the dial on the meter being tested. Repeat the above sequence by doubling the flow rate used in the first test. Compare the first test's accuracy to that of the second test. The difference should not be greater than +/– 5%. If greater, investigate possible causes for the difference, which might be attributed to: (a) A malfunction with the test meter; running low flows may confirm this suspicion. (b) The tested meter may have a badly worn register causing excessive pointer play. Tapping on the register lens and observing the amount of pointer movement might confirm this suspicion.

Source: Expanded version of Section 14.3.4 Control Manual 1st ed.

TABLE **12.9**   Customer Meter Testing: Procedure and Safety Checklist (*Continued*).

	(c) Air may be trapped in the hoses connecting to the meter tester. Flush the hoses and rerun the tests. (d) One or both isolation valves may be leaking, causing inconsistent tests. Check by looking at the tested meter's register low flow indicator for movement over a 1- to 5-minute period. (e) The strainer is clogged with debris, or partially blocked. (f) The test meter may be clogged with rocks or debris, or may have been damaged during flushing. There are many other causes for inconsistency of test data. Inconsistency problems must be resolved before continuing or the validity of the test results will be questionable.
10.	Continue testing of additional targeted large meters by referring to AWWA M6 Manual for test rates on specific meter types. Use the test flow rates from the meter manufacturer's literature if these are available. It is also recommended to perform one additional test at the average customer flow usage rate. This will provide important information on how efficiently the meter is operating in the primary revenue producing flow rates.
11.	When testing compound meters review the consumption rate at the changeover point with the accuracy of the meter at that flow rate. Determining the exact crossover rate requires use of a pressure gage and rate-of-flow display. Slowly open the rate control valve. When a rise in the pressure gage needle is noted, the flow rate indicated by the register is the crossover rate. If crossover is not detected, close the rate valve until the gage drops back again. Repeat opening and increasing the flow until the crossover rate is identified. This process may take practice but is worth the effort.

Source: Expanded version of Section 14.3.4 *Water Loss Control Manual* 1st ed.

TABLE 12.9   Customer Meter Testing: Procedure and Safety Checklist (*Continued*)

Manufacturers of meter test equipment typically provide detailed procedures for conducting accuracy tests. In general, the tester is connected and the line flushed. As a preliminary step, a brief test should be conducted at a relatively high flow to determine if there are any leaks or unknown taps in the pipeline. The flow rate should be set approximately to 50% of the meter's capacity and the test conducted for 10 sweeps of the dial for adequate resolution. After determining the accuracy of the meter, the test should be rerun for half the volume. The second accuracy test's results should be within 1/2% of the first test. If not, a leak or other uncontrolled flow of water may be compromising the test. If the meter in question has a flow indicator, it may indicate water movement as a result of downstream isolation valve leakage.

When conducting tests, it is suggested that no test be less than 1 minute long and that the meter's sweep hand make at least one complete revolution. The residual pressure on the tester should never be less than 20 psi when running a high-flow test. Also, for safety, the tester should not be operated on lines with static pressure exceeding 80 psi unless provisions are made to secure the tester.

The formal testing sequence should then initiate, first in the low flow ranges and progressing to higher flows. Experience has shown that, when most meters begin to wear, accuracy is first impacted at the lower rather than the higher flows. If a meter is performing accurately up through the lower 25% of its capacity, it will normally test accurately through the rest of the range. This is especially true on very large meters.

12.5.2   Evaluating Customer Meter Accuracy Test Results

Evaluating water meter performance requires both experience and confidence in the operator's skill and training in order to correlate testing results with appropriate corrective actions for a given water meter. Included in the following discussion are examples involving Sensus meters, including Table 12.10, which are provided for illustrative purposes only.

Turbo Meter Evaluation			
Meter Size	**Adj. Vane**	**Test Data**	**Possible Cause**
4" W-1000	+15°	90% @ gpm 97% @ 100 gpm 99% @ 700 gpm	• Broken Rotor Blades • Rotor Bearings and/or thrust bearings worn • Debris caught on blade
4" W-1000	+5°	100% @ 10 gpm 103% @ 100 gpm 105% @ 700 gpm	• Jetting from debris (in strainer or caught on flow strainer) • Installation effects • Air entrapped in line • Coating on rotor and/or chamber
4" W-1000	+30°	94% @ 10 gpm 98% @ 100 gpm 99% @ 700 gpm	• Adjusting vane moved to (–) from original test • Installation effects • Improper repair

This meter could be recalibrated by moving the value to 0°.

Compound Meter Evaluation		
Meter Size	**Test Data**	**Possible Cause**
	Low Flow Tests	
3"SRH	105% @ 0.5 gpm 102% @ 3.4 gpm 98% @ 10 gpm	• Leaking downstream isolation valve
	High Flow Tests	
3"SRH	88% @ 25 gpm 94% @ 55 gpm 99% @ 280 gpm	• Damage to propeller • High flow chamber wear • Coordinator wear • Vertical shaft binding and/or bushing wear
3"SRH	95% @ 0.5 gpm 99% @ 3.4 gpm 100% @ 10 gpm 106% @ 25 gpm 108.7% @ 150 gpm 108% @ 280 gpm	• High flow side geared too high • Debris causing jetting • Installation effects

TABLE 12.10   Evaluating Customer Meter Accuracy Test Results (*Continued*)

Fire Line Meter Evaluation		
Meter Size	**Test Data**	**Possible Cause**
		Bypass Meter
6"Compact fire line	192% @ 4 gpm 96% @ 45 gpm 99% @ 500 gpm	• Broken Rotor Blades • Adjusting vane moved to (−) • Rotor Bearings and/or thrust bearings worn • Debris caught in blade
		Detector Check Valve
		• Worn seat • Debris preventing closure
		Large Meter
6"Compact fire line	100% @ 4 gpm 100.3% @ 45 gpm 105% @ 500 gpm	• Jetting from strainer and/or installation • Adjusting vane moved to (+) • Coating on rotor and/or chamber
6"Compact fire line	100% @ 4 gpm 103.3% @ 45 gpm 101% @ 100 gpm 100% @ 500 gpm	• Leaking downstream isolation valve
		Bypass Meter
		• Adjusting vane moved to (+) • Coating on rotor and/or chamber

Source: Water Loss Control Manual, 1st ed.

TABLE 12.10   Evaluating Customer Meter Accuracy Test Results (*Continued*)

To properly evaluate a tested meter, a high level of confidence is needed in the integrity of the test data. It is, therefore, essential that meter testing procedures are followed carefully in conducting the meter accuracy test. Sensus allows a +/− 1½% accuracy spread on Turbo (turbine) meters, fire line meters, and compound meters tested at normal operating ranges. At low flows and crossover flow rates on compound meters, Sensus allows +1½ to −5% accuracy. These limits are more stringent than AWWA standards. See AWWA standards C701, C702, C703, and C704 for additional guidelines.[11–14]

When meter accuracy test results indicate questionable meter performance, be certain to review whether the test process included testing at low, medium, and high flows (manufacturer's or AWWA recommendations), and confirm a minimum duration of one sweep on the tested meter register. If these conditions were not met during the testing, then the test process should be repeated with particular attention paid to these test requirements.

When reviewing the meter accuracy test results, be mindful to look for:

- Normal operating range tests. Is the minimum flow test 95 to 101.5%? If either one is not within the range, the meter should not be geared or adjusted to meet specifications without repair. A complete meter replacement may be required.

- Turbo (turbine) meters show a loss of registration first at low flows due to bearing wear. Be certain that testing was performed reliably at the low flow rates.
- Compound and fire line meters have a crossover flow. Take time when testing to determine this rate. Evaluate each side of crossover as a cause for failure along with valve problems. Do not attempt to isolate measuring chambers and conduct isolated tests.

It is important to remain objective when interpreting meter test results. As long as you are following proper meter testing procedures, let the test results speak for themselves. Attempt to discern any anomalies by explaining the function and application of the meter and be cautious not to quickly dismiss meter test results as poor testing procedure. Again, rely upon your training and expertise to evaluate and diagnose tested meters; never stop looking, listening, and learning. Table 12.10 gives a listing of potential meter problems that can explain variant large meter accuracy test results.

When meter performance is not consistent, the meter should be inspected for any significant change in the customer water consumption pattern and for any meter malfunction. Actions to remedy a malfunctioning meter might include: repair the meter (for large meters greater than 1 in), replace the meter if necessary with consideration to replacement using a different size meter if necessary. When the performance of a meter is in doubt, particularly if a meter has been in service for a number of years, it is best to replace the meter. The meter is the origin of customer consumption data in the water utility and it is very important that the water utility manager have a high level of confidence in the function and accuracy of the customer meter population.

References

1. American Water Works Association. *Manual of Water Supply Practices, Water Audits and Loss Control Programs (M36).* 3rd ed., Denver, Colo.: AWWA, 2008.
2. Davis, S. "Residential Water Meter Replacement Economics." Leakage 2005 IWA Conference, Halifax, Nova Scotia, CA, 2005.
3. Sullivan, J. P. and E. M. Speranza. "Proper Meter Sizing for Increased Accountability and Revenues." *Proceedings, American Water Works Association, Annual Conference & Exposition:* AWWA, 1991.
4. American Water Works Association. *Manual of Water Supply Practices, Sizing Water Service Lines and Meters (M22).* Denver, Colo.: AWWA, 2004. ISBN 1-58321-279-5.
5. Arregui, F., E. Cabrera and R. Cobacho, et al., "Key Factors Affecting Water Meter Accuracy." Leakage 2005 IWA Conference, Halifax, Nova Scotia, CA, 2005.
6. Grothaus, R. "Size Matters: Meters Can Reflect New Standards." *AWWA Opflow*, March, 2007.
7. Rizzo, A. and J. Cilia. "Quantifying Meter Under-Registration Caused by the Ball Valves of Roof Tanks (for Indirect Plumbing Systems)." Leakage 2005 IWA Conference, Halifax, Nova Scotia, CA. 2005.
8. Cohen, D. "UFR (Unmeasured-Flow Reducer): An Innovative Solution for Water Meter Under-Registration—A Case Study in Jerusalem, Israel." *Global Customer Metering Summit:* London, England. July 2007.
9. American Water Works Association. *Manual of Water Supply Practices, Water Meters— Selection, Installation, Testing, and Maintenance (M6).* 4th ed., Denver, Colo.: AWWA, 1999. ISBN 1-58321-017-2.

10. Tao, P. "Statistical Sampling Technique for Controlling the Accuracy of Small Meters." *Journal AWWA*. 1982;74(6):296.

11. American Water Works Association. *Standard C701-07 Cold Water Meters—Turbine Type for Customer Service*. Denver, Colo.: AWWA, 2007.

12. American Water Works Association. *Standard C702-01 Cold Water Meters—Compound Type*. American Water Works Association, Denver, Colo.: AWWA, 2001.

13. American Water Works Association. *Standard C703-96 (R04) Cold Water Meter—Fire Service Type*. Denver, Colo.: AWWA, 1996. (Reaffirmed with no revisions 2004).

14. American Water Works Association. *Standard C704-02 Cold Water Meters—Propeller Type Meters for Water Works Applications*. Denver, Colo.: AWWA, 2002.

CHAPTER 13

Controlling Apparent Losses from Data Transfer Errors by Leveraging Advanced Metering Infrastructure

George Kunkel, P.E.

Julian Thornton

Reinhard Sturm

13.1 The Customer Water Consumption Data Transfer Process

The majority of water utilities in North America provide meters on customer service connections in order to register water consumption from individual customer accounts. Historically, the justification for use of customer meters in water utilities has been to periodically obtain measures of customer consumption that serve as the basis for billing. Linking water consumption volumes to a price also serves as a basic means of water conservation, since consumers are usually more judicious in their water use when its impact on their spending is clear and explicit. Having accurate water meters in place is the first in a multistep process to manage customer consumption data. North American water utilities typically store customer consumption data in a *customer billing system*. Errors can occur in the process used to obtain readings from the customer meter and transfer this data to the billing system. Often such errors result in understated consumption volumes, and represent one form of apparent loss.

Many opportunities for error exist in the customer meter reading and data transfer processes of water utilities. Meters are usually read in one of two manners: manual meter reading or automatic meter reading (AMR). Manual meter reading, with meter reading personnel (meter readers) visiting individual customer premises to visually collect readings, is the traditional approach and, as of 2007, still used by more than 70% of North American water utilities. However, AMR, and a host of innovative end-user capabilities collectively referred to as advanced metering infrastructure (AMI), are being implemented at a rapidly growing pace, giving drinking water utilities highly capable technologies to minimize apparent losses from data transfer error and improve their operational efficiency and level of service to their customers.

13.1.1   Manual Customer Meter Reading

Manual meter reading can work reliably, but in many communities it encounters a number of difficulties that hamper its efficiency and cost-effectiveness. Most notably manual meter readers often find difficulty in gaining access to meters, particularly those located inside customer buildings. A high rate of failed meter read attempts occurs in many water utilities due to this problem. Also, manual meter reading is inherently labor-intensive, with associated high staffing and deployment costs and issues. Because of highly variable field logistics many customer meters cannot be read consistently. In cold climates, water meters are typically located inside customer building premises, often in hard-to-reach corners of basements, boiler rooms, or other subterranean areas. See Figs. 13.1 and 13.2. It is not uncommon for property owners to store items in these areas that block access to the meters. With growing numbers of working couples in families, many properties have no one at home during business hours to let a meter reader into the house. Because of security concerns, many customers are wary of allowing strangers onto their premises at all. The Greater Cincinnati Water Works encountered such difficulties, which led to their decision to install an AMR system, as described in a newsletter account.[1] "Because the utility employed a door-to-door manual read system, employees were bogged down with the management of over 30,000 house keys entrusted to them by their customers. In addition, an increasing number of people were

Figure 13.1   Indoor 3-in turbine meter servicing a 100-unit apartment building in Philadelphia. (*Source:* Philadelphia Water Department.)

Figure 13.2   Indoor 3-in turbine meter servicing a 100-unit apartment building in Philadelphia showing location in underground basement of the building. (*Source:* Philadelphia Water Department.)

unwilling to hand over a key to their home and unable to be there during the day to let the meter reader inside." It is clearly understood that the traditional means of using manual meter reading is fast being outmoded by the more efficient, less labor-intensive capabilities of AMR systems.

In warm climates not subject to freezing and frost, customer meters are usually located outdoors in meter pits, or small, shallow chambers housing the meters. See Fig. 13.3. The pits are usually located midway between the water main and the customer building and often serve as the delineation between the service line responsibilities of the water utility and the customer. Large meters serving industrial customers are typically located in larger, deeper pits or chambers outside of the buildings that they serve; this is common even in cold climates. While outdoor meters generally have less restricted access than indoor meters, outdoor installations also suffer from problems of inhibited access. Many outdoor meter pits are susceptible to flooding. Entrance ways can be buried or covered by debris or parked vehicles. Outdoor residential meter pits often also require access to private property, and the security apprehensions of private property owners. At sensitive sites, such as hazardous industrial buildings or military installations, special security clearances and/or escorts may be required, greatly complicating the process and extending the amount of time to conduct manual meter reading.

Regardless of whether the meter is located indoors or outdoors, meter readers entering private properties encounter safety risks from aggressive dogs, dark or poorly maintained spaces, hostile customers or crime-ridden neighborhoods. The stark rigors of physically visiting dozens to hundreds of customer properties each day in all types

FIGURE 13.3 Typical outdoor meter pit installation. (*Source:* Neptune Technology Group.)

of weather and adversity carry a high potential for monotony-driven inattention, fatigue, illness, and injury, conditions that frequently result in inaccurate or incomplete meter readings and high staffing turnover.

In addition to access difficulties, many meter reading attempts suffer human error of visual misreads of the meter register, or error in transcribing the meter reading to handwritten paper records. Poor handwriting may result in the meter reading numbers being transcribed incorrectly to the billing system. Additionally, less diligent meter readers sometimes abandon all attempts at accessing difficult meters, instead fabricating meter readings and submitting them as actual reads. Occasionally, corrupt meter readers may collude with dishonest customers and intentionally fabricate meter readings to understate consumption and billings, and thereby defraud the water utility. All forms of erroneous or fabricated consumption volumes create distorted consumption records and apparent losses that usually cost the water utility a portion of the revenue to which it is entitled.

Despite the above-mentioned difficulties, manual meter reading is still very common and generally effective in many water utilities, perhaps more often in smaller communities with smaller meter populations, fewer logistical difficulties, and stable demographics. But the improving capabilities of AMR systems continue to make cost-effective business cases in a growing number of water utilities of all sizes.

13.1.2 Automatic Meter Reading

Because of the many difficulties encountered in manual meter reading programs, meter reading success rates have declined in recent years in many water utilities and a rapidly growing number of these systems have installed AMR systems, which are usually more accurate, less labor intensive, safer, and typically more cost-effective than manual meter reading. AMR has greatly reduced the accessibility and safety problems that have plagued manual meter reading programs. Many water utilities have achieved great success in moving from manual meter reading to AMR, such as the account described in Figs. 13.4 and 13.5. AMR has a successful history in the gas and electric utility industries, with implementation in the water industry growing rapidly since the mid-1990s. AMR market penetration in the U.S. water sector stood at greater than 25% of customer accounts in 2007 and is expected to reach over 40% by 2012.[2] This is a good trend for the drinking water industry as AMR offers advantages of improved accuracy, efficiency, and cost-effectiveness.

AMR systems consist of a device that is mounted to the customer water meter. This endpoint device has the ability to obtain a reading from the meter register and transmit it via one of the variety of communication mechanisms offered by manufacturers. The first generation of water utility ARM systems communicated the reading signal to a meter reader walking by the property, either wirelessly or by plugging in a handheld device to a port on the exterior of the customer building. Such *handheld* readings eliminate the need to gain access inside the customer building; yet this method still requires the labor of the manual meter reader patrolling a fixed route. In this approach, meter reading success rate and efficiency is increased while labor costs are little changed or only slightly improved.

A second common form of AMR is the *drive-by* method of communication, whereby meter readers patrol the service area in vehicles to collect meter readings. Meter readers need not leave their vehicle in order to collect readings. Dozens of readings can be quickly collected, virtually at the same time, as the patrol vehicle drives slowly down a street. Equipment in the vehicle sends out signals to *awaken* the AMR endpoint devices attached to the meters and obtain the current meter reading. This drive-by method

---

The Benefits of Automatic Meter Reading Systems

Prior to the start of AMR installation in 1997, Philadelphia's Water Department and Water Revenue Bureau encountered such poor meter reading success that only one out of every seven water bills issued was based upon an actual meter reading; six were based upon estimates. With the installation of over 425,000 residential AMR units by 2000, the city witnessed a meter reading success rate of over 98% in its monthly billing process using a mobile drive-by system. A system of mostly estimates was replaced with a system of mostly actual meter readings. This has greatly improved the confidence of customer consumption data, lessened the number of customer billing complaints and aided the detection of systematic data handling error and unauthorized consumption in the City of Philadelphia. Meter readers were assigned to new duties: no layoffs or terminations occurred, and the project has been highly cost-effective. Philadelphia envisions moving to fixed network AMR as its next generation system.

Figure 13.4   The benefits of automatic meter reading systems. (*Source:* American Water Works Association. "Water Audits and Loss Control Programs." *Manual of Water Supply Practices M36*, 3rd ed., Denuer, Colorado.: AWWA, 2008.)

Figure 13.5   Philadelphia Water Department's AMR system: Typical Itron endpoint "ERT" (encoder, receiver, transmitter) and 5/8-in residential meter from Badger meter. (*Source:* Itron, Inc.)

offers the same advantage of not needing access to customer properties to collect readings as the "handheld" method. However, drive-by AMR offers the additional benefit of needing fewer meter readers since the patrol vehicle can collect many more daily meter readings than individual meter readers on foot. Handheld and drive-by meter reading systems have been the most common form of AMR in use since AMR began widespread penetration in the water utility sector. However, AMR in the water industry is poised to move to the next generation of communication method: *fixed network* AMR. Figures 13.6 and 13.7 show typical ARM endpoint device installations in a meter pit for a fixed network AMR system.

Fixed network AMR refers to AMR systems that use a fixed communication network of established tower, antennae, WIFI, or similar telecommunication networks to send AMR signals when needed. Establishing a fixed network AMR system is certainly more involved than mobile communication systems, since a permanent communication system must be designed and constructed. Initial costs to construct such a system are higher than handheld or drive-by systems. But, fixed network AMR largely frees the water utility from the need to have permanent meter reading personnel in the field, thereby offering a major savings on personnel costs and reduced staffing problems. Fixed network AMR also provides the capability to obtain customer meter readings at any frequency or time of day, since reading schedules don't rely on staffing constraints. Fixed network AMR provides the capability to obtain sufficient data to create customer profiles by obtaining data at hourly intervals (or similar short times) and displaying the

Figure 13.6 AMR endpoint device for residential meter in meter pit. (*Source:* Itron, Inc.)

Figure 13.7 AMR endpoint device being installed in a residential meter pit. (*Source:* Itron, Inc.)

variation of consumption throughout the day, week, season, or year. Meter readings can be gathered and transmitted at short intervals via the fixed communication network. Alternatively, some systems use data-logging AMR devices at the customer endpoint. The data-logging equipment continuously gathers and stores meter readings at the meter reading device. A customer consumption profile is then transferred to the central data collection location on a periodic or as-requested basis. Variations in fixed network or data-logging AMR systems give water utilities a variety of choices to consider in finding the specific type of AMR network that will provide the granular customer consumption that they desire.

Fixed network AMR requires investment in the construction of the fixed communication network, in addition to the installation of user endpoint devices, software, and other standard components of the AMR system. The cost of a fixed communication network is variable, depending upon the nature of the service area: urban versus rural, hilly terrain versus flat, customer density, and other factors that affect the communication system requirements. The network typically requires a number of antennae or collector units (see Figs. 13.8 and 13.9) spatially distributed in a manner that allows AMR signals to be collected and forwarded to a central computer. The fixed communication network in any given service area must be independently evaluated, designed, constructed, and tested to ensure that it meets the service level requirements of the utility

Figure 13.8   AMR fixed network collector antenna. (*Source:* Itron, Inc.)

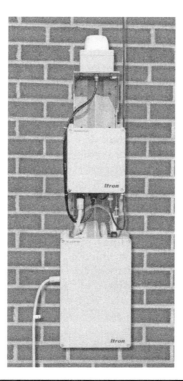

Figure 13.9   Fixed network AMR collector unit—collects and stores up to 10 days of hourly profile data for up to 10,000 customer endpoints. (*Source:* Itron, Inc.)

AMR contract. Whereas fixed network AMR requires notable planning and design effort, notable advantages are gained relative to mobile AMR from the significantly lower annual staffing costs and the more sophisticated customer consumption data that is obtained. Also, as described later in this chapter, the fixed communication network provides the opportunity to gather more than just a meter reading at the customer endpoint, thereby furthering the business case for fixed network AMR.

Water utilities can reduce the likelihood of apparent losses due to data transfer error via the use of AMR systems. AMR systems offer water utilities the current best practice means for cost-effective and efficient collection of customer consumption data. The use of AMR systems in water utilities will continue to grow significantly in coming years, as will the use of fixed network AMR in lieu of mobile AMR.

Detecting and Quantifying Data Transfer Errors

While AMR is less susceptible to data handling error than manual meter reading, both forms of meter reading can incur errors. Meter reading attempts can fail for many reasons. The difficulties of manual meter reading were discussed earlier. AMR attempts can fail due to a malfunction of the automatic meter reading device from causes such as improper installation or calibration, or battery failure. AMR equipment that is improperly installed or configured can result in erroneous readings. This occurrence can be minimized by using a good quality control protocol during system installation.

When a meter reading attempt is unsuccessful in obtaining an actual meter reading, most water utilities bill customers using an *estimated* volume that is calculated based upon a standard estimating protocol or the customer's recent consumption history. While these are reasonable approaches, multiple cycles of meter readings without an actual reading greatly increase the prospect of inaccurate estimates. Over periods of time, buildings are sold and new owners with vastly different water consumption habits may become the permanent occupants. An estimate generated for a household of two may be fine until the house is sold to a family of seven. Water consumption could triple, but understated billings based upon the outdated estimate could continue for some time. When an actual meter reading is eventually obtained, a large billing adjustment will confront the new property owner, a scenario that commonly creates customer's ill will toward the water utility. Clearly, obtaining routine, accurate meter readings is critical in maintaining sound oversight of customer consumption patterns and stable billing and revenue collection functions.

Recognizing that some level of meter reading and data transfer error occurs to a degree in virtually all water utilities, managers should designate staff time to periodically analyze meter reading and billed consumption data in order to detect trends of irregular consumption stemming from data transfer error. A billing analyst should look for trends such as successive cycles of "zero consumption" or other suspicious consumption patterns. Accounts that register zero consumption for several successive meter reading cycles should be sampled and investigated to determine if the zero consumption is valid (which could occur if a building becomes unoccupied) or whether AMR failure or tampering has occurred. The analyst should monitor the meter reading success rate for both residential and industrial/commercial categories of accounts. The number of estimates assigned should also be tracked and an approximation of the error due to poor estimation should be attempted. Estimating protocols should be reassessed if they have long been in use. Other sources of systematic data transfer error can exist in any given water utility. Depending on available resources, investigations can be conducted to assess any occurrences of data transfer error that are unique to the utility.

The auditor should attempt to quantify the major components of apparent loss due to data transfer error and include them in the water audit. By investigating and analyzing a manageable number of suspect accounts, the auditor should be able to identify apparent loss volumes for a valid sample of the customer accounts in the water distribution system. By extrapolating this value of apparent loss volume per account to the entire customer population, the auditor can determine reasonable volumes for various types of data transfer error. One type of potential apparent loss occurrence typical in water utilities are accounts that have not been read for many billing cycles due to access difficulties that prevent the meter reader from obtaining a manual reading. Special efforts will likely be needed to gain access to these meters; perhaps written notices to customers to arrange specific appointment times to allow for a meter read, or a request to remove household items blocking access to the meter. If the local water regulations allow—and the situation warrants it—the utility may need to send to the customer a notice of violation that states that they must provide access to the water meter, or penalties could be enacted, including, if permitted, shutting off water service to the customer. Figure 13.10 gives an example of the calculations used to quantify volumes of apparent loss due to data transfer error in the fictitious County Water Company.

Depending upon the size of the customer population, mode of meter reading, water regulations or policies, and other circumstances unique to each water utility, the number of apparent loss subcategories due to data transfer error could range from as few as

---

Example Calculation of Data Transfer Error in County Water Company (CWC)

The manager compiling the water audit for county water company has suspicions that customer accounts with many billing cycles of estimates are a potential source of data transfer error. He determines to field investigate 50 customer accounts that have not had an actual meter reading in over 2 years. Estimates have been used in billing these accounts over this 2-year period.

The manager's first step was to obtain access to the customer premises in order to obtain current meter readings. After sending violation notices to the customers and making contact during the first month of this effort, CWC was able to gain access and obtain current meter readings in 38 of the 50 properties. For the 12 properties that could still not be accessed, more aggressive steps by the water utility–such as service disconnection–will be needed in order to force access to the customer meter. For the 38 accounts that were accessed, updated meter readings found that these accounts had been collectively under-billed by 360 thousand gallons (kgal) during the 2-year period, or an average of 180 kgal per year. Billing records show that the water utility had a total of 487 accounts that went without an actual meter reading over the past 2 years. Based upon the findings of the 38 accounts

$$\text{Apparent loss (2-year missing reads)} = (180 \text{ kgal}) \frac{487}{38} = 2{,}307 \text{ kgal}$$

The value of 2307 kgal should be included in the water audit as one subcategory of apparent loss due to data transfer error. Any other groups of suspicious accounts, such as zero consumption accounts, should also be investigated and extrapolated over the customer population to obtain a quantity of apparent loss due to data transfer error. Several subcategories might be identified in any water utility and these should ultimately be included in the water audit and totaled under apparent loss due to data transfer error.

FIGURE 13.10   Example calculation of data transfer error in County Water Company. (*Source: George Kunkel.*)

two to as many as eight or more. Subcategories that might be considered for investigation can include

- Accounts without actual meter readings for one year or longer
- Accounts showing zero consumption for three or more billing cycles
- Accounts suddenly evidencing a significant drop in consumption after a stable history of higher consumption levels
- Accounts with confirmed AMR equipment failures
- Accounts known to have suffered from manual meter reading inaccuracy from one or more meter readers confirmed to be inattentive or dishonest
- Accounts known to have suffered data distortion in transferring data from handheld meter reading devices into the *customer billing system*

These are but several possible causes of apparent loss due to data transfer error. It is incumbent on each utility to determine a reasonable extent of cause and volumes of this form of apparent water loss. The key bottom-up activities in this regard are analysis of billing records for unusual consumption patterns or missing meter readings, and auditing/investigation of samples of suspect accounts to confirm actual volumes of apparent loss.

The ability of the water utility to minimize data transfer error also depends upon the strength and clarity of its regulations and procedures. While the use of advanced technology such as AMR certainly can improve effectiveness, necessary improvements might only be gained in updating outdated regulations regarding customer service requirements around the use of estimates, back-billing, entry to private property for meter/AMR repairs and related activities. If regulations have not been reviewed in recent years, it will be worthwhile for the water utility manager to work with community policy managers to ensure that water service regulations are current in meeting the water service requirements for customers, as well as ensuring efficient water utility operations. Similarly, procedures for meter reading and data handling should be clear, current, and monitored for compliance among staff. Training should be conducted on a regular basis for new employees and as refreshers for longstanding employees so that meter reading success and accuracy is maintained at optimum levels.

13.1.3   Advanced Metering Infrastructure

As AMR systems have gained substantial use in the drinking water industry—and appear to be heading toward the most common form of meter reading in the future—manufacturers have come to recognize the potential for significant new customer end-point benefits. With fixed network AMR systems able to communicate water meter readings automatically at established short intervals, the usefulness of the customer endpoint as a data collection location has been greatly elevated.

The historical use of the customer meter—periodic meter readings—has provided a valuable, but singular, purpose: providing a basis for billing based upon consumed water volumes. When water meters are read manually, often considerable difficulties are encountered in collecting meter readings such that conducting a single round of meter readings every 30 or 90 days is a significant challenge. But, with fixed network AMR, meter readings can be collected as frequently as every 15 minutes in an accurate and cost-effective manner. Fixed network AMR provides the ability to collect granular consumption data that can be used to develop customer consumption profiles which show the hourly, daily, weekly, and seasonal variation in consumption flows, as well as allowing for the traditional calculation of the consumption volume for the billing period. Consumption varies in a repeatable pattern for many customers, typically with low consumption during a portion of the day (often nighttime hours) and high consumption peaks at one or more times of the day. Gaining detailed insight into customer consumption patterns can provide benefits to water utilities in a number of ways. Uses of detailed customer profile data are discussed in Sec. 13.2. Since fixed network AMR is capable of obtaining detailed consumption information, why stop there? Other potentially useful water system information also exists at the customer endpoint. Manufacturers have developed the capability to obtain information that includes

- Tampering with metering or AMR equipment
- Consumption trend analysis that sends alerts of leakage on customer piping
- Acoustic leak detection: on customer service connection piping, or leaks in the neighboring water distribution system
- Backflow (flow reversal) detection

Various manufacturers have developed sensing devices for a number of these parameters, and the future might see many manufacturers offering a wider slate of monitoring capabilities at the customer endpoint, possibly including water pressure levels and water quality parameters. The value of the fixed communication network is greatly leveraged since these systems can be designed to communicate data and alarms on a number of endpoint parameters, in addition to the traditional meter reading. This expanded host of capabilities can provide water utilities with efficiencies that go well beyond billing and the water loss control program. These additional capabilities offer a number of operational efficiencies to the water utility and better justify the overall business case for fixed network AMR in a water utility.

Several leak detection capabilities are being developed and refined. Evidence of leaks on the customer side of the meter can be obtained by analyzing the consumption flow profile. An increase in water consumption registered by the water meter can present evidence that a leak has emerged. Having the ability to detect leaks in this manner, and sending an alert to the customer, gives the water utility the opportunity to minimize the run time of the leak, which saves considerable water relative to systems without any leak detection capabilities. This also provides a service to the customer by limiting the likelihood of an *unusually high bill* (UHB) being issued to the customer. Leak detection software that analyzes customer consumption patterns in this manner is becoming a standard feature for a number of AMI manufacturers. Figure 13.11 displays a sample customer bill that includes an alert about a detected leak, based upon the customer's consumption pattern.

The AMR network is also being studied as a means to assist leakage control in water distribution systems. American Water operates water utilities in many states in the United States and has been an industry leader in conducting research on water industry innovations. For several years American Water has been running trials of a technology that utilizes the fixed communication AMR network to transmit meter readings and alerts from leak noise loggers attached to customer service connections near the customer water meter.[3] Spaced at intervals of roughly every 10 customer service lines, the sound patterns generated by leaks—either on customer piping or water distribution system piping—are collected and correlated in order to pinpoint the location of leaks. This approach provides advantages in several manners. Firstly, it leverages the fixed communication network, increasing the benefits of the AMR system and building a better business case for the use of fixed network AMR. Secondly, it minimizes the awareness time in which a leak is detected, relative to manual leak detection. Lastly, it automates the meter reading and leak detection processes by placing permanent, automated equipment in use to replace manual methods. American Water has found success in identifying leaks in its initial pilot projects and is further exploring the use of this technology in several other trail locations. Figure 13.12 shows a typical leak noise logger installation on a customer service line in one of the trials being conducted by American Water.

In addition to endpoint information communicated via fixed network AMR to the water utility, other customer endpoint innovations are being developed to provide enhanced service to customers. A residential water consumption display has been developed to allow customers to view the current water meter reading and consumption at a convenient location in the household.[4] Designed to be mounted at a visible location in the home, the display provides customers with information that helps them understand the water demands of various water-using household fixtures and appliances and helps them to better manage their water demand. Residential water leaks can be detected using software that analyzes the consumption pattern and an alert is issued by in-home display, email, or SMS message. Separate from the water utility metering and AMR system, manufacturers have developed end-use metering devices that display measurements of water consumption at individual customer fixtures such as

OPELIKA UTILITIES
P. O. Box 2587
Opelika, Alabama: 36801-2587
Customer Service: 334-705-5512
Fax: 334-745-3487

Account Number	22546
Due Date	3/15/2003
Service Address	2900 Birmingham Highway, Apt. 235
Amount Due	70, 164, 68

Dan H. Hilyer
502 Geneva Street
Opelika, Alabama 36801

CONTINUOUS LEAK DETECTED

Service ID	Service	From Date	To Date	Meter No.	Previous Reading	Current Reading	Usage	$ Amount
505953	Water	1/23/2003	2/24/2003	46996639	506	511	5	$ 14.33
	Sewer	1/23/2003	4/24/2003				5	$ 17.51
505954	Imgation	1/23/2003	4/24/2003	46996637	1558	1565	7	$ 20.14
	Fire protection	1/23/2003	4/24/2003					$ 11.50
	Deposit							$ 50.00
	Origination fee							$ 25.00
	System development							$ 70,???.00

Period	Days	Water	Irrigation
Current	32	5	7
Last month	28	5	6
Year age	31	3	7

Current charges	$ 70,138.48
Tax	$ 0.57
Past due amount	$ 25.63
Total amount due	$ 70,164.68

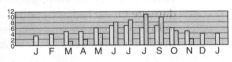

A continuous leak has been detected on your water service. Causes may include a toilet that is running or a faucet that drips continuously. You may have a broken water line under your house or out in the yard. We suggest that you have a licensed plumber inspect your water lines.

Notice: Please bring entire bill if paying in person.
Detach and return bottom partion if paying by mail.

Account no.	Service ID	Cycle no	Route no
22546	505953	2	22
Delinquent date	Past due amount	Total if paid late	Total due
3/16/2003	25.63	$1,980.25	$70,164.68

FIGURE 13.11 Customer billing statement with printed leak alert. (*Source:* Neptune Technology Group.)

FIGURE 13.12 MLOG leak noise logger installed on customer service line and connected to AMR endpoint in tandem with water meter in trials of water loss control project in Connellsville, Pennsylvania by American Water. (*Source:* American Water.)

FIGURE 13.13   LCD display of solid state register for Neptune meter capable of detecting reverse flow (backflow) events and leaks on internal plumbing. The arrow indicators show the direction of flow and the faucet indicator reveals the presence of a leak. (*Source:* Neptune Technology Group.)

faucets and showerheads.[5] One manufacturer also offers an endpoint control device to prevent water wastage from leakage on premise piping. This feature gives the capability to close a valve on the customer service connection piping if a significant leak is sensed by consumption-analyzing software, thereby quickly halting a potentially damage-causing situation. This technology has dual benefits of minimizing water waste from customer premise leaks and minimizing water damage to customer property.[4]

At least one manufacturer provides the capability to detect flow reversal (backflow) events and send alerts to the water utility via the fixed network AMR system. Figure 13.13 shows the LCD of a solid state register with arrow indicators giving the ability to show flow being supplied to the customer as normal, or flowing backwards, as might occur in an unusual back pressure or vacuum condition in the water distribution system. Back-flow conditions can have serious water quality impacts, as water drawn back into the water distribution system during a backflow event may be of questionable quality and risk contamination of the distribution system. As a result of the post 911 vulnerability assessments conducted by water utilities, intentional contamination of a distribution system has been hypothesized as a possible threat to water utilities. Such actions could occur by pumping a contaminant into a building service connection and forcing it into the distribution system. Having a reverse flow detection and quick alert capability can give water utilities effective tools to counter the risks that backflow portends.

Manufacturers are rapidly expanding the functionality of the customer endpoint. Their innovation is giving water utilities many effective tools to better promote accountability and efficiency in their water supply operations. With multiple benefits offered through AMI packages, the business case for fixed network AMR is more persuasive than systems that provide meter reading solely. It is believed that the use of fixed network AMR, and many AMI features, will continue to expand rapidly in the North American water industry. These systems will enhance water accountability and minimize data transfer error as a source of apparent losses, while giving water utilities new capabilities and customers improved service.

13.2   Customer Consumption Profiles—Transitioning from Periodic Customer Meter Readings to Granular Consumption Data

While traditional meter reading programs gather single meter readings every 30 or more days, fixed network or data-logging AMR systems can generate detailed customer consumption profiles by obtaining readings as frequently as every 15 minutes. By collecting more granular data in this manner, these AMR systems utilize capabilities to better address customer billing complaints, quickly identify plumbing leaks, and assist water conservation and loss control efforts.

Data-logging capability for individual customer meters has existed for many years. The first commercial data-logger was designed to attach to an individual customer meter and log pulses from the meter register, thus allowing a detailed profile to be developed (see Chap. 12, Figs. 12.4 and 12.5). Individual data-loggers are now available with communication capabilities similar to those of fixed network AMR systems, and are cost-effective in applications where the business case for full-scale AMR is not yet justified. One example is a small, rural water utility with good success from manual meter reading, but operationally challenged by five large concentrated animal feeding facilities that account for a significant amount of the total customer consumption. Stand-alone data-loggers and communication systems on these five accounts could give the water utility detailed consumption profiles on these significant customers at a relatively minor expense to the water utility. Some water utilities have even negotiated contracts with such large users to sell them detailed profile data on a monthly basis to give the customer information to better manage their water demand, while generating additional revenue for the water utility.

With meter readings taken at close intervals a typical consumption profile can be generated to show the variation in customer consumption throughout the day. A profile generated by a data-logging fixed network AMR system is shown graphically in Fig. 13.14. As

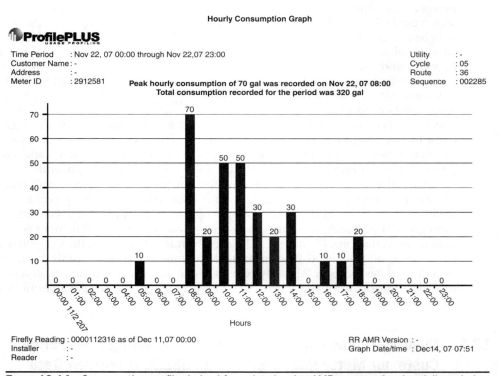

FIGURE 13.14 Consumption profile derived from data-logging AMR system showing daily variation in water consumption for a typical residential customer. (*Source:* City of McKinney, TX, and Datamatic.)

shown, the customer consumed 320 gal on November 22, 2007 but the hourly consumption varies in a typical pattern for a residential customer account. During the night and early morning hours, no consumption is registered at the minimum increment of 10 gal, although it is likely that a smaller rate of consumption occurred during some of these hours. During daytime hours, consumption varies from 10 gal/hr up to a peak hour of 70 gal from 8 a.m. to 9 a.m. This profile is reflective of a typical consumption pattern for domestic usage only, without the larger volumes of consumption related to outdoor irrigation systems used during the warm months of the year in warm climates.

Both water utilities and customers can gain considerable advantages from the availability of customer consumption profiles. For most water utilities the greatest number of telephone inquiries received from customers are related to billing. Customers are quick to call the water utility if they believe that they have been overcharged for water service. UHBs usually cause much concern and frustration for customers. When meters are read only every 30- or 90-days, it may be difficult to ascertain when and why consumption became unusually high for the billing cycle. A customer profile, however, can be used to determine precisely when the consumption accelerated from a normal range to a high range. Knowing the time allows the water utility and customer to relate events—such as filling a residential swimming pool from the household supply—that can explain the high consumption. In the case of the emergence of a leak on the customer premise piping the profile will typically reveal an increase in consumption that remains at a high level. A consumption profile for such an event is shown in Fig. 13.15

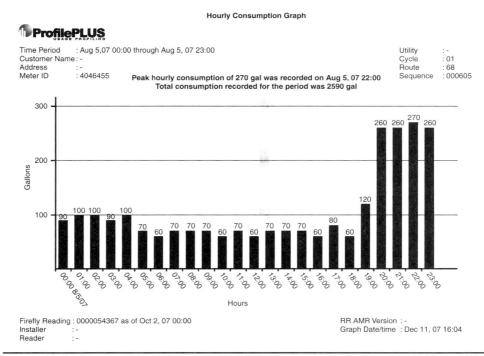

FIGURE 13.15 Consumption profile showing consistently high flow indicative of a significant leak, which erupts dramatically at 7 p.m. on August 5, 2007, giving the basis for an UHB for the customer. (*Source:* City of McKinney, TX, and Datamatic.)

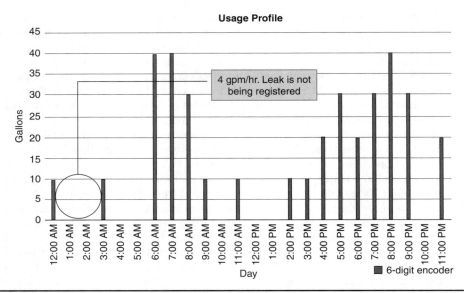

FIGURE 13.16 Traditional competitive 6-digit encoder fails to capture low leakage flow in early morning hours. (*Source:* Neptune Technology Group.)

for a significant leak on customer piping downstream of the meter. Once such a pattern is detected—and many utilities now offer a leak alert feature in their customer billing package—leaks can be more quickly addressed, thereby saving water and preventing an UHB and customer ill will.

Small, *below detectable limits* (BDL) leaks, which trickle at such low rates of flow that they are often unmeasured by many water meters, are perhaps more common and more difficult to detect than the large leak which is profiled in Fig. 13.15. One means to address these leaks is the use of flow modification devices such as the unmeasured-flow reducer (UFR) described in Chap. 12, Sec. 12.4. However, improvements in metering and meter reading technology are also giving water utilities effective capabilities to detect these types of leaks. Figures 13.16 and 13.17 illustrate the value of a high resolution meter in registering a 4 gallon per minute (gpm) leak, where less sensitive metering fails to detect this low leakage flow. The volume impact of a very small leak is shown in Fig. 13.18. The leakage rate of 1/16 gpm is very small, yet results in a significant volume loss over a period of months.

Another example of the use of a customer profile is water conservation tracking. At times water utilities may need to impose water conservation restrictions such as twice-a-week or odd-even day schedules for outdoor irrigation use, which is a highly water intensive use in many dry regions of North America. Figure 13.19 shows a customer profile that clearly displays higher consumption on Wednesdays and Saturdays due to outdoor irrigation flows. Figure 13.20 shows a similar graph where the customer has violated the outdoor water restriction by operating their irrigation system briefly during early morning hours in the belief that this consumption would not be detected. The customer profile can be used as evidence of this unauthorized irrigation consumption and allow the water utility to pursue enforcement action against the customer if it determines that this is warranted.

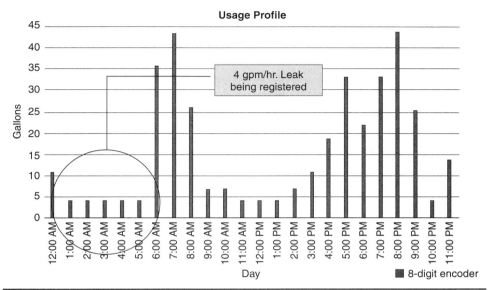

FIGURE 13.17 Neptune's 8-digit encoder captures low leakage flow in early morning hours. (*Source:* Neptune Technology Group.)

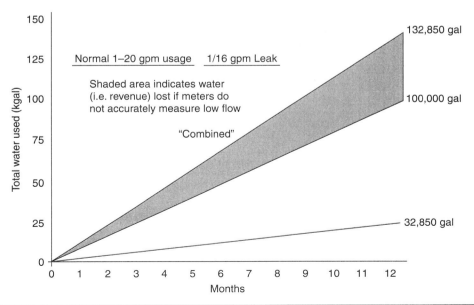

FIGURE 13.18 Total water usage for one residence with and without 1/16 gpm leak. (*Source:* Neptune Technology Group.)

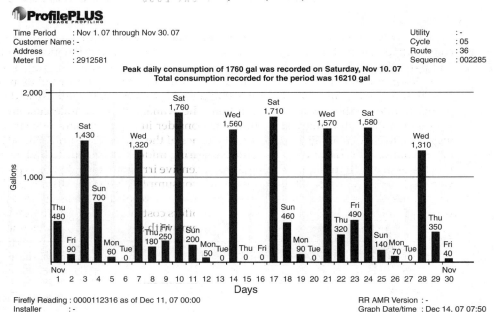

FIGURE 13.19 Consumption profile showing high flows on Wednesdays and Saturdays reflecting twice weekly outdoor irrigation pattern. (*Source:* City of McKinney, TX, and Datamatic.)

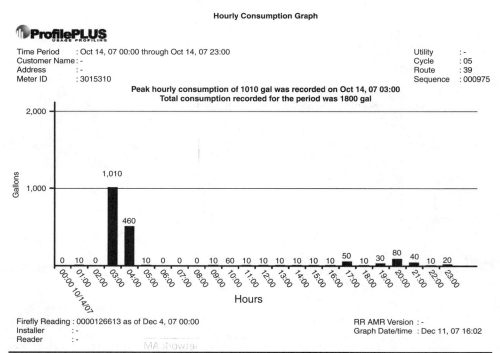

FIGURE 13.20 Consumption profile showing high flows during early morning hours for customer attempting to hide outdoor irrigation system operation. (*Source:* City of McKinney, TX, and Datamatic.)

New ways of encouraging customers to conserve water in dry regions are being constantly developed. In addition to technology that saves water (low flow appliances and fixtures), conservation efforts focus on customer education and incentives to save water. Common financial incentives in use include both rebates for installation of low flow fixtures, and water rate structures that reward conservative water use and charge premium rates for high water consumption. The availability of granular customer consumption data is allowing water utilities to consider innovative water rate structures, including those that assign costs to specific times of the day. In dry, sunny climates outdoor water use for irrigation is less efficient during midday hours due to high evaporation losses in bright sunlight. Shifting water intensive irrigation to night hours promotes water efficiency. With the ability to measure consumption volumes during peak daytime hours, some water utilities have begun to implement time-of-use (TOU) billing, which penalizes peak hour consumption and offers cost incentives to shift high volume practices to evening or night hours. AMR systems with specific software designed with TOU billing capabilities are now available on the commercial market to provide water utilities with this option.[4] Figure 13.21 shows a billing statement displaying water

Itron EE		**Time of Use Report**		07/19/2006 13:05 Page 1

Service point 80018797			**Start time**	05/01/200 00:00
Account 80018797			**Stop time**	05/14/2006 24:00
Reason Billing			**Time zone**	PacificUS
Customer Water customer 2			**UOM**	GAL
Premise 80018797			**Int**	60
Meter			**Roll**	60

TOU ID Simpleonoff   Simple on peak off peak rate
TOU time zone PacificUS
Interval channel 1
TOU set TOU 80018797:1simpleonoff
TOU set channels

	Summer	Total
Onpeak	1,111.0000	1,111.0000
Usage	25.0000	25.0000
Peak	05/04/2006 17:00	05/04/2006 17:00
Peak time	0.6349	0.6349
Load factor	20.34	20.34
% Usage	20.83	20.83
Offpeak	4,350.0000	4,350.0000
Usage	24.0000	24.0000
Peak	05/03/2006 07:00	05/03/2006 07:00
Peak time	0.6814	0.6814
Load factor	79.66	79.66
% Usage	79.17	79.17
Total	5,461.0000	5,461.0000
Usage	25.0000	25.0000
Peak	05/04/2006 17:00	05/04/2006 17:00
Peak time	0.6501	0.6501
Load factor	100.00	100.00
% Usage	100.00	100.00

Summer	4/1–10/31
On peak	Weekdays 12:00–1900
Off peak	All other
Winter	11/1–12/31: 1/1–3/31
Off peak	All other

FIGURE 13.21   Billing statement derived from fixed network AMR meter reading displaying consumption during peak and off-peak hours and used in a TOU water rate structure. (*Source:* Itron, Inc.)

consumption for peak and off-peak periods, as well as the aggregate consumption for the billing cycle. TOU billing is also being employed to mitigate high peak flows and allow water utilities to design smaller capacity into their infrastructure. Wide Bay Water Corporation on the east coast of Australia has installed a data-logging AMR system to provide consumption data in a TOU billing structure to motivate smaller daytime peaks by spreading consumption rates more evenly throughout the day, thereby reducing capacity needs. This is one of many benefits expected from this AMR system.[6]

In addition to TOU billing, some water utilities have begun to develop sophisticated water rates tailored to individual customer consumption patterns. Also known as a water budget rate structure, these rates reward customers that are able to implement water reductions to notable levels below their characteristic budgeted consumption levels. Likewise, the structure requires customers using water well above their budgeted level to pay for the additional water at a premium rate. For each customer an individual consumption pattern is measured and their budget is determined. One approach is to base each customer budget on typical annual indoor use as well as a monthly determination of actual needs for outdoor irrigation. Such an approach, while requiring considerable data management, recognizes that many people will conserve water if a clear financial benefit is available to them. However, there is often a small portion of the customer population—usually those owning expansive, well-cultivated landscapes—that is willing to pay a premium rate for high water consumption. Tailored or budget water rate structures work to promote water conservation in the community while providing equity among ratepayers and a stable revenue base for the water utility. The City of Boulder, CO is a recognized pioneer of budget rates, which it enacted in response to drought and the need for an effective, long-term strategy for water conservation and revenue equity and stability.[7,8]

Another use of granular customer consumption data is to assist distribution leakage assessments in pressure zones or district metered areas (DMA). DMAs are small zones of the distribution system usually encompassing 500 to 3000 customer service connections. A boundary is established that permits the DMA to be supplied by one or more water supply mains that are metered. In this way the supply to the DMA can be monitored directly and the variations in supply flow observed. Granular customer consumption data from AMR systems provide minimum hour consumption volumes that can be compared directly to the water supply input flows to DMAs such that a water balance can be constructed for individual DMAs and water supply tracked specifically. Precision is added to leakage assessments in DMAs when the input supply volume is compared to the customer consumption volume during minimum hours. In regions where nighttime irrigation systems are not in use, minimum hour residential consumption usually occurs during the early morning hours between 1 a.m. until 5 a.m. Night flow analysis for leakage quantification relies on the fact that, during the minimum consumption hours, leakage is at the highest proportion of the supply input to the DMA. Therefore, a reasonably precise measure of DMA leakage can be obtained by subtracting customer consumption from the supply input during the minimum consumption period. In regions where night irrigation flows are common, this analysis may need to be scheduled for an off season (winter) period when irrigation equipment is not in use.

In Philadelphia, the Philadelphia Water Department (PWD) has periodically employed its mobile-read AMR system to obtain nighttime customer meter readings in order to compare them with supply input in a DMA. This technique was initially employed in a temporary DMA and has been used successfully on a number of occasions in PWD's first permanent DMA (DMA5) which was designed and implemented

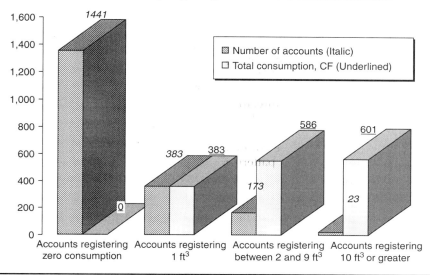

FIGURE 13.22 Customer water consumption in DMA5 during 2 a.m. to 4 a.m. on April 5, 2005; data used as part of minimum hour assessment of leakage in this DMA. (*Source:* Philadelphia Water Department.)

as part of the AWWA Research Foundation project "Leakage Management Technologies."[9] PWD is believed to be the first water utility in the United States to employ minimum hours AMR readings in a DMA setting to provide a more reliable leakage assessment. When gathering customer night meter readings, PWD arranges with its AMR service provider to perform one complete drive-by round of readings in DMA5 around the hour of 2 a.m. A second round of drive-by readings of the +2000 accounts is gathered around 4 a.m. For each customer account, the 2 a.m. reading is subtracted from the 4 a.m. reading and any numerical difference calculated as registered consumption for the minimum night period. A distribution of the results from one set of nighttime AMR readings is shown in Fig. 13.22. As might be expected for DMA5, an area that is largely residential in nature with no industries or irrigation systems using water on a 24-hour basis, water usage is minimal during the 2 a.m. to 4 a.m. period. This is evidenced in Fig. 13.22 showing 1441 of 2020 customers, 71% of customer accounts registered zero consumption and 19% of accounts that registered only 1 cubic foot of water consumption. Seventy-five percent of the total consumption of 1570 cubic feet was registered by only 10% of the customer accounts during the 2-hour period. Five accounts each registered over 40 cubic feet (almost 300 gal) during the 2-hour period, suggesting leakage in these properties on the customer plumbing. The total customer consumption for the 2-hour period of 0.141 million gallons per day (mgd) is subtracted from the supply inflow of 0.55 mgd to infer a difference as leakage in the amount of 0.409 mgd. This represents a significant amount of leakage which is estimated to exist as one half background leakage and one half as unreported leaks. PWD is pursuing this leakage by employing pressure management to address the background leakage and targeted leak detection surveys to locate and abate the unreported leakage.

PWD's work in DMA5 demonstrates how customer consumption profiles can assist leakage management assessments in DMAs. In PWD's case, AMR has been useful in its leakage assessments for DMA5; however, mobile drive-by AMR is not the most efficient means to collect night meter readings across such a short time interval. PWD looks forward to using DMA5 as a pilot area to test fixed network AMR and establish the basis for conversion of its drive-by AMR system to a full-scale fixed network AMR system in coming years.

13.3   Summary: Apparent Loss from Data Transfer Error

Having accurate customer water meters is only the first step in properly accounting for water billed to customers. The water meters must be successfully read on a regular schedule and the measured consumption data accurately transferred to the central data management system, which is typically the customer billing system. There are numerous opportunities for error to be introduced into the customer consumption data throughout the data transfer process. Water utilities should provide analysis and investigation into samples of customer accounts in order to determine the extent and magnitude of apparent losses due to data transfer error. Significant advances in metering, AMR, and a host of new capabilities developed under the label of AMI are providing water utilities unprecedented tools to both optimize the data transfer process and improve operations and customer service. These tools should prove to be highly cost-effective and valuable in saving water for many North American water utilities; and their implementation by a growing number of water utilities reflects an important trend for the water industry in North America and worldwide.

References

1. Neptune Now newsletter. Case Study: Greater Cincinnati Water Works, Tallassee, Ala.: Neptune Technology Group, Fall 2007.
2. Schlenger, D. "Water Utility AMR Systems Begin Transition to Advanced Information Systems." Tulsa, OK.: WaterWorld, PennWell Corporation, August, 2007.
3. Hughes, D. "A Piggyback Ride on AMR—Communicating More than Just a Meter Reading", *Proceedings of the Workshop entitled The ABC's of Apparent Loss Control and Revenue Protection for Water Utilities, AWWA DSS: Distribution & Plant Operations Conference*, Tampa, FL, 2005.
4. Bharat, B. "Elster AMCO's Evolution AMI Empowers Water Utilities in Conservation Efforts." Tulsa, Okla.: WaterWorld, PennWell Corporation, August, 2007.
5. WaterWatch water meters (2007). www.h2owatch.net/more_meters.html.
6. "Hervey Bay taps into high-tech water" Media Release February 19, 2007. [Online]. Available: www.yourwater.com.au/html/19_feb_07_amr.html. [Cited February 19, 2007.]
7. Western Resource Advocates. "Structuring Water Rates to Promote Conservation," 2005. [Online]. Available: www.westernresourceadvocates.org/water/wateruse.php. [Cited April 2, 2008.]
8. City of Boulder, Colorado. "Water Budgets", 2007. [Online]. Available: www.ci.boulder.co.us/index.php?option=com.content&task=view&id=6243&Itemid=2039. [Cited April 2, 2008.]
9. Fanner, V. P., R. Sturm, J., Thornton, et al. "Leakage Management Technologies." Denver, Colo.: AwwaRF and AWWA, 2007.

CHAPTER 14

Controlling Apparent Losses from Systematic Data Handling Errors in Customer Billing Systems

George Kunkel, P.E.

Julian Thornton

Reinhard Sturm

14.1   Compiling Billed Consumption in Customer Billing Systems

There is a tendency of many in the drinking water industry to assume that their system's apparent losses are solely due to customer meter inaccuracy, leaping to the conclusion that replacement of the entire customer meter population is the appropriate remedy. This publication clearly defines that apparent losses occur in a number of different manners. It is important that the auditor first assemble the water audit and identify the nature, quantity, and cost-impact of each of the apparent loss components, and only then develop a rational loss control strategy. Flowcharting the process of the *customer billing system* is a recommended first step. It is a very expensive and inefficient proposition to implement comprehensive customer meter change-out if the bulk of the apparent losses are actually due to billing system data error or unauthorized consumption. Yet, many water utilities have done just this, and are perplexed when, after spending up to millions of dollars on new meters, their apparent loss standing remains unchanged. Conversely, apparent losses in the data handling process of the customer billing system may be addressed by relatively inexpensive computer programming or procedural improvements. In this way, a quick payback can be earned by additional revenue recovery. Planning the apparent loss control strategy based on the results of the water audit is the best way to start, followed by flowcharting, or otherwise investigating, the workings of the customer billing system.

The eminent nineteenth century British physicist, Lord Kelvin, provided the following quote, which has as much relevance to the field of water loss control as to physics:

If you don't measure it, you can't manage it.

A modern corollary of his statement might read:

If we don't properly define it, measure it, data-warehouse it, and report it, we can't manage it.

We exist in the information age and the availability and integrity of the information available to us is of critical importance. A wide variety of information is employed in the provision of safe drinking water. This information is needed by those working in the drinking water industry including utility employees, government officials, regulators, service and equipment providers, and external stakeholders such as business and civic groups, customers, and the news media.

The customer billing system is typically the most significant information warehouse in most drinking water utilities. Revenue is generated via billings to customers for water consumption, typically on a monthly or quarterly basis. For utilities that meter their customers, the billing system stores customer account and meter data, as well as routine customer meter readings, from which consumption volumes are calculated.

Authorized consumption is any water delivered for consumptive purposes that are authorized or approved by the water utility, thereby providing a benefit to the community. The majority of the aggregate customer consumption volume in a community is *billed authorized consumption*, but a small portion is *unbilled authorized consumption*.

Billed authorized consumption may exist as metered or unmetered consumption and represents the collective amounts of water delivered to individual customers that have accounts in a customer billing system. Billed authorized consumption is the primary basis for revenue generation for most water utilities that don't charge based upon flat fees. Billed accounts are customer properties served by permanent customer service connection piping. In North America, most water utilities require customer meters on service connections and bill based upon metered consumption on a monthly or quarterly basis. Metered water can be categorized as residential, industrial, commercial, agricultural, governmental, and other uses. Not all water utilities, however, meter their customers, instead charging a flat billing fee per consumption period, or a charge based upon property or other characteristics. Therefore billed authorized consumption may be metered or unmetered. The American Water Works Association (AWWA) recommends that all customers with permanent service connection piping be metered with billing based upon measured consumption.

Unbilled authorized consumption can also exist as metered or unmetered consumption and describes water taken irregularly in a variety of manners from nonaccount connections that typically do not supply permanent structures. Withdrawing water from fire hydrants is the most common example of such nonaccount consumption. Water utilities often allow water to be taken from fire hydrants for firefighting (their primary purpose), flushing, testing, street cleaning, construction, and other purposes. These uses should be metered to the extent possible, with clear and explicit usage policies in force to protect water quality and public safety. Sometimes unbilled water supplied to government properties is also included in this category although it is recommended that all water continuously supplied to permanent structures be metered and be tracked in a billed account in the customer billing system.

In this way water consumption is monitored even though the property is issued a "no-charge" bill.

Modern metering, automatic meter reading (AMR) systems, and customer billing management technologies offer outstanding capabilities to water utilities to gather and utilize accurate customer consumption and billing data. It is strongly recommended that water utilities measure individual customer consumption via water meters and utilize computerized customer billing systems to store customer account data. AMR systems are being implemented by a growing number of water suppliers because of their cost-effectiveness and accuracy in gathering metered consumption data. For water utilities that utilize these technologies, consumption data is typically accessed via a variety of reports from the customer billing system. Examples of typical reports are shown in Tables 14.1 and 14.2 for the fictitious County Water Company, where consumption is summarized by meter size and customer consumption category, respectively.

All active accounts should include the meter identification number, meter size, and meter type. If an AMR system exists, the automatic meter reading device number and meter reading route number should also be included in the customer billing system, along with any other pertinent information. First, assemble the total (uncorrected) water consumption for all accounts and connections for each size of meter by month (or other billing period) and for the entire study period, as shown in Table 14.2. Remember to use the same unit of measure for *billed authorized consumption* as the *water supplied* value—this will likely require performing a conversion, for example, from cubic feet to million gallons.

> **R**emember to use the same unit of measure for *billed authorized consumption* as the *water supplied* value—this will likely require performing a conversion, for example, from cubic feet to million gallons.

Meter Size, in	Number of Meters	Percent of Total Meters	Percent of Metered Consumption
5/8	11,480	94.1	71.2
¾	10	0.08	0.1
1	338	2.8	2.8
1½	124	1.0	2.8
2	216	1.8	11.7
3	15	0.12	6.6
4	7	0.05	2.2
6	6	0.05	2.6
Total	12,196	100.00	100.0

Source: American Water Works Association. "Water Audits and Loss Control Programs." *Manual of Water Supply Practices M36*, 3d ed. Denver, Colo.: AWWA, 2008.

TABLE 14.1 Water Consumption by Meter Size for County Water Company: January 1 to December 31, 2006

2006 By Month	Residential (million gal)	Industrial (million gal)	Commercial (million gal)	Metered Agriculture (million gal)	Total for All Meters (million gal)
January	146.6	35.8	8.1	0	190.5
February	162.9	35.8	8.1	0	206.8
March	162.9	35.8	8.1	0	206.8
April	179.2	39.1	8.1	24.4	250.8
May	211.8	42.4	8.1	57.0	319.3
June	228.1	48.9	8.1	74.9	360.0
July	260.3	48.9	8.1	57.0	374.3
August	266.5	48.9	8.1	74.9	398.4
September	228.1	45.6	8.1	65.2	347.0
October	162.9	35.8	8.1	0	206.8
November	162.9	35.8	8.1	0	206.8
December	146.6	35.8	8.1	0	190.5
Annual Total	2,318.8	488.6	97.2	353.4	3,258.0
Daily Average, mgd	6.35	1.34	0.27	0.97	8.93

Source: American Water Works Association. "Water Audits and Loss Control Programs." *Manual of Water Supply Practices M36*, 3d ed. Denver, Colo.: AWWA, 2008.

TABLE 14.2   Total Metered Water Consumption by Category for County Water Company in 2006 (Uncorrected)

Data on *Water Supplied* to the distribution system (see Chap. 10) can be matched with customer billing data and the amount of nonrevenue water tracked on a preliminary basis, as shown in Table 14.3 for the Philadelphia Water Department. Using such a report on a monthly basis gives the water utility a snapshot look at its water efficiency status on a more periodic basis than the annual water audit. However, these monthly numbers do not offer the detail or degree of validation of the water audit data, but are still useful for short-term tracking of water efficiency standing.

If computerized billing records or reports do not exist, the water auditor must assemble customer account information from available records. Start by identifying all customer users from permanent structures who should have meters. Accounts should be identified by several descriptors such as account number, property street address, meter size, meter serial number, connection size, assessor's parcel number, and the name and address of the property owner, as well as any tenants. In order to track customer consumption patterns and water conservation impacts it is important to list the consumption category for each account: residential, industrial, commercial, agricultural, governmental, and the like. Any data that is gathered manually in this manner should be input into a computerized format. Ideally, the water utility should move to purchase/install a standard computerized customer billing system. In lieu of this, or during transition to such a system, data might be entered into a desktop database or spreadsheet.

WATER STATISTICS—JUNE 2006

The Philadelphia Water Department distributed an average of 259.3 mgd of water from its treatment facilities during the month of June. This number is lower than June 2005 (268.1 mgd). Water Revenue Bureau customer billing records show that for this June, 174.4 mgd of water was billed to customers to city customers and exported to our wholesale water utility accounts. This figure is higher than last June (170.7).

Nonrevenue water (water supplied minus billed consumption) at this end of Fiscal Year 2006 stands at 76.3 mgd. This is a major improvement over last year's number of 83.6 mgd.

The following table shows water statistic trends for the previous twelve month period.

12-Month Running Period	Water Supplied (mgd)	Billed Consumption (mgd)		Nonrevenue Water (mgd)	Number of Customer Billing Accounts	
		City	Exports		Large	Small (5/8", ¾")
8/04–7/05	260.7	156.9	18.8	**85.0**	13,355	458,339
9/04–8/05	261.3	159.4	19.1	**82.9**	13,332	458,251
10/04–9/05	261.5	160.5	18.8	**82.2**	13,312	458,144
11/04–10/05	261.4	159.9	18.8	**82.7**	13,292	458,056
12/04–11/05	260.9	159.4	18.9	**82.6**	13,274	457,966
1/05–12/05	260.3	159.4	19.1	**81.8**	13,253	457,906
2/05–1/06	258.8	160.6	19.4	**78.8**	13,237	457,922
3/05–2/06	256.9	159.6	19.3	**78.0**	13,217	457,949
4/05–3/06	255.6	158.5	19.3	**77.8**	13,194	457,956
5/05–4/06	254.8	158.0	19.4	**77.4**	13,176	457,946
6/05–5/06	254.5	157.7	19.5	**77.3**	13,156	457,972
7/05–6/06	253.8	157.8	19.7	**76.3**	13,137	458,043

Source: Philadelphia Water Department

TABLE 14.3   City of Philadelphia Monthly Water Statistics Report

14.2   Using the Customer Billing System to Extract Customer Water Consumption Data

Customer Billing Systems historically have been designed with a primarily *financial* purpose: to generate bills that result in revenue collection.

It has become evident in recent years that the value of customer consumption data goes beyond serving as the basis for billings. Consumption data is needed to evaluate water conservation practices. It is needed to realistically size meters and service lines on an individual basis, and to size water supply infrastructure on a community basis. Consumption data is necessary to develop accurate hydraulic models. It is also needed to assist water loss control programs, by separating components of authorized consumption from components of loss. Beyond their financial purpose, customer billing systems have also come to be relied upon for these important *engineering* purposes. Unfortunately, many systems were designed with only the financial function in mind, and water utilities that now also use billing system data for engineering purposes may be doing so without knowing whether adequate controls exist to ensure the engineering integrity of customer consumption data.

It is important that water utility managers understand the workings of the customer billing system with regard to consumption data integrity. Many billing systems—while configured with sound billing intentions—may unknowingly corrupt the engineering integrity of water consumption data. Some systems, when generating a credit to the customer, back calculate the adjustment by changing the actual meter readings or consumption. A monetary credit to the customer is thereby triggered by reducing, eliminating, or creating negative consumption values for the period in question. Frequent adjustments in this manner can greatly distort the true amount of consumption for individual customers or whole communities. Other programming features in customer billing systems—while created with good financial intention—might unintentionally corrupt consumption data in an engineering sense.

It is recommended that sufficient controls be designed into the customer billing systems if the system is to be used for both billing (*financial*) and operational (*engineering*) purposes. This will protect customer consumption data integrity while providing proper billing functions. The primary function of most existing customer billing systems is to accurately account for the revenue received by the utility for services rendered to individual customers. Utility operators embarking upon conservation, hydraulic modeling, or water loss control programs should undertake a careful review of the billing system function and configuration to ascertain that the actual consumption amounts are not unintentionally modified by billing operations, and that the customer consumption amounts recorded as output of the billing system are unchanged from the data generated by customer water meters. The utility should undertake a flowcharting exercise of the billing process, as detailed later in this chapter, in order to identify any impacts to customer consumption integrity, as well as to identify any apparent loss components from the data-handling process. If consumption data is found to be modified by billing operations, the utility manager should consider reprogramming the billing system to record both the *registered* consumption and *billed* consumption as separate fields, thus ensuring that the accuracy of billing functions and customer consumption data are preserved. Until this is implemented, an estimate of the impact of such adjustment activity should be included as a component of the apparent losses.

14.3   Adjusting for Lag Time in Customer Meter Reading Data

Corrections must be made to metered use data when the source-meter reading dates and the customer-meter reading dates do not coincide with the beginning and ending dates of the water audit period, which is recommended to be a 1-year period.

Adjusting for one meter route. For example, a utility is studying one calendar year, January 1 through December 31. Source meters are read on the first day of each month and customers' meters are read on the tenth day of each month. The goal is to calculate the amount of water supplied and consumed for the calendar year.

Source meters: No lag time correction is made for source meters, because their reading usually occurs on the days that the water audit period begins and ends. If the last reading (December 31) was a day late (January 1), then the water supplied for January 1 should be subtracted from the total water supply reading.

Customer meters: Because customer meter readings do not coincide neatly with the study period, a correction must be made. The best way to account for changes in the number of customers and in consumption patterns is to prorate water consumption for the first and last billing periods within the water audit period.

The first billing period has only 10 days that actually occur in the water audit period. Yet the billing information represents 31 days of consumption. If consumption for the December 11 through January 10 period is 33.204 million gal, the amount applicable to the water audit period is:

$$33.204 \text{ million gal} \times \frac{10 \text{ days}}{31 \text{ days}} = 10.711 \text{ million gal}$$

Thus, 10.711 million gal of the consumption read on January 10 applies to the water audit period.

At the end of the water audit period, there are 21 days not included in the billing data collected on December 10. Consumption for the last 21 days in December is obtained from the following month's billing. If sales for that month are 36.66 million gal, the amount applicable to the water audit period is:

$$36.66 \text{ million gal} \times \frac{21 \text{ days}}{31 \text{ days}} = 24.83 \text{ million gal}$$

Thus, 24.83 million gal is added to the consumption read on December 10.

Adjusting for many meter routes. The preceding discussion describes the basic method for correcting lag time in meter reading when all customers' meters are read on the same day. Unless fixed network AMR systems are used (see Chap. 13) that seldom happens since most utilities have such large customer populations that it is impossible to read all of the meters on a single day. Usually, meters are assigned to different routes and read on different days. Therefore, a meter lag correction should be used for each meter reading route, particularly if each customer's meter is read on the same date each month. Figure 14.1 gives an example of this.

A meter lag correction can involve a number of steps. In our example, County Water Company has three meter routes, each with its own reading date. The water audit period is one calendar year, and the consumption is prorated for each meter route or book. Meters are read bimonthly: route A on the first of the month, route B on the tenth of the month, and route C on the twentieth of the month.

The uncorrected total metered consumption is based upon bills issued during the water audit period. But, because of the bimonthly billing schedule, these bills would not include all water consumed during the year. Some water shown as used in the first billing period (issued in February) actually occurred in the preceding December. The last set of bills, issued in November and December, would not include water consumed in December. Two corrections need to be made. First, water consumed in the month preceeding the water audit period must be subtracted from consumption figures. Second, water consumed in the final month of the water audit period must be added.

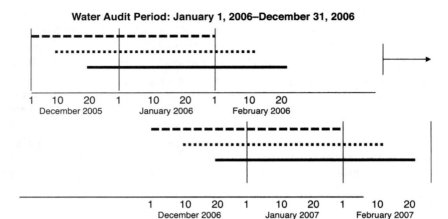

Water Audit Period: January 1, 2006–December 31, 2006

Meter Route A: ▪ ▪ ▪ ▪ ▪ ▪ ▪ ▪ ▪ ▪ ▪

Meter Route B: • • • • • • • • • • • • • • • •

Meter Route C: ▬▬▬▬▬▬▬▬▬

The December through January billing period is 62 days long.

Route	Date Read	Sales	Adjustment
A	2/1/2006	4.0 million gal	31/62 = 2.0 million gal
B	2/10/2006	3.3 million gal	21/62 = 1.1 million gal
C	2/20/2006	3.6 million gal	11/62 = 0.6 million gal

Total adjustment to eliminate 2005 consumption from the audit period = −3.7 million gal. This amount appears on the February billing, but the water was consumed during the previous December.

Route	Date Read	Sales	Adjustment
A	2/1/2007	4.2 million gal	31/62 = 2.1 million gal
B	2/10/2007	3.3 million gal	21/62 = 1.1 million gal
C	2/20/2007	3.9 million gal	11/62 = 0.7 million gal

Total adjustment to add December 2006 sales to the audit period = +3.9 million gal. This amount did not appear on the final bill for the year; it is prorated from the bill on which it appears.

Net adjustment... +0.20 million gal

Figure 14.1 Detailed meter lag correction. (*Source:* American Water Works Association. "Water Audits and Loss Control Programs." *Manual of Water Supply Practices M36*, 3d ed. Denver, Colo.: AWWA, 2008.)

Figure 14.1 shows how to adjust water consumption figures for meter lag time. The result of a net adjustment of +0.20 million gal is shown at the bottom of this figure. Many utilities combine accounting and billing procedures into a computerized format to make this procedure easier and quicker.

14.4   Determining the Volume of Apparent Loss Due to Systematic Data Handling Error in Customer Billing Systems

The majority of North American drinking water utilities meter their customers and bill based upon measured consumption. This is standard practice recommended by AWWA. However, not all utilities meter their customers, instead these water utilities bill customers a flat fee per billing period. Others meter a portion of their customer accounts. This latter scenario can occur if

- The utility is in transition to a fully metered customer population.

- Utility policies dictate that certain accounts, such as municipal properties or fire connections, need not be metered.

- Some of the meters are known to be nonfunctional, highly inaccurate or readings unobtainable, in which case estimates of consumption are used in place of measured consumption.

Without functional meters in place the water auditor must devise an estimate of the water consumed by the unmetered population. A number of means exist to develop reasonable estimates. For instance, in an unmetered system, water meters could be installed in a small sample of accounts (50 or 100) based upon consumption category or meter size. Data from these meters could be used to develop average consumption trends that could be inferred for the entire population in each category. Make certain that any estimating process that is developed is fully documented and based upon current conditions. Unmetered accounts require the use of estimation, an action which can interject a degree of error into the measure of customer consumption. For this reason it is highly recommended that all customer consumption be properly metered, read, and archived.

For water utilities that meter customer consumption, integrity must exist not just with the accuracy of the meter, but also with the processes to transmit, archive, and report customer consumption totals as derived from the meter population. An error at any point in this process potentially represents an apparent loss by distorting the ultimate documented value of customer consumption, causing a portion of the consumption to be understated and possibly missing a portion of revenue. Systematic data-handling error can therefore occur anywhere from the time that the meter reading is registered to the final reporting and use of the consumption data.

As discussed in detail in Chap. 13, considerable error can occur in the customer consumption data transfer or meter reading process. Procedures to quantify apparent losses due to data transfer error, and ways to reduce this form of error, are presented in Chap. 13, with AMR and advanced metering infrastructure (AMI) highlighted as remarkably improving technologies that are greatly assisting water utilities in this realm.

Typically meter readings are transferred to customer billing systems where they are used to calculate the volume of customer consumption occurring since the previous reading. The consumption volumes are archived and the data is used in the process of generating water bills and to assist the varied financial operations involved in the revenue collection process. Systematic data handling errors often occur in this process in a manner that corrupts the integrity of some of the customer consumption volumes.

In the United States consumption is most often recorded in units of cubic feet or thousand gallons. Billing systems often include programming algorithms that assign estimates of consumption if an actual meter reading cannot be obtained. These algorithms often base the estimate on the recent trend of customer consumption, or they may use another method. If a poor or outdated estimation algorithm exists in the customer billing system, underestimation or overestimation of customer consumption can occur, either of which could distort consumption data needed for operational purposes. The water auditor should come to understand the method used to estimate consumption and consider programming refinements if it is determined that the existing method creates inaccuracies. A quantity representing the amount of missed customer consumption due to this occurrence should be included in the water audit.

A significant type of error can occur in the way that billing adjustments affect registered consumption data. An important question: are billing adjustments triggered by modifying actual consumption volumes? As described in Sec. 14.2, billing systems designed with good revenue collection intention may corrupt the operational integrity of customer consumption volumes when generating a credit.

Distortions in customer consumption due to billing adjustments can occur when billing systems do not distinguish between *registered* consumption (from meter readings) and *billed* consumption, listed on the customer bill and archived in the billing records. Billed consumption can differ from registered consumption when the customer is due a monetary credit. If the billing system creates the credit (negative revenue to the utility) by creating negative consumption values, actual consumption data becomes distorted. Billing systems that include separate fields for registered and billed consumption avoid this problem.

Table 14.4 gives an example of a residential customer account that incurred estimates for a 23-month period, during which time the property was temporarily vacant and then sold to a new owner who consumes less water than his predecessors. Beginning in October 2002 the water utility was unable to obtain a reliable meter reading at this property. This may have been due to blocked access to the meter, a failure of AMR equipment or another cause. Unfortunately, the water utility was unable to correct this condition and obtain an accurate meter reading until August 2004. During the period without readings, the water utility assigned an estimate of the consumption based upon the customer's recent history, in this case 885 cubic feet/month.

This estimate, shown in Column D, closely matched the actual consumption (shown in Column G for illustrative purposes) until April 2003, when the property was vacated and placed for sale. The property was vacant until August 2003 and experienced only minimal water consumption during periodic caretaker visits from April to August 2003. Upon sale to a new owner in August 2003, a regular pattern of water consumption resumed, but at a slightly lower rate than the previous owner.

Between April 2003 and August 2004 (17 months) the assigned estimate (885 cubic feet) notably overestimated the consumption for this account. When the water utility was once again able to gain an accurate meter reading, it found that its estimate of the July 2004 meter reading (42477) was overstated by a total of 4132 cubic feet, since the last accurate meter reading in September 2002. This resulting cumulative overestimation error was compounded by

- The lengthy duration (23 months) of the period with no meter readings
- The 4-month period of vacancy of the property
- The lower water consumption habits of the new property owner

Upon obtaining an accurate meter reading in August 2004 an adjustment of negative 4132 cubic feet was necessary and a credit due to the customer in the dollar amount commensurate with the volume of adjusted consumption.

A Year	B Month	C Meter Reading (Estimates Shown in BOLD)	D Billed Consumption (ft³) (Current Minus Previous Meter Reading, Estimated Consumption in BOLD)	E Cumulative Billed Water Consumption (per year)	F Actual Meter Reading	G Actual Consumption (ft³)	H Cumulative Actual Consumption
2001	Dec	15004			15004		
2002	Jan	15838	834	834	15383	834	834
	Feb	16654	816	1,650	16654	816	1,650
	Mar	17496	842	2,492	17496	842	2,492
	Apr	18304	808	3,300	18304	808	3,300
	May	19220	916	4,216	19220	916	4,216
	Jun	20162	942	5,158	20162	942	5,518
	Jul	21130	968	6,126	21130	968	6,126
	Aug	22105	975	7,101	22105	975	7,101
	Sep	23007	902	8,003	23007	902	8,003
	Oct	**23892**	**885**	8,888	23867	860	8,863
	Nov	**24777**	**885**	9,773	24722	855	9,718
	Dec	**25662**	**885**	10,658	25535	813	10,531
2003	Jan	**26547**	**885**	885	26360	825	825
	Feb	**27432**	**885**	1,770	27184	824	1,649
	Mar	**28317**	**885**	2,655	28021	837	2,486
	Apr	**29202**	**885**	3,540	28433	412	2,898
	May	**30087**	**885**	4,425	28513	80	2,978
	Jun	**30972**	**885**	5,310	28578	65	3,043
	Jul	**31857**	**885**	6,195	28633	55	3,098

TABLE **14.4** Distorted Customer Consumption Data due to Customer Billing Adjustments Triggered by the Use of Negative Consumption Values (Ex. 5/8-in Residential Meter Account) (*Continued*)

A Year	B Month	C Meter Reading (Estimates Shown in BOLD)	D Billed Consumption (ft³) (Current Minus Previous Meter Reading, Estimated Consumption in BOLD)	E Cumulative Billed Water Consumption (per year)	F Actual Meter Reading	G Actual Consumption (ft³)	H Cumulative Actual Consumption
	Aug	**32742**	**885**	7,080	29255	622	3,720
	Sep	**33627**	**885**	7,965	30059	804	4,524
	Oct	**34512**	**885**	8,850	30836	777	5,301
	Nov	**35397**	**885**	9,735	31592	756	6,057
	Dec	**36282**	**885**	10,620	32315	723	6,780
2004	Jan	**37167**	**885**	885	33032	717	717
	Feb	**38052**	**885**	1,770	33740	708	1,425
	Mar	**38937**	**885**	2,655	34462	722	2,147
	Apr	**39822**	**885**	3,540	35150	688	2,835
	May	**40707**	**885**	4,425	35884	734	3,569
	Jun	**41592**	**885**	5,310	36686	802	4,371
	Jul	**42477**	**885**	6,195	37520	834	5,205
	Aug	38345	–4,132	2,063	38345	825	6,030
	Sep	39113	768	2,831	39113	768	6,798
	Oct	39811	698	3,529	39811	698	7,496
	Nov	40515	704	4,233	40515	704	8,200
	Dec	41230	715	4,948	41230	715	8,915
2005	Jan	41951	721	721	41951	721	721

Source: American Water Works Association. "Water Audits and Loss Control Programs." *Manual of Water Supply Practices M36*, 3rd ed. Denver, Colo.: AWWA, 2008.

TABLE 14.4 Distorted Customer Consumption Data due to Customer Billing Adjustments Triggered by the Use of Negative Consumption Values (Ex. 5/8-in Residential Meter Account) (*Continued*)

How the customer billing system awards this credit has bearing on both the billing (*financial*) and operational (*engineering*) functions of the system. While money can flow *both* to and from the drinking water utility—via charges and credits, respectively—water flows in *only* one direction, being supplied by the utility to the customer. If the billing system contains only a single field for customer consumption, then the billed consumption value for August 2004 is *negative* 4132 cubic feet. While a negative consumption number is acceptable for use for billing (financial) reasons as it translates into a monetary credit, a negative consumption number is unacceptable for operational (engineering) purposes since the actual consumption for August 2004 was 825 cubic feet (Column G), not negative 4132 cubic feet as shown in Column D.

The distortion of the consumption data is further reflected in the estimated versus actual consumption based upon yearly periods. Water utility analysts reviewing the account data shown in Table 14.4 for conservation or loss control purposes would be in error by 3840 cubic feet (10,620 minus 6780) over the actual consumption in 2003. Conversely, the analysis would be understated for this account by 3967 cubic feet (8915 minus 4948) in 2004. Some may reason that the periods of estimation and adjustment ultimately balance with no net difference over the long term, therefore using a single consumption value is acceptable. However, many analytical and reporting functions are performed over the course of a calendar or business year. If a given account has been poorly estimated for many years, the use of a huge multiyear adjustment in the last year will greatly distort the consumption for that final year. Additionally, in any given drinking water utility many hundreds or thousands of accounts could utilize estimates for varying periods of time. Reliably estimating the net impact of the aggregate overestimation or underestimation of these accounts in a given year is unnecessarily complex. Clearly, while a negative consumption value can be acceptable for billing (*financial*) purposes, it is quite harmful to the integrity of the data for operational (*engineering*) purposes.

For the reasons explained above, it is recommended that water utility customer billing systems include two separate fields for customer consumption: one for *registered* consumption and a separate field for *billed* consumption. Using the same data from the example in Table 14.4, the form of the data with separate fields is shown in Table 14.5.

Table 14.5 includes separate columns for billed consumption (Column D) and registered consumption (Column G). When actual meter readings resumed in August 2004 the consumption adjustment of negative 4132 cubic feet appears as billed consumption in Column D and is used to generate the monetary credit to the customer. However, Column G reflects the revised estimate of consumption for the prior 30-day period, which is based upon the difference between the two most recent actual meter readings (September 2001 and August 2003). This one-time estimate is determined as:

$$(38345 - 23007)/23 \text{ months} = 667 \text{ cubic feet}$$

By September 2004, the second consecutive actual monthly meter reading was obtained, estimates are no longer utilized, and billed consumption once again matches registered consumption. The benefit to the operational integrity of data using separate billed and registered consumption fields is shown by comparing the cumulative consumption for 2004 in Column D and Column G, or 4948 and 9747 cubic feet, respectively. If only a single field is used for consumption the billed value of 4948 greatly understates

A Year	B Month	C Meter Reading (Estimates Shown in BOLD)	D Billed Consumption, ft³ (Current Minus Previous Meter Reading, Estimated Consumption in BOLD)	E Cumulative Billed Water Consumption (per year)	F Actual Meter Reading	G Registered (Actual) Consumption, ft³	H Cumulative Registered (Actual) Consumption
2001	Dec	15004			15004		
2002	Jan	15838	834	834	15383	834	834
	Feb	16654	816	1,650	16654	816	1,650
	Mar	17496	842	2,492	17496	842	2,492
	Apr	18304	808	3,300	18304	808	3,300
	May	19220	916	4,216	19220	916	4,216
	Jun	20162	942	5,158	20162	942	5,518
	Jul	21130	968	6,126	21130	968	6,126
	Aug	22105	975	7,101	22105	975	7,101
	Sep	23007	902	8,003	23007	902	8,003
	Oct	**23892**	**885**	8,888	UNKNOWN	885	8,888
	Nov	**24777**	**885**	9,773		885	9,773
	Dec	**25662**	**885**	10,658		885	10,658
2003	Jan	**26547**	**885**	885		**885**	885
	Feb	**27432**	**885**	1,770		**885**	1,770
	Mar	**28317**	**885**	2,655		**885**	2,655
	Apr	**29202**	**885**	3,540		**885**	3,540
	May	**30087**	**885**	4,425		**885**	4,425
	Jun	**30972**	**885**	5,310		**885**	5,310
	Jul	**31857**	**885**	6,195		**885**	6,195

Year	Month						
	Aug	**32742**	**885**	7,080		**885**	7,080
	Sep	**33627**	**885**	7,965	N	**885**	7,965
	Oct	**34512**	**885**	8,850	O	**885**	8,850
	Nov	**35397**	**885**	9,735		**885**	9,735
	Dec	**36282**	**885**	10,620	R	**885**	10,620
2004	Jan	**37167**	**885**	885	E	**885**	885
	Feb	**38052**	**885**	1,770	A	**885**	1,770
	Mar	**38937**	**885**	2,655	D	**885**	2,655
	Apr	**39822**	**885**	3,540	I	**885**	3,540
	May	**40707**	**885**	4,425	N	**885**	4,425
	Jun	**41592**	**885**	5,310	G	**885**	5,310
	Jul	**42477**	**885**	6,195	S	**885**	6,195
	Aug	38345	−4,132	2,063	38345	667	6,862
	Sep	39113	768	2,831	39113	768	7,630
	Oct	39811	698	3,529	39811	698	8,328
	Nov	40515	704	4,233	40515	704	9,032
	Dec	41230	715	4,948	41230	715	9,747
2005	Jan	41951	721	721	41951	721	721

Source: American Water Works Association. "Water Audits and Loss Control Programs." *Manual of Water Supply Practices M36,* 3rd ed. Denver, Colo.: AWWA, 2008.

TABLE 14.5   Utilizing Separate Fields for Registered and Billed Consumption in the Customer Billing System. Example Data for a 5/8-in Residential Water Meter Account (see Table 14.4) the actual consumption for the year. The registered consumption value of 9747 cubic feet is a much more representative value of the water consumed by this account during 2004.

It is recommended that water utility customer billing systems include two separate fields for customer consumption: one for *registered* consumption and a separate field for *billed* consumption.

In determining the amount of data analysis error occurring in billing system operations the water auditor should determine how billing adjustments are calculated. If adjustments are triggered by changes in consumption, then an approximation of the number of adjustments—both overstating and understating actual consumption—should be attempted. If a significant understatement of customer consumption has occurred, then an estimate of this difference should be included as an apparent loss and entered in the water audit.

14.5   Billing Policy and Procedure Shortcomings

Apparent losses can occur due to policies and procedures that are shortsighted or poorly designed, implemented, or managed. Such occurrences can be subtle and numerous. Flowcharting the customer billing process—with a focus on impacts to customer consumption values—gives insight to the likelihood of these types of apparent losses. Some of the common occurrences to consider are:

- Despite company goals to meter all customers, the installation of meters in certain customer classes is ignored: this is common for municipally owned buildings in water utilities run by local governments.
- Provisions allowing customer accounts to enter "nonbilled" status, a potential loophole often exploited by customer fraud or poor management by the water utility.
- Bureaucratic regulations or inefficiencies that cause delays in permitting, metering, or billing operations.
- Poor customer account management: accounts not initiated, lost, or transferred erroneously.

The degree to which such shortcomings in billing account management exists is largely dependant upon the accountability "culture" that exists in the water utility. If accountability is only casually emphasized, it is likely that numerous opportunities for missed consumption exist. If sound accountability is trumpeted by the utility's leaders and managed down to all levels of staff, then such occurrences are likely to be isolated and of minor significance. The water auditor should consider including an estimate of apparent loss that represents the collective policy and procedure shortcomings of the water utility. During the top-down audit, perhaps only a rough approximation can be ventured. During subsequent audits, bottom-up investigations can give greater insight to such problems, and corrections can be identified. Before a definitive strategy is set, however, the auditor should begin to perform more detailed investigations of the source data and billing functions in order to validate the preliminary loss quantities and obtain a more accurate picture of the apparent losses. The bottom-up process involves detailed investigation or auditing work, similar to detailed financial audits that accountants perform. Bottom-up water auditing functions should consider the following activities:

- **Step 1**: Analyze the workings of the customer billing system to identify deficiencies in the water consumption data handling process resulting in apparent losses. Flowcharting the data-handling pathways is a good way to perform this analysis.

- **Step 2**: Compile listings of basic customer account demographics, including number of meters by meter size, customer type, and consumption ranges. Look for anomalies such as groups of small meters registering large annual consumption volumes or large meter accounts registering unusually small annual consumption volumes.

- **Step 3**: Perform meter accuracy testing for a sample of meter installations in order to establish an understanding of the functional status of the meter population (see Chap. 12).

- **Step 4**: Assess a sample of customer accounts or locations for unauthorized consumption potential (see Chap. 15).

- **Step 5**: Identify accounting policies that have the potential to allow water to be unbilled. Some billing systems have provisions to allow accounts to enter a *nonbilled* status on the basis that the customer is, at least temporarily, not consuming water. A vacant property is a good example. However, such nonbilled accounts have been found to be a ripe source of apparent loss, as many times poor record-keeping results in the customer remaining nonbilled even after water consumption has resumed.

It is recommended that the billing system analysis always be performed as the initial step, since gaps in this process could affect the data that is evaluated in the other steps.

For most drinking water utilities the customer billing system serves as the source of all customer data, including water consumption. Early in the development of the water loss control program, the auditor should develop a detailed understanding of the ways in which consumption data is managed in the customer billing system. Constructing a series of flowcharts that outline the various information handling processes is a systematic approach that can reveal gaps in policy, procedures, or programming that may allow apparent losses to occur. Any such deficiencies that allow customers to exist without billing accounts, without accurate metering and meter reading, or allow metered consumption data to be unduly modified, can create apparent losses.

Figures 14.2 to 14.5 represent several customer billing system flowcharts for the City of Philadelphia. The Philadelphia Water Department and Water Revenue Bureau, an office of the city's Revenue Department, together manage the customer billing process. Figure 14.2 gives a flowchart that represents an overview of the entire billing process. While it displays the major billing functions at a glance, it lacks sufficient detail to identify likely occurrences of apparent loss. Additional flowcharts that display individual sub-processes of the customer billing system are given in Figs. 14.3 to 14.5. In these flowcharts the meter reading sequence for both automatic and manually read customer meters are shown, as well as the meter rotation (replacement) process. Although Philadelphia installed the largest water utility AMR system in the United States from 1997 to 1999, approximately 2% of its customer accounts await AMR due to access issues or the need to address large meter sizing/piping constraints. Therefore Philadelphia utilizes both automatic meter reading and, to a much smaller degree, manual meter reading. Using flowcharts to assess various subprocesses of billing operations

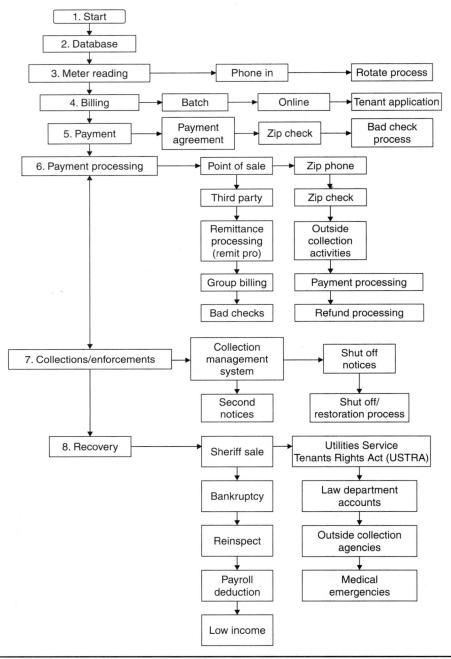

Figure 14.2 Overview of customer billing system for water and waste water services in the City of Philadelphia. (*Source:* Philadelphia Water Department.)

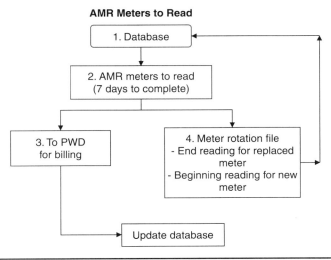

FIGURE 14.3   AMR flowchart for the city of Philadelphia. (*Source:* Philadelphia Water Department.)

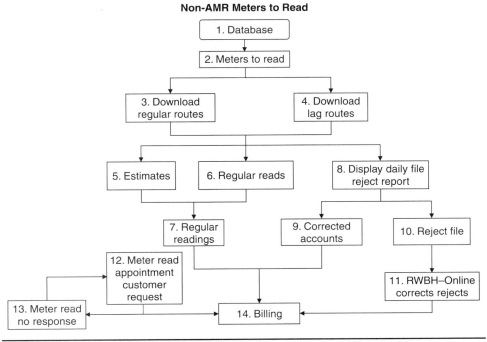

FIGURE 14.4   Manual (non-AMR) meter reading flowchart for the city of Philadelphia. (*Source:* Philadelphia Water Department.)

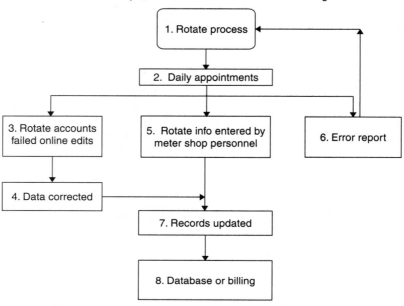

FIGURE 14.5 Customer meter rotation process flowchart for the city of Philadelphia. (*Source:* Philadelphia Water Department.)

allows the auditor to confirm the billing functions that are working properly and identify gaps that cause customer consumption to be understated and the utility to lose revenue.

The billing system flowcharts shown in Figs. 14.2 to 14.5 are given for illustrative purposes only. While they are valid for the process used in Philadelphia, each water utility has a customer billing process that has features that are unique to their organization. Therefore each utility should generate flowcharts that reflect their individual processes.

By outlining the billing data flow paths and documenting information handling policies, procedures, and practices, the auditor can usually establish a highly detailed picture of the billing process and sources of apparent losses due to data handling error. A small sample of several dozen to several hundred customer accounts in various categories should be analyzed to determine if any loss impacts are found to exist, and whether systematic error exists in the procedural or programming aspects of the system. The auditor should analyze samples of accounts in any special billing categories (municipal properties, nonbilled accounts), as well as a sample of the largest water consumers to reveal likely occurrences of apparent losses.

In analyzing customer billing system operations, the auditor should be particularly mindful to assess the integrity of

- Policy: Are policies regarding customer metering, billing, water rates, service line responsibilities, and the like, rational consistent, codified, and well-communicated?

- Procedures: Do written procedures exist? Are procedures used to ensure that consistent metering, meter reading, and billing functions are employed for all customers? Are checks and balances built into the system to flag breakdowns or gaps in the process?

- Practices: Do the actual practices reflect the mandates of the procedures? Does an effective training program exist to ensure that all employees are educated on policy and procedures? Are meter readers, billing clerks, or similar employees properly monitored and supervised to detect and minimize human error and ensure that policies and procedures are being followed?

Additionally, when searching for more specific occurrences of apparent loss:

- Are certain classes of customers, such as municipal properties, exempt from metering and billing? If so, how is their water consumption accounted for by the water utility?

- Can customers enter a nonbilled status for conditions such as property vacancy, delinquent or shutoff accounts, and the like? If so, are these accounts routinely monitored to detect any water consumption in these supposed nonwater using accounts?

- Are estimates of customer consumption employed if meter readings are not available? If so, how accurately does the estimate reflect actual consumption? Do checks exist to validate or periodically update the estimation protocol?

- Does a policy exist for enforcement to deter unauthorized consumption? Can customers have service terminated for nonpayment? If so, are significant numbers of customers illegally reactivating their service? Is there a mechanism to detect and thwart this activity?

- Do programming algorithms incorporate billing adjustments that unduly modify actual metered consumption data, such as shown in Tables 14.4 and 14.5?

- Are metering, meter reading, and billing functions actively tracked and monitored by the issuance of routine management reports that are structured to summarize performance, identify trends, and flag anomalies?

- Are customer consumption and billing trends evaluated on a regular basis to discern specific and overall trends in consumption and loss patterns in response to conservation, loss control programs, or demographic trends such as growth in the industrial sector?

These are just some of the questions that might be posed during the bottom-up audit of the data handling process. For every water utility certain unique processes can exist and should be scrutinized by the auditor.

14.6 Quantifying Systematic Data Handling Errors in the Water Audit and Addressing These Losses

A rough approximation of apparent loss due to systematic data handling error in the customer billing system can be included in the initial top-down water audit. After this, it is recommended to assign moderate staff or consultant resources to analyze the workings of the customer billing system: flowcharting is the recommended approach. Figure 11.7 in Chap. 11 gives an example of the development of a *revenue protection program* to assess and address apparent losses. As shown in this figure, billing system flowcharting is the first and foremost undertaking in launching such a program. While such an effort involves some expense and time, these investments are typically minimal and hold the potential to quickly identify feasibly recoverable apparent losses that can immediately increase revenues, thereby providing a fast payback and successful start to the program.

As suspect customer accounts become identified by revenue protection analysts, individual inspections of customer properties will be needed to confirm water using status in nonbilled accounts, evidence of meter tampering or illegal connections and other occurrences that result in underregistered water and missing revenue. Depending upon the extent of the apparent loss problem in the utility, the utility manager may determine to dedicate full-time staff to the investigative function. This may be accomplished by assigning dedicated personnel to this role, or by cross-training meter reading or other personnel who routinely visit customer properties.

Apparent loss control via revenue protection programs is often a very cost-effective undertaking, particularly in the early phases of the water loss control program. Frequently an almost-immediate impact is realized through increased revenue recovery. Establishing the revenue protection program as outlined in Fig. 11.7 is therefore one of the initial first steps in the bottom-up phase of the water loss control program.

CHAPTER 15

Controlling Apparent Losses—Unauthorized Consumption

George Kunkel, P.E.

Julian Thornton

Reinhard Sturm

15.1 Unauthorized Consumption Occurs in Many Ways

Unauthorized consumption is a label for water that is taken against the policies of the water utility and often occurs from

- Illegal connections to the water distribution system
- Open bypasses, typically around large customer meters
- Buried or otherwise obscured meters that result in water consumption being hidden from the water utility
- Misuse of fire hydrants and fire fighting systems (unmetered fire lines)
- Vandalized or manipulated consumption meters (meter tampering)
- Tampering with meter reading equipment
- Illegally opening of valves or curb-stops on customer service piping that has been discontinued or shutoff for nonpayment
- Illegally opening valves to neighboring water distribution systems that are intended to stay closed except for emergency or special use
- Dishonest actions by water utility employees working in the metering, meter reading, or billing functional groups of the utility
- Failure to notify the water utility to activate a billing account after new customer service connections are placed into service

Unauthorized consumption occurs to some extent in virtually every drinking water utility. It typically occurs through the deliberate actions of customers or other persons who take water from the system without paying for it. The nature and extent of unauthorized consumption occurring in a water utility usually depends upon a combination of the following factors:

- The demographic scale of the community being served
- The economic health of the community being served
- The value the community accords to water as a resource, often as a function of the relative abundance or scarcity of water in the region
- The strength and consistency of the enforcement policies and practices existing in the water utility
- The political will of water utility management and public officials to enact and enforce effective policies to thwart unauthorized consumption

The value that the community and water utility place upon water supply and the management effectiveness of the water utility are often reflected by the amount of unauthorized consumption occurring in a locale. Establishing features of a good accountability and loss control program—water auditing being foremost—will inevitably uncover situations where unauthorized consumption is occurring.

Of the major components of apparent loss, unauthorized consumption creates the greatest impacts to ratepayer equity. When a portion of the customer population under-

> **U**nauthorized consumption occurs to some extent in virtually every drinking water utility.

pays or fails to pay for water service, the paying portion of the customer population effectively pays for the nonpaying portion, since rates are usually set to recover all costs of service. When the need arises to increase water rates, the paying population is forced to shoulder an even greater financial burden while scofflaws remain unchecked. If a water utility does not control unauthorized consumption, it does a disservice to its paying customers and risks a public relations backlash should the knowledge of high unauthorized consumption reach the media or general public.

15.2 Quantifying the Volume of Unauthorized Consumption in the Water Audit

Most instances of unauthorized consumption are attributed to customers who either cannot or will not pay for the services they are rendered. All utility systems are susceptible to the occurrence of unauthorized consumption, and this occurrence is substantial for some. In large, urban systems, occurrences of unauthorized consumption are likely to be more numerous than that of medium or small systems in suburban or rural settings. Yet, in most cases and regardless of system size, the total annual volume of water lost to unauthorized consumption is likely to be a small portion of the water that a utility puts into supply. The water audit should quantify the component of unauthorized consumption occurring in the utility. For first-time water audits, or where unauthorized consumption is not believed to be excessive, the auditor should use the default value of 0.25% of water supplied (WS) as the volume of unauthorized consumption. This percentage has been found to be representative of this component of loss in water audits compiled worldwide. For water utilities with well-established water audits, or those believing that unauthorized consumption is excessive, individual components of unauthorized consumption should be specifically identified, as well as policies and practices that may, unwittingly, create opportunities to obtain water service without making proper payment for the service.

If the auditor believes that unauthorized consumption is significant in his utility, and time and resources are available to investigate, then work to quantify individual components of authorized consumption can be carried out. This work can be tedious, however, and the auditor should use good judgment to determine whether the extra effort to obtain specific estimates of unauthorized consumption is worthwhile compared to merely applying the default value. Once the water auditing process matures over a period of years, some reasonable effort should be dedicated to investigation of specific occurrences of unauthorized consumption in the water utility. Still, in most cases, the default value gives a reasonable quantification of unauthorized consumption.

15.3   Controlling Unauthorized Consumption

As previously discussed, unauthorized consumption can occur in many ways. For virtually all water utilities, some portion of the customer population will attempt to obtain water service without properly paying for it. However, the extent to which such occurrences exist also depends upon the policies, practices, and oversight of the water utility. Water utilities exert control over unauthorized consumption via

- Detection—the ability to become aware of unauthorized consumption in its various manners

- Enforcement—means to halt such consumption and invoke appropriate penalties

Water utilities should have mechanisms in place to detect trends of unauthorized consumption. As an example, the auditor might review opportunities for the unauthorized use of fire hydrants and ensure that a rational policy regarding fire hydrants exists. Flowcharting the processes of the customer billing system (as described in Chap. 14) can give the auditor insight into loopholes that allow unauthorized consumption to occur and go unnoticed by the water utility. Once identified, loopholes can often be expeditiously closed by procedural, programming, or permitting corrections, realizing a quick return of additional revenue. Billing data should be reviewed for suspicious trends that might reflect unauthorized consumption. For instance, active accounts registering unchanged meter readings (zero consumption) for consecutive billing cycles might be an indication of meter tampering. Household inspections can be conducted on select zero consumption accounts to determine whether actual consumption is occurring. Boundary valves to neighboring water systems should be inspected periodically to ensure that they are in the proper position. If utility policy allows customer service to be terminated due to payment delinquency, follow-up random inspections should be conducted on a sample of accounts to ensure that customers have not reactivated their service illegally. Customer meter tampering can be cost-effectively controlled by locking devices that are now commercially available at competitive prices for all sizes and configurations of customer meters. All of these actions are typical of the bottom-up activities utilities can undertake to control unauthorized consumption.

For control of unauthorized consumption on a long-term basis, the water utility should employ effective policies and enforcement capabilities. This may require changes in existing regulations, statutes, or codes and the creation of new ones. Implementing such change in these instruments can be politically sensitive and requires skilled effort over potentially long periods of time to implement, however, a strong legal framework will ultimately allow the water utility to operate with enforcement powers to keep unauthorized consumption to an economic minimum.

> **R**ecognizing that a portion of customers in any region live with real economic hardship, the water utility may choose to operate programs offering appropriate discounts, grants, or similar services to qualified customers in order to keep essential water service affordable. Having such a program working in tandem with aggressive unauthorized consumption enforcement is the best policy.

Recognizing that a portion of customers in any region live with real economic hardship, the water utility may choose to operate programs offering appropriate discounts, grants, or similar services to qualified customers in order to keep essential water service affordable. Having such a program working in tandem with aggressive unauthorized consumption enforcement is the best policy. It is never justified for a customer to take water service in an unauthorized manner based upon their purely subjective statement as to economic hardship. However, it is appropriate that water utilities recognize the limitations of certain customers in justifiable need and offer them an avenue to legitimately purchase water service at affordable rates.

15.3.1 Successfully Managing Fire Hydrants

Many water utilities lose an appreciable amount of water from fire hydrants that are opened without authorization or knowledge of the water utility. This activity not only results in an apparent loss but also frequently results in damage to fire hydrants from improper operation. In addition to loss concerns, fire hydrants are viewed more seriously as a security issue in the post 9/11 world. Hydrants could potentially be used as an entry point to intentionally inject contaminants into the drinking water supply, therefore, having strong oversight of fire hydrants is now viewed as more critical as in the past.

The primary purposes of fire hydrants are fire fighting and water distribution system testing and maintenance, including flushing water mains. In many water utilities, however, the use of fire hydrants—for both authorized and unauthorized purposes—goes far beyond these basic functions. Unauthorized consumption from fire hydrants, which is classified under apparent losses, occurs when water is drawn illegally from hydrants to fill tank trucks for landscaping or construction purposes, to wash cars, or to use recreationally for personal cooling in hot weather, such as shown in Fig. 15.1. Many water utilities have policies that permit water to be drawn from fire hydrants for a variety of community-spirited purposes. This water typically falls under unmetered, unbilled authorized consumption in the water audit and includes water used in street cleaning, filling public swimming pools, providing transient supplies (such as nonpotable supply to a traveling circus), community gardens, and constructions sites. Some allow hot weather cooling relief from fire hydrants via the use of spray caps. These varied uses of fire hydrants pose potential problems for water utilities and customers, including

FIGURE 15.1   Unauthorized fire hydrant openings waste water and pose a security threat. (*Source*: Philadelphia Water Department.)

- Water taken from fire hydrants is often unmetered. The more numerous the openings of hydrants, the greater the amount of water that must be metered or estimated to quantify this consumption in the water audit.

- Water taken continuously from fire hydrants should include backflow protection to prevent contaminants from entering the distribution system during a negative pressure event. Often no backflow protection is used.

- Water drawn from a fire hydrant could pose a health risk if used for human consumption since water quality degradation can occur as the water passes through the barrel of the hydrant.

- Cooling off from the spray of a fire hydrant is a significant safety risk as fire hydrants are usually configured to face the street, placing the public (often children) in the roadway to compete with traffic while being pushed by water under high pressure.

- Widespread unauthorized openings of fire hydrants can result in greatly reduced pressure in the distribution system, crippling firefighting capability and greatly increasing the risk of backflow contamination.

FIGURE 15.2   Unauthorized fire hydrant openings often cause damage to fire hydrants from the use of improper tools. (*Source:* Philadelphia Water Department.)

- Allowing a variety of people to operate fire hydrants increases the likelihood of damage occurring to hydrants due to lack of familiarity with operating procedures or use of improper tools to operate the fire hydrant. (See Fig. 15.2)

- Allowing multiple uses of fire hydrants sends a poor public relations message that water is free for the taking to those who can manage to open a hydrant. This is a precarious position particularly due to the need to secure drinking water systems and preserve water resources.

For the above reasons it is recommended that water utilities keep the number of permitted uses of fire hydrants to a minimum, and such usage should be carefully regulated and overseen. Utility managers should vigorously maintain control of their fire hydrants and resist requests for sundry uses of hydrants. It is important that utility managers establish a sound policy for fire hydrant usage that is supported by fire departments and political leaders. Procedures for permitting and tracking allowable uses should be put in place and enforced. Commercially constructed bulk water sales stations are available on the marketplace to provide water utilities with a means to supply water to permitted users, typically via tank trucks, rather than allowing the use of fire hydrants. This is one step of a good policy on fire hydrant use. Water utility managers should work to educate public officials, contractors, customers, the media, and other stakeholders on the need to maintain strict utility control over fire hydrants. The Loudoun County Sanitation Authority in Loudoun County, Virginia developed a comprehensive policy and detailed procedures for fire hydrant usage which has allowed them to better balance the need for access to water supply versus protection of the water distribution system and drinking water quality.[1]

15.3.2   Unauthorized Consumption at the Customer Endpoint

Unauthorized openings of fire hydrants are often visibly evident to the general public, water utility, and law enforcement personnel, thus aiding the detection of such occurrences. Unauthorized consumption at the customer endpoint is not nearly so obvious.

Unlike water visibly spewing forth from a fire hydrant, water obtained illegally at the customer endpoint occurs at meter/piping locations inside buildings or meter pits, and is likely to escape detection of all but those with a trained eye. The good news is that the array of new capabilities offered under the label of advanced metering infrastructure (AMI) gives water utilities effective new tools to detect signs of illegal endpoint consumption (see Chap. 13, Sec. 13.1.3).

The most common ways to violate water service provisions of a water utility include

- Tampering with customer meters
- Tampering with meter reading equipment
- Making illegal connections into building piping upstream of the water meter
- Illegal use of (often unmetered) fire connection lines for routine water supply
- Illegally reopening a curbstop/valve on customer service connection piping after it has been closed as an enforcement action for nonpayment by the customer
- Opening valves that should remain closed: bypasses around large meters, control valves to neighboring water utilities, etc.
- Any other means to corrupt the utility metering and billing process to illegally obtain water service at partial or zero cost

Meter tampering has been a common illegal action virtually as long as water utilities have employed meters to register consumption volumes to serve as the basis of customer billing. Perhaps the most common form of tampering has been *jumping* the meter. This is accomplished by closing the service connection supply curbstop/valve, removing the meter, and installing a straight piece of pipe ("jumper") in place of the meter, as shown in Fig. 15.3. The water thief routinely places the meter back into the line

Figure 15.3 Bypassing a water meter is not complicated. This photo shows a "cheater" or "jumper" pipe (top) that was illegally used to replace the meter in the meter setter. In the lower part of this meter pit, is a typical 5/8-in residential meter. (*Source:* Morgantown Utility Board.)

before and after the meter reading in order to register some consumption and not draw scrutiny as a zero consumption account. If the water utility is lax in gathering meter readings—and the customer thief is aware of this—the jumper may stay on the line indefinitely while the customer pays a reduced fee based upon an erroneously low estimate. In a similar guise, the Philadelphia Water Department documented an instance of over 100 residential meters being swapped into varying customer addresses in a small neighborhood of the service area. Apparently, one person had taken on the role of full time meter-jumper in order to generate reduced water bills for his customers.

Depending on the brand and age of individual water meters, tampering may be attempted in order to change the register reading on the meter. Any attempt to corrupt the registration of actual consumption at the meter is a form of meter tampering.

With the same motives in mind, some customers have attempted to disrupt the billing process by tampering with meter reading equipment of automatic meter reading (AMR) systems. This is more readily detected than meter tampering, since most AMR systems include tamper detection capabilities that send an alert to the water utility when tampering has been detected and the meter reading process is interrupted. This is one of the most basic features now offered by manufacturers to help thwart unauthorized consumption.

Detecting unauthorized consumption that occurs via illegal piping connections or valve operations is more difficult than water taken from fire hydrants or water meters, and typically relies on physical inspection of building piping and meter pits by the water utility. An illegal pipe installed into an unmetered fire line may be plainly visible in the basement of a building, but personnel must be directed to any suspicious property and be trained in identifying an illegal connection. In extreme cases (at least for developed countries) the water thief may excavate to install an illegal piping connection, and then backfill the excavation, covering the new illegal pipe. Unless the excavation and illegal pipe connection are observed when carried out, such a connection will not be detectable from above ground. If such a connection is suspected, evidence of illegal piping might be gathered by using pipe locators or performing test shutdowns of various segments of the customer service connection to identify the water source. New valves may need to be installed at different points in the service line to perform the conclusive shutdowns. Water utilities can benefit from cross-training employees—particularly meter technicians, meter readers, and backflow technicians—to observe and identify occurrences of meter or meter reading equipment tampering, illegal connections, or illegally opened valves.

> The Philadelphia Water Department documented an instance of over 100 residential meters being swapped into varying customer addresses in a small neighborhood of the service area. Apparently, one person had taken on the role of full time meter-jumper in order to generate reduced water bills for his customers.

Requirements for continuous provision of water service vary across utilities or political jurisdictions. Some utilities or communities prohibit water utilities from halting water service to customers under any circumstances, least of all nonpayment of water fees. The American Water Works Association (AWWA) believes that water utilities have the right to terminate water service if payment is withheld. The AWWA policy statement on discontinuance of water service for nonpayment is given in Fig. 15.4. For water utilities that do discontinue water service, a hardened portion of this customer population will make strenuous attempts to illegally restore their water service. Many water utilities terminate service to residential customers by closing the curbstop valve

The American Water Works Association (AWWA) believes that water utilities must have the right to discontinue water service for nonpayment to maintain self-sustaining utility operations.

AWWA realizes the importance of the nondiscriminatory billing and collection procedures to ensure that each customer pays for the services rendered by the utility under its rates and tariffs. Failure on the part of the customer to pay a water bill necessitates that other customers bear the burden of paying for the service.

AWWA recognizes that certain circumstances may require some flexibility because water service is a necessity in maintaining sanitary conditions in the home, and may be required for life-sustaining equipment. It may also be a vital part of industrial and commercial operations. Discontinuance of water service for nonpayment is considered a final phase of a collection procedure and is never to be instituted without sufficient notification and until all other reasonable alternatives have been exhausted.

FIGURE 15.4 American Water Works Association policy statement: Discontinuance of water service for nonpayment. (*Source:* American Water Works Association.)

located between the customer premise and the water main in the street or right-of-way. While a distinct valve key is used to operate curbstops, it is not difficult to manipulate a closed curbstop to illegally restore service. Some water utilities have policies in place that suspend meter reading and billing for customer accounts that have been discontinued for nonpayment. A customer illegally restoring their service under such a policy structure encounters the opportunity to reactivate their service while the utility no longer monitors the customer accounts. Fortunately, with the advent of AMR and AMI technology, many water utilities continue to monitor discontinued accounts for signs of metered consumption or tampering, thus indicating that the customer has illegally restored their service connection.

The discussion in Chap. 13, Sec. 13.1.3 gives considerable detail on the wide array of astounding technical capabilities that meter and AMR manufacturers are developing under the heading of advanced metering infrastructure (AMI). With a trend in the AMR industry to move toward fixed communication networks, it has become evident that the communication network can communicate more than just water meter readings. It can also collect data on meter tampering, reverse flow events, leak noises, high flows, and other potential parameters of interest. Water utilities that install fixed network AMR systems effectively achieve a means for almost continuous monitoring of customer endpoint devices. With such capability, water utilities will be able to quickly detect many incidents of meter tampering that have often gone unnoticed in the past. Also, by analyzing customer consumption profiles that are developed by fixed network AMR, utilities can interpret and explain unusual flow patterns that have historically confounded both the customer and water utility.

15.3.3   The Future of Unauthorized Consumption Control for Water Utilities: Prepayment Structures and Endpoint Controls

Water resources are being stressed at ever-growing rates by climate change, growing populations, and pollution. Water utilities act as stewards of their water resources but must also deal with the daily realities of maintaining service and meeting regulations and the long-term reality of upgrading deteriorating infrastructure. Water utilities must recover the full cost of service, including the long-term costs, and rightfully expect that their customers should pay for this service.

Because a portion of the customer population in any community will strive to illegally obtain water service, utility managers must have in-place programs to detect and contain the occurrence of unauthorized consumption to economic levels. Ever-improving technology gives water utilities outstanding tools to manage supplies and track authorized and unauthorized consumption. But water utilities must also have appropriate policies and regulations to clearly define the roles and responsibilities of the customer and service provider. Appropriate regulations should exist to give the water utility enforceable rights to take action against customers who willfully take water illegally.

Technology and policy are being merged in both the water and energy industries (electricity, gas) in a number of settings around the world. Merging improved technology and policy gives opportunity to improve the balance between providing a population access to clean water, the utility's right to receive fair payment for the service rendered to customers, and the joint responsibility of all concerned to preserve precious water resources. In the energy industry, technology has been developed and projects are starting that include the use of prepayment regulations for utility services. Prepayment requires customers to pay for service prior to receiving the service, unlike the traditional business model of postpayment that is prevalent throughout most of the world's utility industries.

A large-scale energy (gas, electricity) prepayment project is launching in Azerbaijan in attempt to improve revenue collection so that the utilities can increase their investment in infrastructure renewal. The prepayment structure is set to launch in the City of Ganga in 2008 and features smart card technology linked to banking institutions.[2] The smart cards feature two way communication capability, carrying credit to the meter and meter readings back to the utility.

The complexities of managing prepayment structures are being addressed by the development of "smart meters," or metering and related infrastructure which includes the capability to communicate and enact a variety of functions. "The market for smart meters is estimated at nearly 1.28 billion units worldwide—1 billion of which are outside of the North American market."[3] Currently, most of this market is projected for the energy sector, but many of the capabilities being developed for the energy industry have potential for future use in the water sector. Some of the capabilities being offered as smart meter technology in the energy sector include: two-way automated meter reading, multitiered billing, time-of-use and real-time pricing, remote electrical disconnect and reconnect, distribution system asset optimization, electricity outage detection and restoration management, blackout and brownout elimination, revenue protection, real-time direct load control, power quality management, and tamper detection capabilities. Many of these features could be considered directly, or for parallel functions, in the water industry.

A prepayment structure for water has been implemented in South Africa by Johannesburg Water in order to address significant problems of unauthorized consumption, poor revenue collection, strained water resources, and growing populations. The water supply scheme employed in this structure provides many users with a free basic allotment of water. Once they have consumed this monthly volume, they must have a credit registered in the prepayment meters to continue to receive water service for the remainder of the month. A shutoff feature is included in the program. The program seeks to improve on the revenue collection of the water utility and better manage supplies. The previous structure provided unlimited water for a flat fee. Such programs offer water utilities much stronger and more direct methods to provide reasonable service, but only with the guarantee of receiving payment for the service. In a structure such as this, the water utility has a much stronger level of control over unauthorized consumption. In fact, such control represents a proactive stance by the water utility to optimize its revenue stream and guard against unauthorized consumption. This is a dramatic departure from the purely reactive controls that most water utilities in the world employ against unauthorized consumption.

The experience of Johannesburg Water is not without controversy, however as several advocacy groups have joined to support several customers in legal action against the plan, largely on the basis that the volume determined for the free allotment is too small for large, poor households who are ultimately suffering several weeks of no water service each month once their free allotment is consumed.[4] This project will serve as an interesting early test case in an attempt by a water utility to institute stronger, integrated technology and policy that balances the economic, social, and environmental concerns surrounding the provision of safe drinking water.

Still it is notable that smart technology exists to provide utilities with more control over their services than ever before. This technology gives utilities the tools to operate efficiently, collect appropriate revenue, and provide good customer service. Water is unique in the utility world in that it is the only utility service ingested in the human body and is therefore essential as life-sustaining. Given this, it cannot be regarded in the same vein as other utility services, as all people must have water service. It is up to the managers of water utilities to proactively control losses and optimize revenue capture, but to also appropriately recognize those segments of society that are truly in need, and offer the appropriate discounts or other accommodations that ensure life-sustaining service.

References

1. Villegas, Samantha. "Hydrant Use: Balancing Access and Protection," AWWA *Opflow*. Denver, Colo.: October, 2006.
2. Itron, Inc., Media Release. "Itron Announces Prepaid Metering Contract," 2007. Available Online: www.itron.com/pages/news_press_individual.asp?id=itr_016305.xml. [Cited: 19 December 2007.]
3. Echelon Media Releases. "Echelon Announces World's Most Advanced Residential Utility Meters," 2006. Available Online: www.echelon.com/Company/press/newmeters.htm. [Cited: 31 January 2006.]
4. Right to Water, Media Summary. "Legal Challenge over Water Policy in Poor Community in Phiri, Soweto," 2006. Available Online: www.righttowater.org.uk/code/legal_6.asp. [Cited: 12 January 2006.]

CHAPTER 16

Controlling Real Losses in the Field—Proactive Leak Detection

Reinhard Sturm

Julian Thornton

George Kunkel, P.E.

16.1   Introduction

Chapters 7 and 9 provide guidance in the steps of assessing the volume of real losses and calculating the economic optimum volume of real losses for any water utility. Once the nature and value of real losses have been identified, quantified, and economic sustainable limits calculated, realistic targets can be set. Once the targets and budgets for intervention have been identified then the most suitable methodologies for economically reducing and controlling the real losses can be selected. This chapter presents some of the most common technologies and practical methods used for proactive leakage detection. Figure 16.1 shows the four arrows representing interventions against real losses (Chaps. 17 to 19 will address the other three arrows in detail).

The practices that water utility managers employ to become aware of leaks in their distribution system can be categorized as occurring in one of the two following operational modes:

1. **Proactive Leak Detection**: also known as active leak detection (ALD) is an operational mode in which the water utility deploys resources and equipment in order to actively detect leaks that are currently running undetected (also called *hidden losses*). Proactive leak detection has various benefits:

   • Reducing leakage reduces the production costs to treat and energize the water.
   • Can reduce the amount of treated water that is entering the sewer system—adding unnecessary loading to the sewage treatment process.

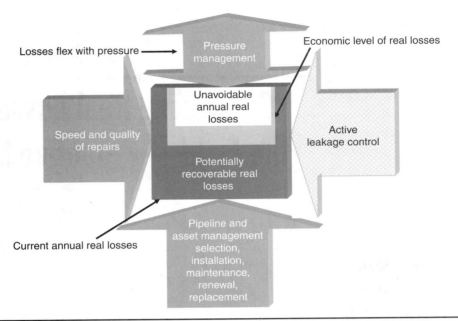

Losses flex with pressure

Pressure management

Economic level of real losses

Unavoidable annual real losses

Speed and quality of repairs

Active leakage control

Potentially recoverable real losses

Current annual real losses

Pipeline and asset management selection, installation, maintenance, renewal, replacement

FIGURE 16.1   Four potential intervention tools of an active real loss management program. (*Source*: IWA Water Loss Task Force and AWWA Water Loss Control Committee.)

- Reducing leakage may help to avoid or defer capital expenditure needed to develop new resources for water supply to meet the needs of a growing service area.
- Helps prevent damage to the infrastructure if leaks are found and repaired before they can cause a catastrophic failure.
- Reduces the liability to the utility.
- Increases supply standards and reliability.
- Has positive impact on the public perception of the water utility.

2. **Reactive Leak Detection:** also known as passive leak detection, this mode is practiced by most North American water utilities—whether economically justified or not. Reactive leak detection means responding to leaks only when they are brought to the attention of the water utility, typically when they become visible on the surface or they are causing a drop in pressure to a customer. Under this operational approach the utility does not seek to actively identify leaks that are not visible or causing supply problems. Under normal circumstances, the overall volume of leakage will continue to rise when only reactive leak detection is used to control the volume of real losses.

In order to schedule field activities properly, it is first necessary to prioritize intervention against real losses. Most utilities have limited budgets, so the methods of intervention with the shortest paybacks are usually the ones that are put into place at the start of the program. In this way the programs start to self-fund out of savings after a certain period of time.

16.2   Mapping

The first thing which must be done when considering tackling real losses in the field is to ensure that the maps and plans of the system and its components are as accurate and current as possible. The media on which water company plans are kept vary widely, from distribution systems with the latest geographical information system (GIS) software, to systems with up-to-date paper plans, to systems with an up-to-date picture in someone's head, to systems in which no one has any idea where anything is! Obviously, the cost of updating such systems will vary greatly.

Systems with good plans and organized, structured background data tend to be more efficient, as the managers responsible for day-to-day decisions have tools at their fingertips with which to make decisions about the performance of their organization. Systems with very little background data find it very hard to set a realistic objective for leakage reduction performance. Even after launching a project to improve performance, these systems find it hard to justify the results, as they do not have reliable baselines from which to measure. Water utilities in this position should consider putting the data in order and improved mapping in place prior to beginning a structured leakage control program.

GIS software is a highly effective tool for managing system plans and provides a very user-friendly graphical interface. GIS also brings other benefits, as it can integrate with other management information systems such as financial and billing databases, telemetry and SCADA systems, and work order management systems. A GIS can also be linked with a hydraulic model, which is a decision-making tool used by many water utilities. As the GIS is linked with the model, there is less need for costly model updates—the distribution system asset data in the model is automatically updated as edits are conducted in the GIS.

Global positioning systems (GPS) are being used by many water utilities to automatically register or locate system components and major features within the system through the use of satellite positioning. GPS coordinates are often used within the GIS environment. While GPS sounds like it might be complex and difficult to operate, it is very user-friendly. GPS data can be downloaded automatically into most GIS databases. The cost of GPS systems varies widely with the resolution and capabilities required, but in most countries, are not prohibitive. Figure 16.2 shows GPS being implemented in the field in Pietermaritzburg, South Africa, as part of an overall upgrading of plans and water loss management program funded by the federal government.

If a water utility lacks reliable system maps, or intends to make a major upgrade, GIS is a very well-recommended route. There are many packages available on the market today, although it is important to use a package suited to the water utility personnel, whether or not the system support is provided by in-house staff or by contracted services. The software should be well supported and easily upgraded. Figure 16.3 shows a layer from a GIS system put into place in SABESP, São Paulo, Brazil, during implementation of a leakage management program. This particular figure shows municipal blocks, roads, pipes, and number of service connections per block. This plan was used to determine areas where pressure management might be implemented. Figure 16.4, from the same company, shows another zone where pressure management and leak detection and repair are being carried out. The GIS system is being used to map reported and unreported leak locations for repair and monitoring of leak frequency.

After making the decision as to whether a software package or paper plans will be used, necessary functionality should be determined to allow the system to be utilized

Figure 16.2 Using GPS to locate fittings.

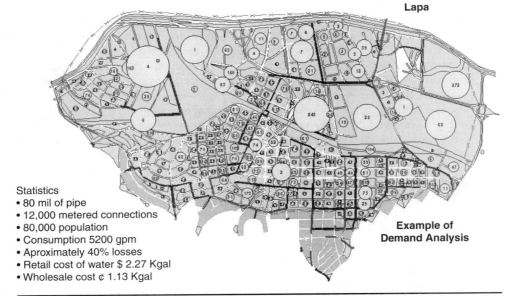

Lapa

Statistics
- 80 mil of pipe
- 12,000 metered connections
- 80,000 population
- Consumption 5200 gpm
- Aproximately 40% losses
- Retail cost of water $ 2.27 Kgal
- Wholesale cost ¢ 1.13 Kgal

**Example of
Demand Analysis**

Figure 16.3 GIS system used to determine potential areas for pressure management Contract No.66.593/96.

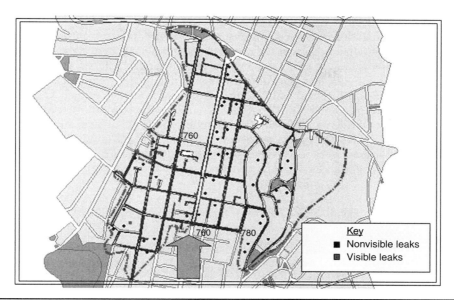

FIGURE 16.4   Using GIS to map reported and unreported leaks Contract No.4.134/97.

and updated easily. In addition to the requirements listed below, the operator should consider the size and scale of the plans to be generated or updated. If GIS is available, the operator can select areas and manipulate them to the required size and scale by using zoom tools. This is not true for paper plans, however. Obviously, there must be sufficient detail to be able to make accurate decisions, but the scale should be sufficiently small to review an area in its entirety. Many utilities use a scale of 1:2000 for urban areas where complex detail and a high density of piping interconnections are encountered. Rural systems or areas often use plans to a scale of 1:5000, as piping density is lower and it is preferable to see a larger area in a single view.

In general, to allow a thorough loss management strategy to be implemented, plans must be available with the following basic information:

- Roads with road names
- City or municipal blocks
- Customer meter reading book routes
- Water distribution system features including pipe diameters, pipe material, and, where possible, age of pipe (entire system transmission included)
- Clear identification of major consumers
- Ground levels and contours to at least 20-ft intervals
- All water sources, wells, treatment plants, pumping stations, transfer points, and above-ground and below-ground storage facilities
- All valves, control valves, source, and master meters into zones or district metered areas
- Clear identification of any zones within the system and their function (pressure control, zone flow analysis, step testing, billing, municipal land use, and the like.)

Once a list of necessary features has been created, a careful desktop review must be undertaken to determine which features are present and usable from existing information sources, what is present but out-of-date, and what is missing altogether. Once this information is known, teams can be assembled to collect the necessary data.

Most fieldwork to collect data on distribution system features is conducted using pipe and cable locators, metal detectors, and recording devices such as ground-probing radar (GPR) instrumentation, but even a simple clipboard and paper will sometimes suffice, depending on the level of detail required for the project, and available budget. Figure 16.5a illustrates pipe location work being undertaken in the field using a pipe locator. Location work is also often done with GPR, as shown in Fig. 16.5b. All data collected must be stored in a manner which is easily accessible to all members of the team, as well as the utility staff in general.

FIGURE 16.5a Using a pipe locator.

FIGURE 16.5b Using GPR.

When launching a system plans update project, careful thought must be given to the planning of a process that ensures that the new mapping system is reliably updated as system changes occur. It is also important to have buy-in from all departments within the water utility, such as finance and maintenance.

16.3   Leakage Fundamentals

It is important that the utility's water loss management personnel understand the fundamental characteristics of leakage occurrences and appropriate means to control them. The *basics* of leakage are given below

16.3.1   Leak Types

Just like there are many different types of infrastructure used in a distribution system there are also many different types of leaks occurring in a distribution system as given herein

Main Break or Pipe Fracture

This term is widely used in North America to describe a catastrophic pipe failure caused by pipe deterioration, fluctuating or excessive pressure, ground movement or a combination of these factors. Breaks, or bursts, in water mains are relatively easy to locate as water released in these failures usually becomes quickly and visually apparent at street or ground surface level, particularly in areas of high pressure. However, on occasion, main breaks are not visually evident from above ground as the water finds an escape channel underground. This can make detection difficult, as paradoxically a notable rupture may not necessarily produce a loud leak noise. This is due to the fact that a large amount of leaking water often results in dramatically reduced pressure, thereby resulting in low noise level. A water pocket may also quickly form around the leak, further diminishing the quality of leak noise. The leak noise from a main break is normally characterized by a low frequency rumbling rather than a high frequency hissing and may therefore be difficult to detect audibly by an inexperienced leakage inspector. In the case of a large volume escaping underground and a reduced pipeline pressure, evidence of the break in a general area may be detected if the water utility monitors water pressures across the distribution system and notes a detectable reduction in pressure. A note regarding terminology: while the term "main break" is widely recognized and utilized in the North American water industry, it is not applied consistently as the terms "main break" and "leak" are interpreted inconsistently by utilities. This makes performance comparisons among water utility main break and leak data difficult. Terminology established during the development of component analysis models using "reported" and "unreported" leakage better defines "main breaks" and "leaks," respectively, and the reader is referred to Chap. 10 and the Glossary for the basis of these definitions.

Crack

This term is used to describe a pipe failure mechanism occurring as circumferential or longitudinal failure that usually results from pipe deterioration or ground movement. They may go undetected for some time and eventually deteriorate to become a reported main break or fracture. The quality of the leak noise depends on factors such as pressure and pipe material, but usually is distinct and of high audible frequency.

Pinhole

Pinhole leaks are small circular failures in a pipeline usually caused by corrosion or stress by stones after poor backfill procedure during installation. Steel pipe installed in a corrosive environment without proper corrosion protection is particularly susceptible to the development of pinhole leaks, which can develop very quickly—as short as several months time in extremely corrosive environments. Pipelines should always be placed in a layer of sand as a minimum protective measure, but often more robust protections are needed, particularly if steel pipe is being specified for the pipeline material. The quality of leak noise varies depending on the pressure, pipe material, and backfill but is usually distinct and of high audible frequency.

Seepage

Most commonly found on deteriorated asbestos cement (AC) pipes where the pipe wall becomes semiporous and water escapes slowly. These types of leaks are extremely difficult to locate as leak noise is minimal. They are therefore normally classified as undetectable background leakage. Losses caused by seepage can be minimized by use of pressure reduction and/or infrastructure replacement.

Leakage on Packing Glands of Pumps or Valves

Caused by deterioration over time and usually occurs when a valve is used after a long period of inactivity. These are relatively easy to detect visually at pumps or by a valve chamber that is full of clear potable water and a good audible noise detectable by direct sounding at the valve spindle. Newer types of valves have a more resilient gland and/or no packing at all, effectively reducing the occurrence of such a common leak problem.

Pipe Joint Leaks

These are common points of leakage, particularly on older cast iron and AC pipes where the caulking or joint gasket deteriorates over time. Many older couplings are not corrosion protected and therefore deteriorate long before the pipe itself. When ground movement occurs, pipe joints bear most of the strain, often resulting in leakage and, eventually, a fracture.

Welded joints on steel pipes are actually stronger than the pipe itself but are seldom corrosion protected after jointing and therefore a vulnerable point of corrosive attack. Joint leakage is reasonably easy to detect on metallic pipes as these pipes usually create a clear leak noise. However, they can be difficult to locate on AC and plastic pipes due to leak noise attenuation that occurs on these materials.

Leaking Service Connection Pipe

Service connection pipe leaks are the most common type of leak that occurs in water distribution systems. From the ferrule connecting the service connection to the water main, to the customer water meter there is often more than one change of pipe size and/or material, which necessitates joints which frequently are a weak point in the connection pipeline. Service connections are also often laid very shallowly; very close to the road surface. They are therefore vulnerable to weakening by movement caused by traffic load. The ferrule connection is a susceptible point of leakage due to corrosion combined with frequent fluctuation in pressure. Service connection leaks are usually easy to detect, as there is normally access to the pipe via a curb stop or meter to enable close direct sounding for leak noises.

Leaking Fire Hydrants, Air Valves, and Scour Valves

Leaks also occur at distribution system appurtenances such as fire hydrants, air valves, and scour valves. These are relatively straightforward to detect as they are usually visible or able to be detected by direct sounding.

16.3.2   Leak Detection via Leak Noise Sounding

Leak sound frequencies vary depending on the type of leak, the type of pipe, backfill material and density, and whether a water-filled cavity has formed around the leak. In general, there are three types of situations that generate leak sound frequencies:

- *Friction sound* is the sound created by water forcing its way through the pipe wall and making vibrations along the pipe. This tends to be a high-frequency leak sound, ranging between 300 to 3000 Hz. In general, high frequency leak sounds are easy to recognize but do not travel very far along the pipe. Note that pipe vibration against surrounding material may cause noise overlaying the leak sound, especially metallic service connection pipelines (see Fig. 16.6).

- *Fountain sound* is the sound of water circulating around the leak site and tends to be lower frequency, in the range of 10 to 1500 Hz (see Fig. 16.7).

- *Impact sound* is the sound of a leak impacting on the walls of the hole around the leak and the sound of the impact of rocks, which often are thrown around the leak. This sound also occurs in the range of 10 to 1500 Hz (see Fig. 16.8).

> **S**ound generated by leaks is continuous, unlike transient ambient noise.

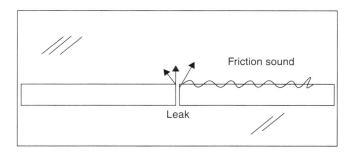

FIGURE 16.6   Friction sound.

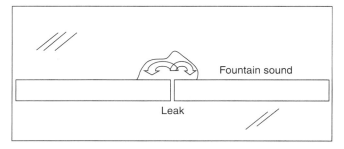

FIGURE 16.7   Fountain sound.

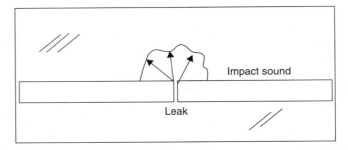

FIGURE 16.8   Impact sound.

16.3.3   Factors Influencing the Quality of Leak Sound

Pressure

The audible volume, quality, and propagation of leak noise are pressure dependent. The higher the pressure, the better the quality of the leak noise and vice versa. There needs to be at least 30 psi (~21 mH) in a zone for effective direct sounding using a listening stick. In many distribution systems pressure during the day is less than pressure at night due to higher consumer demand during the day. If day pressure is less then 30 psi (21 mH) sounding should be carried out at night when consumption is reduced and pressure in the system is higher. The most effective time for sounding in densely populated urban areas is generally expected to be between the hours of 2:00 and 4:00 a.m. when consumption is at its lowest and pressure is at its highest. This is also a time when traffic and other above-ground noises are at a minimum.

Pipe Material and Pipe Size

The audible volume, quality, and propagation of leak noise are also dependent on pipe material and pipe size. Generally, the harder the pipe material (i.e., steel) and the smaller the diameter, the better the quality of leak noise and the further the noise will travel along the pipe wall. Conversely, those pipes with softer material (i.e., PVC) and larger diameter, will attenuate leak noise. The following is a guide to noise propagation values on varying pipe material:

Cast iron
Steel   } Good (for *leak noise sounding*)
Copper

Ductile iron
Asbestos cement   } Average

PVC
MDPE
HDPE   } Poor
Internally lined
Externally wrapped

Some corrosion protection applied to metallic pipes can diminish quality and propagation of leak noise as well. Steel pipes usually have an internal concrete lining and are externally wrapped and/or coated in bitumen paint, both absorbing leak noise. Smaller lengths of pipe and fittings such as flanged tee's, bends, and valves may also have a coating hardened epoxy resin. This too will reduce quality and propagation of leak noise.

Types of Leak Noises

Smaller fittings, valve packing, and pinhole leaks tend to produce higher frequency leak noise than cracks, breaks, and some joint leaks. Smaller leaks often have a "hissing" sound and larger leaks have more of a "rumble."

Surface Covering the Pipe

Sandy soils and asphalt conduct sound quite well as opposed to clay and concrete. Backfill soils that have cavities diminish the transmission of leak sound. Sounding on unpaved surfaces is more difficult and a probe rod may need to be rammed through the ground, if possible, to make content with the pipe to transmit the sound. If this is not possible a device known as a "thumb tack" (round metal plate with a rod attached transversely) can be placed in the ground as a sounding surface.

Soil Moisture

The variation of soil moisture and the altitude of the water table effect the propagation of leak noise. Increased water in the soil diminishes the leak sound transmission. A saturated soil creates a backpressure against the leak origin.

Sources of Interference with Leak Noise

Noise generated by traffic, pressure reducing valves (PRVs), consumer consumption, partially closed valves, aircrafts, leaf blowers, air conditioners, generators, trains, cable cars, compressors, transformers, and the like, all can compete with leak noise in a given location and make leak detection more difficult. Many of these noises, however, occur at frequencies outside of the typical leak noise frequencies. Today's modern electronic sounding equipment (leak noise loggers, leak correlators) have filters that separate sound outside of typical leakage frequencies, making the existing leak noise detectable. Still, the existence of interfering noises should always be closely considered by the leakage inspector in attempting to discern and pinpoint leakage.

16.4   Leak Detection Equipment

Leak detection equipment is available in a wide range of technologies, capabilities, and prices. Hence a good understanding of the nature and occurrence of real leakage losses enables the water utility operator to select the most appropriate technology. It is very important to note that even highly sophisticated and expensive leak detection equipment does not represent an appropriate remedy for a utility's leakage problem if the utility does not yet understand the real extent and nature of leakage occurrences in its distribution system. A reliable benefit to cost analysis should be conducted before making a major investment in leak detection equipment.

The most important factor for success in detecting leaks using any type of equipment is the experience of the leak detection team in using the available equipment and interpreting the results received from the equipment. The operator must be trained to understand not only how to use the acoustic sounding equipment but also its limitations. Only then will he successfully tailor the leak survey procedure to match the capabilities of his equipment to the distances that leak sounds typically travel in his distribution system. For example, if sounding is undertaken every 656 ft (200 m) but the pipes are plastic with poor sound propagation, there is a very good chance of missing detection of an existing leak unless the leak happens to be next to a fire hydrant or fitting which is being sounded. Every water distribution system, and every distinct section of the system (pipe materials, pressure levels), must be treated on its own merit. Sounding may be a simple technique, but the planning must be done by someone who understands the capabilities and limitations of the equipment and personnel, and the characteristics of the leakage occurrences in their distribution system. Appendix B provides further information on leak detection equipment and leak detection techniques.

16.4.1   Acoustic Leak Detection Equipment

Mechanical and Electronic Listening Stick

The listening stick, probe rod, or similar name describes a traditional instrument used to systematically sound all mains fittings and service connection pipes. There are various designs, the most common having an earpiece attached to a steel shaft. Alternatives have a mounted ear piece housing a diaphragm that amplifies sound. However, these tend to give a "seashell" sound effect, which can be misleading to an untrained ear. Use of the listening stick is by placement on a fitting, whereby any leak noise is transferred from the pipe, through the steel shaft and is heard at the earpiece.

The electronic listening stick is used in the same way as the mechanical version, but has a battery powered sound amplifier attached so that the leak noise is enhanced and then heard through headphones. The electronic listening stick is utilized in areas of low pressure, where leak noise is weak and requires amplification. It is also useful for direct sounding in areas where there may be high noise interference from passing traffic.

Ground Microphone

Ground microphones (Geophones) are listening devices mostly used to listen for leaks from the surface where contact points such as valves, hydrants, service connection curbstops, and the like are far apart. Ground microphones are also used to pinpoint the exact location of a leak. Mechanical listening devices have an appearance and work on the same principle as the physician's stethoscope. Today's electronic devices have signal amplifiers and noise filters to attenuate the leak noise signal. Ground microphones (see Fig. 16.9) are usually used in conjunction with other leak detection equipment, although it can be used alone, especially in areas with few fittings and predominantly plastic pipe.

Leak Noise Correlator

Just like traditional sonic equipment a leak noise correlator relies on the leak sound generated by a leak. A leak noise correlator typically consists of a receiver and processor (correlator unit) unit and two sensors equipped with a radio transmitter. The two sensors are placed on valves or hydrants on each side of the suspected leak. The leak noise detected by the sensors is converted into electrical signals and then transmitted via the radio transmitters to the correlator unit. Leak sound travels along the pipe with a constant velocity depending on the pipe diameter and pipe material. The leak noise will first arrive at the sensor closer to the leak. The correlator uses the time difference between the two arrival times, information about the pipe material and size, and the distance between the two sensors to calculate the location of the leak. The calculation principle is set out in the formula $L = TD \times V + 2I$, where L is length, TD is the time

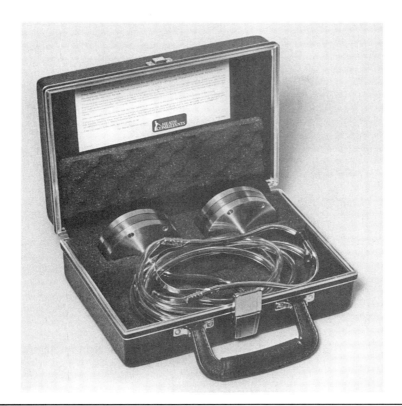

FIGURE 16.9 Mechanical geophones.

delay for the signal to reach the farthest sensor after reaching the first sensor, V is the speed at which the leak sound can travel either in the pipe wall or in the water, and I is the leak position from one sensor and fitting. This calculation principle is shown more clearly in Fig. 16.10. The calculation is then reworked to give us a way to calculate the leak position, which is $I = L - (TD \times V)/2$.

The ability of the correlator to precisely determine a leak position is very much dependent on the necessity to detect the leak noise at both sensors, and on the accuracy of the information input by the operator. Upsets in the process usually occur when correlating over long distances when the leak is quite close to one end. This creates a high value of the time delay, TD. Since, in many cases, the velocity is estimated, a longer time delay multiplied by an incorrect velocity value distorts the pinpointing projection from the actual leak position.

Many correlators have a velocity calculation feature which should be used to address this condition. However, a precaution to ensure a reasonably accurate pinpointing location is to keep the suspected leak area fairly close to the middle of the distance between the sensors. This can be achieved by running a quick correlation to locate the leak roughly, and then move the sensors to centralize the suspected leak point. As previously discussed, no single piece of leak detection equipment is infallible and use of a combination of tools is often required to detect and pinpoint leaks reliably. It is recommended, after locating a suspected leak location with a correlator to employ the ground microphone or geophone and listen over the suspected point.

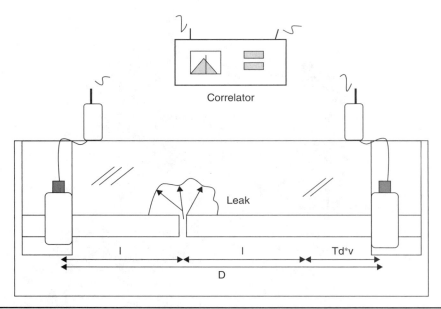

FIGURE 16.10   Principles of leak noise correlation.

Operators must gain experience in using correlator equipment and many situations exist in the field that challenge the correlation process. For example, confusion can occur when correlating on a length of main with an actual leak on a tee or branching line from the water main. Most correlator models are not able to distinguish whether the leak is on the water main or branching line, and often indicates that the leak is on the tee fitting on the main. It is up to the operator to know the system and the correlation process sufficiently to check the line branching from the tee fitting (see Fig. 16.11).

A good way to avoid errors is to perform three or more correlation runs with varying lengths of pipe between the sensors, and then do a linear regression of the data as shown in Fig. 16.12. This has the effect of averaging the errors in each velocity calculation and giving a more precise leak location. Some correlators have this facility built in, but any experienced operator can use this method by manually plotting the results of varying length and changing time delay.

Leak Noise Loggers

Noise loggers are installed at fittings and programmed to automatically turn on at night to monitor system noise and listen for signs of leakage. The usual logging period is between 2:00 and 4:00 a.m. Nighttime logging has the dual benefits of increased intensity of leak noise due to higher pressures and the absence of interfering ambient or consumption sound. It is important to bear in mind that nighttime irrigation will seriously affect the usefulness of results generated by the leak noise loggers. Leak noise loggers can be installed permanently or be moved from location to location depending on the leakage management practice of the utility. Unlike listening sticks and leak noise correlators, noise loggers do not pinpoint the location of a leak. Leak noise loggers only give an indication that there is a leak present within the vicinity of the logger. Hence, the leak pin-pointing needs to be carried out by an experienced operator using a leak correlator, ground microphone and/or listening stick.

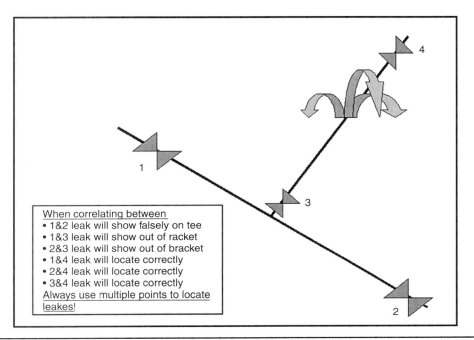

When correlating between
• 1&2 leak will show falsely on tee
• 1&3 leak will show out of racket
• 2&3 leak will show out of bracket
• 1&4 leak will locate correctly
• 2&4 leak will locate correctly
• 3&4 leak will locate correctly
Always use multiple points to locate leakes!

FIGURE 16.11   Tee connection rule.

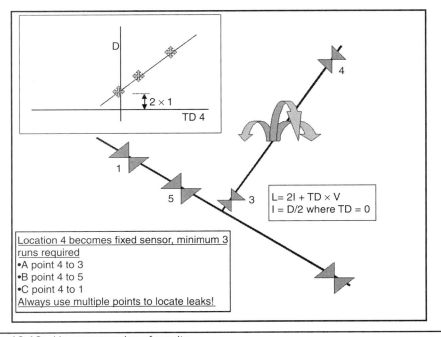

$$L = 2I + TD \times V$$
$$I = D/2 \text{ where } TD = 0$$

Location 4 becomes fixed sensor, minimum 3 runs required
•A point 4 to 3
•B point 4 to 5
•C point 4 to 1
Always use multiple points to locate leaks!

FIGURE 16.12   Linear regression of results.

Digital Correlating Leak Noise Logger

The next step in noise logger technology was the devolvement of digitally correlating leak noise loggers, which combine acoustic noise logging and leak noise correlation. This technology has the advantage of reducing the time span between identification of a leak noise and localization of a leak. Nevertheless, it is still highly recommended that the exact location of the leak be verified by a trained leak detection specialist using a ground microphone before excavating for the leak repair.

16.4.2 Nonacoustic Leak Detection Equipment

Tracer Gas

Water insoluble gas such as helium or hydrogen is inserted into an isolated segment of a water pipe. The gas escapes at the leak and permeates to the surface where it can be detected by using a highly sensitive gas detector. Solid surfaces such as concrete slow the process of gas permeating to the surface. Tracer gas is an option used for transmission mains, on low pressure mains where the acoustic sounding is difficult and for leaks on small plastic pipes on house connections. This technique is also used to validate watertight construction of new water mains before they are commissioned into service. A disadvantage of this technique is that water pipelines must be removed from active service in order to apply this method.

Ground Penetrating Radar

Ground penetrating radar identifies water leaks by detecting cavities around the pipe created by the leak, detecting the presence of water around the pipe stemming from the leak, or through the observation of disturbed ground caused by the leak. This technology is not in wide use because of its relatively high logistical requirements and related cost. However, it can be a highly effective tool in situations such as low pressure or plastic pipe leaks where very little sound is generated and sonic leak detection is not possible.

16.4.3 Leak Detection Equipment for Transmission Mains

The general difficulty one faces when trying carry out a leak detection survey on a transmission main is the long distance between fittings that can be used as sounding contact points and the fact that the leak sound decreases with increasing pipe diameter and increasing distance from the leak.

Sensors Inserted into the Transmission Main

One type of leak detection equipment developed for transmission mains uses the principle of a sensor (different manufacturers use different types of sensors) being inserted in the transmission main, which then travels along with the flow in the pipe picking up any noise generated by a leak. The use of inline transmission main leak detection service is proving to be very accurate. It is still a new technology in North America but has a well-established history in the United Kingdom. This technology is expected to be more widely embraced in the North American water utility industry in the near future.

Fiber Optics

Another type of technology utilizes acoustic fiber optics for managing and monitoring large diameter mains. A continuous fiber optic cable is installed in the pipeline and the fiber optic cable is then connected to a data acquisition system that allows permanent real-time acoustic monitoring.

Infrared Technology

Infrared thermography can be used as a method of testing for leaks which do not surface. The underlying principle of this technology is that the water escaping from a leak is of a different temperature than the surrounding ground and can therefore be detected by a thermographic camera. The method is quite expensive and in many cases is undertaken by flying over the areas to be tested. This method has been used successfully for testing transmission mains in rural areas, but is not practical for dense urban areas, where interference from other underground utilities, such as sewers, would unduly complicate the process. Some operators are using this method to detect reservoir leakage.

16.5 Leak Detection Techniques

The monetary value of water lost through leaks plays an important role in deciding which leak detection technique or combination of techniques is most suitable for the water utility. When deciding on the right techniques it is also important to consider age, condition, and material of the distribution system and the skill level of water utility personnel carrying out the leak detection effort.

District metered areas (DMAs) are used to monitor flow into discrete zones of the distribution network in order to determine the leakage level and to monitor any rise in inflow due to newly formed leaks. The use of DMAs in tandem with active leak detection can serve as the basis for a comprehensive leakage management strategy. A detailed discussion of DMA technology is given later in this chapter.

16.5.1 Visual Survey

The most basic form of leak location is the visual survey. A visual survey consists of walking the lines looking for either leaks which appear above the ground or, in very dry countries or regions, areas that have suspicious green growth patches above the water lines. Figure 16.13 shows a leak that could easily be located by visual survey. This particular leak is on an above-ground air valve. Other leaks, which are not quite so obvious, are also often picked up.

While the visual survey is not the most sophisticated technique but it should not be underestimated, particularly by utilities which have suffered from lack of good and frequent maintenance.

16.5.2 Acoustic Leak Detection Survey

The acoustic leak detection survey is probably the most common and familiar leak detection methodology which has been around for many years. Different types of acoustic sounding equipment are used in two distinctly different levels of detail.

General Survey

This survey method is often referred to as a hydrant survey in Canada and the United States. This survey method generally listens only to fire hydrants and valves on distribution system mains in order to detect any leak sound, no service connections are sounded. Fire hydrants can be found at more or less constant distances providing a good coverage of most areas. In this survey mode geophones and leak noise correlators are generally

Figure 16.13   Visible leakage from an air valve.

only used for pinpointing a leak. It is a time-saving leak detection methodology which has one shortfall. Service connection leaks often go undetected in this mode, especially if the area mainly consists of nonmetallic mains and service connections.

Comprehensive Survey

This survey method listens to all available fittings on the mains and service connections. Geophones are used to sound above the mains in case contact points are far apart. Once a leak sound is detected, geophones and leak noise correlators can be used for pinpointing the leak. Even though this leak detection method is time consuming it is the most effective way to detect all detectable leaks in the system, including service connection leaks.

16.5.3   Step Testing

Step testing involves isolating sections of the water distribution system into small zones and measuring the supply to the zone. This is often done on a temporary basis and portable flowmeters are used to measure flow into the zone. Every time a section with a leak is isolated, a marked drop will be seen on the flow graph as shown in Fig. 16.14. This drop represents the leak volume, which represents valuable information for cost-benefit calculations and program tracking, and also saves time by directing leak pinpointing crews only to those sections of the water main where leakage has been proven to be occurring.

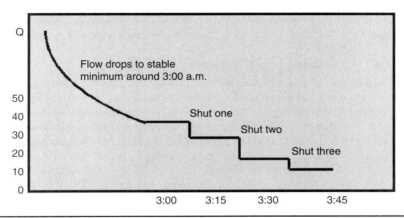

Q

Flow drops to stable
minimum around 3:00 a.m.

Shut one

Shut two

Shut three

50
40
30
20
10
0

3:00   3:15   3:30   3:45

FIGURE 16.14   Step test.

When undertaking step testing, it is very important to execute the test in a manner that does not cause undue interruption to customer supply. For this reason, step testing is usually carried out at night when customer consumption is at a minimum.

Step testing can be carried out in other forms too. Many rural water systems do not have zones, but they do have sections of plastic pipe with many fittings, making it difficult to find leaks. One interesting way of outfitting these systems for sustainable leakage management is to install either temporary or permanent flow measuring sites in the distribution system. Flow data gathered from these points are then analyzed to assess night-time flow patterns, when there is relatively little customer consumption. Any sections showing significant flow are marked as having leakage and intervention is scheduled. This method saves costly surveys over miles and miles of water main that may not have any leakage and helps the operator to focus on the sections with significant leakage. When the volume of leakage is known, the operator can justify higher levels of effort to locate leakage than he would be able to do if he were unsure about the existence of leakage. In some cases, the cost of excavating additional test holes can be justified to track a leak when its volume is known to be significant. If the operator is unsure that the leak is actually on that section he would not wish to spend the additional time and money.

There is a fundamental problem with step tests which needs to be stated. As each "step" is closed, the total flow into the zone is reduced. This reduces the frictional head-loss in the mains in the "open steps," that is, those steps that have not yet been isolated. This in turn causes an increase in pressure in the open steps and, as a result of the direct relationship between leakage and pressure, causes an increase in the leakage rate in the open steps. In theory the reduction in flow, seen when a step is closed, reflects the flow reduction caused by the closed step. However, it is possible that the reduction in flow is less than it should be because of an increase in flow in the open steps, due to increasing pressure in the areas still supplied. In worst case scenarios it is possible for the increase in flow in the open steps to completely mask the drop in flow when a step is closed. This scenario is commonly found in zones where several leaks are present. In these situations the presence of a leak in a step can be masked and because the leak pin-pointing crews are not directed into that step, the leak remains undetected. To carry out step tests, there is a need to operate a large number of valves and in general there is a reasonable chance of using valves that are not "shut-tight." If a step valve or circulating valve is not fully shut-tight, then the drop in flow will not be correct. Therefore it is necessary to carefully evaluate if a step-test is the appropriate method to be used.[1]

16.5.4   Leak Noise Logger Survey

Leak noise loggers are a relatively new technology compared to other acoustic leak detection instruments. They have been used in various forms for about the last 15 years. Noise loggers are installed on pipe fittings such as valves and hydrants through the use of a strong magnet. They are programmed to listen for noise generated by leaks. Noise loggers typically record at 1 second intervals over a period of 2 hours during the night, when background noise is likely to be lower. By recording and analyzing the intensity and consistency of noise, each logger indicates the likely presence (or absence) of a leak. Noise loggers can either be permanently installed in the distribution network, or temporarily for a user definable period (mostly one or two nights). Noise loggers only indicate the presence of a leak and do not provide the exact location of the leak. Therefore, it is necessary to follow up on an indicated leak with a pin-pointing exercise to exactly determine the location of the leak for excavation and repair.

16.5.5   Leak Noise Mapping

Noise mapping is an improved form of regular acoustic survey pioneered by the Halifax Regional Water Commission. Locations for sounding are predetermined (mainly hydrants), and the level and type of leak noise (reading taken from the leak detection equipment) is written down on a map. In addition, the presence or absence of noise is entered into a spreadsheet with standardized information on date, location, general description, leakage inspector, and noise sound legend. The next step is to validate the recorded noise and document the results in the same spreadsheet. This process is only completed when all recorded noises have been validated. The leakage manager can now easily control the work of the inspectors and can compare actual noise levels to previous ones and thus, easily identify areas where more detailed leak detection activities are needed. This is a simple but very effective improvement of regular sounding surveys which can best be applied in distribution systems with a very high hydrant density where the sounding points can be identified easily.[2]

16.5.6   Summary of Leak Detection Techniques

Understanding the nature of leakage occurrences in a distribution system and the relative capabilities of leak detection equipment, techniques and staff training and experience is important to the successful control of leakage levels. Tables 16.1a and 16.1b[1] provide a summary of specific characteristics of each leak detection technique mentioned above.

Every leak survey needs preparation in order to achieve good results. The practitioner should consider the following points when planning a survey and evaluating the results.

Presurvey Checklist

- Prepare good system plans at workable scales.
- Clearly mark limits of zone(s) to be tested.
- Locate unknown pipe lengths.
- Identify a suitable distance between testing points per main type.
- Identify large users who could interfere with sounding.
- Identify locations of pressure reducing valves and their inlet/outlet pressure ranges.

Technique	Localization/ Pinpointing	Used in Combination with	Type of Leaks Found	Manpower Intensive	Well Trained Personnel Required
Visual survey	Localization	Pin-pointing technique	Surfacing leaks	No	No
Temporary noise logger	Localization	Pin-pointing technique	Mostly mains leaks	Moderate	Moderate
Permanent noise logger	Localization	Pin-pointing technique	Mostly mains leaks	Moderate	Moderate
Step testing	Localization	Pin-pointing technique	Mains leaks	Yes	Yes
General survey	Localization and pin-pointing	Noise loggers and step testing	Mostly mains leaks	Moderate	Yes
Comprehensive survey	Localization and pin-pointing	Noise loggers and step testing	Mains and service line leaks	Yes	Yes
Leak noise mapping	Localization and pin-pointing	Pin-pointing technique	Mains and service line leaks	Yes	Yes

Source: Ref. 1.

TABLE 16.1A   Matrix of Leak Detection Techniques—Part One

Technique	Impact on Awareness Duration	Impact on Location Duration	Most Appropriate for
Visual survey	If it is the only technique applied, awareness times (AT) will be very long.	Additional pin-pointing required. Depends on availability of staff and the priority given to the leakage event.	Initial run through in utilities with huge backlog of leaks and poor infrastructure.
Temporary noise logger	Positive impact on awareness time. Dependent upon frequency of area surveyed by temporary deployment of noise loggers, e.g., average AT for an area covered once a year is 183 days, and for twice a year AT is reduced to 92 days etc.	Requires additional pin-pointing. Therefore highly dependent on availability of staff and priority given to the leakage event.	Areas with high background noise and to avoid night work of leak detection crews.

Source: Ref. 1.

TABLE 16.1B   Matrix of Leak Detection Techniques—Part Two (*Continued*)

Technique	Impact on Awareness Duration	Impact on Location Duration	Most Appropriate for
Permanent noise logger	Reduces AT down to a couple of days— AT depends on how often noise logger data is retrieved, e.g., AT can be 1/2 day if noise logger data is transmitted or collected every day.	Requires additional pin-pointing. Therefore depends on availability of staff and priority given to the leakage event.	Due to high capital cost it is difficult to justify permanent installation.
Step testing		Requires additional pin-pointing. Therefore depends on availability of staff and priority given to the leakage event.	Rural distribution networks with high level of excessive leakage.
General survey	AT depends on frequency of general survey.	Leak is pinpointed immediately after it is localized—very short location time.	Areas where leaks occur mainly on the distribution network and the mains and services are of metallic material.
Comprehensive survey	AT depends on frequency of comprehensive survey.	Leak is pinpointed immediately after it is localized—very short location time.	Areas where high level of service connections leaks exist and where a significant portion of nonmetallic pipework exists. Most appropriate technique to detect all leaks and remove the backlog of hidden leaks.
Leak noise mapping	Awareness time depends on frequency of surveys.	Leak is pinpointed immediately after it is localized—very short location time—plus all other noise picked up during survey by equipment has to be noted down and followed up and verified.	In areas with high density of fire hydrants.

Source: Ref. 1.

TABLE 16.1B   Matrix of Leak Detection Techniques—Part Two (*Continued*)

- Prepare protective clothing.
- Prepare a suitable leak location form.
- Charge batteries for electronic equipment.
- Check sensors against a reference sound such as a tap running to ensure sensitivity.
- Carry appropriate identification badges, since access to private properties is periodically required.
- Take the necessary signs and cones to warn traffic.

Postsurvey Checklist

- Clearly record all suspected leak points on prepared sheets.
- Clearly identify the points on the maps.
- Attempt to rank the leaks by severity of loss and potential damage to life or property.
- Prepare a repair work order.
- Identify a realistic time frame for repairs to be undertaken, ensuring that the worst leaks are repaired first.
- Where possible, visit the leak site during repair to make a photographic record of the leak.
- Attempt to make volumetric measurements for larger leaks, to assist in preparing the annual balance.
- Prepare a leak report card (see Fig. 16.15).

16.6   Zoning and District Metered Areas

The use of discrete zones or DMA can form an integral part of their leakage control strategy for many water utilities. DMAs have the benefit of combining two of the four tools against real losses (see Fig. 16.1). DMAs help reduce leak awareness times by identifying newly occurring leaks through minimum hour or nighttime flow analysis. DMAs also improve proactive leak detection efforts thorugh prioritization of leak detection efforts to areas where DMA analysis has shown that leakage levels are highest. DMAs can be designed for permanent installation or can be established for temporary measurements. The use of DMAs is standard practice in some countries such as the United Kingdom where many thousands of discrete DMAs are in service. This technique is relatively new to North America. However, the technique was successfully piloted and investigated in the North American setting as part of the American Water Works Association Research Foundation project "Leakage Management Technologies." Readers are strongly encouraged to review the final report of this project as valuable information on DMA technology was brought forth in this undertaking.

By dividing the distribution system into smaller, easier to manage and monitor areas, leakage levels can be quantified for each DMA and leak detection activities can be directed to those DMAs where leakage levels are highest. Also, the DMA approach

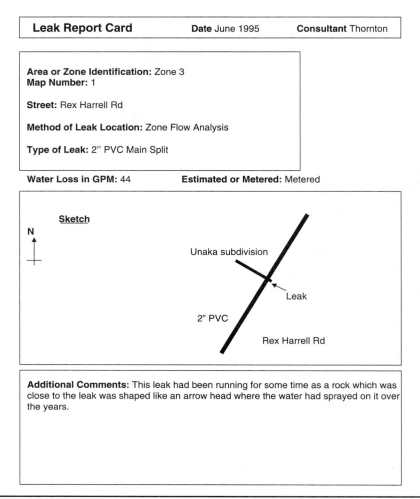

| Leak Report Card | Date June 1995 | Consultant Thornton |

Area or Zone Identification: Zone 3
Map Number: 1

Street: Rex Harrell Rd

Method of Leak Location: Zone Flow Analysis

Type of Leak: 2" PVC Main Split

Water Loss in GPM: 44   **Estimated or Metered:** Metered

Sketch

N

Unaka subdivision

Leak

2" PVC

Rex Harrell Rd

Additional Comments: This leak had been running for some time as a rock which was close to the leak was shaped like an arrow head where the water had sprayed on it over the years.

FIGURE 16.15   Sample leak report form.

offers the advantage that, once the leakage levels are reduced to an economic optimum level, it is possible to closely monitor the subsequent rise in leakage in the DMA. Leak detection personnel need not be sent into the DMA until a preset threshold (economic) level of leakage is reached. The threshold level is determined by factoring the cost of lost water and the personnel/equipment costs of the leak detection crew.

16.6.1   DMA Principles and Effectiveness in Leakage Management

Depending on the characteristics of the distribution network a DMA is a hydraulically discrete area supplied by a single or multiple feeds. The water supplied to the DMA is monitored by flowmeters and in certain circumstances a DMA may cascade into an adjacent DMA (see Fig. 16.16).

By subdividing the distribution network into small hydraulically discrete areas known as the district metered area (DMA), input supply flows can be continuously

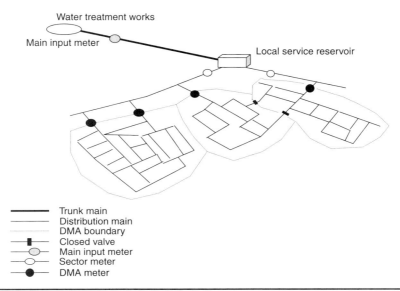

Trunk main
Distribution main
DMA boundary
Closed valve
Main input meter
Sector meter
DMA meter

FIGURE 16.16   Typical DMA layout. (*Source*: Ref. 3.)

tracked and minimum hours or nighttime flow rates assessed to reveal leakage trends. This technique has been found to be one of the most successful approaches for reducing the run time of unreported or hidden leaks, and therefore reducing the overall volume of real losses. There are two main benefits related to the installation of DMAs in water distribution systems:

1. They allow the network to be divided into smaller sections, each with a defined boundary and equipped with flowmeter(s) to monitor the total supply (with special focus on the minimum or nighttime flows), which enables the utility to identify the presence of unreported breaks and leaks. The minimum nighttime flow information is used to identify the occurrence of new breaks and leaks and also to prioritize leak detection efforts toward the DMA where the volume of leakage has risen above the economic optimum threshold.

2. DMAs provide the utility with the opportunity to manage pressure on a microscale assuring that each DMA is operated at the optimum level of pressure (see Chap. 18).

DMA minimum hour or nighttime flow analysis is also used in conjunction with the validation of real losses derived from the top down annual water balance. DMA measurements serve to field verify the calculated volumes of real losses based on the top down water balance, with the actual leakage volumes of real losses found in the DMA. Utilities without permanent DMAs can establish one or more temporary DMAs, representative of the entire network, to assess leakage volumes through bottom-up DMA measurements.

16.6.2 DMA Design

The research project "Leakage Management Technologies,"[1] investigated the applicability of the DMA technique in North American water distribution systems. The findings clearly suggest that DMAs are applicable to North American networks when certain design criteria are properly addressed. Two DMA design elements of note in the North American water industry are fire flow capabilities and water quality. However, extensive field testing in North American water utilities revealed that, by following some simple design rules, it is possible to create and operate a DMA and still provide the necessary fire flows while maintaining adequate water quality. The design rules discussed in this chapter draw from the research findings of the above project[1] and can be applied to the use of the DMA technique in any setting, and not just the North American water industry.

The most important factors that need to be taken into consideration when designing a DMA are

- The economic level of leakage or the economic intervention frequency needs to be considered since it will impact the optimum size of the future DMA.

- The types of consumers (industrial, multifamily, single family, commercial, and critical customers such as hospitals, and the like.) need to be assessed and considered during the design phase.

- Existing pressure control zones should be assessed and if possible converted into DMA. This is the easiest and most economic way of creating a new DMA.

- Variations in ground elevation need to be assessed thoroughly.

- DMAs should be designed in a way so that new boundary valves are located on smaller mains.

- Existing check valves and closed PRVs should be used as DMA boundary valves providing additional back up in case of fire flow emergencies.

- A boundary should be designed not only to fit the broad design criteria for the DMA, but also to cross as few mains as possible. The boundary should follow the "line of least resistance" by using natural geographic and hydraulic boundaries. The aim is clearly to minimize the cost of installation, operation, and maintenance. A hydraulic model is particularly useful to identify existing hydraulic balance points where a DMA boundary valve can be closed without modifying the existing operation of the network, thus limiting potential pressure or water quality problems.

- Transmission mains, service reservoirs, or tanks should not be included in a DMA.

- Water quality considerations have to be addressed and water quality should be monitored prior to and after the installation of the DMA.

- The targeted final leakage level should be defined to make sure the DMA meter and PRV are not oversized once the backlog of leaks was removed.

- Minimum flow and pressure requirements for fire flow and insurance purposes need to be assessed.

- Minimum and maximum pressure at the critical zone point should be assessed.

- Looping and redundancy requirements need to be assessed.

- System changes required for DMA installation, like the number of new valves required, installation of meter point(s) and chamber(s), and and so on. should be considered.

- The configuration of the distribution network pump system and location of pumping stations need to be carefully assessed and included in the design stage.

- When selecting the meter locations, it is necessary to consider the size of the *feeder main* through which the DMA will be supplied. Feeder mains with larger diameters will experience very low flow velocities during the minimum nighttime flow period. In many cases, those velocities might be below the accuracy limits of the flowmeters that are to be installed. The minimum nighttime flow into the zone is the crucial information for DMA monitoring and analysis. Therefore, it is important to locate feeder mains with a smaller diameter, which still can meet all necessary flow requirements or to install a bypass around a closed valve on which the DMA inflow meter is installed (see Fig. 16.17).

Several design criteria are addressed more specifically below

DMA size: The smaller the size of a DMA the quicker new breaks will be identified through the minimum nighttime flow monitoring and analysis. For example if a DMA is larger than 1000 properties/service connections it becomes difficult to discriminate small leaks (e.g., service line leaks) from customer consumption volumes. However, the DMA size depends ultimately on the economic level of leakage. If economic analyses have shown that it is economic for the utility to quickly identify and repair new service leaks then the DMA size needs to be less than 1000 service connections. However, in most cases the DMA size should be somewhere between 3000 and 5000 service connections.

Water quality considerations: Creating a DMA involves closing valves to form a boundary, which creates more dead-ends than would normally be found in a fully open system. Hence the potential for water quality degradation from flow disturbance (initially) and stagnation (eventually) exists. The greater the number of closed valves in a DMA, the greater the care that should be exerted in designing

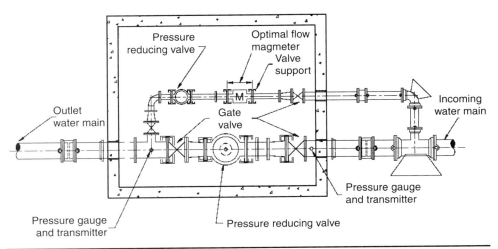

FIGURE 16.17 Typical DMA meter chamber design used by Halifax Regional Water Commission—providing redundancy for fire flow requirements. (*Source*: Halifax Regional Water Commission.)

water quality safeguards. Conversely, the creation of a DMA allows the water utility to focus more specifically on valves, fire hydrants, pressure levels, and water quality than in a typical open system. Water utilities are often hard-pressed to actively manage system valves, and many valves are overlooked for maintenance, hence, failing to operate in times of emergency such as water main breaks. Good valve exercising and management practices can be incorporated into DMA efforts to provide proactive management of these often neglected assets. Water utilities operating multiple DMAs often have better valve management than those not employing DMAs. Water quality sampling and assessment should be conducted during the planning and implementation phases of the DMA, as well as routinely during the DMA operation. This will give the utility operator the opportunity to proactively build any needed water quality controls into the design of the DMA. Good water quality can be maintained by properly configuring the boundary or performing periodic flushing.[4]

Minimum flow and pressure requirements for fire flow and insurance: During the design phase of a DMA it is important to properly assess the impact the creation of a DMA has on the ability to provide sufficient flow and pressure in case of an emergency.

There are several design options to meet fire flow and insurance requirements. Multiple or redundant feeds have proven to be a successful method of designing DMAs, where only the primary feed (or feeds) is equipped with a DMA meter and the stand-by feed (or feeds) is equipped with a PRV that only opens up in case of an emergency (see Fig. 16.17). The lead and stand-by feed can be located in the same chamber or at different feed points on the DMA boundaries. Another method of meeting the fire flow requirements while accurately measuring the DMA inflow is to use check valves or pilot-controlled hydraulic valves in place of closed-gate boundary valves. When fire flows are required, the system pressure will drop within the DMA causing the check valve or hydraulic valve to open thereby introducing additional flow as required.

16.6.3 Initial DMA Installation and Testing

Following the initial design phase the DMA needs to be set temporarily and field measurements gathered in order to verify the integrity of the DMA and to gain the data necessary for the DMA chamber design. The DMA needs to be set up by closing all identified boundary valves and verifying the status of already closed valves. The supply into the DMA through the selected feeder main/s needs to be monitored by using temporary flowmeters (e.g., electromagnetic insertion flowmeters, or clamp on ultrasonic flowmeters).

Next the integrity of the DMA boundaries should be assessed by conducting a "pressure drop test." During this test the pressure is dropped within the DMA in various steps by operating the valve or PRV controlling the inflow to the future DMA. Such tests should be conducted during the minimum nighttime flow period (between 1 and 4 a.m.) in order to avoid customer consumption disruption that would generate complaints. This period needs to be adjusted to any local differences in demand patterns. The steps in pressure reduction should be in the range of 10 psi (7 mH) to 15 psi (11 mH) down to the pressure level where the minimum required pressure at the critical zone pressure point is set. In order to monitor if the DMA is hydraulically discrete or not, several pressure loggers need to be installed outside the DMA boundaries prior to the test. These boundary loggers will record any change in pressure related to pressure drops created within the DMA in case the DMA is not hydraulically discrete. In addition to the boundary loggers, it is also necessary to install loggers inside the DMA. If any of the pressures recorded by the boundary loggers have the same pattern as the pressures from loggers located inside the DMA, then the DMA is not hydraulically discrete, and an unidentified cross connection to adjacent areas or a passing boundary valve exists.

Once the integrity of the DMA has been confirmed it is necessary to measure the total inflow to the DMA over several days to gain the necessary data to calculate the existing volume of leakage and to estimate future leakage target volumes. This stage should also be used to simulate fire flow emergencies to see if the selected feeder mains have the capacity to provide sufficient flow during such an event. If it's found that the selected feeder main does not provide sufficient supply capacity during an emergency then it is necessary to redesign the DMA and either change the boundaries or to include an additional feeder main in the design.

16.6.4   DMA Meter Selection

The selection and installation of the DMA meters are key components when designing and creating a new DMA. There are several key issues related to DMA metering that need to be considered, such as the sizing of the meter, the ability of the meter to record accurately at maximum and minimum flow rates, and the necessity to meet peak demand and fire flow requirements. Fire flow demands for a DMA are dependent upon the customer building demographics, since fire flow requirements vary notably between residential structures and industrial, commercial and institutional customers such as factories, shopping malls, schools, airports, and the like. Seasonal fluctuations and demand changes are also factors that need to be considered when specifying the DMA flowmeter.

The choice of meter size and type depends upon:

- Size of main
- Flow range
- Head loss at peak flows
- Reverse flow requirements
- Accuracy and repeatability
- Data communication requirements
- Cost of the meter
- Cost of ownership and maintenance requirements
- Water utility preference

When selecting the appropriate meter size and type, it is critical to assess the current proportion of leakage to customer demand and to project the future reduced leakage rate anticipated to occur after leakage reduction controls are established. The estimate of future leakage will affect the future minimum nighttime flow range. The flow measurements conducted during the initial DMA installation and testing, in conjunction with analysis of seasonal demand fluctuations and leakage, can then be used to finalize the design of the DMA inflow meter(s) and meter chambers. Utilities with calibrated hydraulic models may use their models to calculate expected flow ranges at the DMA metering point instead of temporary flow measurements.

If initial leakage levels are very high it is recommended to conduct a thorough leak detection and repair campaign to remove the majority of the leakage backlog before finalizing the meter design. This will allow the design to be based upon flow characteristics representative of the desired low leakage operation, and avoid oversizing the meter.

A simple rule of thumb is to limit the number of metered inlets and outlets (if any), since multiple supply and pass-through locations can give rise to misleading leakage levels due to the compounding of errors from multiple flowmeter.

16.6.5 DMA Data Monitoring

The economic optimum volume of leakage is a driving factor influencing the selection of DMA monitoring and data transfer capabilities. In utilities where the cost of water is relatively low it is very likely that there is no financial incentive in detecting small-sized leaks instantly. This means that it will not be necessary to have real time transmission of DMA data. The data from the DMA might be transferred and analyzed once a week. If several leaks occurred over this period, the minimum night flow might only reach the level of intervention after a number of days. Only then does it become necessary to send a leak detection team into the DMA to conduct a leak survey. However, there are several options to consider when selecting the optimum interval to collect the DMA flow and pressure data.

> *Real time data transmission*: Supervisory control and data acquisition (SCADA) systems are commonly used in North American water utilities to provide real time monitoring and to control pumping stations, remote treatment facilities, reservoir sites, pressure reducing chambers, and any other desired water supply facilities. In recent years, the role of the SCADA system has expanded to include security, video transmission, water quality monitoring, and other parameters not directly associated with water distribution. If real time data collection is desired, the use of an existing SCADA system is a viable option for many water utilities. With an existing SCADA system in place, the cost to outfit and individual meter/PRV site is basically the cost of one additional SCADA endpoint device, or remote terminal unit (RTU), and related instrumentation. If a SCADA system does not exist, the possibility of outfitting multiple DMAs in a SCADA system might be one of a number of benefits to help justify the cost of a complete SCADA system in the water utility. Real time monitoring is appropriate if the water utility needs to respond immediately to an emerging leak or main break in the DMA. However, most leaks emerge slowly and are initially small in volume and are not identified until the next minimum nighttime flow period. Therefore, it is not essential to receive DMA data in real time. For distribution systems with mostly slowly emerging leaks, the greatest data collection frequency that can be economically justified is once per day, ideally early in the morning after the minimum nighttime flow period. Monitoring DMA data through a SCADA system is the most comprehensive option to monitor DMA data, but is not likely to be cost-justified strictly based upon DMA use alone. Usually the business case for SCADA systems is based upon the multiple benefits of monitoring and control of many parameters at many sites and facilities.
>
> *Data transmission through GSM telemetry*: Another option of monitoring the DMA flow and pressure data is by transmitting data through global system for mobile

communications (GSM) short message service (SMS). Several manufacturers provide loggers able to transmit the recorded and logged flow and pressure values on a regular basis using SMS. These loggers can transmit the data to a host computer on a daily, weekly, or monthly basis. The cost of installation for this option is very low. However, it is necessary to assess the cost for the SMS messages since this service cost is set by the mobile phone provider in the local area. Dial up connections using telephone lines or low power radio can be used as well for the transmission of flow and pressure data. This option does not require power supply.

Manual data collection: Another option is to manually download the data recorded by the logger on a regular basis depending on the intervals set up as economically justifiable. This approach requires personnel to visit the DMA equipment on a regular basis for data downloads. It has the advantage of frequent visual checks on the equipment, but the disadvantage of high staff time required to make the regular visits. This option involves the lowest installation costs since no automated communication system is required, but the operational costs are high, since ongoing staff and transportation costs are necessary.

16.6.6   DMA Data Analysis

The concept of DMA monitoring is to measure flow into a discrete area with a defined boundary and observe typical variations in flow. The estimation of the real loss component via minimum night flow analysis is carried out by subtracting an assessed or measured volume of legitimate night consumption for each of the customers connected to the water mains in the DMA. The minimum nighttime flow in urban areas usually occurs between 2:00 and 4:00 a.m. This flow value is the most meaningful data used in determining the leakage rate in the DMA. During this period, authorized consumption is at a minimum and, therefore, leakage is at its maximum percentage of the total inflow. In regions where customer landscape irrigation makes up a significant part of the demand during the minimum nighttime flow period, the accuracy and the confidence in the calculated real loss figures will diminish. The result obtained by subtracting the legitimate night consumption from the minimum nighttime flow is known as the net night flow (NNF) and provides an estimation of the volume of real losses during the MNF period. The leakage volume can be modulated over the whole 24-hour period using the fixed and variable area discharge paths (FAVAD) concept (see Chap. 18, or where we discuss the FAVAD principle).

$$NNF = MNF - \text{legitimate nighttime consumption}$$

The NNF is mostly composed of real losses from the distribution network and the service connection piping between the water main and the customer meter. However, it may also include leakage on the customer side of the meter and consumption through unauthorized connections. Figure 16.18 shows the results of a MNF analysis.

Data for DMA Minimum Nighttime Flow (MNF) Analysis

In addition to the inflow measurements, pressure measurements at the zone inlet point(s) and at the average zone pressure point the following data is also required to be able to conduct a MNF analysis.

- Length of mains
- Number of service connections

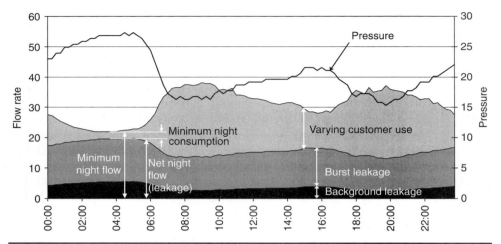

FIGURE 16.18 Twenty-four-hour leakage modeling based on minimum night flow analysis. (*Source:* IWA Water Loss Task Force.)

- Number of household properties
- Number and types of nonhousehold properties
- Legitimate nighttime consumption (can be estimated or obtained by measuring a sample of customers and inferring for the entire population, or measuring the entire customer consumption by a fixed network AMR system)

Legitimate nighttime consumption is generally composed of three elements:

- *Exceptional night use*: Some public, commercial, industrial, and agricultural customers will have significant water use during the nighttime period due to the nature of their business processes. Such uses can be large in relation to the minimum night flow into the zone. These customers have to be identified through discussions with local operational staff and analysis of consumption data from the billing system. Where a customer in a zone is thought to have a significant night use, consumption readings or recordings of this customer need to be taken during the MNF measurement in order to accurately deduct this component of legitimate consumption from the total inflow.

- *Nonhousehold night use*: Nonhousehold customers who are not identified as exceptional night users may, nevertheless, consume some water at night, for example, in automatic flushing urinals. Some allowance for this night consumption has to be made. This is often accomplished by making estimates based upon the type of industry and typical published consumption volumes for such users. Where necessary, this data can be supplemented by local short-term data-logging of specific customer meters.

- *Household night use*: Household or residential customers also consume some water during the minimum nighttime flow period. Consumption occurs due to toilet flushing, automatic washing machines, time-programmed dishwashers, and outdoor landscape irrigation. Ideally, night customer consumption measurements of typical household customers can be gathered for the proposed DMA in order to determine an appropriate level of household night water consumption to include in the night flow analysis. Household night water use can be determined by gathering data through manual or fixed network automatic meter readings during the minimum nighttime consumption period, or through deployment of data-loggers recording the household night use. Alternatively, data from the literature can be used, for example, the U.K. Managing Leakage Series provides details of assessed night use studies in the U.K. between 1991 and 1993 and the AWWA Research Foundation project residential end uses of water (REUWS) provides data on North American night use volumes. If there is significant nighttime landscape irrigation consumption at certain times during the year, it is recommended to undertake DMA MNF analysis during periods of no irrigation or when irrigation is at a minimum, usually during the winter period.

16.6.7 Prioritizing DMA Leak Detection Efforts

DMAs allow assessment of leakage volumes in a hydraulically discrete zone. If multiple DMAs are established in the service area, leakage volumes can be assessed for each of the DMAs on a regular basis. The results gained from the DMA measurements allow a utility to prioritize its leak detection efforts, targeting the DMA(s) with the highest leakage volume, where the leak detection efforts bring the best results in real loss reduction in relation to the work effort required. Consequentially, targets can be set to decide which DMA needs to be addressed and in what order by the leak detection team. A simple starting point for prioritization of leak detection efforts is to rank DMAs according to their volume of real losses per service connection. This applies to utilities in urban areas, while rural utilities should consider expressing the volume of real losses by length of main. The use of DMAs results in a strategic scheduling of leak detection crew activities. This is more efficient that the historic practice of crews canvassing portions of the service area based upon fixed time intervals.

Ideally targets, or thresholds, for leak detection intervention are set based on analysis of the economic optimum volume of leakage in each DMA (see Fig. 16.19).

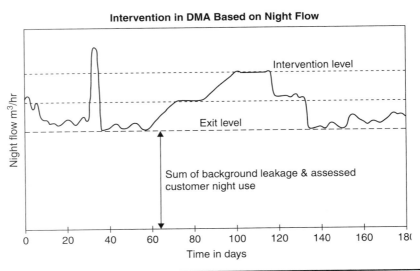

FIGURE 16.19 Example for leak detection intervention level based on economic optimum analysis of real loss volumes in a DMA. (*Source:* IWA Water Loss Task Force.)

Figure 16.19 depicts an example of continuous DMA measurements and the rise of leakage volume in this DMA. Based on economic optimum analysis of real losses for this particular DMA an optimum threshold level for intervention against real losses was set. Once this level is reached, a leak detection team is sent to the DMA to detect the running leaks and bring down real losses to the exit level where further leak detection efforts would not be cost-effective.

16.6.8   DMA Management

Like any other part of the distribution system, DMAs need to be managed and maintained in order to achieve the expected results. DMA related equipments such as DMA flowmeters and PRVs should be properly maintained to ensure that high quality DMA data is collected. It is vital for the success of the DMA that the integrity is maintained by keeping all boundary valves closed and by ensuring that boundary valves are not leaking or passing water into a neighboring zone. However, boundary valves can be opened temporarily for operational purposes as long as they are properly closed again afterward and normal operating conditions are confirmed. It is a good practice to keep records about events, locations, and durations of boundary valve operations. Boundary valves should be clearly labeled on maps and in the field so that they are not operated inadvertently. This information will assist the leakage management team in interpreting high flows as leaks and breaks rather than an open boundary valve.

For each DMA a file should be established containing key information including number of all types of customers, location of sensitive customers and their contact information, number of hydrants and fire sprinkler systems, pressure information, and assessed minimum nighttime consumption volumes. Creating a map with as much information as possible is another important part of DMA management. Records should be kept for future component analysis on all leaks found, their location, type of pipe failure or defect, material, and size of the pipe on which the break was detected.

16.7   Testing for Reservoir Leakage

Large amounts of leakage can be lost through either leakage from the structure of the reservoir or from reservoir overflow. Leakage from the structure itself is probably more common in older underground brick or masonry reservoirs which have not been lined, but leakage can occur in other forms of storage too.

The easiest way to check for leakage is to isolate the reservoir from the system by closing the inlet and outlet valves. This is usually done at night. Once the reservoir is isolated, a depth test over time can be performed either by simply measuring carefully the drop in level over time or by installing a high-resolution level data logger to measure the drop over time. It is then just a matter of calculating the area of the reservoir, calculating the volume per area times the drop measurement, and calculating the volume of loss.

> **A** level drop test can be performed to check whether reservoir leakage is present.

Care must be taken to ensure that the outlet valve is not letting some water escape. Calculations are more difficult when the shape of the structure is not prismatic, as the unit volume per increment of level changes as the level drops. Most water utilities should have accurate as-built drawings showing exact measurements.

If the reservoir is found to be leaking, then one way of finding the actual leaks is to send in a diver with fine sand. The fine sand is sprinkled along the walls of the reservoir and the base and is drawn into patterns where the suction of the leak takes effect. In many cases, though, if the leakage is significant, the reservoir should be programmed to be lined, as long as the basic structure is still structurally sound. Determining the presence of leakage will help to justify the cost the reservoir lining project.

Storage overflow losses are more common where storage is in a remote location and water is not obviously visible or evident from above-ground, as it would be in an urban situation. Overflows usually happen at off-peak times when head losses and demand are low in the system and storage is filling. Overflows occur most typically from malfunctioning level instruments or control equipment and/or inattentive operators who fail to halt a filling operation at the proper level.

Overflow pipes should be inspected to see if there are obvious markings on the ground or wet patches where water has been ejected. Another simple method is to wedge a ball or object into the pipe during the day. If the object moves, it is likely that there has been an overflow situation. A more detailed analysis can be undertaken by using a high-resolution level logger. When the level of the overflow pipe is reached, loss starts to occur. Coupled with a temporary meter at the inlet to the tank, it is easy to calculate the volume and value of the loss. Once the value of the loss is calculated, a suitable and cost-effective method of intervention may be installed. The simplest forms of level control are mechanical float valves or altitude valves, which are discussed in Chap. 18. However, utilities often use remote control systems and SCADA systems to control tank levels. In some cases tank overflows occur because this equipment malfunctions due to lightning strikes or other causes.

Sometimes the problem is no control or inefficient manual control, and sometimes the problem is lack of maintenance on simple mechanical controls. In all cases the loss should be resolved in a cost-effective fashion.

Utilities with SCADA or telemetry systems in place can utilize these systems to periodically read zonal meters and analyze the condition of the losses through periodic modeling and assessment.

16.8   Summary

In this chapter we have discussed the methods of cost-effective leakage management. The intent of this chapter is to provide guidance on effective and innovative methods and technologies to control water distribution system leakage, particularly underground, nonvisible (unreported) leakage. A variety of methods can be considered, in developing the leakage management strategy, and each distribution system should be evaluated individually before a commitment is made to one methodology or approach. However, without active leak detection, leakage in a water distribution system will only worsen!

References

1. Fanner, V. P., R. Sturm, J. Thornton, et al. *Leakage Management Technologies.* Denver, Colo.: AwwaRF and AWWA, 2007.
2. Fanner, V. P., J. Thornton, R. Liemberger, et al. *Evaluating Water Loss and Planning Loss Reduction Strategies.* Denver, Colo.: AwwaRF and AWWA, 2007.
3. *"District Metered Areas Guidance Notes,"* IWA Water Loss Task Force, 2007.
4. UK Water Industry Research Limited. *A Manual of DMA Practice.* Report Ref. No. 99/Wm/08/23, UKWIR: UK 1999.

CHAPTER 17

Controlling Real Losses—
Speed and Quality
of Leak Repair

Reinhard Sturm

Julian Thornton

George Kunkel, P.E.

17.1 Introduction

The speed of leak repair comes into the overall four-component picture of reduction of real losses as shown in Fig. 17.1.

We have discussed various methods of locating leaks in Chap. 16. It is very important to rank leak repairs for severity of loss or danger to life or property and schedule them to be repaired as soon as possible. The annual volume of real losses stemming from reported and unreported leaks depends on the number of leaks, their magnitude, the operating system pressure, and probably most importantly the total time the leak was permitted to run. All leaks are pressure dependent—more pressure equals greater leakage rates—and Chap. 18 discusses all aspects of pressure management used to reduce leakage volumes.

17.2 Leak Runtime Reduction

The total run time of a leak is comprised of three elements (Chap. 10 discussed the component analysis of real losses in more detail than provided in this chapter):

> **Awareness Time:** This is the time needed for the operator to become aware that a leak exists, a parameter strongly influenced by the presence or absence of an active leakage control program.

> **Location Time:** This is the time taken to pinpoint the location of the leak once the operator is aware of its existence.

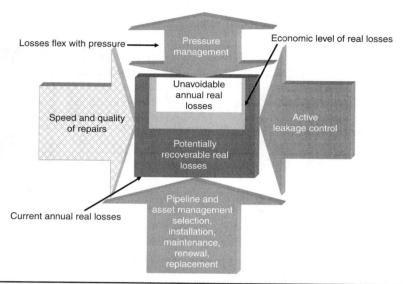

FIGURE 17.1   Four potential intervention tools of an active real loss management program. (*Source:* IWA Water Loss Task Force and AWWA Water Loss Control Committee.)

Repair Time: This is the time to affect a repair that halts the leakage flow, once the location of the leak has been identified. This is not just the time of the shutoff or repair action, but all time needed to route the repair work order, schedule the repair, notify customers, and other activities, which can take days or weeks depending upon the policies of the water utility.

Figure 17.2 depicts the impact the awareness, location, and repair time duration have on the total volume lost from a leak.

Figure 17.2[1] provides the results from a study carried out in England in order to investigate the impact leak run times of various size leaks have on overall system leakage volume. It clearly shows that mains breaks, which due to their disruptive nature are

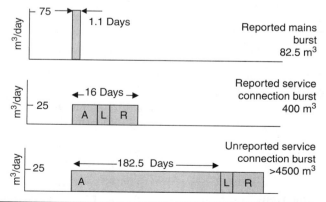

FIGURE 17.2   Impact of awareness, location, and repair time on total leakage volume. (*Source:* Ref 1.)

easy to locate and are repaired quickly, actually only contribute a small volume to the annual real loss figure. The reason for this is that, even though they usually have a high leakage flow rate, the utility responds quickly to such an event and the pipe section of the break is promptly shut down, therefore the leakage loss volume is relatively small.

Conversely, small leaks, especially on service lines, generally contribute the greatest volume to the overall real loss volume due to their average long run time. Small leaks can run for periods of weeks, months, and even years before being discovered and repaired.

17.2.1   Reducing the Awareness Time

Some leaks are reported and others are unreported and whether a leak is reported or unreported has an impact on the total leak run time, especially the awareness time. The time it takes to become aware of an unreported or a reported leak is significantly different. Reported leaks have a short awareness time since they either become visible on the street or ground surface (sometimes in form of catastrophic failures) and are reported to the utility, or they cause a drop in supply pressure and again are reported quickly to the utility.

Unreported leaks, however, can run for very long periods of time (up to years) before they become big enough to surface, cause a catastrophic failure, and the like, and therefore become reported leaks.

There are two activities that help reduce the awareness time of unreported leaks.

1. **Active Leak Detection**: Conducting an active leak detection campaign covering the entire distribution system once a year reduces the awareness time of an unreported leak to on an average 6 months. Doubling the intensity of active leakage control effort, that is, completely sounding the system every 6 months instead of every year, would reduce the duration for which the unreported leak runs to an average of about 95 days, reducing the leakage volume related to unreported leaks by half. However, a reduction in activity to sounding the system every 2 years would allow breaks to run for an average of 365 days before their location and repair, doubling the losses resulting from sounding the complete system annually. This illustrates why the detection of unreported breaks can be of such importance to a water service provider. The various frequencies of leak detection sounding carry differing levels of cost—personnel, equipment, and materials—to implement, and these costs must be compared with the value of the water that would be either saved by a greater frequency of active leakage control activity or lost due to a lower level of activity (see Chap. 9).

2. **District Metered Areas (DMA)**: Dividing the distribution system into small hydraulically discrete zones where the total inflow is monitored continuously allows the water utility to become aware of a new leak shortly after it emerges in the DMA. With a DMA in place the utility can analyze the minimum nighttime flows (MNF) on a daily or weekly basis to identify the emergence of new leakage as indicated by an increase in MNF. The size of the leaks that can be identified by DMA analysis depends on the size of the DMA—the smaller the DMA the smaller the leakage events that can be discerned. A leakage management strategy including both DMAs combined with active leak detection is usually a more efficient approach than regular sounding alone, but a combined approach incurs a higher capital cost to create and install the DMA.

17.2.2   Reducing Location Time

The time it takes a utility to locate a known leak depends on the tools and skills of the leak detection crew in pinpointing the exact location of the leak source. The location time can be reduced by making more crews available for leak pinpointing and by making sure that the crews are well-trained, motivated, and equipped. Leak detection crews that utilize state-of-the-art leak correlators will typically respond faster and more accurately in pinpointing leaks than crews that only utilize mechanical sounding equipment. The policy decision on how fast the utility responds to a leak it is aware of should be evaluated using benefit to cost analysis. The higher the value of the water lost through running leaks the quicker the utility should respond in locating a leak.

17.2.3   Optimizing Repair Time

The time it takes a utility to repair a leak depends upon a number of factors. The number of repair crews available to address leaks, their level of training and motivation, and how well they are equipped are primary factors. Water utility policy also has a strong bearing on the average repair time to address leaks. Water utilities may have set performance targets for how fast a service line leak, a mains leak, and the like need to be repaired. Nevertheless, there are significant differences in leak repair time from utility to utility. Top performing water utilities repair leaks on an average within 12 to 24 hours once they are aware of the leak. Other utilities may take weeks to months to repair leaks that are not causing supply disruptions or infrastructure damage.

A significant factor in overall leak repair time is the type of policy that water utilities accord to customer service line leaks. It is standard practice in many water utilities worldwide to require customers to own and maintain at least a portion of the service line that extends from the water distribution main to the customer premises. A small number of utilities require customers to own the entire length of service line, while a majority of systems require the customer to own the section between the property line or curb stop to the customer premise. Requiring customers to arrange for repairs of known leaks on their service lines has been found to be a highly inefficient leakage management policy, since customer inherently responds in much slower fashion than utility crews would in effecting leakage repairs. It is very feasible for water utilities to operate service connection repair programs that efficiently implement repairs in 2 to 4 days after a leak is discovered. For most customer-arranged repairs, response time typically averages several weeks. The longer leaks run, the greater the leakage losses.

In order to operate efficient leakage control programs, and to save customers the effort and aggravation of arranging leak repairs, many water utilities operate service connection insurance or warranty programs. For a small additional fee included in their regular billing, customers can rely upon the water utility to make all arrangements for service connection repair or replacement when leaks arise, and pay no additional costs. These approaches generally handle service connection leaks more efficiently than customer-arranged repairs, and help to improve customer relations. Water utilities should track response and repair times and, if they require customers to arrange repairs, consider reevaluating this approach as a means to reduce the duration of customer service leaks occurring in their system.

Another important factor to consider in leak repair efficiency is the effectiveness of the work order management in the water utility. Water utilities should employ a robust work order management information system to track and archive information on customer complaints and utility-generated work orders. Sound work order tracking is essential for water utilities to provide excellent service, particularly for large water utilities that encounter thousands of complaints and work requests every year. The average time to repair leaks can be increased if poor work order tracking results in a delay in forwarding leakage information from the leak detection crew to the designated repair crew. For utilities that employ paper work order tracking, it is not uncommon for paperwork to be lost, resulting in a failure to respond to suspected leaks that have been identified during leak detection surveys. Water utility managers should review the structure and effectiveness of their work order tracking process in order to ensure that unnecessary delays are not injected into the leak repair process due to poor work order information management.

Component analysis of real losses, as discussed in Chap. 10, is a powerful tool to analyze the impact various repair time policies have on the overall real loss volume of a water utility. Again, leak repair policies and leak repair time targets should be based on sound benefit to cost analysis.

17.3   Quality of Leak Repair

The quality of leak repair work plays a significant role in the overall leakage management effort. Quality of materials and quality of workmanship are two main factors influencing the overall quality of leak repairs. If the quality of leak repair is poor, then there is a good chance that leaks will recur at the location of a previous leak repair. In the worst case poor repair quality might even result in the creation of new leaks.

17.4   Summary

Leaks must be repaired in timely and effective fashion to ensure that loss volumes are kept to a minimum. Surprisingly, many utilities do not always repair known leaks! This may in some cases be due to an economic decision or one based on distribution logistics, but in some cases it is just a lack of awareness of the impact on annual loss volumes. Chapter 10 discusses in detail some of the methods available to model annual losses and the impact of various interventions on those losses. The impact of an improved repair program can be easily modeled. Not only must leakage be repaired, it must be done in a manner which will ensure that this particular leak will not recur in the short term. Unfortunately, quality of repair is an area which is sometimes overlooked. The time until leak repair is carried out will almost always have a large effect on the annual volume of real losses, whether it is leak repair from surfacing reported leaks or unreported leakage which is located during a routine leak survey. Many small leak volumes soon add up to one large leak volume!

Reference

1. IWA Water Loss Task Force. "Leak Location and Repair Guidance Notes," 2007.

CHAPTER 18

Controlling Real Losses— Pressure Management

Julian Thornton

Reinhard Sturm

George Kunkel, P.E.

18.1   Introduction

System optimization is in many cases far more cost-effective than system expansion and most certainly always has a much more positive environmental impact. Many water systems are designed considering the minimum level of pressure required for the demand types, but in many cases no consideration is made for maximum pressure levels. If no consideration or only basic consideration was made at the time of installation then there may very well be room for optimizing the pressures within a system. Pressure management is one of the most basic and cost-effective forms of optimizing a system and can in many instances provide fast paybacks on large investments.

Figure 18.1 shows where pressure management fits into the four-component scenario of real losses management.

Pressure management has been around for many years in various forms, however it is only in the last few years that, advanced pressure control has been used on a wide basis in system optimization and loss reduction and management programs.

This chapter takes the reader through all planning stages of a pressure control scheme from deciding whether or not it is necessary for his or her system and to what degree, to cost justification and practical field installations.

Many systems are designed with minimum pressure requirements in mind but not maximum pressure limitations therefore many systems have areas, which are grossly over pressured.

Pressure management is one of the most basic tools available for total real loss management.

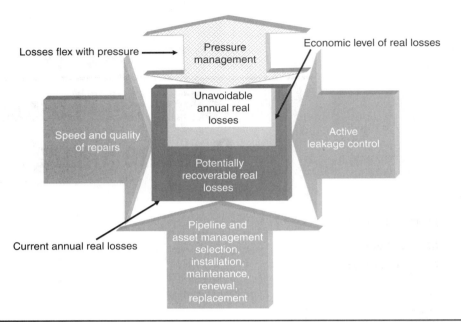

Figure 18.1 Pressure management component of real loss management. (*Source:* IWA Water Loss Task Force/AWWA Water Loss Control Committee.)

This chapter is not designed to replace an all-encompassing valve manual. Covering every aspect of hydraulic control, valve manuals are available from most manufacturers and a good manual is available from the ISA. This is rather a very practical "hands on" guide to using pressure management (pressure-reduction, level-control, flow-control, and pressure-sustaining valves) as one of the many tools to reduce losses and run water distributions systems in a more efficient manner.

18.2   Why Undertake Pressure Management Schemes?[1]

18.2.1   Positive Reasons

Leakage Reduction
The reduction of leakage is a subject, which is on the minds of most water utility engineers and managers throughout the world. In other chapters of the book we have discussed various types of leakage reduction programs of which pressure reduction is one. As with all of the other techniques for reducing leakage, pressure management is just one tool, which should be used where applicable in conjunction with other technologies and methodologies.

Recent studies and research have shown that both leakage volume and new leakage frequency is reduced greatly by the reduction and stabilization of pressure within a distribution system. Obviously not all systems can tolerate pressure reduction and indeed many systems suffer from lack of pressure, however there are still many, utilities that are operating pressures in excess of those required, who would benefit greatly from a pressure management scheme. When considering reduction of leakage practitioners usually think of pressure reducing functions, however, in many cases, in particularly pumped systems, leakage can be reduced greatly by surge anticipation.

Water Conservation

In direct pressure use situations, see Fig. 18.2 pressure reduction can be an effective way of controlling unwanted demand. A simple example is somebody cleaning their teeth for 5 minutes at a high pressure or 5 minutes at a low pressure. If the tap is left on for the duration, much less water will be consumed at the lower pressure.

This is not the case in tank-fed residential situations as in Fig. 18.3 as the head controlling the demand is a function of the height above the equipment being used not the incoming pressure. (Work is being undertaken by practitioners in areas with residential tanks to better understand the role that pressure management may play in the reduction of ball valve leakage, which often goes undetected as meters have trouble reading these low flows. It has been noted that below certain pressures the ball valves stop leaking with no further intervention needed. This may in effect mean that pressure management can also have a positive effect on apparent losses).

While many utilities may not want to reduce demand, because they will have a negative impact on their billing, many other utilities have found that it is much more cost-effective to reduce demand than to implement costly capital expansion programs to increase supply or meet excessive demand peaks. Utilities with direct feed systems should carefully analyze the demand types within the residences and commercial industrial customers, as many demands are volumetric and therefore will not be affected by pressure reduction other than fill time changes.

Nonpayment

Some utilities are faced with a nonpayment situation, which is difficult to resolve due to political, or social pressure meaning that they have to continue to supply water even though the customers are not paying. In these situations pressure management to

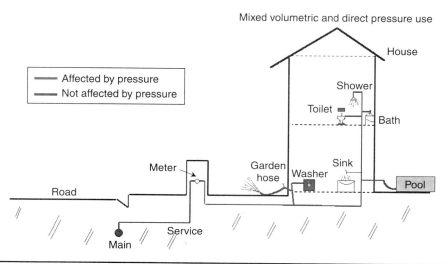

FIGURE 18.2   Residential demand direct feed.

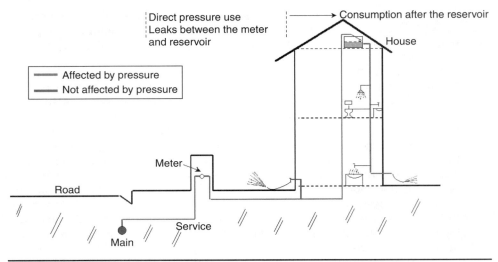

FIGURE 18.3   Residential demand tank-fed.

reduce consumption while maintaining a minimum level of supply is of the utmost importance to optimize losses and conserve resources.

Some utilities are faced with the situation where they are not permitted to increase their supply, due to environmental restrictions. Pressure reduction can be performed on a zonal basis or indeed on an individual customer basis, as the situation requires.

Pressure reduction can also be used as an emergency measure for drought control; levels of demand and leakage can be drastically cut until reserves return to normal.

Efficient Distribution of Water

Many water distribution systems have problems supplying some customers, while others enjoy a constant source of water. The reason may be due to aging infrastructure, poor design, geographic constraints, or demographic layout. Pressure management using not only pressure reducing techniques, but also pressure sustaining techniques, boosters or flow control, can ensure that the system distributes its resource as evenly as possible, ensuring required volumes for a majority of the customers.

> **P**ressure management is not only about pressure reduction but also in some cases pressure increase, pressure sustaining, and surge control and level management.

Guaranteed Storage

The implementation of pressure management schemes can assist the utility operators in ensuring that reservoirs and storage tanks remain at realistic levels to meet demands. This may be done using a mixture of pressure-reduction, pressure-sustaining, and flow-control valves. See Fig. 18.4. Level control also ensures that storage is not allowed to overflow during off peak hours when system demand and head loss is low and pressures are at there highest. Reservoir overflows can form a large part of a utility leakage if not properly controlled and calibrated.

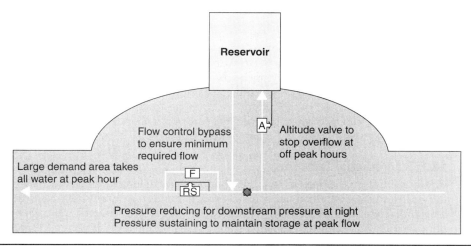

FIGURE 18.4   Pressure management often uses a mixture of valve types.

Reduced Hydraulic Impact

Hydraulic impact, surge and transient waves are caused by quick changes in system condition. Unfortunately most systems have situations where an operator closes a valve too fast or the opposite. Maybe a hydrant is operated quickly in an emergency or a large consumer suddenly stops drawing water. Without valve control in the system transient waves are allowed to travel backward and forward within the system, causing damage at any weak point. While pressure-relief valves and surge-arrestor valves are the tool for this type of situation see Fig. 18.5, simple pressure management schemes limiting pressure to those required are also effective in reducing the negative impact of transient waves. Simple pressure-reducing valves installed to maintain lower pressures would also damp the potential negative effects.

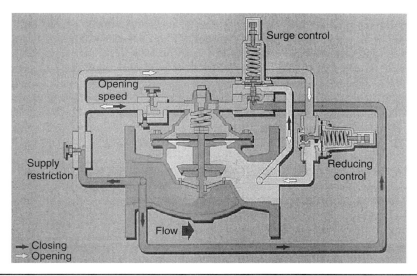

FIGURE 18.5   Surge arrestor valve diagram.

Reduced Customer Complaints

Pressure management schemes are designed not only to reduce pressure but also to provide a constant supply of both water pressure and volume. Some customers experience periods of the day with low pressures caused by high head losses in the system. High velocities, some of which may be due to uncontrolled demand downstream of the customer, cause high head losses. Other customers complain of pressures, which, are too high and cause either discomfort or damage to equipment in the home. Uncontrolled leakage can also cause lack of supply for customers.

Contrary to belief pressure management can increase customer satisfaction.

18.2.2   Potential Concerns

It may seem from the preceding pages that pressure management is the answer to all of the utility problems! However a poorly implemented program may also cause problems of its own. When discussing a pressure management scheme for a utility that does not currently have control or is intending increasing the level of control, the usual concerns are as follows:

- Fire flow concerns
- Loss of revenue
- Reservoirs not filling at night

Fire Flow Concerns

Where fire flows are a concern, sectors can have multiple feeds, controlled by PRVs with flow-modulated capability. Therefore, if there is a fire, the system has sufficient hydraulic capacity to maintain pressures and flows for fire fighting, as required for example in the National Fire Protection Association (NFPA) regulations, in the United States and Canada. The valves will automatically regulate pressure as determined by the demand requirement plus the minimum safe operating limit at residual conditions.

Systems, which do not have the benefit of the more efficient flow-modulated valves often, have a large sleeper valve either in parallel with the operational valve or at a strategic entrance to the sector. This valve will open when the system pressure drops due to additional head loss created by the fire flow. In many cases this large valve will remain closed unless an emergency situation is encountered. The use of a large dormant valve may in many cases not be cost-effective when calculating against the more modern efficient demand modulated options, however in some situations the range of demands dictates that a second parallel valve be used.

The NFPA basically states that systems should have an available residual pressure of 20 psi while the hydrant is flowing and 40-psi static head. The hydrants are then coded based on testing as discussed in Chap. 22, as to their flow capacity at these standard reference pressures.

> **W**hen setting up pressure management zones, fire codes must be respected.

Obviously when setting up potential pressure controlled sectors, these limits along with insurance regulations for the types of property in the sector, should be taken into account.

Most countries have some kind of fire code, which should be followed when planning a pressure management scheme.

Loss of Revenue

As far as the loss of revenue is concerned, systems with high leakage will almost always see a positive benefit from pressure management, even when stacked against a potential loss of revenue, due to reduction of pressure in the residence or industry.

Any lost revenue is included in the cost to benefit calculations as a cost against the project just as installation and product costs are included.

This is also true for systems with lower losses and high costs to produce or purchase water. In situations where a loss of revenue cannot be tolerated, pressure management can be limited to nighttime hours, when legitimate consumption is at its lowest and system pressures are at their highest.

Remember also that many systems are enforcing water conservation programs. Pressure reduction, is also a water conservation program.

A large portion of water use within a household is from the toilet; tank type toilets use a fixed volume, which does not change significantly when pressure is reduced. There are, also many other fixed volume uses within a residence, which will not vary significantly with pressure, see Fig. 18.2 previous.

When considering pressure management for a sector, we must consider the per capita use and if this is excessive. Sample per capita uses can be found in Table 18.1.

If it is excessive then pressure management will become a natural part of a conservation program. If it is not, we must decide on the components of consumption within the sector (residential, commercial, industrial), the volumetric consumption and the consumption directly tied to pressure. We can then analyze the potential benefits of loss reduction over reduction in revenue.

> **F**or a detailed breakdown of usage in the United States the AWWA has recently undertaken an excellent residential end use study, which can be purchased through the AWWA web site.

Reservoir Filling

Regarding reservoirs not filling at night because of reduced system pressure, many pressure reduction programs concentrate on the smaller mains, therefore allowing reduction of losses in selected areas, while allowing normal system pressure in the larger trunk or transmission lines. (As with the example in Fig. 18.4, a complete pressure management project can in some cases actually improve reservoir-filling characteristics). This is particularly important in pumped systems where the storage tanks balance on the system pressure. Gravity systems are less affected.

Reservoirs are usually connected with the larger pipes, so there should not, in many cases be a problem. Most utilities find that nonvisible leakage tends to be on the smaller pipes and service connections, so the effectiveness of a potential pressure management program should not be reduced significantly, by the exclusion of larger pipes in the control area. See Fig. 18.6.

18.3 Various Types of Pressure Management

Pressure management comes in various forms from the basic sectorization of a gravity system to dynamically controlled automatic control valves (ACVs) or pump speeds. Every distribution system in the world may have different requirements or indeed multiple

State	L/per capita/per day	Gal/per capita/per day
Alabama	379	100
Alaska	299	79
Arizona	568	150
Arkansas	401	106
California	556	147
Colorado	549	145
Connecticut	265	70
Delaware	295	78
District of Columbia	678	179
Florida	420	111
Georgia	435	115
Hawaii	450	119
Idaho	704	186
Illinois	341	90
Indiana	288	76
Iowa	250	66
Kansas	326	86
Kentucky	265	70
Louisiana	469	124
Maine	220	58
Maryland	397	105
Massachusetts	250	66
Michigan	291	77
Minnesota	560	148
Mississippi	466	123
Missouri	326	86
Montana	488	129
Nebraska	435	115
Nevada	806	213
New Hampshire	269	71
New Jersey	284	75
New Mexico	511	135

Source: Soley et al. In *Water Distributions System Handbook*, Larry W. Mays., Ed. © 2000 by McGraw-Hill.

TABLE 18.1 Estimated Per Capita Consumption in the United States (*Continued*)

State	L/per capita/per day	Gal/per capita/per day
New York	450	119
North Carolina	254	67
North Dakota	326	86
Ohio	189	50
Oklahoma	322	85
Oregon	420	111
Pennsylvania	235	62
Rhode Island	254	67
South Carolina	288	76
South Dakota	307	81
Tennessee	322	85
Texas	541	143
Utah	825	218
Vermont	303	80
Virginia	284	75
Washington	522	138
West Virginia	280	74
Wisconsin	197	52
Wyoming	617	163
Puerto Rico	182	48
Virgin Islands	87	23
United States Total	397	105

TABLE 18.1   Estimated Per Capita Consumption in the United States (*Continued*)

requirements. Some of the most common forms of pressure management are discussed in the following sections of this chapter.

18.3.1 Sectorization

Sectorization is one of the most basic forms of pressure management, which is still very effective. Subsectors are divided either naturally or by physical valving. The sectors are usually quite large and often with multiple feeds, therefore they do not usually develop localized hydraulic problems because of valve closures. Systems with gravity feeds usually sectorize by ground level and systems with pumped feeds usually sectorize depending on the level of elevated tanks or storage.

One of the hardest parts about controlling pressure solely by using sectorization is enforcing boundary valve control. Nowadays telemetry devices are available which transmit valve status to a central control every time the valve is operated, therefore allowing managers to control the integrity of the sectors and ensuring that they are returned to normal after either an emergency or maintenance procedure.

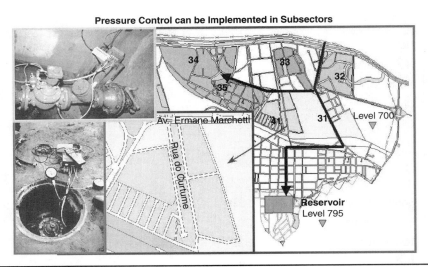

FIGURE 18.6   Pressure management in subsectors.

Sectorization in its simplest form does not require the implementation of costly ACVs and controllers, however without them, in many cases it is not completely efficient. Many systems that have had sectorization in place for many years are finding that it is very cost-effective to implement more advanced control in addition to the basic control already in place.

18.3.2   Pump Control

Many utilities use pump control as a method of controlling system pressure. Pumps will be activated or deactivated depending upon system demand. This method is effective if the reduced level of pumping (usually at night) can still maintain reservoir levels. With recent energy conservation concerns this methodology should be carefully reviewed, as to the efficiency of energy use. The pump(s) may operate outside of the designed profile if subjected to upstream valve throttling or demands outside of the design limits. Inefficient pumps can cause huge increases in electricity consumption and in some cases expensive fines for over use during peak times.

Properly controlled pumps in particularly with variable speed drives can however provide very effective system pressure control.

18.3.3   Throttled Line Valves

Many system operators recognize the need for reducing system pressure and partially close a gate or butterfly valve to create a head loss and reduce pressure. This method is the least effective, as the head loss created will change as system demand changes. At night when a distribution system needs the least pressure, the pressure will be higher and during the day when the distribution system needs the most pressure to supply demand, the pressure will be lower. This creates a classic case of an upside down zone.

> **T**hrottled system valves are the least effective way of controlling pressure.

FIGURE 18.7   Pressure-reducing valves working in parallel.

18.3.4   ACVs—Fixed Outlet

ACVs are a traditional method of control and use a basic hydraulically operated control valve. See Figs. 18.7 and 18.8. Later in the book we will be discussing controllers and varying profiles. The fixed outlet control valve method is effective for areas with low head losses, demands, which do not vary greatly due to seasonal changes and areas with uniform supply characteristics.

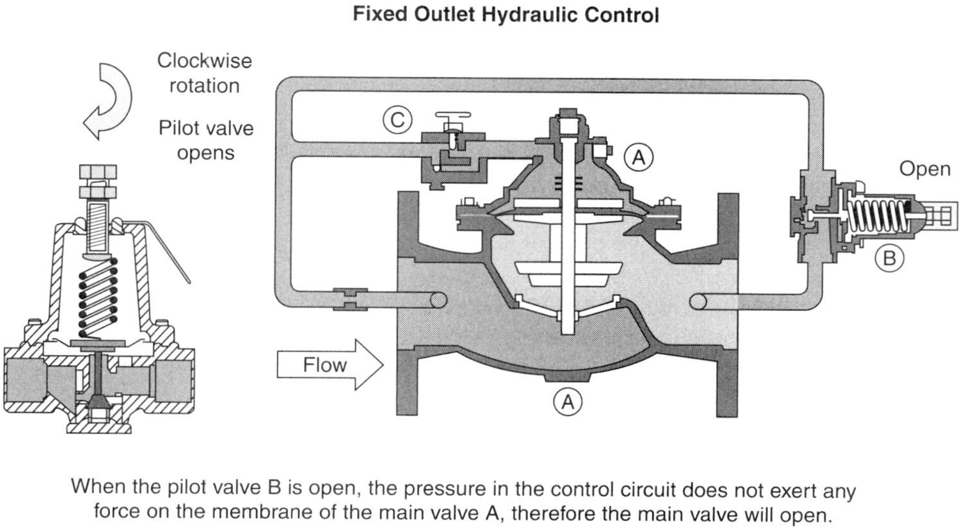

Fixed Outlet Hydraulic Control

Clockwise rotation

Pilot valve opens

Open

Flow

When the pilot valve B is open, the pressure in the control circuit does not exert any force on the membrane of the main valve A, therefore the main valve will open.

FIGURE 18.8   Pressure-reducing valve diagram.

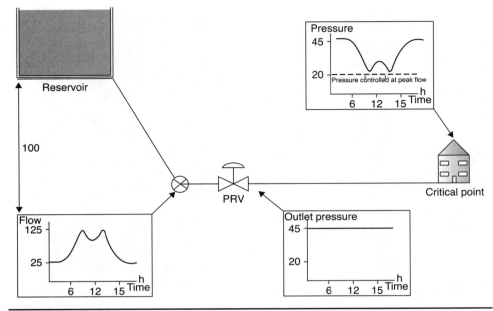

Figure 18.9   Effects of fixed outlet control.

Fixed outlet control in areas other than previously quoted may be inefficient, as outlet pressures have to be set high enough to meet minimum pressures during peak demand. As system demands reduce usually at night the head losses in the system reduce and system pressure returns toward the static pressure, which in many cases is far in excess of that required to meet nighttime demand plus fire demand. See Fig. 18.9.

18.4   Leakage Control—Pressure Leakage Theories

As discussed in Chap. 10 it is now proven that the relationship between leakage and pressure is not merely related to the square root of the pressures in question, but rather an expanding power law. As well as PVC leaks, many other types of leaks, in particularly joints, are subject to a change in area as pressure changes. This means that potential benefit in pressure reduction, on the volume of these leaks has a much greater impact, as not only the velocity of leak flow changes, but also the leak area.

18.4.1   Traditional Calculations

Traditional calculations for reduction in leakage through reduction in pressure assumed a fixed area leak. The calculation for this type of situation was

$$\frac{L_1}{L_0} \text{ varies with} \left(\frac{P_1}{P_0}\right)^{N1\,=\,0.5}$$

and still is as follows:

When the pressure is changed from P_0 to P_1, the leakage rate changes from L_0 to L_1.

Therefore: $L_1 = L_0(P_1/P_0)^{0.5}$

An example could be:

Q. A zone with fixed area leakage has a leak rate of 500 gpm at 80 psi. If the pressure were reduced to 50 psi, what would the savings in leakage rate be?

A. $L_1 = L_0(P_1/P_0)^{0.5}$ = New leakage = $500(50/80)^{0.5}$ = 500 − 395 = 105 gpm

18.4.2 Fixed and Variable Paths

Leakage can be described in either fixed or variable paths. Fixed area leakage could be pinholes in galvanized service line or a hole in a cast-iron pipe. This type of leakage follows the traditional calculation shown in the last paragraph. Savings through reduction in fixed area leakage are usually more conservative than in areas with variable area leakage.

Variable area leakage normally occurs in systems with some kind of PVC or plastic based pipe, systems with joint leaks (often found in systems with AC piping or old hydraulic couplings) and systems with high background leakage.

Variable area leakage is not calculated using the traditional square root power but rather a power which is very much system dependent. $N1$ values range from 0.6 to 2.5 and should be calculated on a zone-by-zone basis. International research has however identified an $N1$ of 1.15, as being representative of large zones with varied materials.

Calculating the $N1$ is quite simple and can be undertaken in the field with either data loggers or manually by flow and pressure readings. This type of testing is commonly referred to as step testing.

To calculate the correct $N1$ the pressures and flow should be read at night during stable demand conditions. The pressure should be lowered by either reducing the pressure on an existing PRV or by throttling a gate valve. The corresponding drop in flow will dictate the $N1$ as discussed in Chap. 10. Usually the $N1$ used for estimation is an average of three or more drops or steps. See Fig. 18.10 for a sample step test result.

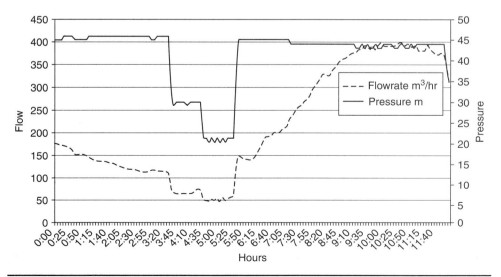

Figure 18.10 Reduction in pressure provides a reduction in leak flow rate.

Using the internationally accepted average N1 of 1.15, our sample calculation of reduced leakage changes too:

$$L_1 = L_0(P_1/P_0)^{1.15} = \text{New leakage} = 500(50/80)^{1.15} = 500 - 291 = 209 \text{ gpm}$$

It can be seen that we have an additional saving of 104 gpm. This additional saving is a function of the changing area of the leak(s) in the second example. We must therefore conclude that other than for systems with 100% fixed area leakage (which is very hard to find), the traditional method of calculating potential savings from reduction of pressure, is to say the least very conservative and misleading.

18.4.3 Background Leakage

While many utilities are undertaking very efficient leak assessment, detection, and repair, there still remains an element of leakage, which is undetectable. This is often also referred to as background leakage. This leakage is made up of many small pinhole leaks, joint leaks, drips, and the like, which cannot be detected by traditional means. The only efficient way of reducing the impact of background leakage, (other than infrastructure management interventions such as mains and service replacement program, as discussed in Chap. 19), is to efficiently control pressure.

High background leakage will often be found in systems with high service density, high hydrant density, or systems where maintenance is difficult because of a highly urbanized situation.

18.4.4 Reduction of New Leak Frequency

Pressure management helps to reduce not only the volume of leakage and background leakage, but also reduces the frequency of new leaks occurring. It should be noted that pressure is not the only influencing factor in the frequency of new leakage, however, it is often a significant one. Other factors may include ground conditions, traffic conditions, pipe material and condition, stray currents, temperature, and backfill. A method for estimating the reduction in break frequency due to reduction in pressure is shown in Chap. 10.

18.5 Overflow Control

When discussing pressure management and its impact on water loss it is important to also discuss level management in reservoirs, tanks, and storage.

> **P**ressure management includes the management of reservoir and tank levels, which can often be the source of considerable annual losses.

Water loss from overflows in storage facilities is too often overlooked, as it is deemed not to be significant and often tanks are in out of the way places, so overflow is not always evident.

Overflows usually occur at night (when pressure conditions are often at their highest due to lack of demand and head loss on the system) and are caused by either lack of level controls or malfunctioning controls. Level control can be performed manually by pump control, by SCADA, which involves automatic control by computer-linked software, or by simple hydraulic control, using either altitude valves or ball valves. Sometimes a utility will have a sophisticated series of automatic controls, however, external forces such as lightening may affect them. A simple hydraulic backup is often cost-effective. We will be discussing how the hydraulic solutions work later in this chapter.

Most tanks and reservoirs have an overflow pipe. If a utility wishes to discover if overflow is occurring, it is a simple task to inspect the point where the overflow pipe dumps water. If there is recent evidence of water being discharged, then either level should be data-logged and compared with the overflow level, or if data-logging technology is not available a simple solution is to locate a ball in the overflow pipe and inspect the position of the ball each day. If the ball has come out of the pipe then there has been an overflow situation.

Pressures and levels should be monitored and the level of loss analyzed. A simple cost to benefit exercise will identify if a new system of control is warranted.

18.6   Fundamental Monitoring Points

For any pressure management project, it is first necessary to monitor as a minimum the following points:

- Supply nodes
- Storage nodes
- Critical nodes
- Average zone point (AZP) nodes

Supply nodes could be considered as any point, which supplies a system or subsector of a system. A supply node could also be an outlet point from one zone to another. In some cases it may be necessary to monitor bidirectional flows.

Storage nodes would be any reservoir, tank, standpipe, or location where water is stored.

A critical node is a location point where supply may be at its weakest, for example, a high level within the system or a point where there is high head loss in the supply pipe. Alternatively it could be a point where a user cannot be left without water, for example, a production plant or hospital.

An **AZP node** is a location, which is chosen to be representative of the average condition, (ground level, pressure, head loss, and the like.) within the system or zone. Methods for properly identifying AZP points are discussed in Chap. 10.

18.7   Flow Measurements

In general flow measurements should be taken at any supply or exit point as discussed above. A supply point may be a pumping station, treatment plant, storage facility, well or bulk transfer point to the system or zone. It may be deemed necessary during a demand analysis, to measure demands from large consumers, if they are considered to be large nighttime water users.

Measurements should be taken for a minimum of 24 hours, but preferably for 7 days or more, the decision on how long to measure for usually comes down to cost.

> The longer the measurement period the better, however measurement periods are usually limited by the cost.

Care should be taken when measuring flows to ensure that these flows are easily related to changing seasonal trends. Obviously the best situation in areas of changing demands is to monitor for 1 year, however this is very rarely possible. The next best thing is to normalize annual demands and to relate the week of flow monitoring to the normalized curve for security.

Flows should be measured accurately with calibrated equipment, however an accuracy of ±10% is usually acceptable, as the valves to be installed have quite a wide range.

18.8   Pressure Measurements

Pressure measurements should be taken at all of the node points mentioned above. Pressure should be measured with a reasonably high-resolution logger (±0.1% full scale), which should be calibrated for accuracy and drift before and after the field installation. Further information on the measuring process, recommended above can be found in App. B.

18.9   Using Hydraulic Computer Models to Identify Ideal Locations for Installation

It is not necessary to have a computer hydraulic model to select areas for pressure control. However if one is available then it can be used, to identify areas with high pressures and also to identify areas of high head loss, where the more advanced dynamic controllers could be used beneficially. See Fig. 18.11. Modern modeling techniques can be used to identify the optimum number and location of control zones. In general the model should be reasonably calibrated and include any extreme demands necessary, such as fire flows or seasonal adjustments. A model, which is calibrated to only ±15%, is acceptable for this type of work.

Using a model is a very nice way of quickly identifying potential areas, although it is still necessary to go into the field and make field measurements, as often the situation in the field changes, valves get left closed, new leaks occur, and so on.

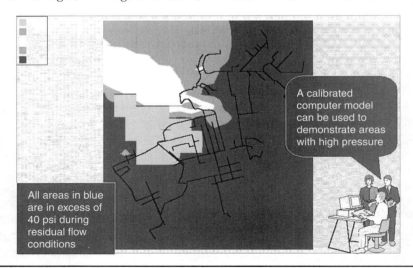

FIGURE 18.11   Hydraulic models can be used to show areas of high pressure.

18.10 Understanding the Hydraulics of Your System Prior to Implementation

In addition to using hydraulic models for the location of pressure control stations and field measurements to locate critical data, it is also very important to understand exactly how the system functions hydraulically.

This analysis is normally undertaken in the demand analysis phase of the project and should identify:

- Percentage of direct pressure consumption
- Percentage of consumption from individual storage tanks
- Distribution feeds, by pump or by gravity
- Breakdown of consumer categories, residential, commercial, and industrial
- Level controls for elevated storage
- Pump shut off controls

The results of this research will form the basis of the control scheme, providing limits of control and cost to benefit assumptions.

18.11 Using Statistical Models to Calculate the Potential Benefit of a Scheme

Once we have identified an area, made field measurements, and identified how the water is used within the sector, we can proceed to the decision-making stage. During this phase we identify how much control we can effect without disrupting normal supply and what benefit this control will have on reduction of leakage volume, reduction of new leak frequency, deferral of new source schemes, and in some cases water conservation.

A simple model can be constructed by most users of excel following the guidelines shown in Chap. 10, however there are also various commercial models available for purchase. The decision to purchase a model or construct one should really lie with the type of staff a utility has and time that they have available. While the calculations are not really complex, it can in some cases be false economy to try to build your own model when a small investment will buy a tried and tested version.

Most commercial models are flexible however care should be taken to ensure that the model purchased takes into account the hydraulic characteristics of the utility system in question. As discussed earlier there are significant differences between hydraulic characteristics of demand for a system, which uses residential storage, and a system, which has direct pressure feed.

18.12 Calculating Cost to Benefit Ratios

Once the data has been entered into the model and calibrated to a certain degree of confidence, the model can be used to analyze the cost of a potential project and its estimated benefit. The components and the diameters to be installed, the type of bypass and chamber, the ground type, the type of control to be effected, and the type of maintenance program to be put into action after installation dictate the cost. An additional cost can be a small reduction in revenue from the direct pressure user component. This cost should only be used if there are no ideas for water conservation within the utility. If the utility is trying to get their consumers to reduce consumption then pressure control will form a very efficient part of this program and reduction of consumption will become a benefit not a cost.

The benefit is calculated from the reduction in leakage volume, reduction in maintenance costs, deferral of costs to build a new source of water if water is scarce, and reduction of supply to nonpaying customers and increased storage management.

The cost to benefit is calculated as a function of the cost divided by the benefit and is usually displayed as a ratio and also as the number of months required to payback the initial investment. In most utility cases a good payback is somewhere inside of 24 months. In many cases advanced pressure control is giving paybacks of less than 12 months due to the huge impact on leakage and the simplicity of installation.

18.13   How Do ACVs Work?

There are various types of ACVs available on the market. Some use a diaphragm, some a piston and some have a collapsible sleeve arrangement.

> **M**ost valve bodies can be piloted for a number of different pressure management activities.

However most of the valve bodies for each manufacturer are interchangeable for type of control. For example:

A valve which was designed originally as a pressure-reducing valve, can easily be changed into a level-control valve, a pump-control valve, a flow-control valve, or any number of other functions, by changing the way that it is tubed and the type of pilot control fitted.

Hydraulic ACVs basically work by using the upstream force of the pressure to either open or close the valve by water entering or leaving the head of the valve, in function to the pilot setting. This can be seen in the previous Fig. 18.8. A good rule of thumb is to size the ACV to work within a range which is 20 to 80% open.

When considering a pressure management scheme for the first time it is a good idea to talk to various valve manufacturers to ensure that their valves can be altered into various functions, by altering the tubing and piloting. This will ensure the optimization of your investment as systems change character.

For example a utility may install a 6-in valve for pressure control in a zone this year, correctly sizing all parts, however in 2 years a construction company may construct a large condominium, therefore changing the demand conditions. The 6-in valve may now be undersized, however it is a simple job to replace this valve for a more representative one and reuse the 6-in valve in another location. Care should be taken when sizing the bypass assembly to allow for flexibility.

Another important point is maintenance. For ease of maintenance most utilities only use valves from one or maybe two different manufacturers. This saves on having to stock the same size parts for various different makes of valve. Obviously price is a concern for most projects but local support and product and spares stocking should also be taken into consideration. After all in any ongoing project the initial investment is often only a small part of the overall investment on the life of the project.

Pressure management as with any of the other loss control tools shown in this manual is not a static concept but rather a constantly changing project, which follows the ever changing needs of the utility.

18.14   Pressure Reduction

Pressure reduction is probably one of the most common forms of pressure management being practiced today, with very positive impacts on leakage. A valve not dissimilar to the diagram shown in the previous Fig. 18.8 undertakes pressure reduction hydraulically.

Placing more or less tension on the control spring changes the pilot stem position and the pilot valve opens or closes. As the available orifice size in the pilot changes, more or less water is forced into or out of the head of the valve, making it modulate either toward the open or closed position. Pilots can be adapted to be fitted with a controller as explained later in this section.

In most cases pressure control will be undertaken in a zone, which has excess pressure throughout the zone. It is however the case in some larger zones where cost to benefit ratios are good, that it may be necessary to boost water to certain high critical locations. While this may seem ridiculous, it is a simple matter of performing the cost to benefit calculations to see if the ratio is good.

Usually large areas with only a small amount of potential for pressure reduction will still give good results.

It is not uncommon to find that in addition to smaller sectors, which allow a large amount of pressure reduction, that is, a valley, other larger zones with the potential for only a very small amount of pressure control, will also give very good paybacks. Figures 18.12 and 18.13 show an example of this situation from one of the SABESP installations in Sao Paulo Brazil.

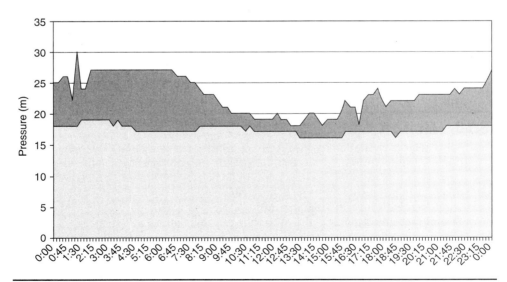

Figure 18.12   Pressures before and after management, contract No.69.502/96.

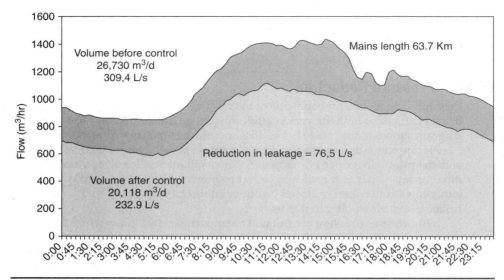

FIGURE 18.13   Flows before and after pressure management, contract No.69.502/96.

18.15   Locating Installation Points in the Field

Once the potential sector has been chosen and control points identified either on paper or in a computer model, it is very important to go into the field and locate the exact spot where the valve assembly will be installed. Other underground utilities should be carefully located prior to excavation.

Care should also be taken when setting a valve on an inclined section of road, to ensure that inlet pressure always exceeds the required maximum outlet pressure, plus a few extra pounds which are required head loss across the valve to make it function.

Once the spot has been located it is a good idea to make a location diagram, ensuring that the valve housing is constructed exactly in the right location and that this location can be easily located at a later date if it has been asphalted over for example. See Fig. 18.14 for a sample location diagram.

18.16   Multiple Valve Sectors

Some sectors cannot be hydraulically fed only from one point. This may be due to fire flow volume requirement, or it could be due to high head losses during peak demand periods or any number of other reasons, such as water quality concerns. This does not necessarily mean that the zone is not viable.

Zones with several feeds are quite viable and reasonably easy to set up, as long as careful thought is given to the hydraulic reactions of one valve against its counterpart. It is important to rank valves in order of importance and ensure that the control set points reflect the order of ranking. For example some valves may be required to function only during periods of high head loss, that is, during peak or emergency demands. These valves would then remain closed during the rest of the day. Hydraulic grade lines (HGLs) can be used to ensure that the valves are balanced out.

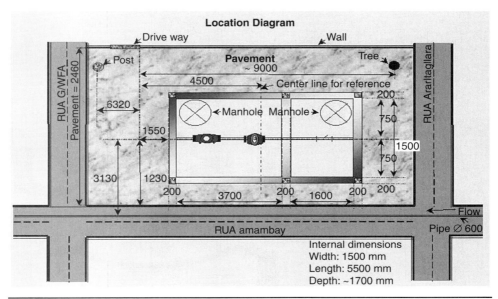

FIGURE 18.14   Valve location diagram.

Normally the valves on larger feeds will be set to respond quicker to changes in demand, while the other valves feeding the system may be set with a slightly longer response time.

18.17   Reservoir and Tank Control

As discussed previously there are various ways of controlling tank levels and many utilities may already have this under control. We are therefore only going to discuss two simple hydraulic solutions to level control, ball valve control and altitude valve control. These two methodologies are probably the simplest and most maintenance free solutions to reducing water loss through overflow. Some utilities with SCADA systems, which have lightening problems, may also wish to consider this as a backup system, which will operate hydraulically and independently to the automated system they may have.

18.17.1   Ball Valve Control

Ball valves operate very simply by a floating ball on the surface of the water. The newer units have a ball connected to a pilot system, which in turn operates the main valve, as per the diagram in Fig. 18.15.

It is important in reservoirs with turbulence, to make sure that the ball assembly is installed in a stilling well or a calm location as per the picture in Fig. 18.16. This ensures that the turbulent surface does not affect the control, making the valve open and close. The ball valve assembly is ideal for storage facilities, which fill from the top, as opposed to bottom filling tanks and storage.

18.17.2   Altitude Valve Control

The altitude valve uses a column of water, which equals the level of the tank to control a pilot valve, which in turn opens and closes the main valve as per the diagram in Fig. 18.17.

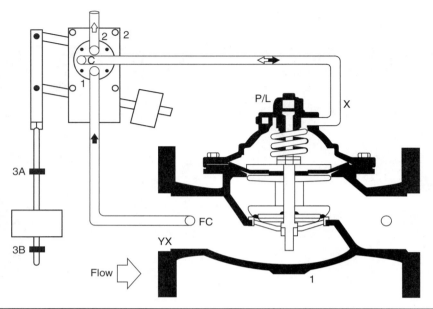

Figure 18.15   Ball valve control.

Figure 18.16   Calibrating a ball valve.

Altitude valves are usually installed on bottom filling tanks and storage, as can be seen in the photograph in Fig. 18.18, however, they can be installed on top filling tanks if the sense line is connected to the outlet pipe.

Care should be taken not to install the altitude valve too far away from the tank, as this will create delayed reactions and poor control. Manufacturers supply good installation diagrams, which should be adhered to. Altitude valves can be installed on unidirectional pipes and on bidirectional pipes and can be set for on–off response or relational response.

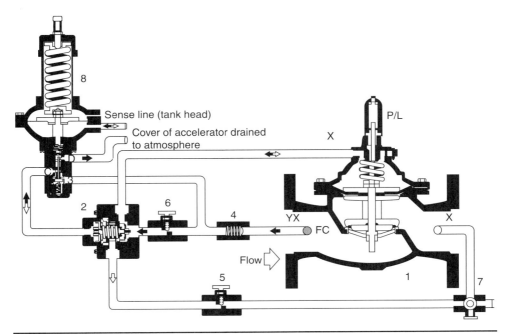

FIGURE 18.17   Altitude valve diagram.

FIGURE 18.18   Altitude valve installation.

18.17.3   Demand Control Using Flow and Sustaining Valves

Some systems find that during peak flow conditions certain parts of the system are hydraulically deficient. In many cases when this occurs, certain consumers will use most of the water while leaving only a small volume for others. This is often so in the case of very large consumers, who create large, localized head losses. It is also often seen in the case of irregular settlements in developing country situations.

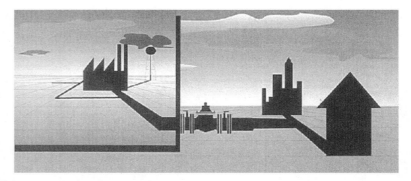

FIGURE 18.19   Sustaining valves can be used to protect supply.

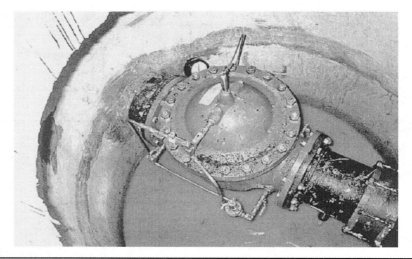

FIGURE 18.20   Sustaining valve installation.

Flow-control valves and pressure-sustaining valves can be used to reduce the impact of these situations and ensure a constant supply of water for all consumers. These valves in conjunction with pressure-reducing valves in areas of excess pressure help to ensure that an even supply pressure is given to all parts of the system.

Sustaining and flow control features can also be added to pressure-reducing valves, see Fig. 18.19, and the photograph in Fig. 18.20, making the valve an efficient tool for directing water around the system and ensuring a good turn over of reservoir water, which in turn ensures good water quality. This type of control is often necessary when considering protecting reservoir volumes during times of high demand, as hydraulically some reservoirs will empty quicker than others will. In this situation without control, some reservoirs can always be virtually empty, while others never empty.

18.17.4   Sectors with Large Industrial Customers

In sectors with very large consumers, care should be taken to make sure that flow and pressure profiles used for sizing valves are representative of the highest and lowest demand periods. In some cases it may be advantageous for both the utility and the customer to install an on-site storage facility, if one is not already installed. Once the large consumer has some storage, either flow or pressure-sustaining valves can be used to control the demand of the large consumer and minimize the peak impact caused through high-localized head losses.

Site visits for all large consumers are a must to ensure that their demand needs and emergency needs are taken care of properly. The cost of these surveys and in some cases retrofitting fire sprinkler systems or providing storage may sometimes be included in the project cost. This would be so in the case of a customer who didn't want to change their system and was holding up the whole project. Obviously this would only be done at the utilities expense in favorable payback situations.

18.17.5   Sectors with large seasonal variations

In sectors with large seasonal variations it may be necessary to install multiple feeds, or valve installations in parallel. The parallel installation would consist of a large valve, which will provide high flows during peak conditions, usually at weekends or holiday periods in the case of tourist areas. The smaller valve would then function most of the time. In many cases where a controller is installed, the controller need only be installed on the valve, which is most active. In certain cases the larger valve will function most of the time with the small valve operating just at night during periods of minimum night flow. See Fig. 18.21 for an example setup.

18.17.6   Sectors with Weak Hydraulic Capacity

In sectors with weak hydraulic capacity it may not be uncommon to find potential for pressure reduction at night, whereas during the day there is not sufficient pressure. Pressure reduction at night however can often still be justified and would depend on the cost to benefit analysis for final decision. As mentioned earlier, often pressure-sustaining valves

> **P**ressure management can be undertaken at off peak hours only, if systems are weak during peak demands.

FIGURE **18.21**   Parallel installation can be used to extend flow range.

in conjunction with pressure-reducing valves will be used. An alternative if possible is to install a small tank, which will pick up lost pressure only during the peak hours when the system is most stressed and pressures uncontrollable.

18.18   Valve Selection and Sizing[2]

Valve selection and sizing is often done using average values of flow and pressure and as valves are fairly forgiving, in most cases the valves work. This however is not recommended practice!

In the case of pressure reduction for leakage control, it is strongly recommended that flows and pressures be measured in the field. As previously discussed the impact of reduced pressure on leakage is often critical to the operation of the valve and without accurate data, valves may be installed incorrectly and operate erratically.

> **S**elect the flow range that meets your system requirements and correct valve(s) size(s). Note: Maximum flow rates in this table allow for continuous flows at velocity of 20 to 22 feet per second

Field measurements are also beneficial when seasonal corrections need to be made and ensure that the valves can cope with the top end flows, without creating too much head loss. This is also true when calculating the effects of emergency water use such as fire fighting.

All valve manufacturers provide valve-sizing charts, an example is shown in the chart in Table 18.2.

18.18.1   ACV Types—Diaphragm, Piston, Rolling Diaphragm, Sleeve

All pilot operated hydraulic ACVs operate use similar principles, however the mode of control changes significantly. Each manufacturer will quote his/her type of benefit and will attempt to justify why his/her valve is better. Full bore valves such as the rolling diaphragm and sleeve will quote low head losses at high flows, while globe style diaphragm manufacturers will quote stable modulation and control. At the end of the day it is important for the utility engineer to understand what it is that he or she wishes to achieve and then to select the best valve for the job.

As well as technical benefits the engineer should also consider two other very important points, local support and ongoing maintenance costs. One of the biggest problems facing a utility when they have many valves installed is the cost and availability of quick and fast maintenance. A utility should try to avoid installing many different make of valve as the cost of stocking parts increases significantly and the availability of local support drops drastically.

18.18.2   Valve Sizing and Limits—Qpmax and Qpmin, Cavitation, Head Loss

In many cases control limits are set as a function of the maximum pressure controllable at the valve, while providing a constant minimum pressure at the critical node(s). If a substantial amount of pressure is to be controlled, the manufacturer's cavitation chart should be consulted to ensure that the valve is operating within its limits. See example in Fig. 18.22.

System Flow Range (gpm)	Size Range (gpm)	Size Range (gpm)	Size Range (gpm)
Single valve installations			
1–100	1–1/4″		
1–150	1–1/2″		
1–200	2″		
20–300	2–1/2″		
30–450	3″		4″ 30680
50–800	4″	ACV 6000	6″ 501025
115–1800	6″	Series	8″ 1152300
200–3100	8″	Valves	10″ 2004100
300–4900	10″		
400–7000	12″		
500–8500	14″		
650–11000	16″		
Parallel installations			
1–400	1–1/4″ (1–100)	2–1/2″ (20–300)	
1–800	1–1/4″ (1–100)	3″ (30–500)	
1–1000	1–1/2″ (1–150)	4″ (50–850)	
1–2000	2″ (1–200)	6″ (115–1800)	
1–3800	1–1/4″ (1–100)	3″ (30–500)	8″ (200–3100)
30–3800	3″ (30–500)	8″ (300–1800)	
1–5400	1–1/4″ (1–100)	3″ (30–500)	10″ (300–4900)
30–5400	3″ (30–500)	10″ (300–4900)	
1–8000	1–1/2″ (1–150)	4″ (50–850)	12″ (400–7000)
50–8000	4″ (50–850)	12″ (400–7000)	
1–9500	1–1/2″ (50–850)	4″ (50–850)	14″ (500–8500)
50–9500	4″ (50–850)	14″ (500–8500)	
1–13000	2″ (1–200)	6″ (115–1800)	16″ (650–11,000)
115–13000	6″ (115–1800)	16″ (650–11,000)	

TABLE 18.2   Quick Sizing Chart

Valve Sizing—Pressure Reducing

Selection of the correct size pressure reducing valve is a relatively simple process. Criteria for selection are minimum flow, maximum flow, and pressure drop across the valve. Following are explanations of the three types of PRV installations.

These also apply to any functions combined with the reducing function such as reducing/check function such as reducing check and reducing/solenoid valves.

Single Valve Installation

A single reducing valve can be applied if operating flow requirements are within the capacity of one size valve, and pressure drop is outside the cavitation zone.

1. Select the valve size from sizing chart, that is, within the range of flow to high flow. (Consider requirements of lowest demand equipment.)
2. Check pressure drop (inlet-outlet) to confirm desired outlet pressure is above the recommended lowest outlet setting to avoid cavitation conditions. (Check cavitation chart.)

Parallel Installation

If flow requirements fall outside the capacity of a single valve, an additional smaller valve installed in parallel may be required. In parallel installations, the larger valve handles the requirements for maximum flow down to its low flow capacity. The small valve extends the low flow range. Total capacity of this installation is equal to the sum of the maximum flow of both valves.

1. Select the valve size combinations from sizing chart, that is, within low to high flow system range.
2. Check pressure drop (inlet-outlet) to confirm if desired outlet pressure is above index psig, or check cavitation chart.

Series Installation

If pressure drop requirements cause the outlet pressure to be below the index psig, or fall in the cavitation zone, then two valves in series may be required. Each valve will function outside the cavitation zone to safely drop the high inlet pressure, in two steps, to the desired outlet pressure. Valve size is based upon the minimum-maximum flow ranges previously explained.

Isolation Shutoff Valves

Butterfly or similar type valves should be installed in the line upstream and downstream of the automatic control valve to allow for maintenance service without draining the system or exposing service personnel to the pressure.

Installation Recommendations and Requirements

Avoid mounting valves 6 in and larger in a vertical discharge position (valve stem horizontal or cover pointing sideways). If your installation requires this mounting position, consult the factory or specify at time of order.

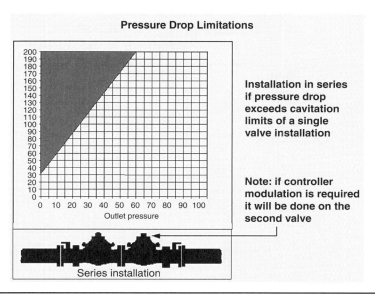

Pressure Drop Limitations

Installation in series
if pressure drop
exceeds cavitation
limits of a single
valve installation

Note: if controller
modulation is required
it will be done on the
second valve

Series installation

Figure 18.22 Cavitation chart.

If the valve is found to potentially be operating within the cavitation zone, then installation of two valves in series should be considered as discussed in the next sections. See Fig. 18.22.

Care should be taken when sizing the valve, to check the potential head loss through the total valve assembly (gate-valves, filter, meter, control-valve, and pipe fittings), especially when the pressure during the peak hours is already low and modulated control is only desired during off peak times. If care is not taken, supply may be reduced during peak hours resulting in no water complaints. See Table 18.3 for a reference spreadsheet which can be used to check head losses across all of the fittings within the bypass assembly.

As a consequence of the pressure control, the existing flow profiles will be reduced, in particular when a high level of leakage is present. See Chap. 10 for more details on calculating the effects of reducing pressure on system leakage.

Care should be taken when selecting valve sizes, so that the flow cannot fall below the minimum acceptable flow for the valve, after leakage has been reduced. If this happens the valve may control erratically, as it is controlling at the almost closed position, therefore any small modulation affects the flow and pressure more than when it is modulating at the nominal position. This may result in either higher maintenance costs or increased leakage. Most hydraulic valves are designed to operate from about 20% open to about 80% open.

In situations where a valve has to deal with both high and low flow conditions it is often common to install a small bypass valve around the main control valve to ensure smooth hydraulic control.

18.18.3   Parallel Installations—Fire Control, Large Flows, Variation in Flow Pattern

In situations where flow patterns vary greatly or there is a requirement to meet safety or emergency periodic demands, it is normal practice to install valves in parallel. These types of installations usually take the form of one large valve and one small valve working

---

ESTIMATED HEADLOSS
BY-PASS

Company: Date:
Site: Site Identity:
Sector   Zone:

DATA TO BE SUBMITTED in: [   ]

Ø Inline piping : [   ] m   Q : [   ] m³/h
                  A = [   ] m²   v : [   ] m/s
Ø By-pass piping : [   ] m   Length [   ] m
                   A = [   ] m²

HEADLOSS DUE TO FRICTION CAUSED BY PIPING: (see Moody diagram)

$hf = f \dfrac{L \times V2}{D \times 2g}$ Re =   ∴ f = [   ]   hf =   m
                        Roughness Coef.=

HEADLOSS DUE TO FRICTION CAUSED BY COMPONENTS:

$hf = K \dfrac{V2}{2g}$ From manufaturers catalogs

[   ] Reducer   0,00 m   [   ] Y filter [   ] m

[   ] Amplifier   0,00 m

[   ] Tee(lat. exit) (K = 1,30) :   0,00 m   [   ] Meter [   ] m

[   ] Bends 90° (K = 1,20) :   0,00 m

[   ] Bends 45° (K = 0,40) :   0,00 m   [   ] PRV [   ] CV

[   ] Gate valve   0,00 m $\Delta p = \left(\dfrac{Q}{3, 6\ CV}\right)^2$ 0, 702: 0,00 m

TOTAL HEADLOSS : ========= m

Obs. :

Calculations made using liquid = H_2O at 20°C;

Formulas used: Reynolds, Darcy, Moody diagram,

Mechanical joint included in piping

TABLE 18.3   Sizing Consideratons

FIGURE 18.23 Parallel installation unequal diameters, contract No.3.066/98.

together to increase the effective flow range. Figure 18.23 shows an example of two valves of different diameters working together. In this situation the large valve takes account of most of the daily flow and the small valve just deals with the relatively low nighttime flow. In this case if a controller were fitted it would be fitted to the large valve and the smaller valve would have a fixed outlet pressure just slightly higher than the minimum modulated pressure of the larger valve. As the larger valve modulated toward closed position reducing the outlet pressure the effect of the smaller valve kicks in. A difference in outlet settings of around 5 psi gives a nice smooth change over.

In a situation where the larger valve is fitted to meet fire flow conditions and stays closed most of the time, the opposite is true. If a controller were fitted it would be fitted to the smaller valve and the outlet of the larger valve would be set just below that of the smaller valve. As the demand increases and the small valve starts to create excessive head loss therefore forcing outlet pressure down, the larger valve will modulate open and feed the system with a fixed outlet pressure and large volume.

Parallel installations can also be undertaken in situations where a large flow dictates that two large valves are installed in parallel. The picture in Fig. 18.24 shows an example of a parallel equal diameter installation. It also shows how a controller would be hooked up to one pilot, which would then control two valve head chambers. This is normal practice and functions well as long as valve opening and closing speed is calibrated equally for both valves. In some cases (usually dual large diameter valves) a pilot with a larger diameter nozzle should be used to allow a larger volume of water to pass through while still maintaining control and not being effected by localized head loss.

18.18.4   Series Installations—Large Pressure Drops

In situations where a large amount of pressure needs to be cut and a single valve would enter into the cavitation zone, then two equal diameter valves operating in series can be fitted. In the case of a controller being fitted, it would be fitted to the second or downstream valve allowing the first valve to cut pressure from upstream to required maximum

FIGURE 18.24 Parallel installation equal diameters.

FIGURE 18.25 Parallel and series installation with controller.

and the second flow-modulated valve from required maximum to required minimum. The picture in Fig. 18.25 illustrates a complex example of two valves in parallel and series with a controller.

18.19 Using Controllers to Make Your Hydraulic Valves More Efficient

The advent of intelligent and cost-effective controllers now allows us to use conventional fixed outlet hydraulic valves in a more efficient manner. The controller effectively allows multiple set points for downstream pressure depending on either time based or system demand based requirements. In many cases just altering the set point by the amount of head loss difference between daytime and nighttime conditions can affect huge savings in leakage.

In addition to standard leakage control, controllers can be used to make valves function for emergency situations like earthquake control for example. When an area is hit by earthquake, it is very possible that a major transmission line could rupture, leading to possible depletion or complete loss of storage. By calibrating controller profiles, the controller can reduce the amount of reservoir loss or shut down the line completely, therefore saving valuable resource and the headache of trying to refill storage under emergency conditions.

18.19.1   Time-Based Control

Can be affected by using a controller with an internal timer. Control is affected in time-bands in accordance with demand profiles. This methodology is very effective for areas with stable demand profiles and head losses and is usually used where cost is an issue, but advanced pressure management is desired. Time-based modulation controllers can be supplied with or without data-loggers and or remote links. Some manufacturers connect the controller to the pilot valve and alter the set point of the pilot valve by introducing a force against the existing force of the pilot spring, as shown in Fig. 18.26.

Other manufacturers use a timer and a solenoid valve to reroute control through preset pilots.

18.19.2   Demand-Based Control

This is the best type of control for areas with changing conditions, head loss, and fire flow requirements and the need for advanced control. This type of control is affected by controlling outlet pressure in relation to demand by connecting the controller to a metered signal output. Modulation of outlet pressure is achieved by altering the force against the pilot spring. The controller is normally supplied with a local data logger and optional remote communications see diagram in Fig. 18.27.

Control can be affected with a preset profile, which shows the changing relationship of demand and head loss in the sector. Alternately a direct communications link can be

FIGURE 18.26   Time-based controller fitted to valve pilot.

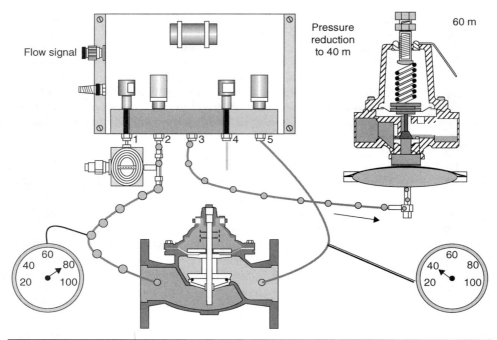

Flow signal

Pressure
reduction
to 40 m

60 m

FIGURE 18.27 Demand-based control diagram.

made between the controller and the critical point. Obviously the second option involves communications and therefore higher costs, which are not always necessary.

In general installation costs are higher for this type of control, however, additional savings and guaranteed fire flows due to more intelligent control usually make this type of control more desirable.

Demand-based or remote-note control to key critical points can also smooth out pressure fluctuations at weak points in the system and reduce break frequencies especially in system which are very fragile.

18.20 SCADA

Some utilities will have existing SCADA systems. Many SCADA systems are designed to run the transmission level system and are not designed at distribution level, due to the cost involved. However, utilities are starting to cost-justify system optimization and SCADA should not be ruled out as an excellent, although costly, means of managing pressure within the distribution system. (The author has seen installations and presentations from Canada, Australia, Japan, and the United States where this type of control is affected.) Generally speaking, this type of control is the most expensive to install but is obviously by far the most efficient. It may not always however be necessary, and cost to benefit calculations may not justify, the installation of a full SCADA system, just for loss control. If a utility already has distribution level SCADA in place, then it may be very cost-effective to add on modules.

18.21   Valve Installation

Once the valve has been sized and control limits identified, it is time to decide how we are going to install the valve. Valves can be installed in a number of ways at varying costs. We will be discussing various options in the following paragraphs.

18.21.1   Where to Dig the Hole

This may seem like an obvious answer, the easiest location! However sometimes decisions can be made quickly and on paper without the proper site investigations. The onsite investigation is probably one of the most important tasks and should not be taken lightly.

Prior to selecting a final location it is necessary to locate all other underground utilities, see Chap. 16 on locating underground utilities. Identify an area where traffic will be easiest to contend with in the case of an underground chamber and consideration should be given to properties, which may be affected by the excavation.

18.21.2   Mainline or Bypass

Once a reasonable location has been located it is necessary to decide if the valve assembly will be installed on the mainline, allowing for a smaller diameter bypass which is cheaper to install, or if the valve assembly will be installed on the bypass allowing the valve chamber and access manhole to be installed in the verge for easy access, if the mainline is under the road. See Figs. 18.28 and 18.29 for both options. While a bypass

FIGURE 18.28   Valve being installed on main line, contract No.69.502/96.

FIGURE 18.29 Valve being installed on bypass, contract No.3.066/98.

facilitates maintenance of the main valve some utilities prefer not to install a bypass as there is always the opportunity for the bypass to be opened outside of planned maintenance activities which effectively takes the pressure control out of service and can cause significant increases in volumes from leakage as well as causing increased leak frequencies. If a bypass is installed then bypass control also needs to be instilled into the utility operators.

18.21.3 Head Loss Concerns

When sizing a bypass and fittings for valve installation it is important to consider the head loss, which will be created at peak flow plus emergency, demand, not only through the valve, but also through all of the fittings. Remember that if the valve will be modulated to various set points with a controller, then it may be desirable in some instances to almost open the valve completely making it transparent during high demands and only controlling during low demand periods. If the bypass or adjoining fittings are downsized to reduce cost, head losses may be created in excess of the minimum control desired. Careful consideration of fittings sizing should be given at this stage. In some cases economy is justified and in others it is not, each case is site specific and will be analyzed during the cost to benefit stage.

18.21.4 Hydraulic Connections

When considering the type of connections to use it is important to consider the existing type of pipe work. If the existing pipe work is flanged cast iron or ductile then anchorage for horizontal movement is not so critical a problem. If however, the existing pipe work is made up of asbestos cement or to a lesser degree bell and spigot PVC type pipe then consideration should be given to potential horizontal movement during periods of control. In all cases vertical movement should be considered.

18.21.5 Anchorage

Depending on the size of pipe work and valve, each case will be considered independently and calculations should be made to ensure no vertical or horizontal movement of the installation is possible. It is normal to put thrust blocks on the 90° bends of the bypass, but additional thrust restraint should be calculated if the valve itself would be installed on the bypass.

18.21.6   Chamber or Above Ground Installations

In some cases it is desirable to install the valve assembly above ground. See the photograph in Fig. 18.30 for an example. Above ground installations a very practical and allow the operator to avoid confined space entry requirements, potential flooding problems (especially in areas with high groundwater conditions), and generally working in cramped conditions. Obviously there are negative points to above ground installations, in that the installation takes up a lot of space, in particular in the case of a large diameter installation. Above ground installations also attract more attention and could be the focus of vandalism, in certain environments.

Each site should be considered on an individual basis and the merits of each type of installation considered.

18.21.7   Valve Commissioning

In all cases it is better to have a skilled and experienced operator perform valve start up. Sometimes for a number of reasons the valve will not function as it should and without the necessary knowledge an unskilled operator could create serious problems in the system, if the valve controls erratically and at the very least could loose a lot of time and effort attempting to resolve simple problems.

> In all cases it is better to have a skilled and experienced operator perform valve start up.

18.21.8   Start up Procedures

Start up procedures are, obviously site specific, however it is a good idea to make a checklist prior to start up. A sample start-up procedure for a pressure-reducing valve, which also has rate of flow control and a sustaining function, can be found in Table 18.4.

FIGURE 18.30   Above ground assembly, contract No.3.066/98.

RATE OF FLOW/PRESSURE REDUCING/SUSTAINING VALVE

Installation/Start-up

Start-up of an automatic control valve requires that proper procedure be followed. Time must be allowed for the valve to react to adjustments and the system to stabilize. The objective is to bring the valve into service in a controlled manner to protect the system from damaging overpressure.

- Clear the line free of slag and other debris.
- Check to ensure the orifice plate is installed in the valve inlet flange and that the inlet sensing port is not covered by the retainer ring. If so, rotate until the space aligns with the port.
- Install the valve so that the *flow arrow* marked on the valve body/tag corresponds to flow through the line.
- Close upstream and downstream isolation valves.
- Open ball valves or isolation cocks in the control tubing, if the main valve is so equipped. Failure to open these will prevent the valve from functioning properly.

Step 1   Preset pilots as noted:

Rate of flow: Adjust *out*, counterclockwise, to start valve at a lower flow rate.

Pressure sustaining: Turn sustaining control adjustment screw *out*, counterclockwise, backing pressure off the spring, to allow it to stay open while adjusting other controls.

Pressure reducing: Adjust *out*, counterclockwise backing pressure off the spring, preventing possible overpressuring of the system.

Step 2   Turn the adjustment screws on the closing speed and opening speed controls, if the main valve is to be equipped, *out*, counterclockwise, $1\frac{1}{2}$ to $2\frac{1}{2}$ turns from full closed position.

Step 3   Loosen a tube fitting or cover plug at the main valve to allow air to vent during start-up.

Step 4   Pressure the line, opening the upstream isolation valve slowly. Air is vented through the loosened fitting. Tighten the fitting when liquid begins to vent.

Setting the Rate of Flow Control

Step 5   Slowly open downstream isolation valve until valve is full open.

Step 6   With a demand for flow on the system, the valve can now be adjusted for the proper flow rate. This requires a meter to read the flow that the valve is providing.

Step 7   While reading the meter register, adjust the rate of flow control:

- Turn the adjustment screw *in*, clockwise, to increase the flow rate regulated.
- Turn the adjustment screw *out*, counterclockwise, to reduce or lower the flow rate regulated.

TABLE 18.4   Calculating Head Loss in the Installation (*Continued*)

RATE OF FLOW/PRESSURE REDUCING/SUSTAINING VALVE

Setting The Pressure-Reducing Control

Note: Reducing control is set higher than the sustaining control.

Step 8   Fine tune the pressure-reducing control to the desired pressure set point by turning the adjustment screw *in*, clockwise to increase or *out*, counterclockwise, to decrease downstream pressure.

Step 9   Opening speed flow control adjustment: The opening speed flow control allows free flow into the cover and restricted flow out of the cover of the main valve.

If recovery of pressure is slow upon increase downstream demand, turn the adjustment screw *out*, counterclockwise, increasing the rate of opening.

If recovery of downstream pressure is too quick, as indicated in a rapid increase in pressure, probably higher than the desired set point, turn the adjustment screw *in*, clockwise, decreasing the rate of opening.

Step 10   Closing speed control adjustment: The closing speed needle valve regulates fluid pressure into the main valve cover chamber, controlling the valve closing speed. If the downstream pressure fluctuates slightly above the desired set point, turn the adjustment screw *out*, counterclockwise, increasing the rate of closing.

Setting The Sustaining Control

Step 11   Setting the sustaining control requires lowering the upstream pressure to the desired minimum sustained pressure.

Step 12   Leave the downstream isolation valve full open and close the upstream isolation valve until the inlet pressure drops to the desired setting.

Step 13   Adjust the sustaining control screw *in*, clockwise, until the inlet pressure begins to increase, or *out*, counterclockwise, to decrease, stopping at the desired pressure.

Step 14   Allow the pressure to stabilize.

Step 15   Fine tune the sustaining setting as required as detailed in step 13.

Step 16   Open upstream isolation valve to return to normal operation.

TABLE 18.4   Calculating Head Loss in the Installation (*Continued*)

Air

When starting up a new installation or restarting one that has been subjected to zero pressure, it is common to get air entrapped in the head of the valve. The effect of this is that the valve will not control properly. Usually in this case the valve will not close or modulated toward the closed position. It is common practice to install a small air release valve on the head of the valve as can be seen in the picture in Fig. 18.31. Alternatively one can be installed on the line. The later is usually done when the utility wishes to have an indicator stem installed on the head of the valve to show the valve position. A line air valve will also handle a larger capacity and can be used as part of the systems air release feature.

FIGURE 18.31   Air valve installed on PRV head, contract No.3.066/98.

Modulation Speed

Valve modulation speed is always an issue whether a controller is fitted or not. The hydraulic speed controls should be set to allow smooth controlled modulation. In addition when changing the outlet pressure in the case of advanced modulation then the controller reaction speed should be matched with the system needs. In general as a rule of thumb, larger valves need to be controlled faster as their hydraulic reaction time is longer due to the larger amount of volume in the head of the valve.

> **M**odulation speed is a critical issue for all types of control valve and should be addressed according to independent system conditions.

Smaller valves should be modulated slower. A reasonable band for modulation by the controller would be between 10 and 25 seconds per control pulse. In some cases where the head volume is very large in the case of two valves running from one controller in parallel then it may be desirable to change the pulse volume, or the time the solenoids are open, to allow a more forcible control. Most manufacturers provide detailed manuals explaining how to do this with their equipment. However the operator should have a feel for the type of reaction he needs for his system. System needs may include fire fighting response or large consumer draws. Obviously a valve should not be modulated too quickly otherwise it will set up a very negative hydraulic reaction. If unsure how the system will react to control, it is a good idea to put out pressure data-loggers in the system logging very fast, then experiment with the pulse size and frequency on the controller(s) to see which combination gives the smoothest control.

Stability

Prior to fitting a controller it is a good idea to log the system pressures to see if the hydraulic valve(s) are controlling in a stable manner without the controller. If speed controls are incorrectly set or in the case of a multiple feed system the outlets are incorrectly set then the valves will hunt. This should be hydraulically corrected prior to trying to establish control through a more advanced controller type regime.

18.22   Maintenance Concerns

After the valve has been installed and properly commissioned and calibrated, it is very important to put into place a periodic maintenance schedule to ensure ongoing efficient operation of the valve. The time between maintenance visits is something, which is usually defined by water quality, the location of the installation (if it is going to be vandalized) and the variability in demand requiring changed modulation profiles.

> **V**alves like any other equipment need regular maintenance to ensure ongoing efficient operation.

Maintenance should include but not be limited to the following items:

18.22.1   Valve Maintenance

- Clean principle filter and secondary filter
- Check tubing for leaks or kinks
- Check operation of control isolation valves
- Check pressure gauges
- Check smooth modulation of valve

18.22.2   Controller Maintenance

- Check battery
- Check input cables
- Check logger functionality
- Check modulation speed

18.22.3   Subsector Maintenance

- Check boundary valves
- Check night flows
- Check critical node pressures
- Check critical node validity
- Repair new leakage

18.23   The Chamber

The chamber should be periodically checked for leakage and seepage, air quality, and general usability. Chamber manhole covers should also be periodically checked and greased to allow easy lifting.

18.24   Nonhydraulic Pressure Control

Other options for pressure control are available and should be installed where justified by local conditions[3]. One example of an electrically actuated valve is shown in Figs. 18.32 and 18.33.

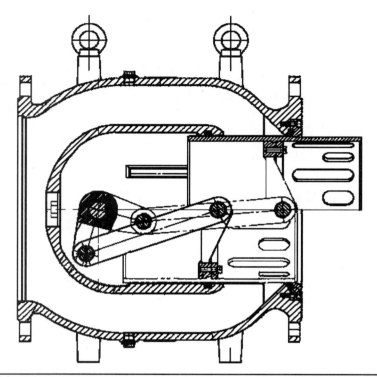

FIGURE 18.32   Plunger valve assembly.

FIGURE 18.33   Plunger valve installation in Sao Paulo SABESP MS.

18.25   Summary

Pressure management is one of many tools, which may be used by leakage management practitioners to combat either leakage volumes or increasing leakage frequencies. It may also be used in conjunction with demand reduction programs. Water efficiency programs are discussed in more detail in Chap. 20 by Bill Gauley.

Pressure management is a suitable means of controlling water losses in all areas of the world from the highly industrialized nations to developing countries.

References

1. Thornton J. *Pressure management.* AWWA Publication: *Opflow*, Vol. 25, No.10: USA, October 1999.
2. Thornton J. *Correct selection sizing and advanced operation of PRV's*, ABES national congress: Salvador, Brazil, 1998.
3. Thornton J. *New tools for precision pressure management—a case study in SABESP, Sao Paulo, Brazil*. IWA World Water Congress, Beijing, 2006.

CHAPTER 19

Controlling Real Losses— Infrastructure Management

Julian Thornton

Reinhard Sturm

George Kunkel, P.E.

19.1   Introduction

Underground piping is one of the largest investments a utility can have and the cost to maintain and or replace old piping is often prohibitive due not only to the physical costs of the pipe-work itself but also the excavation and reinstatement in dense urban situations.

Unfortunately maintenance is often overlooked as the problem is out of sight and therefore until an emergency situation occurs it can be out of mind. However any good proactive loss management program should address ongoing maintenance as one of the key issues. Figure 19.1 shows where maintenance, rehabilitation, and replacement figures in our four arrows concept of real losses control.

Pipe maintenance can come in many forms and can be undertaken over varying time frequencies, depending on the nature of the problem, the attitude of the operator, and the seriousness of the situation. However some of the more frequent maintenance programs encountered to counteract losses are corrosion control and pipe lining and replacement. In the case of pipe replacement new technologies are being used to undertake trenchless replacement. It is important to note at this point that many pipes whether mains or services suffer from high break frequencies from early days of being laid particularly if material's quality or installation quality is neglected during installation. Any utility considering replacement as an option needs to consider the reason for failure and ensure that the new pipe does not quickly fail at the same frequency as the old one.

The following chapter touches on some of the problems and methodologies currently encountered in the market today, although both subjects are sciences in their own right and have been widely discussed and published in other documents and publications.

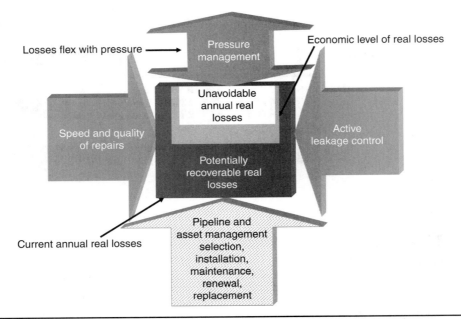

FIGURE 19.1   Maintenance, rehabilitation, and replacement play an important role in any loss management program. (*Source:* IWA Water Loss Task Force and AWWA Water Loss Control Committee.)

19.2   Pipeline Corrosion

There are many forms of corrosion some of which could be

- Galvanic
- Oxygen concentration cell attack
- Bacteriological
- Stray current
- Pitting
- Crevice
- Selective dissolution
- Stress related
- Erosion
- Fatigue related
- Impingement
- High temperature

However as water operators we usually have to deal with the following types:

- Galvanic
- Bacteriological soil born

- Bacteriological water born
- Stray currents

As with any of the other components which make up a systems water loss, the reasons for corrosion in any one particular water system are varied and often complex and should be studied on a one on one basis, however some of the standard methods of corrosion control are as follows:

- Protective external coatings
- Pipe relining
- Insulation of pipe joints
- Water treatment using corrosion inhibitors
- Cathodic protection
- Stray current drainage bonds

Corrosion control can be a complex issue involving a very proactive approach to system maintenance, however in most cases paybacks on this type of service are very fast, with severe reduction in the number of new breaks and leaks in very short time periods.

19.3   Pipe Rehabilitation and Replacement

Underground pipelines have a limited life and often need to be rehabilitated or replaced for a number of reasons, some of which may be:

- High break or leakage rate
- High occurrence of joint leaks
- Encrustation or corrosion (internal or external)
- Hydraulic carrying capacity
- Structural reinforcement
- Threat to life or property

In this manual we are focusing on water loss management. Rehabilitation and periodic maintenance can effectively add years to the life of a pipeline, however different methods will be more or less effective and costly in different situations. Recently the IWA water loss task force pressure management team has confirmed that pressure management can have a significant impact on the annual number of breaks in the system—thus increasing system useful life.

19.3.1   Pipe Replacement and Rehabilitation Methods

In general pipe replacement methods such as the first options mentioned below will be more effective for reduction of leakage, in particular, if the pipe is seriously structurally damaged. However there are many case studies, which do show good results in water loss reduction from spray linings.

Some of the methods of rehabilitation and replacement are discussed below:

Main and Service Replacement

Obviously pipes can be replaced by laying new pipe and discarding or removing the old one, however this is often extremely costly and in some cases completely impractical, as in the case of dense urban cities. In many utilities a service replacement program will resolve large volumes of loss, as in many cases the largest annual volumes of real loss lie in the smaller leaks on service lines which run for longer periods undetected or unreported. Additionally replacing mains or services most often reduces the new break frequency and therefore reduces annual maintenance costs and reduces frequency required for leak detection survey activity.

Trenchless Technologies

Other methods of pipe replacement can be undertaken using no dig or trenchless technologies, which are usually cheaper and almost always less disruptive.

Some of the methods of trenchless pipe replacement are discussed below:

> **T**renchless technologies for pipe replacement repair and maintenance are often more cost effective especially in dense urban situations.

Slip Lining   Slip lining is probably one of the simplest of no dig replacement techniques. In this case the old pipe is cleaned out and a new smaller diameter pipe is drawn through or pushed through the old one. The new pipe is of a smaller diameter and usually made of polyethylene (PE). Once the new pipe is in place the service connections are usually excavated and reconnected.

Slip lining does reduce the original diameter of the pipe and care should be taken that enough hydraulic carrying capacity remains for the job in hand. However, in many cases, in particular with old cast iron pipe, the old pipe; although a larger diameter may have corroded to give a much smaller effective diameter.

Close fit lining is another type of slip lining where a deformed liner is inserted into the pipe and then restored to its original size once in place.

Pipe Cracking or Pipe Bursting   In situations where the hydraulic carrying capacity needs to be maintained or indeed increased, pipe cracking can be undertaken. The old pipe is prepared and then a conical wedge is drawn through ahead of the new pipe. In this way it is possible to use the old pipe as a guide for the new pipe, however the new pipe is actually larger than the old one.

The methods mentioned above will in all cases assist in reducing leakage as well as providing other benefits such as increased hydraulic capacity and clean safe water supply conditions.

19.3.2   Rehabilitation Methods

In most cases where structural integrity is not found to be a problem, pipes can be cleaned and lined. The liners tend to be either cement or epoxy and are not in most cases designed to be structural or reduce leakage but rather, provide a clean smooth environment, to ensure a healthy water supply and a lower friction factor.

Pipe Cleaning

Before any kind of relining intervention can be undertaken it is important to properly clean the pipe to ensure that the lining can bond with the pipe wall, without pockets of debris or corrosion, which could later form problem areas.

Pipes with internal corrosion can be cleaned in a number of ways, however some of the most common are:

- Air scouring
- Rotating chains, rods, and scraper trowels
- Pigging

Air Scouring   Air scouring is undertaken using an air compressor to inject air pressure at a slightly higher pressure than the water pressure. As the air is introduced into the line and then let out downstream, a surge is set up which has the effect of ripping the corrosion off the pipe walls. Air is usually injected and purged through selected fire hydrants while the main is under pressure, although it is a good idea to close surrounding distribution valves to limit discoloration of the water in surrounding areas. Air scouring should be properly supervised and after any program a mains flushing exercise should be undertaken to ensure that there are no health hazards, dirty water, or entrained air complaints. One of the benefits of air scouring is that it is not necessary to excavate to undertake the cleaning work. This methodology is often used when lining is not going to be undertaken but an improved hydraulic capacity is required.

Rotating Chains, Rods, and Scraper Trowels   When a section of main has been identified for relining then access pits will have to be excavated prior to application of the lining. These same pits are used to pull rotating rods or chains through. Alternatively scraper trowels can be pulled through the line. After the main has been scraped then it should be flushed and often will be pigged prior to lining.

Pigging   Pigs come in various shapes and sizes and can be used for the initial clean or to clean up after a rodding or scraping exercise as discussed above. The pigs are inserted into the main through the pits, which will be used for the relining process and retrieved at the end of the section. Some pipelines have "pig traps" which allow regular pigging of the line even when a relining exercise is not warranted.

Spray Linings

Epoxy   Epoxy linings have been approved by many environmental agencies throughout the world, however all are not approved and care should be taken when considering their application. Epoxy lining is sprayed onto the pipe wall through the use of a towed centrifugal pump gun. Epoxy linings have a benefit that they are usually quite thin and therefore have less negative impact on internal pipe diameter and effective hydraulic capacity. Epoxy linings also tend to dry quickly allowing the mains to be put back into service quickly.

Cement   Cement linings are also widely approved and provide an excellent way of improving internal pipe condition and improving "C" factors. Cement lining is often the lining of choice for new pipelines too.

Cement linings are applied by centrifugal spraying and also by trowelling, depending on the pipe diameter. Cement linings tend to take a little longer to dry than epoxy and as the lining is thicker, care should be taken to ensure that effective hydraulic capacity is not reduced below acceptable limits due to reduction in diameter.

19.3.3 When to Replace or Rehabilitate

From a water loss reduction perspective the decision to replace or rehabilitate a pipeline would often be made on a cost-to-benefit basis, although other factors such as those shown below will often influence the decision:

- Environmental considerations
- Health concerns
- Structural problems
- Emergency hazards
- Demand growth
- Reduced hydraulic capacity
- Lack of alternative supplies

The cost of not replacing or rehabilitating the pipe can be evaluated using the following components:

- Average historic break frequency
- Cost of volume of lost water per incident
- Cost of damage caused by blow-out
- Cost to repair the main
- Cost to reinstate the surrounding area

This cost should then be compared to the cost-to-replace or rehabilitate the main or service in question and the life span of the proposed intervention.

19.4 Summary

Most utilities today will have older less effective pipes, which are coming to the end of their useful life. As technology advances trenchless replacement and rehabilitation options are becoming very attractive options to traditional mains replacement. Careful tracking of reported leak and break frequencies along with hydraulic and camera inspections will allow the operator to quickly identify those sections of main which can no longer be maintained cost-effectively in their current condition.

CHAPTER 20

Water Efficiency Programs

Bill Gauley, P.E.

20.1 Introduction

Developing programs to improve water efficiency is fast becoming a preferred alternative for municipalities faced with a need to expand their water supply or wastewater infrastructure. Improving water efficiency is almost always more environmentally responsible and can often be considerably more cost-effective than expanding capital works.

This chapter is intended to help those planning to implement a water efficiency program (WEP) to focus on elements that will be important to the overall success of the project. Success here is defined as achieving the maximum cost-effective water savings through implementing publicly acceptable measures. The material in this chapter should help program designers to establish specific goals for water demand reduction, as well as to understand that if the goals are not specific it will be impossible to quantify program success.

Capital expansion is extremely expensive and in some cases virtually impossible. In these cases systems will first reduce their system real losses, while maintaining current billing levels. If the reduction in real losses is not significant enough to defer the capital construction, then demand reduction is undertaken.

This chapter will also explain that it is only after the program's overall goals have been established that it will be possible to identify which demand components should be targeted and, ultimately, which water efficiency measures will be best suited to achieve these goals.

The section dealing with water saving targets will explain the importance of knowing both the maximum potential water savings and the target water savings associated with a water efficiency plan, and why there is usually a difference between these values.

Important implementation issues are identified later in the chapter; some of these issues are often overlooked or misunderstood, and are therefore described more fully.

The final section outlines the importance of monitoring and tracking program results. It describes some of the tools often used to assess program performance as well as some of the more common monitoring misconceptions.

Understanding the material outlined in this chapter should help both program designers and implementation staff to have a better understanding of some of the more basic elements involved in implementing a successful water efficiency program.

20.2   Why Plan a Water Efficiency Program

Since the 1980s it seems that an increasing number of municipalities and agencies are implementing water efficiency programs. Some even require that the potential for water demand reduction be determined and evaluated before approval to expand the water or wastewater infrastructure will be granted. Even when it is not mandated, many municipalities are showing fiscal responsibility by considering the economical and environmental benefits associated with demand-side management (water efficiency) versus supply-side management (infrastructure expansion).

> **S**ome municipalities are mandating conservation before granting system expansion or extraction rights.

Unlike the old adage, "art for art's sake," there should be very clear and well-defined reasons for undertaking a water efficiency program. Fortunately, in today's environment there are generally a myriad of reasons for doing so. Some of the more common reasons include

- A need to expand water or wastewater treatment plants or infrastructure
- Nearing the capacity of water source (e.g., reservoir or aquifer)
- An interest in being environmentally responsible

Whatever the reason, for the program to succeed it is important that the overall goal is understood and accepted by all involved parties—politicians, works department, the public, and the like. After all, it is the program goal that will dictate which demand component (as described in the next section) should be targeted, and it is the target demand component that will dictate which water efficiency measures should be included.

It is essential, therefore, for the program designer or implementation team to understand the different system demand components and how they relate to the various water efficiency measures that are commonly implemented as part of a water efficiency program.

20.3   System Demand Components and How They Relate to a WEP

Throughout the year, most water supply systems experience a range of water demand rates*—often changing with the season. Figure 20.1 illustrates the various demand components commonly experienced by water supply systems. These demand components

---

* Note that this statement refers to demand rates, not billing rates.

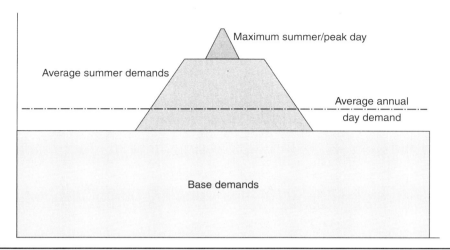

FIGURE 20.1 Typical demand pyramid. (*Source:* Bill Gauley.)

tend to form a demand pyramid, with the base of the pyramid comprising the system's average daily base demands, and the top of the pyramid representing the system's peak day demand.* Each of these demand components is described in detail later in this section.

Because most water efficiency measures target a specific demand component, it is important that these measures are properly selected based on the program's goals. Improperly chosen measures may not only be ineffective, but even worse and may actually have a negative impact on the program (e.g., they may reduce system revenues).

20.3.1 Base Demands

Generally, an assortment of demand types contribute to a system's overall base demand. Those that are related to residential indoor water use, such as toilet flushing, showering, clothes washing, and the like, generally experience little variation from season to season. Distribution system leakage, and most nonirrigation and noncooling water demands in the industrial/commercial/institutional (ICI) sector, are also fairly constant throughout the year. Base demands form the largest component of average winter[†] day water demands. As seasonal temperatures rise, however, irrigation and other seasonal demands increase. In fact, in the heat of summer, as much as 50% or more of a system's total water supply can be related to irrigation and cooling.

Base demands are affected by changes in population size, number of employees, and demographics. However, since base demands are not generally affected by changes in the weather, they tend to be reasonably constant from year to year.

Commonly, a significant percentage of a system's base demand is discharged to the sanitary sewer system. Therefore, water efficiency programs targeting reductions in wastewater flows (i.e., to defer wastewater infrastructure expansion) should focus on reducing base demands (as well as other inflow and infiltration, explained in the next section).

---

* Typically, water treatment facilities are designed to meet peak day demands, while system storage is utilized to meet peak hour demands.

†The term "winter" is used here to describe any nonirrigation season.

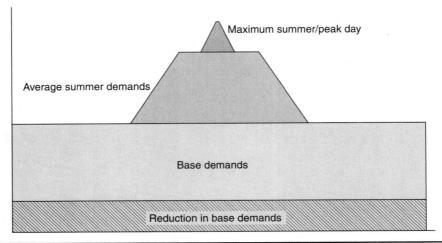

FIGURE 20.2 Peak demand rate lowered. (*Source*: Bill Gauley.)

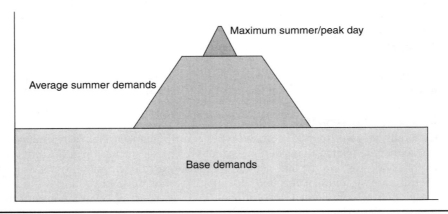

FIGURE 20.3 Entire demand pyramid lowered. (*Source:* Bill Gauley.)

Although reducing base demands does not decrease the demand volumes related specifically to irrigation and other outdoor water uses, it does lower the peak demand rate by taking a "slice" off the bottom of the pyramid (see Fig. 20.2) and lowering the entire demand pyramid (see Fig. 20.3). Note also in Fig. 20.3 that the peak demand is reduced by the same demand rate (not the same percentage) as the base demand reduction.

A sometimes overlooked, yet critical, aspect of a water efficiency measure is the sustainability of the water savings. Are the savings maintained in subsequent years, or must the reduction measure be repeated or reinforced? Generally, water savings that are not sustained are of little value to a municipality.*

---

*The exception to this comment is in the use of temporary emergency measures, such as watering bans during periods of drought.

20.3.2   Base Sewage Flows

Sanitary sewage flows are, in general, relatively constant throughout the year. Flows in systems with high levels of inflow* or infiltration† (I&I) will vary depending on changes in groundwater levels or precipitation. Generally speaking, when rainfall events are eliminated, there is relatively little variation in sanitary sewage flow rates from season to season.

Typical sewage reduction programs generally involve water efficiency measures targeting base demands (replacing toilets, showerheads, clothes washers, etc.), or reducing the levels of I&I.

Municipalities that maintain combined sewer systems, where both sanitary sewage and storm water are collected by the same system, often include I&I reduction measures in their water efficiency program.

20.3.3   Average Annual Day Demand

Some system operators calculate average annual day demand (AADD) by dividing the total annual water production by 365 (i.e., the number of days in a year). This value actually represents the average annual day production and includes water lost through system leakage and other unaccounted-for water demands. This volume can be divided by the total population‡ serviced by the system to determine the average daily gross§ per-capita water demand (or, more accurately, the average daily gross per-capita water production).

It should be noted that the average daily net per-capita water demand, that is, the average volume of water attributed specifically to personal use, is generally determined by dividing the total volume of water billed to residential customers by the total residential population. This value is usually presented as a demand rate, typically gallons per capita per day (gcd) or liters per capita per day (lcd). Demand rates can be determined for population subsets as well (single-family households, multifamily apartment buildings, industrial facilities, commercial sites, etc.) and can also reflect seasonal demand variations (average summer day single-family household water demand, average winter day commercial site water demand, etc.).

The AADD is an academic value that changes from year to year (differences in summer irrigation demands, for instance, can have a significant effect on the AADD). Since they are a blend of the various seasonal demand components, AADD values generally do not provide sufficient data to design water efficiency programs that target either base demands (affecting both water and wastewater treatment infrastructure) or peak demands (affecting only water supply infrastructure). For example, two systems could have identical AADDs and yet have completely different operational characteristics (see Fig. 20.4).

For this reason, it is not usually practical to base a water demand reduction target or a water efficiency program on AADD demands. In fact, programs that reduce AADD

---

* Infiltration: groundwater seeping into sewers through cracks and joints.
† Inflow: surface water being directed into the sewer.
‡ The term "population" generally refers to the residential population, that is, those persons living in single-family and multifamily households within the community. Care should be exercised when evaluating municipalities where significant portions of their population work outside the community, or where a significant number of their employees actually reside outside the community.
§ Gross per-capita water demands include residential water demands, ICI water demands, fire-fighting demands, mains flushing, as well as all unaccounted-for demands.

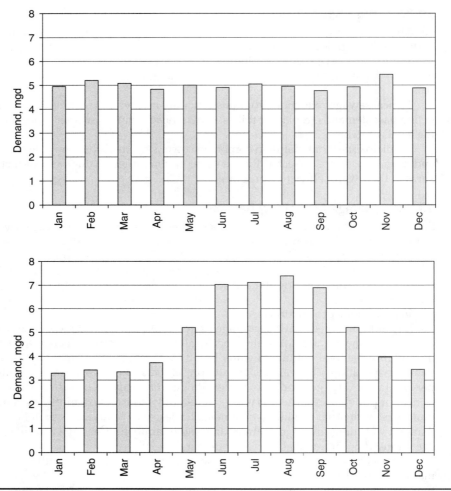

FIGURE 20.4   Two systems could have an identical AADD (in this case 5 mgd) and yet have completely different operational characteristics. (*Source:* Bill Gauley.)

while not reducing either base flow demands or peak demands may accomplish nothing more than reducing system revenues.

Example 1   A community has determined that their wastewater treatment plant is nearing capacity and decides to implement a water efficiency program to extend the life of the facility. They set their goal as a 10% reduction in the average annual day demand.

A year later they are surprised to find that although they did achieve a 10% reduction in AADD, their wastewater flows were unchanged. After some investigation they determine that the reduction in AADD was related entirely to reduced summer irrigation (perhaps the result of a cooler than average summer with higher than average precipitation rates) and not to their water efficiency program.

Although reduced irrigation demands did reduce the community's AADD, it had no effect at all on the wastewater flows.

The community decides that future programs aimed at extending the life of wastewater treatment plants will focus on reducing base flows rather than AADD.

Example 2 A community has determined that their water treatment plant is nearing capacity and decides to implement a water efficiency program to extend the life (capacity) of the facility. They set their goal as a 10% reduction in the average annual day demand.

A year later they are surprised to find that although there was no reduction in AADD, they actually achieved their target of reducing the system's maximum water demands, thereby extending the capacity of the plant. After some investigation they determine that because of a warmer than average April and September, the total volume of summer irrigation demands was slightly higher than average. However, their landscape irrigation reduction program, which involved providing customers with informative bill stuffers, radio and TV ads, and soon had the desired effect of reducing customer irrigation demands during the hottest and driest part of the summer.

Although the program did not reduce AADD at all, it did achieve the goal of extending the capacity of their water treatment facilities.

The community decides future programs aimed at extending the life of water treatment plants will focus on peak demands rather than AADD.

20.3.4 Maximum Summer/Peak Day Demands

The peak day demand is usually defined as the highest water demand recorded during a single 24-hour period in any calendar year and, as such, it changes from year to year. Although technically the peak day demand occurs only on a single day, in reality there can be several peak-type days (maximum summer demands) within a year, and they may occur sequentially (e.g., during a hot, dry period) or at several times throughout the year.

A system's peaking factor is a mathematical value determined by dividing the peak day demand by the average annual day demand and, as such, also changes from year to year. Dissimilar systems can have the same peaking factor. The largest peaking factor experienced over a period of several years is often used as a design peaking factor* when planning new water supply infrastructure components.

It is important to note that although peak day demands are generally related to outdoor irrigation demands and usually occur after long periods of dry, hot weather, they may also be the result of large water main breaks, fires, or industrial demands, or a combination of any of these factors.

There can be significant benefits associated with reducing peak day water demands, for example, deferring the need to expand water treatment facilities or distribution infrastructure, to enable the current infrastructure to service an expanding population, and the like. As a result of these benefits, many municipalities implement at least some type of program targeting outdoor irrigation, for example, watering restrictions (odd/even day watering, time-of-day, for example.), bill stuffers, radio/TV/newspaper articles or advertisements, irrigation audits, and so on.

Although it can be very important to reduce peak day demand, it is important to note that water efficiency programs that reduce average summer day demand but do not reduce peak day demand will not achieve the program's goal, but will reduce water sales revenues. See Fig. 20.5.

Peak day demands are usually determined by monitoring daily water production volumes at water treatment plants or in well fields. Changes in weather patterns (e.g., hot

---

* Design peaking factors that are higher than actual historical values contain an additional margin of safety.

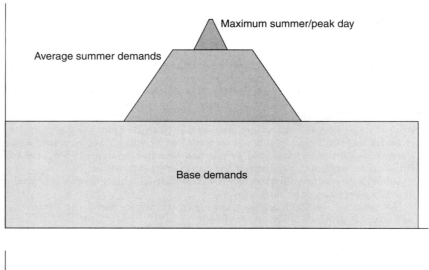

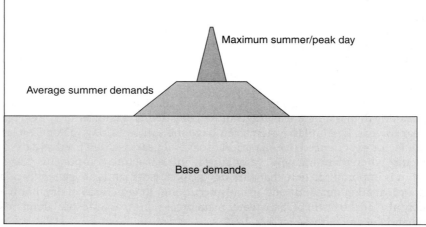

FIGURE 20.5   Water efficiency programs that reduce average summer day demands but do not reduce peak day demands will not achieve the program's goal. (*Source:* Bill Gauley.)

and dry versus cool and wet summers) can cause large variations in peak day demands from year to year.

Agencies or municipalities implementing water efficiency programs targeting peak day demands must be aware that any savings achieved through their program may be either exaggerated or eclipsed by the generally significant demand changes related to weather.

Ideally, the program should reduce peak day demands alone (thus extending the service life of the water treatment or distribution infrastructure) while not affecting average summer day or base demands (thus not reducing revenues). For this reason, some municipalities are implementing pilot programs to quantify specifically the peak day water savings achieved as a result of implementing their irrigation reduction measures.

These pilot programs involve bulk monitoring* of a study area and a control area. The water efficiency measure (or intervention) is implemented only in the study area. Bulk monitoring must be used for this type of program because it is not possible to quantify savings using water billing data.[†]

Similar to base demand reduction programs, the sustainability of the water savings is critical. Since the target of many peak day reduction programs involves changes to customer irrigation habits rather than changes to fixtures or equipment (e.g., installing new toilets or showerheads), maintaining peak water demand savings over time may be more complicated than sustaining base demand savings.

20.3.5   Summary

Although many water efficiency measures affect more than one demand component, they are generally intended to target a specific goal such as peak or base water demand reductions, or wastewater flows reductions. It is important that the proper measures be selected to address the program's specific demand component target. Table 20.1 relates commonly implemented water efficiency measures to the type of demand component they most effect.

20.4   Water Saving Targets

Once you have determined whether your water efficiency program will focus on peak day demands or base demands (or both), and which measures you will implement, you must determine how much water your program can realistically be expected to save,

Measure	Primary Impact
Toilets	Base demand programs
Showerheads	Base demand programs
Clothes washers	Base demand programs
Landscape irrigation	Peak water demand programs
Seasonal pricing	Peak water demand programs
Watering restrictions	Peak water demand programs
Cooling water reduction	Peak water demand programs
Gray water reuse	Peak water demand and base demand programs
Public education	Peak water demand and base demand programs

Source: Bill Gauley.

TABLE 20.1   Impact of Water Efficiency Measures

---

* Bulk monitoring involves recording the water demands of a large group of customers, for example, an entire subdivision, simultaneously, by installing water meters directly in the supply water mains. The use of bulk monitoring eliminates the Hawthorne effect, which is described later.
† Water billing data, even when bills are issued every month, do not provide the details necessary to identify demand parameters on individual days, nor do these data account for changes in weather conditions.

that is, how effective your WEP will be. The first step in this process is to determine the theoretical maximum savings of your program.

20.4.1   Theoretical Maximum Savings

The theoretical maximum savings (TMS) is a calculated value; it assumes that the measure is implemented perfectly and achieves 100% market penetration. The TMS establishes the upper boundary of the water savings target—a program cannot save more water than the TMS.

It is important to note that the TMS does not consider any of the program delivery elements that would be required to achieve 100% participation. For example, the TMS does not require any knowledge of incentive amounts, installation criteria, how removed fixtures will be disposed of, how marketing will be performed, the cost-effectiveness of the measure, and so on.

The TMS is later used as a tool when determining the target WEP water savings. Some examples of calculating TMS values are illustrated in the following examples.

Example 3   A municipality with a population of 50,000 has a sewage treatment plant that is nearing capacity and decides to implement a toilet replacement program to extend the life of their plant.

They decide to use the information from the AWWARF* Residential End Use Study (REUS) to establish the approximate TMS for this measure in their municipality. The study states that nonefficient toilets are used 4.92 times per person per day, with 4.1[†] gal (15.5 L) per flush. The study also states that efficient toilets are used 5.06 times per person per day with 1.9 gal[‡] (7.2 L) per flush.

A sample household survey has identified that only an insignificant number of existing toilets are currently water efficient. The approximate TMS for their measure is determined as follows:

- Water demand related to toilet flushing—existing nonefficient: 50,000 population × 4.92 flushes/capita/day × 4.1 gal/flush = 1,008,600 gal/day
- Water demand related to toilet flushing—projected efficient: 50,000 population × 5.05 flushes/capita/day × 1.9 gal/flush = 479,750 gal/day
- TMS = 1,008,600 gal/day − 479,750 gal/day = 528,850 gal/day

In other words, if the entire population replaced their existing toilets with water-efficient models and achieved savings similar to those stated in the REUS, the municipality would save 528,850 gal/day (gpd).

Example 4   A community with a population of 50,000 needs to reduce peak water demands to postpone a planned expansion to the water treatment plant. The peak demands in the system occur in the summer and are the result of extensive landscape irrigation. The town decides to implement a water efficiency program targeting residential irrigation to reduce the system's peak demands. There are about 14,000 single-family households in the community.

The utility has completed a billing data analysis and determined that the average household water demands during the nonirrigation season (i.e., winter) is about 200 gpd, while the average household summer day demand is 280 gal, and the peak summer day demand is 350 gal. Since they are trying to postpone expanding their treatment plant they decide to focus on reducing the peak summer day demand.

---

* American Water Works Association Research Foundation.
† The REUS states that nonefficient toilets are flushed an average of 4.92 times per person per day and account for 20.1 gal of water. Average flush volume is, therefore, 4.1 gal per flush.
‡ The REUS states that efficient toilets are flushed an average of 5.06 times per person per day and account for 9.6 gal of water. Average flush volume is, therefore, 1.9 gal per flush.

The average "additional" household demand (i.e., demand in excess of the winter day demand) on the peak summer day equals 150 gal. The town is aware that some of this additional demand is related to vehicle washing, filling swimming pools, and the like, and they estimate* that approximately 65–70% of additional demand, or about 100 gal per household per day, is related to landscape irrigation.

In this case, the TMS depends on the type of measure used to address the goal.

Scenario A   Program designers know that virtually all of the irrigation demand could be eliminated if a mandatory watering ban was enforced[†]; this would mean that about 100 gpd per household could be saved. The TMS is 14,000 households × 100 gal/household/day = 1,400,000 gpd.

This type of restriction, however, is generally not popular with customers or politicians and is usually used only in an emergency situation, such as a severe drought.

Scenario B   Program designers review the results achieved in other jurisdictions and estimate that about 25% of the irrigation demand could be saved through voluntary restrictions or through the free distribution of hose timers, rain gauges, brochures, and so on, to residents by temporary employees, or by utilizing detailed and informative bill inserts, and the like.[‡] Although this type of program will have a smaller TMS, it is expected to be much more acceptable to the residents. Assuming that 100% of the population participated and each home saved 25% of its irrigation, the TMS is 25% × 100 gpd × 14,000 households = 350,000 gal.

20.4.2   Realistically Achievable Savings Target

A program's realistically achievable savings target (RAST) is usually somewhat less than the TMS value for several reasons, such as

- Actual customer participation rates will be less than 100% especially if participation is voluntary.

- Not all measures will achieve 100% of their potential water savings, especially when the measure requires changes to customer water using habits.

- Not all measures can be implemented cost-effectively; for example, a water efficiency measure would not be considered cost-effective if the cost per unit of water was greater to implement it than it is to expand the water supply.

- Not all measures can be implemented on schedule, especially when water savings are required quickly.

- Water savings may not be sustainable, especially when changes to customer water-using habits are involved.

- Some measures may not be publicly applicable, for example, although the use of gray water offers the potential for substantial water savings, it may not be popular with customers.

---

* Although it is almost always necessary to estimate certain values and make certain assumptions when determining the TMS, they should be based on sound engineering judgment and properly referenced.
[†] Seattle, Washington, issued a mandatory ban on all summer lawn watering during a 1992 drought and, essentially, eliminated the normal peak water demand. The ban was, however, unpopular with both customers and elected officials.
[‡] Savings estimates are for illustration only and are not intended to reflect actual probable savings.

The RAST is generally established for each individual water efficiency measure included in the program; the sum of the various RASTs determines the overall WEP water savings target. Once the RAST is established for a measure, it will be applicable only to the specific set of circumstances (population, demographics, program schedule, public support, etc.) from which it was developed. The RAST associated with a measure will vary from municipality to municipality and, as such, there is no single "cookie cutter" approach to calculating these values—RAST values must be determined for each application on an individual basis.

Following is a list of some of the aspects that must be considered when establishing a measure's RAST.

- Cost/benefit ratio* of water efficiency measure to the customer
- Availability of incentives
- Public attitude toward water efficiency
- Household demographics
- Water rates and/or structures
- Expected building code requirements for new fixtures
- Water use by law enforcement protocol
- Expected advancements in plumbing fixture technology
- Expected changes in the costs of water-efficient fixtures

Once the RAST is established, the next phase in developing a WEP involves designing the implementation plan, described in the next section.

20.5   Implementation Plan

The implementation plan considers and describes exactly how the WEP will be delivered and how the program's goals will be met. In a manner similar to the RAST, implementation plan will also vary from measure to measure and from municipality to municipality.

Although describing all of the potential elements that must be considered when designing an implementation plan is beyond the scope of this chapter, the following list identifies some of the more important elements.

- What is the implementation schedule, that is, how quickly are the savings required?
- What capital works projects will be deferred or eliminated because of the WEP?
- How will the cost-effectiveness of the program be determined?
- If water-efficient fixtures are involved, will they be installed by professionals or self-installed?
- Will incentives be offered?

---

* Cost/benefit ratio: the cost of implementing a measure divided by the value of the resulting savings. A cost/benefit ratio of less than 1.0 indicates a cost-effective measure.

- How will you ensure that new fixtures are installed properly and achieve the maximum savings?
- What criteria will be used to approve plumbing fixtures and appliances?
- How will you ensure that water savings are maintained?
- How will removed plumbing fixtures be disposed?
- How will customer complaints be managed?
- Will additional staff be required?
- Are pilot programs required?
- Is monitoring required?
- Will public education be included?
- Will newspaper, radio, or TV promotion be included?
- What contingency plans are in place if savings are not achieved or exceeded?
- What effect will natural replacement of fixtures have on projected savings?
- What about the effect of "free riders" on the program costs?

As demonstrated by the list above, determining exactly how the WEP will be implemented is usually considerably more complex than determining why the WEP should be implemented. However, as stated earlier in this chapter, the most important aspect of any WEP is not the program itself, but the successful implementation of the program.

Many implementation issues are common to most plans. Four of the most important issues are described in the following paragraphs.

20.5.1 Natural Replacement

In time, all plumbing fixtures and appliances become old or their performance begins to deteriorate and they are replaced with newer units—though not necessarily with units that operate more efficiently. This replacement occurs even without incentives. By offering incentives or rebates, water efficiency programs often try to influence customers to select water-efficient fixtures or appliances. Knowing natural replacement rates allows a WEP designer to better estimate the water savings that can be accredited directly to the WEP implementation.

> **Example 5** If a residential toilet has an average life cycle of 25 years, the natural replacement rate for this fixture equals 4% per year (i.e., $1/25 = 0.04$) and all toilets will theoretically be replaced with new units in 25 years.*
>
> If water-efficient toilets are the only type of units available in the marketplace, then virtually all toilets will be water-efficient in 25 years even without offering incentives. If nonefficient toilets are still available, then there is a possibility that without incentives no water-efficient toilets will be installed within 25 years.

Free Riders

Customers who receive program incentives or rebates even though they would have participated in the program through natural replacement are considered "free riders,"

---

* In practice, some toilets will be replaced before and some toilets will be replaced after the average life cycle is reached.

that is, these participants increase the costs associated with implementing the program but do not increase the program effectiveness. For example, all participants in a program offering an incentive toward the purchase of a water-efficient toilet in an area where only water-efficient toilets are available are free riders.*

Incentives

Many water efficiency programs rely on the use of incentives to accelerate the adoption of a measure. Determining the optimum incentive amount, however, can require some research†—if the incentive is too low, the program will fail to meet its required participation targets; if it is too high, the program will cost more than it should. Incentive amounts are often based on targeted customer participation rates, overall program cost-effectiveness, and the urgency for water savings.

Be aware that changing the value of an incentive partway through a program may cause customer complaints, that is, reducing incentives may offend later participants ("Why did they get more then than I will get now?"). On the other hand, increasing incentives may offend early participants ("Why did I get less than they are getting now, when I supported the program from the onset?").

Pilot Programs

Pilot programs are small-scale programs generally implemented immediately before a full-program rollout to verify design, implementation methodology, participation rates, and so on. Because pilot programs usually include a significant level of monitoring, analysis, and evaluation, the unit costs are often considerably more than those of a full-program rollout.

It is important that the design of the pilot program reflects that of the full-program rollout, for example, rebate amounts, marketing, product types and qualities, and so on.

20.6   Monitoring and Tracking

One of the most important aspects of any implementation plan is establishing the protocol that will be used to monitor and track the program results. Properly conducted program monitoring and tracking will determine if the efficiency measures are achieving their water savings targets (i.e., the RAST), if the program is on schedule, if program costs are on track, and if the water savings are being sustained. If proper monitoring is not performed, it will be impossible to assess the effectiveness of the WEP.

Because no two municipalities are exactly alike, monitoring programs must be designed to suit the specific conditions associated with each individual WEP. It is not possible in this chapter to outline all the parameters that should be considered, but some of the more common elements associated with program monitoring are outlined in the following section.

20.6.1   Water Audits

Some water efficiency programs attribute water savings to the conducting of water audits. Although a water audit can be an excellent tool to evaluate site conditions or even to determine the potential for reducing water demands, the audit itself does not

---

* In this situation, however, the incentive may be intended to accelerate the natural replacement rate of the nonefficient toilets.
† This research is often completed as part of a pilot program.

save any water. It is quite possible to conduct an extensive and expensive water audit without saving any water. Perhaps the customer did not implement any of the measures identified through the audit, or perhaps there were simply no water saving opportunities identified. Water savings that do occur after a water audit is completed are the result of changes in water-using practices or equipment or both.

20.6.2 Water Meters

Water meters are important because

- They help ensure equity (i.e., each customer pays for the water they receive).
- They can be used to help identify opportunities for water savings.
- They can be used to promote awareness.
- They provide a mechanism for measuring and tracking the effects of change.

However, water meters do not save water!

Based on the results of some poorly analyzed case studies, many water efficiency programs have mistakenly assigned water savings directly to the installation of water meters in homes that were previously billed on a flat-rate basis. All of the water savings realized in these homes are the result of other actions—such as changes in the customer's water-using habits, the installation of more efficient fixtures, or both. No water savings are attributable specifically to the meter.

Programs that include water savings because of meter installation run the risk of "double counting." For instance, a water efficiency program that has already considered the effects of installing efficient toilets, showerheads, and faucet aerators, as well as from improving customer habits (turning off faucets when not in use, using full loads for clothes and dishwashers, avoiding overwatering landscapes, etc.), would mistakenly overestimate the potential for water savings if they also include savings from installing a water meter.*

20.6.3 The Use of Percentages Versus Hard Values

Although percentages are commonly used to describe the distribution of a data set, the results can sometimes be misleading, as illustrated in the following example.

Example 6 The water demands of a single-family household are "premonitored" as part of a water efficiency program, and the data shown in Table 20.2 are collected.

The homeowner decides to take advantage of a municipal rebate and installs a new water-efficient clothes washer that uses only 60% of the water of his existing machine. No other changes are made to the home's water demands!

> The use of percentages as a performance indicator can be misleading.

The home is again monitored and the data shown in Table 20.3 are collected.

The percentages illustrated in the tables seem to indicate the absurd conclusion that installing a water-efficient clothes washer will somehow increase the water demands associated with toilets, showers, and faucets. In reality, of course, the actual volumes of water associated with the other plumbing fixtures were not affected by the installation of the new clothes washer—only their percentage contribution to the reduced overall demand.

---

* Consider that no water savings would be expected from secretly metering a customer's water demands, nor would additional savings be expected by installing more than 1 m on a customer's service.

Item	Demand (gpd)	Percentage
Toilet	60	30
Clothes washer	40	20
Shower	50	25
Faucet	50	25
Total	200	100

Source: Bill Gauley.

TABLE 20.2   Premonitoring

Item	Demand (gpd)	Percentage
Toilet	60	32.6
Clothes washer	24	13.0
Shower	50	27.2
Faucet	50	27.2
Total	184	100

Source: Bill Gauley.

TABLE 20.3   Postmonitoring

Example 7   For several years a small municipality was struggling with a high level of unaccounted-for water. Although they produced about 1.2 mgd (million gallons per day) of potable water, they only billed for about 1.0 mgd—a difference of about 17%.

Later, a brewery with an average demand of 0.8 mgd moved into town. Then the municipality produced 2.0 mgd and billed for 1.8 mgd. Their level of unaccounted-for water, at only 10%, seems to have been reduced.

This type of reasoning indicates the absurd conclusion that having a brewery relocate to your municipality will improve your distribution system's performance. In reality, of course, the actual volume of unaccounted-for water was unaffected by the relocation of the brewery.

20.6.4   The Hawthorne Effect

The Hawthorne effect is an initial improvement in a process caused by the obtrusive observation of that process. You should be aware of the Hawthorne effect when you are conducting a monitoring program. The Hawthorne effect occurs when program participants change their normal behavior due to the knowledge that their actions are being monitored. For this reason, monitoring programs often indicate actual conditions if they are implemented without the participants' knowledge.

Monitoring programs should be designed in such a way as to reduce or eliminate the Hawthorne effect; that is, bulk metering or other methods of blind testing should be used where possible.

Example 8   Information collected by data logging water meters that are located in outdoor meter pits (where participants may be oblivious to the monitoring) may provide a more accurate reflection of actual field conditions than data logging water meters located inside the home (where the participants may be aware of the monitoring program and, therefore, alter their behavior).

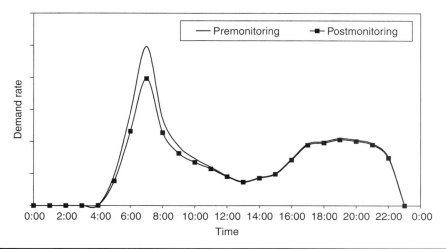

FIGURE 20.6   Diurnal curve of household water demands. (*Source:* Bill Gauley.)

20.6.5   Diurnal Demand Curves

A diurnal curve is often used to illustrate demand rates versus time over a 24-hour period. The shape of the curve depends on the type of facility being monitored. Diurnal demand curves can be produced for individual homes, apartment buildings, subdivisions, industrial parks, or entire municipalities.

The data used to create the curves are obtained by data logging the building's water meter. Data are often collected for periods ranging from 24 hours to several days. The data are usually analyzed and then plotted to show average or "typical" results. It is important to remember that the amount of detail illustrated by the curve will depend on whether the demand data were data logged as instantaneous values or as average values and the frequency of collection.* Generally, the frequency of collection is based on the variability of the data, that is, data that fluctuate significantly should be logged at a higher frequency.

When both "pre" and "post" data are collected, the water savings can be clearly illustrated by comparing the characteristics of the two diurnal curves. Figure 20.6 illustrates what might be expected when monitoring the water demands of a household before and after water-efficient toilets and showerheads are installed. The reduction in the morning's peak demands relates to the resulting water savings.

Diurnal curves can be used to illustrate information such as when people use water (bathing, showering, toilet use, etc.), the flow rate and duration of irrigation, the difference between weekday and weekend water demands, and the difference between summer and winter demands.

20.7   Lost Revenues

Some municipalities may be concerned that implementing a water efficiency program will result in a reduction in water sales revenues. In fact, many water efficiency programs are

---

* Using a data logging frequency of 1 h, for example, some data loggers will "turn on" every hour and log the parameter's value at that instant, while other data loggers will continuously monitor the parameter's value throughout the hour and then log the average of those values.

implemented to allow a greater population to be serviced with the existing infrastructure; a larger population generally means a larger tax base, which will benefit the municipality.

Programs that are implemented without a need (e.g., with no need to defer capital expansion projects) may, however, result in a reduction in revenues. The following points address this concern.

- Most water efficiency programs achieve demand reductions gradually over several years, providing ample time to implement small changes in water rates.

- Changes in summer weather patterns (i.e., cool wet summers vs. hot dry summers) may have a more significant effect on annual water sales than the implementation of water efficiency programs.

- Reducing water production volumes will reduce costs associated with water treatment and pumping.

20.8   Conclusion

It is hoped that the material in this chapter will assist the growing number of persons developing, implementing, and monitoring water efficiency programs to better understand some of the important concepts associated with completing a successful program.

By developing a clear understanding of your WEP goals, the measures and methods that you will implement to achieve these goals, and the protocol that you will employ to monitor your results, it is hoped that water efficiency can become an even more important element of your future water demand planning.

Water is too valuable to waste; let's use it wisely.

CHAPTER 21

Using In-House Staff or a Contractor and Designing a Bid Document

Julian Thornton

Reinhard Sturm

George Kunkel, P.E.

21.1 Introduction

In this chapter we discuss how to prepare for interventions necessary in the field to resolve real and apparent loss problems located through the auditing and modeling phases discussed earlier in the book.

21.2 Using In-House Staff or a Contractor

Most larger utilities have some kind of in-house expertise with knowledge of how to undertake a water audit or intervention against loss on either apparent or real losses. However, often these people have other duties, which make it hard for them to concentrate on the specialized tasks in an ongoing manner.

Many smaller utilities or industrial/commercial/institutional (ICI) systems do not have the in-house expertise and equipment necessary for a full audit and analysis of the water system. If either the people cannot be dedicated to the job or they are not available, the decision is easy—a consultant or contractor should be employed to undertake the audit and subsequent intervention.

> **O**nce the team has been identified and trained, they must identify where the faults in the current system lie and how best to resolve them. Usually the best way is the most cost-effective way! If in-house staff is not available or cannot be fully committed, then a consultant or contractor should be used.

If staff is available, the following steps should be taken to organize a dedicated crew for loss control.

- Identify a team leader who will be full-time loss control supervisor.
- Identify the necessary test equipment for taking field measurements.
- Ensure that the equipment can be periodically tested for accuracy locally and is supported by a local supplier who can undertake repairs on a timely basis.
- Identify either a full-time or part-time team who can assist the team leader.
- Undertake detailed training on the methods and technologies chosen for the audit and intervention methods.
- Be prepared to give authority to the team.
- Be prepared to give a budget to the team for annual testing and intervention.

Most audits and intervention tasks include the following:

- Master meter testing and repair
- Telemetry testing and repair
- Updating of system plans
- Sample testing and replacement of sales and revenue meters
- Selection of performance indicators
- Statistical analysis, modeling, and audit completion
- Leak detection and repair
- Reservoir and storage testing
- Pre- and postintervention monitoring

Other tasks which may arise as a result of the auditing and testing may include

- Pressure management
- Level control
- Mains relining and rehabilitation
- Mains and service replacement

Some of the tasks listed above are quite time consuming and detailed, and have to be repeated frequently to ensure a sustainable and economic level of losses. Therefore in certain cases it may be preferable to use a specialized contractor or consultant to assist the in-house team or undertake the work in place of an in-house team.

The decision as to whether to utilize an in-house staff or a contractor will really come down to time and money, as with most things! The other aspect to consider is the fact that a consultant or contractor will be specialized in the most up-to-date methodologies and techniques of the field.

Water company operators do not always have the opportunity to be exposed to latest technology if budgets do not allow them to attend conferences or seminars, and travel to see other utilities and discuss success and failure with others.

21.3   Designing a Bid Document

21.3.1   Introduction

If a decision has been made to call in a specialized contractor, careful planning must be undertaken prior to going to bid or negotiating directly with a contractor to ensure that both client and contractor understand exactly what the requirements and deliverables of the contract are. If everything is clearly spelled out in the beginning, it is much easier to select the best offer from several in the case of a bid, or to resolve any dispute which may later arise. If the bid document is not there is room for speculation and uncertainty, which will inevitably waste time on both sides—the client and the contractor.

> The bid document must be clearly and carefully written to ensure that both the client and the contractor fully understand the project methodology, goals, and objectives.

21.3.2   Important Factors to Consider

Obviously, each utility will have different requirements for a specialized contractor. Some utilities may require a full service and others may require specific tasks to be undertaken to complement the skills available from in-house personnel.

If the job is to be quite large and encompassing, it is a good idea to call a bidders meeting and present the overall situation clearly. Bidder packs should be distributed, with system condition and information reports.

System condition reports may contain some of the following information:

- System overview schematic if one is available.
- System plans (or a sample if there are many plans).
- Number of supply meters, type, and age.
- Number of sales meters, type, and age.
- If the above information is not available, clearly state that one of the objectives is to acquire this information.
- Topographical information.
- Storage information.
- Average system pressures.
- Schematic of supply zones if applicable.
- If zoning is not in place but is one of the desired deliverables, clearly state so.
- Length and type of mains and services.
- If the above information is not available, clearly state that this must be provided as one of the deliverables.
- Information on previous audits and water loss intervention.
- Estimated water losses.

A simple example follows:

XYZ utility comprises 800 mi of main supplying approximately 35,000 metered connections. The mains are primarily old cast iron mains laid around 1950, with some newer areas of PVC estimated to have been laid around 1980. Meters are between 10 and 30 years old. Meter maintenance has been sporadic. System plans are available in a 1:2000 scale but are not reliable in many cases. There are two main storage reservoirs, one old brick ground reservoir and one newer elevated concrete tank.

Water is supplied through the treatment plant in ZYZ Road. The treatment plant takes water from the XYZ River. The system is a direct pumped system with the two tanks balancing on the system. There are two main pressure zones, A and B. Pressures in zone A range from 50 to 70 psi and in zone B from 40 to 100 psi. In addition to the water treatment plant, zone B receives water through a supply meter from ABC utility. Supply meters are Venturi type with 4 to 20-mA output to a telemetry system, which reports back to central control. Topographical information is not available. Losses are unknown but are estimated at around 25%, of which it is felt that 70% is system leakage.

21.3.3   Project Goals

In all cases the bid document should clearly state the final objective of the contract. For example XYZ utility wishes to undertake a detailed water audit following the third edition of the AWWA M36 guidelines or IWA audit guidelines. The main goal of the audit is to identify and rank, by cost to benefit, the best ways of reducing lost water or lost revenue in the system.

The audit will be complete, covering all aspects of potential loss, both real and apparent. Decisions as to the levels of loss and the potential benefits and costs of intervention will be based on real field measurements taken during the audit.

All data collection, testing, analysis, and recommendations for water loss control intervention will be the responsibility of the successful contractor. Additionally, all system plans shall be updated, with separate 8.5 × 11 sheets for each pipe junction clearly showing valve positions triangulated from three fixed points. The new system plans shall be in ABC GIS format. GIS layers will include pipes, hydrants, valves, and detailed elevation contours.

The contract shall be conducted in three phases:

- *Phase one*: Audit, system measurements, sample meter testing, and updating of plans
- *Phase two*: Leak detection and repair
- *Phase three*: Meter change-out and automated meter reading (AMR) system

The successful contractor shall go out to bid for leak detection services for phase two and meter replacement services for phase three of the project.

The contractor shall provide monthly progress reports and a detailed report at the end of phase one identifying all cost-to-benefit scenarios. The contractor shall provide a fixed sum for supervision of phases two and three.

Phase One Tasks

As a minimum, the contractor shall

- Test all master and supply meters (see sample bid document below). Testing shall be to AWWA M6 recommendations.

- Perform demand analysis to identify classes of consumer and identify where most of the water supplied is being used, that is, residential, commercial, industrial, institutional, agricultural, municipal, and the like.

- Select a statistically accountable sample of sales meters and test it to allow analysis of the potential cost to benefit of meter change-out and meter resizing. The contractor shall comment upon and analyze the benefits of an AMR system.

- Update system distribution plans to an ABC GIS system as stated in the introduction.

- Perform hydraulic measurements to ascertain the level of real losses in the system and the benefits of leakage control. Hydraulic measurements will be defined as a minimum of two-hundred fifty 7-day pressure measurements at locations to be agreed at the start of the contract and fifty 7-day flow measurements. Flows should be accurate to ±5% of real flow and pressures to 1% of real pressure. Equipment must be calibrated to a national standard volume or weight at the beginning and end of the contract and at two separate random occasions during the contract. Data will be manipulated in function to any drift recorded during the equipment testing.

- Identify the benefits of pressure management as a means of reducing and controlling real losses further (this task should include at least one pilot installation).

- Perform drop tests on the storage tanks to see if they are leaking.

- Identify potential losses from overflows.

- Provide a complete audit to the required guidelines, with accompanying ranking of loss recovery measures.

- Provide a complete cost-to-benefit analysis of all recommended measures, including an analysis of the potential benefits of AMR.

- Provide monthly progress reports.

- Provide a final report, including bids for leak detection and repair and meter change-out and AMR if applicable.

- Provide detailed training for utility staff on the measures taken during the audit.

The following is a sample bid document for a contractor to test supply meters, which are primary measuring devices such as orifice plates, Venturi, Dall, and Pitot tubes, as in the case of our example above. Other tasks such as leak detection or meter replacement could be structured similarly.

Inspection and Testing of Primary Devices

1. The contractor shall physically inspect and report on the visual condition of each primary device, the mechanical and hydraulic connections, and the suitability of the environmental housing, that is, chamber, and so on.

2. Using portable equipment provided by the utility, the contractor will measure the differential pressure (DP) in relation to the flow existing at the time of the test and will relate this back to the individual specifications for each device, which will be provided. The contractor will write a report pertaining to the accuracy of this DP.

3. If the DP does not match the flow/velocity, the contractor will suggest possible methods of recalibrating the primary device, that is, cleaning, rodding, and so on. If the primary device is past rehabilitation, the contractor will state this with detailed reasoning.

Inspection and Testing of DP Sensor and Converter

1. The contractor shall physically inspect and report on the visual condition, approximate age, and suitability of the electronic equipment, stating serial number, make, and type of process, that is, square root extractor, linear, and the like.

2. After ascertaining the accuracy of the primary device, the contractor shall check both zero and span calibrations of the electronic equipment. A detailed report will be compiled identifying the conditions in which the calibration was found to be. The report will state the impact on potential metering error and what must be done to rectify a potential error.

3. If the device is found to be in error and can be calibrated, the contractor will perform the necessary calibration of both zero and span. The contractor will report on the calibration technique and values used. The contractor will then arrange with the utility to have the site reevaluated with an insertion meter for a final accuracy evaluation.

4. If the equipment is found to be in error and cannot be satisfactorily calibrated, the contractor will report on the reasons and make suggestions for the replacement of this equipment with other, more suitable equipment.

Transmission and Data Collection

1. The contractor will inspect all radio equipment, electrical connections, and the like, to ensure that this equipment is transmitting and receiving proper data. The contractor will report on any faults or potential problems found.

2. In all cases, the contractor will collect data from the utility's files in the central control for the period of testing and in the case of recalibration, during this period. The contractor will analyze the archived data to ensure that it matches the findings in the field and that in the case of calibration; the data being collected after calibration is a true reflection of the real flow conditions in the field.

Replacement of Faulty Equipment

1. The contractor may provide a separate quotation for supply and services required to replace faulty equipment with new calibrated equipment.

2. If the contractor is called upon to install new equipment, it will be calibrated and tested in situ and a detailed report generated to identify settings, and the like.

Contractor Experience and Requirements

1. The contractor must provide resumes of the instrument technicians who will undertake this service work. The technicians must have a minimum of 10 years relevant experience and be conversant with various makes and models of the above-mentioned equipment.

2. The contractor will be responsible for all test equipment required to undertake this work other than the portable DP meter and insertion meter, which will be supplied by the client. (If this is not the case, then the contractor should supply the insertion meter, etc.)

3. If the contractor is called upon to supply and install replacement equipment, the contractor must be prepared to provide all necessary equipment, fittings, connectors, and the like, to ensure that the old equipment can be taken out of service and the new equipment fitted the same day, without delay or "downtime."

4. In addition to full reporting, the contractor will be required to provide "hands-on training" for client staff during all stages of the testing and or replacement and calibration.

21.3.4 Selecting a Contractor

Once the bid specification has gone out and responses received, it is necessary to grade the responses, in a fashion that allows the best contractor to be selected for the job.

Many utilities put out a bid specification and then select the cheapest bidder. For a simple service this may be OK, but often for more detailed services it is not always the cheapest bidder that gives the best value for money. One way of being more selective is to grade the responses on technical merit and price. To do this it is necessary to apply weighting to the skills and personnel of the bidding contractors and weighting to their financial bid. The bids are then compared on both technical and financial merit. Another popular way of financing a contract is on performance; this will be discussed later in this section.

> **E**nsure that you have a mechanism in place to allow you to select the best contractor for the job. The best is not always the cheapest.

Technical and Price Bids

Technical and price bids are usually made up of components similar to the following:

- Understanding of the problem
- Bidder's experience with similar projects
- Personnel experience with similar projects
- Additional innovation brought to the project
- Equipment to be used on the project
- Price

Each technical topic is assigned a weighted value; this value can then be divided by the price to give an overall weighted score. Tables 21.1 to 21.3 show three scenarios. Bid one is the most technically competent in all respects, and in Table 21.1 Bid one is the winner.

Table 21.2 shows how much cheaper the second bidder needs to be to win even though he is deemed to be less competent. Table 21.3 shows how much cheaper bidder three has to be to win even though he is much less competent than the other two bidders. Technical and price bids are good ways of ensuring that the best bidder for the job gets the project; however, care should be taken to carefully analyze the data submitted

A Project with a Ceiling of $150,000.00 (Bid One Wins)	Max Weight Score	Bid One Allocated Score	Bid Two Allocated Score	Bid Three Allocated Score
Understanding of the problem	30	29	25	20
Bidders experience with similar projects	25	25	24	20
Personnel experience with similar projects	25	24	23	19
Additional innovation brought to the project	5	3	3	0
Equipment to be used on the project	15	15	15	10
Total	100	96	90	69
Price	150,000	150,000	145,000	142,000
	Maximum			
Weighted score	66.67%	64.00%	62.07%	48.59%

TABLE 21.1   Bid One Is the Winner

A Project with a Ceiling of $150,000.00 (Bid Two Wins)	Max weight Score	Bid One Allocated Score	Bid Two Allocated Score	Bid Three Allocated Score
Understanding of the problem	30	29	25	20
Bidders experience with similar projects	25	25	24	20
Personnel experience with similar projects	25	24	23	19
Additional innovation brought to the project	5	3	3	0
Equipment to be used on the project	15	15	15	10
Total	100	96	90	69
Price	150,000	150,000	140,000	142,000
	Maximum			
Weighted score	66.67%	64.00%	64.29%	48.59%

TABLE 21.2   How Much Cheaper the Second Bidder Needs to Be to Still Win

A Project with a Ceiling of $150,000.00 (Bid Three Wins)	Max Weight Score	Bid One Allocated Score	Bid Two Allocated Score	Bid Three Allocated Score
Understanding of the problem	30	29	25	20
Bidders experience with similar projects	25	25	24	20
Personnel experience with similar projects	25	24	23	19
Additional innovation brought to the project	5	3	3	0
Equipment to be used on the project	15	15	15	10
Total	100	96	90	69
Price	150,000	150,000	145,000	105,000
	Maximum			
Weighted score	66.67%	64.00%	62.07%	65.71%

TABLE 21.3   Shows How Much Cheaper Bidder Three Has to Be to Win

and to be accountable for the decisions made. It is certain that somebody will complain and ask for justification!

It is a good idea when using a technical and price bidding structure to clearly state the weightings to be used in the bid documents. It is also a good idea to place a statement to the effect that "XYZ utility reserves the right of final judgment in the assignment of points for the technical criteria. By submission of the bid the bidder agrees to waive any rights to pursue financial claim for loss of earnings resulting from the loss of this bid."

Performance-Based Bids

Another way of ensuring quality for money is to go out to bid on a performance basis. In this scenario the bidder basically becomes a partner of the utility, sharing the gain from reduced overheads or increased revenue streams. Obviously, if the bidder does not perform, there will be no payment for services rendered, or reduced payment if the risk is shared with the utility. Appendix A contains a paper titled "Performance-Based Non-Revenue Water Reduction Contracts.[1]" This paper is excerpted from: "The Challenge of Reducing Non-Revenue Water (NRW) in Developing Countries—How the Private Sector Can Help: A Look at Performance-Based Service Contracting," WSS Sector Board Discussion Paper #8, World Bank, 2006, by William D. Kingdom, Roland Liemberger, and Philippe Marin. This paper and the full WSS Sector Board Discussion Paper provide an excellent source for information on performance-based contracts.

Keep Contracts Simple

In most cases the best agreements and contracts are the simplest ones. In all cases the contractor and the client need to agree on a baseline for payment and a means of measuring increased efficiency over the baseline over time. An agreed value for reduced real (or apparent) losses and an agreed value for increased revenues need to be established.

The length of the agreement needs to be tailored around the amount of investment the contractor will make and the payback periods envisaged. Obviously, these must be realistic or nobody would bid. It is in the utility's interest to allow the contractor to make money so that he will continue to provide good service.

A sample performance-based project structure follows:

1. Phase one (fixed rate, upfront payment, repaid from savings later): Perform source meter testing and a system water audit to an agreed format.

2. Phase two (payment by performance).

Set up temporary district meters and data log the district demands.

- Identify the minimum night flow period and establish reasonable nonrevenue water and real loss levels.
- Identify areas where leakage is evident and undertake "step testing" to quantify the leaks accurately.
- Pinpoint the leaks using sonic and correlation methods and report on the locations for a directed repair program.
- Meter testing and downsizing program.

Detailed engineering report on the findings of the program with procedures for ongoing leakage control practices to be undertaken by utility staff. In this simple case the baseline for recovered leakage could be the minimum night flow before and after leakage location and repair, assuming that consumption and pressure stays constant. A simple check would be for the contractor and utility supervisor to estimate each leak repaired for volume of losses. This could be compared back to the change in night flow. An agreement would have to be made if significant differences were encountered. Care should be taken to deal with backlog leakage, which may distort figures the first time.

Likewise, to calculate the gain in metered sales after testing and correct sizing, the agreement could be to use a weighted rolling average of the last 3 months of sales for the year of change-out against the previous year. This ensures that consumption differences from one month to the other do not confuse the issue and result in conflict. In the event that clients consume less after change-out because they start to instigate conservation measures, an agreement could be made whereby the utility pays a fixed charge to the contractor for a period of time to cover his costs. In this case the value of leakage recovered could be either

- The purchase cost if water is being imported from a water supplier
- The variable production cost if water is being produced locally

The value of recovered revenue should be the sales cost of the water less any fixed charges.

21.4   Summary

The ideas identified above are just an example of the myriad of possibilities for negotiation between utility and contractor. However, in summary, the best way to ensure a successful project for both parties is to state as clearly as possible, up front, the requirements and specifications of the contract. It also always helps to go into a project situation in a position of trust. As was the case in the Metro Water Services contract, an "agree to agree" basis is the healthiest one.

21.5   Checklist

- Bid documents must be clear and to the point.
- Bidders' packages should be prepared with background data, however sketchy it may be.
- A bidders meeting is a good idea.
- Gray areas will cause confusion and can lead to dispute.
- Budgets must be realistic.
- Performance-based options can be negotiated.
- Realistic time frames must be negotiated to allow the contractor to take his money out.
- Good baseline data and performance indicators should be used to clearly identify the situation before and after intervention.
- Have an "agree to agree" clause, which should be adhered to by both parties.

Reference

1. Kingdom, B., R. Liemberger, and P. Marin. "The Challenge of Reducing Non-Revenue Water (NRW) in Developing Countries—How the Private Sector Can Help: A Look at Performance-Based Service Contracting." WSS Sector Board Discussion Paper Series—Washington, DC:World Bank, 2006.

CHAPTER 22

Understanding Basic Hydraulics

Julian Thornton

22.1   Introduction

We are about to embark on a major mathematical exercise to identify losses and their economic value, so we had better go back to school for a few pages and review some of the calculations, which we have either been bypassing or have forgotten! The following sections deal with some of the most frequently used calculations, tables, and transformations used in the water loss management and field-testing world.

22.2   Pipe Roughness Coefficients

All water pipes have a roughness factor, which plays a part in creating friction between the water running in the pipe and the pipe wall. Think about a glass pipe, which would be very slippery because of its high coefficient, as opposed to old rusted and encrusted cast iron pipe, which would have a very low coefficient. This roughness factor is very important when considering either computer modeling or pressure zoning and management. In cases where fire flows are critical, the roughness factor may indicate that a zone may not be shut in, due to poor hydraulic conductivity.

Additionally, the roughness factor can indicate pipes in poor condition and is often used by water system operators to identify which pipes should be earmarked for either replacement or some form of rehabilitation. Earlier in this book we dealt with the various types of rehabilitation techniques available and in use today, where and when to use them, and how they are best applied. We also discussed when to replace and when to repair or rehabilitate.

Continuity: $Q = V \times A$
Hazen-Williams: $V = 1.318 \times CX$
$\times R_{0.63} \times S_{0.54}$

There are various methods of calculating roughness in a closed pipe, such as the Manning, Darcy-Weisbach, and Colebrook-White equations. However, the most common method for pressurized water pipes is through the use of the Hazen-Williams C factor. The Hazen-Williams formula can be found above.

In order to measure the C factor we must measure flows and pressures, levels and elevations, and pipe length and diameter. The following section identifies detailed methodology statement for this type of testing in the field.

A table showing average C factors for various pipes of various ages and diameters may be used as a first estimate when applying values for decision-making models, but such a table is not a substitute for actual values measured in the field.

22.3   C-Factor Testing in the Field

Accurate measurements of flow, pressure, level, diameter, and length must be undertaken. Figure 22.1 shows an example of a C-factor test laid out with two hydrants being used for pressure measurements and one for flow measurement. The spreadsheet in Table 22.1 shows a sample calculation using commercial software called Flowmaster, from Hastead Methods.

It is important to use calibrated equipment when undertaking this kind of testing, as the results are very sensitive, particularly to pressure. Further information on the equipment and calibration procedures used in this type of testing can be found in Appendix B.

It is important that the operator has a basic feel for the results which will come out of the testing, so that unnecessary returns to the field can be avoided. After all, the idea of reducing losses and undertaking rehabilitation is to make a more efficient system! Fieldwork is a very important part of this process and can often be quite expensive when done properly either by contractor or by in-house staff. Proper planning and a good "ballpark" feel for what the results should be can help keep unnecessary work to a minimum.

While it is possible to have C factors higher than 130 and lower than 75, these cases are less likely, so by using this as a safety band the operator can query anything outside of these numbers while still on site and retest while the equipment is still set up. If the numbers are repeatable, then as long as they are not a long way off of the above recommendations, they may be realistic.

> **C** factors should fall somewhere between 75 and 130.

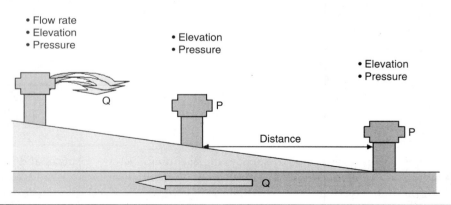

FIGURE 22.1   C-factor test. (*Source:* Julian Thornton.)

Input	U.S. Units	Output	U.S. Units
Elevation @ 1	605.86 ft	Velocity	1.65 fps
Pressure @ 1	25.2 psi	Headloss	2.11 ft
Elevation @ 2	540.65 ft	Energy Grade @ 1	664.04 ft
Pressure @ 2	52.55 psi	Energy Grade @ 2	661.93 ft
Discharge	2325 gpm	Friction slope	0.0004 ft/ft
Diameter	24 in		
Length	4710 ft		
Hazen-Williams Coef.	124.41		

Results calculated using Flowmaster.

TABLE 22.1   Sample *C*-Factor Calculation

Age and pipe material play a large part in the C factor, with the worse conditions often being found on old, corroded, and untreated iron pipe. It is common to find that very old untreated iron pipes can be almost closed off by tuberculation, corrosion, and debris.

22.4   Firefighting Regulations

When considering changing pressures either by zoning or pressure management, as discussed in Chap. 18, it is important to remember that many countries—including the United States and Canada—have mandatory flow and pressure levels for firefighting demand. Although all countries do not have the same regulations, the author has had the opportunity to study several regulations and there is quite a lot of similarity among them. Most countries require a minimum flowing pressure of 20 psi or 15 m.

Before changing system pressures, it is also necessary to check the needs of local insurance underwriters in many cases. Often the insurance rating for a particular type of property is determined partially on local firefighting capability, which is in many cases stated in terms of flow. In many cities the hydrant caps and bonnets are painted different colors to indicate the flow capacity of the hydrant.

In addition to understanding how to define hydrant needs for firefighting, it is also important to undertake a demand analysis in any proposed area where zoning or pressure management may be undertaken, to identify the needs of internal building sprinkler systems. These systems can often be reset to accept lower inlet pressures, but consumers must be made aware of any potential change so that volumes may be properly calculated. Usually the benefit from pressure management is great enough that the cost of recalibrating these systems can be borne by the contract, which also ensures that the clients are happy.

22.5   Flow Terms

The term *flow* is used to describe the amount of water passing a point in pipework, perhaps a meter, in a certain time period. Flow is actually a moving volume of water (or any other substance). Flow can be recorded in many different units, usually depending on whether the country uses metric, imperial, or U.S. units of measurement. The most common units for flow measurement are gallons per minute (gpm, either imperial or U.S.), cubic feet per second (ft³/s), and, in the case of large flows such as those found in bulk mains, transfer stations, and treatment plants, millions of gallons per day (mgd). Corresponding metric units are liters per second (L/s), cubic meters per hour (m³/h), and megaliters per day (MLD).

During a water audit, it is often necessary to convert between different units of flow and also velocity. *Velocity* is the speed at which a liquid moves. Velocity itself does not tell you anything about how much water is flowing—only how quickly it is moving along the pipe.

Flow volume is calculated by multiplying the average velocity of the fluid by the cross-sectional area of the pipe in which it is flowing. The formula for calculating flow can be found to the left.

> $Q = V \times A$
> Where: Q = flow
> V = velocity
> A = cross-sectional area

When calculating the cross-sectional area of the pipe we must make sure that we are using the same units for area and velocity, that is., square feet (ft²) and feet per second (ft/s) or square meters (m²) and meters per second (m/s).

Finally, to calculate flow, multiply the area by the velocity, or speed, of the liquid, which must be measured in the field.

Let's do some calculations of flow and velocity.

22.5.1 Example

A 6-in pipe has a velocity of 1 ft/s. How much flow does this represent in gallons per minute?

We use the equation $Q = V \times A$.

We fill in the missing information: $V = 1$ ft/s and $A = 0.785 \times 0.5 \times 0.5 = 0.196$ ft². So flow $Q = 1 \times 0.196$ ft³/s.

However, we want to express the flow in gallons per minute (gpm). We must multiply the volume in cubic feet per second by 7.48 to get a figure in gallons. Then we have to multiply by 60 to turn the time frame into minutes: $Q = 0.196 \times 7.48 \times 60 = 87.96$ gpm.

Let's now do the same calculation in metric units. We convert the above calculations as follows: $V = 0.3048$ m/s and $A = 0.785 \times 0.15 \times 0.15 = 0.0176625$ m². So flow $Q = 0.3048 \times 0.0176625 = 0.0053835$ m³/s.

We discussed earlier that metric units of flow are usually expressed as either liters per second or cubic meters per hour, so we should change our answer: $0.0053835 \times 1000 = 5.3835$ L/s, or 0.0053835×3600 (60 min of 60 s each in 1 h) $= 19.3806$ m³/h.

There are 3.78 L in a gallon, so, just to check our two calculations: $(5.3835/3.78) \times 60 = 85.45$ gpm. (The difference of around 1% is due to the rounding up or down of the decimal places.)

22.5.2 Types of Flow

Now that we have done some basic flow calculations, let us discuss the different types of flow which may be encountered. It is important to understand the flow conditions in a pipe, so that a suitable place for monitoring or testing can be selected.

Many problems in water system calculations are caused by incorrect siting of meters or test equipment. This is often the case because the individual who is responsible for the installation of the meter does not understand basic fluid dynamics. Other problems are caused by incorrect conversion of units, which is why it is important to understand the relationships between various types of units.

It is a good idea to write down the types of units currently in use in the water system you are going to audit and their respective conversions, before starting any field work or data analysis. (More information on data handling can be found in Chap. 8.) This way you can avoid costly mistakes later.

Although it is not necessary to understand fluid dynamics on a high level, a few basic ideas will come in handy. If we understand the basics, then when it comes time to troubleshoot we will be much better equipped to track down a problem.

If we do not understand the basics of fluid dynamics, we may end up fitting our test equipment at an unsuitable site. In this case our test data might be incorrect and we might input the wrong data to what might be a very important calculation.

The following sections cover the various types of flow. Although it is unusual to have to deal with all of these types of flow in a single water system, it is important to know that they exist.

Steady-State Flow

Flow is considered steady if, at a certain point in the pipe, the velocity of the water does not change with time but remains the same. Steady-state flows are not often found in field situations, but are sometimes used in simple modeling calculations.

Uniform Flow

Flow is considered to be uniform when the velocity does not change speed or direction from point to point. This condition can be found in transmission mains with long lengths of equal diameter and few restrictions such as butterfly valves or control valves. Most distribution-level mains experience nonuniform flow.

At the distribution level in the field, mains size is always changing, making velocity and pressure change constantly. In other situations water is passing through meters and control valves.

For our purposes, when measuring bulk flows on transmission systems, we should always record flow at a point where the flow is reasonably uniform. This is less likely to be possible when measuring distribution system flows.

Many manufacturers of portable equipment suggest a minimum of 10 times the pipe diameter upstream and 5 times the diameter downstream of any fitting or restriction. If in doubt, the 30/20 rule is a good bet; see Fig. 22.2. It is not good to install test equipment next to a permanent meter or valve, as the test equipment may then be less reliable than the equipment being tested.

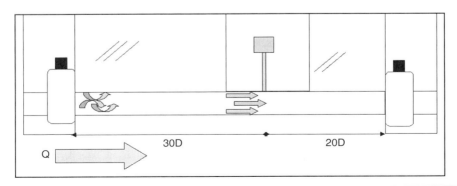

Figure 22.2 30/20 rule for installation of temporary measuring equipment. (*Source:* Julian Thornton.)

Laminar Flow

In a laminar flow condition the water moves along straight parallel paths, in layers or streamlines. The water velocity is not the same. Inside the pipeline, the layers of water closest to the pipe wall rub against the pipe walls and therefore travel more slowly. We discussed friction factors earlier in this chapter. We now know that by measuring flow and pressure we can actually calculate the internal condition of the pipe.

Laminar flow is not very common in water pipes and occurs only at very low velocities. Most utility situations have nonlaminar or turbulent flow. A term called a Reynolds number is used to calculate if flow is laminar or turbulent. Many pieces of flow monitoring equipment quote Reynolds numbers above or below which the equipment may be used. A Reynolds number is a function of velocity, diameter, and viscosity: $R = (\text{velocity} \times \text{diameter}) / \text{viscosity}$.

Most utility pipeline applications have Reynolds numbers in the hundreds of thousands. A Reynolds number of less than 2000 would indicate a Laminar flow condition.

Turbulent Flow

In a turbulent flow condition the water tumbles along in a more confused fashion, although this is the usual state for water flow in a water supply system. Figure 22.3 shows the difference between laminar and turbulent flow.

Although most water system flows are turbulent, the flow does still move in a forward direction. When monitoring flows we usually see that the velocity is greater toward the center of the pipe than it is at the sides. This is very important to understand, as we use different techniques to monitor flow in the field. Sometimes we use a single-point velocity meter, which requires finding the average velocity in the pipe and multiplying this by the cross-sectional area to find volumetric flow. At other times we use equipment which averages the velocity across the pipe diameter and calculate the average velocity for us. Either way, if we understand the basics we are better equipped to deal with anomalies.

It is important to note the law of conservation of mass, which states that material is neither created nor destroyed. So whatever flow enters a system must leave it (at either a consumer point or feed to another system, or leakage) or accumulate inside it (as in the case of storage tanks and reservoirs). Since water is incompressible, it cannot accumulate inside the system or individual pipes. This is why we have to have tanks or reservoirs inside a water system—to accumulate water. We usually refer to this as *storage*. The conservation of mass is the basis behind the water audit or balance. The water that enters a system is delivered to a customer, delivered to storage, or delivered to a loss situation such as leakage or theft. It cannot simply disappear (even though in some audits it may seem that way for a while).

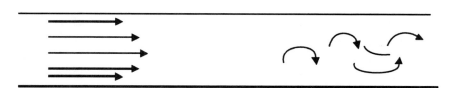

Figure 22.3   Difference between laminar and turbulent flow. (*Source:* Julian Thornton)

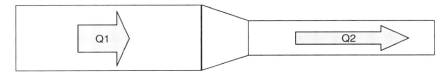

FIGURE 22.4   Conservation of mass. Q1 = Q2; however, the velocity will have increased. (*Source:* Julian Thornton.)

A gas system, however, is different. Storage is achieved by "line packing" or increasing the pressure of the gas in the line. Gas is compressible, so this is possible. This should be remembered if undertaking a tracer gas survey. Due to the fact that gas is compressible, a line break will cause a lot more damage when it is filled with gas than with water.

Figure 22.4 illustrates the law of conservation of mass.

Often it is necessary to measure velocity in a particular pipe. One of the main reasons is to decide if a site is suitable for a monitoring position. Sites with low velocities are not considered good places to monitor, as the equipment tends to stall out. Most flow monitoring equipment will state the stall-out velocity. This is often around 0.3 ft/s or 0.1 m/s. Flows calculated with velocities below these figures should be scrutinized for instability and potential error using the manufacturer's error curve and calibration certificate.

For example, we may know the flow and velocity leaving a treatment station or tank, and we want to undertake a flow balance to assess a system for leakage. Obviously, we want to reduce the inherent error in our monitoring equipment, so we try to locate the pipes with the highest velocity to monitor. Other pipes may be temporarily shut down during the period of testing. Larger pipes are often subjected to lower velocities, especially in newer systems where they may have been sized for future population growth.

22.6   Pressure Terms

To understand pressure we need to think of containers of water resting on the ground. The weight of the water is 62.4 lb for every cubic foot; if we divide by 7.4 gal/ft³ we find that we have 8.34 lb for every gallon. In metric units the corresponding weights and measures are 1 L = 1 kg, and 1 m³, which is 1000 L, weights 1000 kg.

This weight resting on a surface exerts a force on that surface. That force is what we call pressure.

22.6.1   Example

One cubic foot of water resting on the ground exerts a force of 62.4 lb on that bottom square foot. If there are 2 ft³ of water resting on a 1-ft² area, the pressure is 124.8 lb, and if there are 20 ft² resting on that bottom foot, then the pressure is 1248 lb. This is obviously important when we start thinking about storage and measuring pressures at tanks.

The same is true in metric units: 1 m³ of water resting on 1 m² of area exerts 1000 kg of pressure on that square meter, and 2 m³ of water resting on 1 m² of area exerts 2000 kg of pressure. If there are 20 m³ of water in a column resting on 1 m² of area, then the pressure is 20,000 kg.

- **Pressure is force per unit area.**
- **Pressure = weight × height**

During analysis of a water system, we often need to know the pressure at a particular point. Some of the reasons we may need to know are

- To calculate the amount of leakage occurring through a hole of a known size
- To calculate the amount of leakage through a varying hole or split
- To calculate the amount of water flowing from a fire hydrant during a test
- To figure out the condition of the main, when used in conjunction with a flow test
- To calibrate a computer model
- To see if a customer has sufficient pressure for supply

We just saw that there are 62.4 lb/ft² per square foot; however, water systems usually use pounds per square inch, or psi, as the denominator. We must therefore split our

Many systems that use metric units also use psi.

square foot into inches. There are 12 × 12 = 144 in in a square foot, so there are 62.4/144 = 0.433 lb per square inch. Therefore, for every foot of height (head), there is a pressure of 0.433 psi. Most people remember it the other way round, in terms of how many feet of water yield a pressure of 1 psi. So let us calculate this: 1/0.433 = 2.31 ft.

Metric calculations of height and weight (pressure) are generally easier, as the weight or pressure is stated as meters column of water. Therefore, if you have 20 m of water resting on 1 m² of area, you have a pressure of 20 m head. Sometimes this is stated in bar. One bar is 10 m head, so in this example 20 m head would be 2.0 bar. The metric system is quite easy to use, as the multipliers and dividers are all factors of 10 (e.g., 1 m of water = 1.42 psi; 1 bar of water pressure = 14.2 psi).

Static system pressure is due to the depth of water above the point of measurement; it has no relationship to the size of the water pipe or storage tank. Storage tanks of equal height but different shapes all provide the same pressure. See Fig. 22.5. Static system pressure within the distribution system is measured by taking the tank height plus the difference in elevation at the point of measuring in the system to the bottom of the tank. Static system pressure occurs only when there is very little flow in the system, which is usually at night. In some water systems the static pressure is never reached, because of high demands. As we discussed earlier in the chapter, the water flowing in a pipe is subjected to friction losses or headlosses. These headlosses change as flow conditions change within the system.

22.6.2   Gravity-Fed Systems

Often a water system will operate from static head in a tank and will gravity-feed into a system. Water is either pumped from a surface reservoir to a tank, so pumping costs do occur, or, as in the case of some mountain cities, water comes from a high-level natural reservoir or spring. See Fig. 22.6. This second type of system theoretically has the lowest operating costs and often the highest unaccounted-for water levels. The high unaccounted-for water levels are often the product of attitude, as the water is so cheap and pressures from a high-level reservoir run either unchecked or badly calibrated. These systems, however, often experience trouble in treating the large quantity of water passing through the station, particularly when turbidity is high, therefore making treatment costs high.

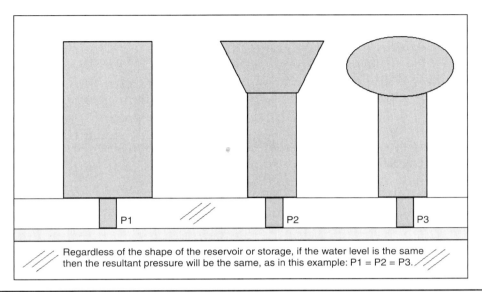

Regardless of the shape of the reservoir or storage, if the water level is the same then the resultant pressure will be the same, as in this example: P1 = P2 = P3.

FIGURE 22.5 Storage tanks of equal height but different shapes all provide the same pressure. (*Source:* Julian Thornton.)

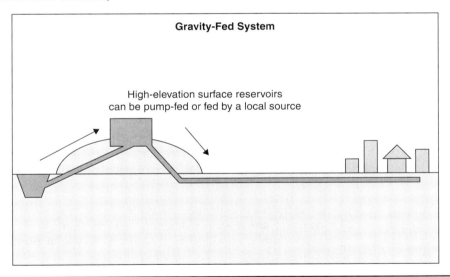

Gravity-Fed System

High-elevation surface reservoirs can be pump-fed or fed by a local source

FIGURE 22.6 Gravity-fed system. (*Source:* Julian Thornton.)

In many cases the only way to cost-justify a loss control project in a mountain community gravity-fed without booster pumps is on deferral of capital costs for distribution system upgrades.

22.6.3 Pumped Systems

Pumps can also provide system pressure. The pumps actually provide lift or head. Pumped pressure is measured in the same way as static pressure.

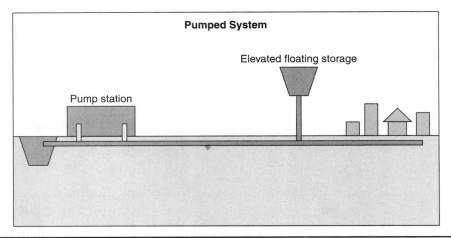

FIGURE 22.7   Pump-fed system. (*Source:* Julian Thornton.)

Pumps are often used to lift water from a surface reservoir or well to a holding reservoir, standpipe, or water tower. Some systems, however, use direct pumped systems. See Fig. 22.7.

Most pumped water systems tend to have higher water prices than gravity-fed systems, particularly if the marginal cost of water is only calculated using the power and chemicals calculation. This of course could be untrue in cases where a system has extremely high treatment costs. An example is a system that uses reverse osmosis for treatment.

22.6.4   Pressure Measurements

Pressure measurements can be made in a number of ways. Some of the more common are

- Piezometers (level tube)
- Pressure gauges
- Pressure loggers

Most pressure measurements in water systems are made using *gauge pressure,* which is the difference between a given pressure and atmospheric pressure. *Absolute pressure* is the reading of pressure including atmospheric pressure. This type of reading may be used, for example, at a weather station. All of our measurements will be done using gauge pressure.

22.6.5   Effects of Pressure

Water hammer or *hydraulic shock* is the momentary increase in pressure which occurs in a dynamic (moving) water system due to sudden change of direction or velocity of the water. Water hammer is often caused by incorrect operation or calibration of valves, pumps, or fire hydrants. A less severe form is often referred to as *surge.* This can be due to pressure fluctuations caused by natural changes in demand.

> **W**ater hammer and surge are often responsible for recurring leakage in a system.

FIGURE 22.8 Surge anticipator valve. (*Source:* Watts ACV, Houston, Texas.)

Many systems fit surge tanks, or surge anticipator valves (see Fig. 22.8), which provide pressure relief for a system. Reservoirs and storage tanks also help to vent unnecessary pressure. Reservoir storage is also often found to be an area of high leakage if not controlled properly. Most utilities are starting to fit pressure control systems, often consisting of fixed-outlet pressure-relief valves or altitude valves (see Fig. 22.9). In addition, pumped systems with high leakage levels often see benefits from installing surge anticipator valves.

FIGURE 22.9 Altitude valve. (*Source:* Watts ACV, Houston, Texas.)

Some utilities are also seeing increased benefits from modulated pressure control. We dealt with this subject in more depth in Chap. 18.

22.7   Summary

In this chapter we have covered some of the very basic concepts of flow and pressure and some of the effects of headloss. The following references were used during this research and are recommended reading for those who may wish to learn more about this topic.

References

General references regarding flow and pressure terms:

1. Hauser, B. A., *Practical Hydraulics Handbook,* Chelsea, Mich.: Lewis Publishers, 1991.
2. Giles, R. V., Evett, J. B., and Liu, C., *Schaum's Outline of Theory and Problems of Fluid Mechanics and Hydraulics,* 3d ed., New York: McGraw-Hill, 1994.

Case Studies

Case Study A.1: The Philadelphia Experiences

George Kunkel, P.E.

A.1.1 Philadelphia's Water Supply: A History of Firsts

The City of Philadelphia has been a leader in water supply technology in the United States for over 200 years. In 1801 the fledgling city became the first in the young nation to construct two steam-driven pumping stations to bring Schuylkill River water to wooden tanks at the city's "Centre Square" where water was piped to 63 private homes, four breweries, and one sugar refinery. In 1815 a larger, improved system was commissioned with the settling reservoir at "Fair Mount" to supply water to the growing city. By 1822 a dam, water-driven turbines, and Greek Revival architecture were incorporated into the Fairmount Water Works. The site entertained many visitors; being widely recognized not only as an engineering marvel but also a place of architectural splendor and beauty.

The first distribution piping of these early systems consisted of bored wooden logs joined end-to-end by iron bands and caulking. The city's first water loss problem was realized immediately as these pipes leaked badly and constantly. Philadelphia soon began to import British-made cast iron pipe to expand its water distribution system and this material became the norm by 1832. The longevity of iron pipes—in use in Europe for hundreds of years—has been confirmed in Philadelphia, where several thousand feet of pipe segments installed in the 1820s still provide reliable service to this day.

While recognized for its historical significance as a center of government during the United States' birth as a nation, Philadelphia's emergence as a major American city actually occurred during the industrial revolution of the nineteenth century as it evolved into a significant manufacturing center and bustling port. By 1900, however, the city's population of almost 1.3 million had begun to degrade its two major water sources: the Schuylkill River and the Delaware River. Philadelphia demonstrated innovation by becoming one of the first large cities in the nation to construct water filtration plants, with five such facilities of various sizes commissioned between 1903 and 1911. At the time, Philadelphia's filtration system was the largest in the world. Philadelphia's readiness to apply emerging technology continued as it adopted cleaning and cement lining rehabilitation of water mains (1949), use of an analog computer, the McIlroy Fluid Network Analyzer (1956), and use of telemetry control of pumping stations—the forerunner to today's modern Supervisory Control and Data Acquisition (SCADA) systems

(1958). More recently, the city of Philadelphia installed the largest water utility automatic meter reading (AMR) system in the United States, with over 400,000 residential units outfitted in an initial phase between 1997 and 1999, and almost 487,000 AMR-capable accounts as of mid-2007.

Philadelphia continues to meet today's complex challenges by providing a full range of water and wastewater services to a discerning public while maintaining a delicate balance with the natural environment. Faced with increasing water quality and stormwater regulations, the Philadelphia Water Department (PWD) and Water Revenue Bureau (WRB) are further challenged by a contracting customer base, mounting infrastructure needs, and the fact that water loss in the city has traditionally been perceived as relatively high. The city has responded to these needs by developing a comprehensive capital planning and rehabilitation program focused upon optimizing its assets and adopting best management practices for efficient water supply operations. At the start of the new millennium Philadelphia continued its tradition of firsts, by becoming the first water utility in the United States to explore the use of the progressive water loss management methods and technology developed internationally during the 1990s. PWD became the first water utility in the United States to employ the water audit methodology published in 2000 by a team from the International Water Association (IWA) and the American Water Works Association (AWWA). In 2004, PWD became the first known water utility to utilize its AMR System to gather nighttime customer consumption readings to assist leakage assessments in a District Metered Area (DMA). PWD continues to be a pioneer in promoting new technologies for improved operations and service to their customers.

A.1.2 Water Loss in Philadelphia

In most communities in the early days of the United States, engineers focused upon building and the development of the industrial potential that was evident in the young country. Water was critical to this end and the coastal regions of the first states have always been blessed with abundant water resources. Engineers exploited these resources and were highly successful in creating a reliable water supply infrastructure. As growing communities or industry required more water, new wells or pumping stations were constructed. However, when John C. Trautwine was Philadelphia's Bureau Chief in 1898 water charges were assessed according to the type and number of plumbing fixtures in a business or home rather than the actual quantity of water used. Greatly concerned about the enormous waste of water in the city, he felt that water meters would be the most successful way to encourage conservation. As a demonstration, he constructed and displayed the "Trautwine Tank" which held 250 gal of water—the amount used by each Philadelphian every day at the time—in order to emphasize the significance of water use to the public. Accountability of water was again promoted with the installation of customer meters starting after World Water II. Despite these early displays of water conservation acumen, it is likely that the City of Philadelphia historically did not maintain a high level of water accountability in its operations. With water relatively available and inexpensive, PWD's primary water supply goals were to provide a safe, sufficient supply of water for industrial, residential, and fire protection needs, and the city has continuously met these goals for over two centuries. Philadelphia completed the expansion of its distribution system and modernization of its water treatment plants by the mid-1960s, with an infrastructure capacity easily able to supply over 400 million gal of high quality water each day.

Philadelphia's population peaked at roughly 2.1 million people in the mid-1950s with the average annual water delivery topping-out at 377 million gallons per day (mgd) in 1957. Then began a slow, subtle shift in the demeanor and demographics of the city. Industry entered a gradual decline as heavy manufacturing migrated away from major northeastern United States cities. The relocation of citydwellers to suburban areas furthered the decline in the city's population, which stood at 1.46 million as reported in the year 2005 mid-term census estimate. Yet its infrastructure size— three water treatment plants and 3100 mi of piping— remains largely unchanged. With water export sales at 7.5% of its production, and only moderate additional sales potential believed to exist, the change in size of the customer base is projected to remain stable at best, or in continued slow decline for the near-term future. The city's volume of water supplied to its distribution system has also declined, reaching a record low in the city's modern history of just under 254 mgd for the fiscal year ending June 30, 2006 (FY2006). While water charges in Philadelphia have not been high by relative standards, the new demographics—with a larger portion of urban poor—have resulted in political pressure to keep water rates affordable.

Philadelphia assessed its water loss condition in limited and detailed fashion in 1975 and 1980 studies, respectively. A cursory review of delivery and billing data in 1975 asserted that the level of "unaccounted-for" water in the city appeared to be high and warranted further attention. Considerable attention was then given to the issue in 1980 when an Unaccounted-for Water Committee undertook a comprehensive, year-long study to identify sources of lost water in the city and propose actions to reduce losses and recoup revenue. A number of initiatives, including master meter calibration, expanded leak detection and meter replacement, came about in the years following this endeavor. However, non-revenue water— defined as the difference between the volume of water supplied and customer billed consumption— remained at levels well above 100 mgd in the decade following this work.

Water loss took on a greater prominence for the city government in 1993 after a proposed 30% water rate (tariff) increase was roundly criticized and eventually reduced to single-digit increases totally 7% over 3 years. The city's water loss standing was scrutinized and resulted in the formation of a permanent Water Accountability Committee to pursue water loss reductions. Further expansion of the main replacement and leak detection programs and a switch from quarterly to monthly billing were implemented shortly thereafter. These efforts made headway in bringing the city's excessive water losses under an initial degree of control. Figure A.1.1 reveals a notable decline in Non-revenue water after 1994. Non-revenue water averaged almost 126 mgd from fiscal years 1990 to 1994, but realized a steady drop to a level of 77 mgd for the fiscal year ending June 30, 2006. This success in cutting water loss is attributed to reductions in both real losses (leakage) and apparent losses (missed billings, meter error, unauthorized consumption). It is believed that real losses have been reduced by applying several new technologies (District Metered Area, inline leak detection probes), as well as refining existing efforts via a combination of stepped-up leak detection effort, improvements in leak repair job routing and pipeline replacement. Apparent losses have been reduced by the use of new residential meters (installed with AMR), large meter right-sizing, missed billing recoveries and metering/accounting of city-owned properties. A new customer billing system, targeted for implementation in 2008 will provide the city with additional capabilities in monitoring consumption trends and identifying losses. While these improvements are significant, it is still understood by city managers that non-revenue water of over 77 mgd and an infrastructure leakage index of roughly 10.0 represents a large amount of water that is not being recovered.

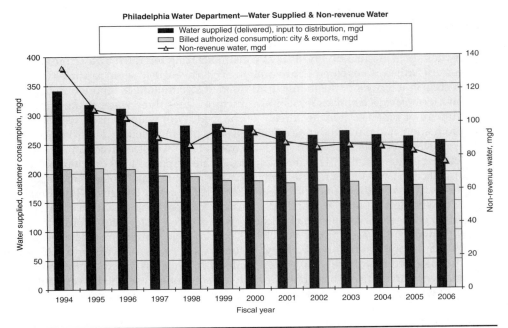

Philadelphia Water Department—Water Supplied & Non-revenue Water

- Water supplied (delivered), input to distribution, mgd
- Billed authorized consumption: city & exports, mgd
- Non-revenue water, mgd

FIGURE A.1.1 Philadelphia Water Department non-revenue water reduction trend. (*Source:* Philadelphia Water Department.)

A.1.3 In Search of Best Management Practices for Water Loss Control

During the 1990s Philadelphia's Water Accountability Committee sought to gain a better understanding of the nature of the seemingly high level of water loss occurring in Philadelphia's water supply system. The committee determined to identify current industry best practices and in 1995 began participation in the Water Loss Control (Leak Detection & Water Accountability) Committee of the American Water Works Association (AWWA). The Philadelphia committee followed the recommendations of the AWWA committee in assembling a water audit in a manner close to that recommended in the first edition of the AWWA publication M36, *Water Audits and Leak Detection.* After several years of gathering detailed information, the city produced its first distribution system water audit for the fiscal year ending June 30, 1996 (FY1996). In retrospect, a very basic water audit could have been issued 1 to 2 years earlier and the author recommends this approach to any first-time auditors, particularly since they now have opportunity to utilize AWWA's Free Water Audit Software. The city continued with the modified M36 format through FY1999, but moved to use the IWA/AWWA methodology upon its publication in 2000.

The years 1998 to 2001 were instrumental for the city's gaining awareness of the rapidly developing water loss technology and policy that was occurring internationally during the 1990s. During this time, a number of motivated researchers and engineers from the IWA and AWWA collaborated to raise awareness of water loss and explore the applicability of international water loss methods in North America. The IWA's Water Loss Task Force published its new water audit methodology in 2000 as a "best practice" approach to water auditing and benchmarking on an international scale. With these important developments occurring, Philadelphia again demonstrated its willingness to implement new technology by contracting with international experts to conduct its Leakage Management Assessment (LMA) project in 2001. Research and development has continued in recent years and Philadelphia has had an active role in the development of the AWWA Free Water Audit Software (2006) and two research projects sponsored by the AWWA Research Foundation (AWWARF): "Evaluating Water Loss and Planning Loss Reduction Strategies" (Project 2811, 2007) and "Leakage Management Technologies" (Project 2928, 2007).

A.1.4   The Importance of the Annual Water Audit

As part of the LMA project, the consultant team guided the PWD in converting its water audit into the IWA/AWWA format, making it the first water utility in the United States to apply this method. The summary of the water audit for the city's most recent water audit report is shown in Table A.1.1.

PWD strongly supports the use of system water audits by water utilities as both a standard business practice and a means by which regulatory agencies can assess the efficiency of water suppliers. The water audit is best compiled on an annual frequency, either on a calendar year or business year basis. PWD conducts its water audit on a business (fiscal) year schedule and accords the water audit report the same status as its other business year reporting functions.

The PWD has also played a strong role in the water industry by advocating the use of the IWA/AWWA water audit methodology as the fundamental practice necessary to assess water efficiency status in water utilities and motivate better control of the high levels of water and revenue loss that are believed to exist in drinking water utilities.

A.1.5   Evaluating and Controlling Real Losses

The City of Philadelphia operates one of the oldest water distribution systems in the United States. Approximately 60% of its pipeline is unlined cast iron installed between 1880 and 1930 with 6-in diameter being the most common water main size. The city's 3100 miles of distribution system piping extend across its 129 mi^2 area and provide water to approximately 4,90,000 customer accounts. Over the past three decades the city averaged 840 reported water main breaks, or bursts, per year, with half of the annual breaks occurring in the cold weather months of December through February. Additionally, in FY2006 the city documented 4301 leaks, with 3621, or 84% of the total, occurring on customer service connection piping. Accompanying Philadelphia's population decline of four decades has been a growing number of abandoned private properties. Left unmaintained, deteriorating service connection piping at many such properties have aggravated the trend of service connection leaks in the city. In Philadelphia customers are responsible to arrange for maintenance and repairs of leaks on their entire service connection piping. This is known to be a highly inefficient policy in terms of leakage control and ways to address this policy shortcoming are being studied. Water pressure levels in most of the distribution system vary between 40 and 70 psi with an average city-wide pressure of 55 psi. Small areas of the city are provided pressures in excess of 100 psi, and some hold potential for improved pressure management.

The PWD has operated a focused leak detection and repair program since 1980. It maintains a leak detection squad of roughly 20 employees with crews performing leak surveys in search of "unreported leaks" on both day and night schedules. Leakage survey progress has typically been measured by the amount of system pipeline mileage

Water Supplied	Volume, million gal	Average Volume, mgd	Costs/Year
System input	**92,931.5**	**254.6**	
Minus correction for source meter error	294.2	0.8	
Corrected system Input	92,637.3	253.8	
Minus exports	6971.5	19.1	
Water supplied (City only)	85,665.8	234.7	
Authorized Consumption			
Billed metered	57,633.5	157.9	
Billed unmetered	0.0	0.0	
Unbilled metered	0.3	0.0	$1176
Unbilled unmetered	892.5	2.4	$191,084
	58,526.3	**160.3**	**$192,260**
Water Losses	**27,139.5**	**74.4**	
Apparent losses			
Customer meter Inaccuracies	114.6	0.3	$520,206
Unauthorized consumption	1579.0	4.3	$3139,437
Systematic data handling error	3826.4	10.5	$16,616,968
Apparent loss totals	**5,520.0**	**15.1**	**$20,276,611**
Real Losses			
Tank overflows/Operator error	0.0	0.0	$0
Reported & Unreported leakage*			
Transmission main leaks/Breaks	5.7	0.0	$916
Distribution main leaks/Breaks	927.5	2.5	$148,850
Customer service line Leaks	9,003.5	24.7	$1,444,858
Hydrant & valve leaks	474.0	1.3	$76,065
Measured leakage (DMAs)	1,094.3	3.0	$175,606
Background leakage	10,114.5	27.7	$1,623,154
Leakage liability costs			$759,198
Real loss totals	**21,619.5**	**59.2**	**$4,228,646**
Water losses -total cost	27,139.5	74.3	$24,505,257

Fiscal Year 2006 Financial Data

$4791	Apparent losses per MG-small meter accounts (5/8" & 3/4")
$4143	Apparent losses per MG-large meter accounts (1" and larger)
$4070	Apparent losses per MG for city property accounts
$4500	Apparent losses per MG—overall average customer rate
$160.48	Real losses—marginal cost per mg
$759,198	Real loss Indemnity costs—added to total of Real Losses
$190,162,000	Water supply operating cost for Fiscal Year 2006

Fiscal Year 2006 Infrastructure Data

13,137	Number of large meter accounts, 1-in and greater
458,043	Number of small meter accounts, 5/8 & 3/4 in (also includes some large meter accounts)
80,779	Number of actual connections in nonbilled account population
3014	Miles of transmission and distribution pipeline
25,199	Number of fire hydrants
12	Ave. length of service connection: curbstop to customer meter, ft.
14.7	Average length of fire hydrant leads, ft.
55	Average operating pressure, psi

*The breakdown of leakage categories is approximate and should not be interpreted literally as most of these components are based on estimates rather than measured nightflows. It is believed, however, that the overall estimate of leakage is reasonably representative of aggregate system conditions.

Performance Indicators for Water Supply System Losses

Water Resources Performance Indicator

Inefficiency of Use of Water as a Resource = Real Losses over system input volume, %

= 21,619.5 MG divided by 85,665.8 MG × 100% = 25.2%

Operational Performance Indicators

	Million Gal	
Water losses	27,139.5	
Apparent losses	5520.0	
Real losses	21,619.5	
Unavoidable annual real losses (UARL)	2185.2	6.0 (see next page for calculation)
Real losses normalized	107.3 Gallons/Service Connection/Day	
Apparent losses normalized	27.4 Gallons/Service Connection/Day	
Infrastructure leakage Index (ILI) = Real losses over UARL		

= 21,620 MG divided by 2185.2 MG = 9.9

	mgd
	74.4
	15.1
	59.2

Financial Performance Indicator for Non-Revenue Water

Non-revenue water = Unbilled authorized consumption + apparent losses + real losses

= 0.3 + 892.5 + 5,520.0 + 21,619.5

= 28,032.3 million gal

= 76.8 mgd

Non-revenue water by volume = Non-revenue water over water supplied, %

= 28,032.3 MG divided by 85,665.8 MG × 100% = 32.7%

Non-revenue cost ratio is the annual cost of non-revenue water over the annual running costs for the water supply system, in %

Non-revenue water costs	$1176	Unbilled metered
	$191,084	Unbilled unmetered (authorized consumption)
	$20,276,611	Apparent losses
	$4,228,646	Real losses
	$24,697,517	Total

Non-revenue cost ratio = $24,697,517 divided by $190,162,000 × 100% = 13.0%

TABLE A.1.1 City of Philadelphia Annual Water Audit Summary—IWA/AWWA Water Audit Method Fiscal Year 2006—July 1, 2005, to June 30, 2006 (Continued)

Calculation of Unavoidable Annual Real Losses (UARL): IWA/AWWA Water Audit Method

Unavoidable annual real losses (UARL) is a reference value that can be calculated for any water distribution system and is used in calculating certain performance indicators. It is not an actual measure of any leakage component, however. The IWA/AWWA calculation for UARL is powerful since it is determined on a system-specific basis. The UARL is the theoretical minimal level of leakage that would exist in a distribution system after all possible leakage management actions are implemented, using the best of today's available technology.

The IWA/AWWA calculation includes leakage allowances based upon the number of customer service connections, length of service connection piping between the curbstop or property line and the customer meter, and average system pressure, all of which are key factors in the rate of active leakage in a water distribution system.

Calculation of Unavoidable Annual Real Loss (UARL) for Philadelphia Water Department Fiscal Year 2006, July 1, 2005–June 30, 2006

Infrastructure component	Quantity	Unit Rate for Unavoidable Annual Real Losses	Average Pressure, psi	Unavoidable Annual Real Losses, million gal	Unavoidable Annual Real Losses, mgd
Total pipeline mileage, including pipeline total & sum of fire hydrant leads	3,084	5.40 gal/m/d/psi	55	334.3	0.916
Number of service connections (includes active and nonbilled connections that remain in place)	551,959	0.15 gal/service/d/psi	55	1662.1	4.554
Service connections, curbstop to meter	551,959 X 12 ft./5280 ft/mile	7.5 gal/m/d/psi	55	188.9	0.517
	Unavoidable Annual Real Losses			2,185.2	6.0

Source: Philadelphia Water Department.

TABLE A.1.1   City of Philadelphia Annual Water Audit Summary—IWA/AWWA Water Audit Method Fiscal Year 2006—July 1, 2005, to June 30, 2006 (*Continued*) covered per year. An annual survey goal of 1300 pipeline miles is used, which translates to roughly one-third of the total mileage each year, or a total system survey interval of 3 years. The leak detection squad also consults to repair crews who have difficulty in pinpointing "reported leaks" assigned to them for location and repair. The leak detection squad utilizes leak correlators, leak noise loggers, and other leak noise sounding equipment to provide locations of leakage sources. All suspected leaks are then referred to various repair crews, or customers if leaks are determined to exist on private service connections. Costs to operate the 20-person leak detection squad are approximately $1million per year including personnel, vehicles, equipment, and training. The PWD employs in excess of one hundred other employees engaged in making routine leak repairs in the water distribution system. The costs to employ this staff are roughly $5 million per year, although only a portion of their workload is leak repair, as they also perform general maintenance and replacement work on valves and fire hydrants, install new connections, and provide a variety of support functions.

In addition to the large maintenance staff repairing several thousand main breaks and leaks each year, the PWD manages a significant capital program for infrastructure replacement, with an annual goal of replacing 25 mi of pipeline, much of which is over 100 years old. This rate of replacement represents approximately 0.8% of the city's total pipeline mileage renewed each year. The primary criterion used to designate sections of pipe for replacement is the recent break or burst rate of the pipeline, although it is believed that this criteria can be refined by including leakage and environmental data.

The LMA project was completed in 2001 and was successful in converting Philadelphia's water audit to the IWA/AWWA format and evaluating PWD leakage and distribution system asset management. The actual work of the project ran for approximately 3 months at a cost of $60,000 in consultant fees and $30,000 of PWD activity. The primary intention of the LMA was to gather information and critique the city's leakage management conditions with respect to the best practices being applied in water loss management throughout the worldwide water industry. Secondarily, the project provided opportunity for city personnel to become educated in the progressive leakage management methods in use internationally. The major steps of the project included

- Convert the Philadelphia Water Audit into the IWA/AWWA format.
- Obtain field measurements of water flow and pressure for night flow analysis in several test District Metered Areas.
- Provide two presentations and a workshop on the methods of progressive leakage management technology to Philadelphia stakeholders.
- Assess PWD's leakage control and distribution system management practices and offer recommendation for areas of improvement opportunity.

Flow measurements and studies were conducted on four temporary DMAs selected based upon varying distribution system attributes. The findings suggested that Philadelphia's water leakage does not occur homogeneously throughout its system, but instead is concentrated in certain areas. The LMA confirmed the feasibility of applying the DMA approach for continuous monitoring and nightflow analysis of leakage in Philadelphia. This led to PWD's decision to pilot a full-scale, permanent DMA as part of AWWARF Project 2928 starting in 2005.

Leak survey records were also evaluated as part of the assessment of Philadelphia's active leakage control practices. The LMA project consultants analyzed coverage frequency and records of survey (unreported) leaks located and repaired to determine whether the leakage survey frequency used by PWD is optimal. However, assessment of PWD's repair tracking system evidenced gaps in the routing of jobs to repair crews creating the potential that a certain portion of suspected leaks identified during leak surveys were not being repaired in timely fashion, or at all. Also, leak repair time suffers from the weak city policy that assigns repair responsibility for service connection leaks to the customer/owner.

The consultant team also made recommendations to refine PWD's capital program strategies for water system rehabilitation. Options for improvement could include replacing the entire length of service connection piping instead of just a portion, pressure management refinements to reduce main breaks by lowering average pressures and eliminating transients, and a strategy to reduce the total pipeline mileage in the distribution system by abandoning unnecessary piping. Finally, the PWD might better leverage its limited capital funding by incorporating the use of "trenchless technology" methods into its infrastructure management tool kit.

The LMA project was highly successful in providing to the PWD a clear understanding of its water loss standing relative to worldwide best practices. It also provided new tools and technology to proactively manage leakage to reduced levels.

Since the completion of the LMA project, the PWD has continued its application of new leakage control technology on several fronts. Most notably PWD served as one of 10 participating water utilities in the AWWA Research Foundation Project "Leakage Management Technologies" (Project 2928), which was completed in 2007. PWD was one of half of these utilities to construct full-scale leakage controls into their existing water distribution system. PWD created its first permanent DMA, labeled DMA5, in the Germantown neighborhood of the city, an area of older infrastructure, high pressure, and high leakage levels. A primary intention of this work was to demonstrate the feasibility of full scale DMA monitoring and advanced pressure management in the United States water utility industry. Figures A.1.2, A.1.3, and A.1.4 show photos of the primary supply equipment which includes piping, a pressure reducing valve, and flowmeter in chambers. Flow measurements taken as part of the planning phase for DMA5 found very high leakage as evidenced by a high minimum hour flow. This is shown in the graph in Fig. A.1.5 with a high leakage rate of 1.29 mgd or 639 gallons per service connection per day. The leakage assessment also found that approximately one-half of the leakage existing in DMA5 was background leakage indicative of infrastructure in poor condition. Background leakage cannot be detected sonically, but can be reduced by improved pressure management and/or pipeline rehabilitation. The long-term effects of the advanced pressure management capabilities designed into DMA5 will be observed carefully to evaluate the impact of this feature in reducing the relatively high level of background leakage in this area. Although PWD operates leak detection crews to seek unreported leakage, the survey frequency across the city service area is approximately once every 3 years. The high leakage rate initially uncovered in DMA5 occurred between leak survey events. With a DMA in place and flows continuously monitored, readings of high nighttime flows now prompt leak detection crews to conduct a survey immediately, addressing leakage before it reaches rampant levels as was found during the preliminary measurements. Once the DMA equipment was installed a standard leak survey was conducted and more than 10 unreported leaks, some quite large, were located and repaired. The graph in Fig. A.1.6 shows the reduced flow profile that resulted from the leakage reduction work; note leakage quantified at 0.23 mgd, or 114 gallons per service connection per day. Since this time, nighttime minimum flow

FIGURE A.1.2   Installation of pressure reducing valve and chamber for DMA5 in Philadelphia. (*Source:* Philadelphia Water Department.)

FIGURE A.1.3   The author, Julian Thornton, performs startup calibration of the electronic controller for Philadelphia's DMA5. (*Source:* Philadelphia Water Department.)

Figure A.1.4   Pressure reducing valve on primary supply water main for DMA5 in Philadelphia. (*Source*: Philadelphia Water Department.)

levels have been retained at the lower level observed in Fig. A.1.6. Efforts to drive leakage levels down further are continuing via additional leak surveys, pressure management, and select infrastructure replacement.

Figure A.1.6 also represents conditions with pressure management in place. In this graph a constant fixed outlet pressure scheme was employed. More recently, it has been determined that a flow modulated pressure control mode is the most appropriate scheme (see Chap. 18 on pressure management). The flow modulated scheme has advantages of providing slightly higher pressure during higher demand periods of the day such as peak morning periods. Perhaps more importantly, this scheme provides lower pressure at low demand periods of the day such as minimum night hours when customer consumption is low and leakage is at its greatest proportion of the DMA input flow. Further refinements of the pressure settings in the flow modulated scheme are expected in DMA5 and the long-term effect on background leakage and main breaks will be observed. PWD also envisions using DMA5 to pilot fixed network AMR technology to serve as a pilot for the envisioned full-scale fixed network AMR conversion of the entire service area several years in the future. (See Chap. 13 for information on fixed network AMR Systems). DMA5 is perhaps the most advanced use of new technology occurring in PWD in attempt to control leakage levels. So far, this technology has proven feasible in the Philadelphia distribution system and its ultimate benefit will be known once the system has been in operation for several years.

In addition to piloting of DMA technology, PWD also initiated use of an inline transmission main leak detection technology that has quickly proved to be very accurate and useful to detecting leaks in hard-to-access sections of large-diameter piping. In 2007 PWD initiated a contract with Pressure Pipe Inspection Company for use of its

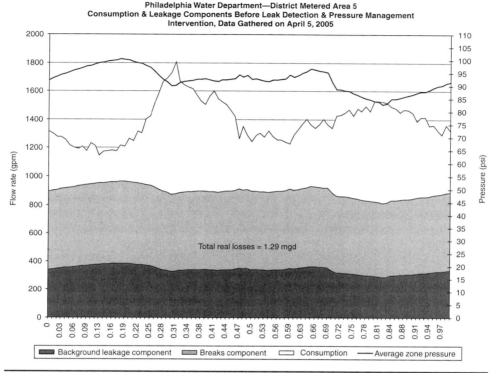

Philadelphia Water Department—District Metered Area 5
Consumption & Leakage Components Before Leak Detection & Pressure Management
Intervention, Data Gathered on April 5, 2005

Total real losses = 1.29 mgd

▨ Background leakage component   ▤ Breaks component   ▢ Consumption   — Average zone pressure

FIGURE A.1.5 Philadelphia Water Department—DMA5—consumption and leakage components before leak detection and pressure management intervention—Data Gathered on April 5, 2005. (*Source*: Philadelphia Water Department.)

SAHARA service. This technology inserts an electronic listening sensor into an active transmission main. The sensor is propelled by the flow of the water and listens for leak noises. Since it is inside the pipeline, extraneous noises are at a minimum. For a number of reasons, traditional above-ground leak detection on large diameter pipelines is more difficult than smaller distribution mains (see Chap. 16) and PWD has encountered the common limitations encountered on this type of pipe. However, in three rounds of scans the SAHARA system has pinpointed 18 unreported leaks on a total of 13.6 miles of various segments of important transmission piping. Two large leaks were quickly pinpointed on a 48-in diameter steel pipeline that traversed under an interstate highway. Numerous attempts and hundreds of hours of standard leak detection crew time had been previously expended but could not pinpoint this suspected leakage; but the SAHARA located the leaks within minutes of insertion into the pipeline. Scenes from this location are shown in Figs. A.1.7. and A.1.8. PWD looks forward to continued use of the SAHARA system in order to obtain a clear status of the condition of its transmission pipelines.

PWD's distribution system rehabilitation methods historically included an active phase of cleaning and cement lining of pipelines from 1949 through the mid-1980s, and water main replacement since the mid-1960s. PWD has opportunity to expand its rehabilitation options by investigating the array of "trenchless technologies" that have been developed in recent years. This technology is believed to have good potential for

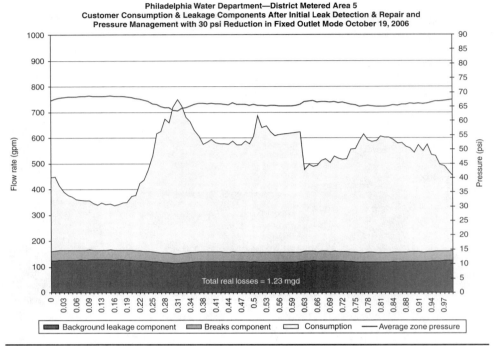

FIGURE A.1.6   Philadelphia Water Department—DMA5 Customer consumption and leakage components after initial leak detection and repair and pressure management with 30 psi reduction in fixed outlet mode. October 19, 2006. (*Source:* Philadelphia Water Department.)

FIGURE A.1.7   Bridge over interstate highway in Philadelphia; leaks exist on 48-in steel main traversing under this highway. (*Source:* Philadelphia Water Department.)

FIGURE A.1.8 Equipment setup for SAHARA inline transmission main leak detection service in Philadelphia on 48-in steel main traversing under interstate highway. (*Source:* Philadelphia Water Department.)

rehabilitation of PWD's large transmission pipelines. An initial project is planned to launch in 2008 using neopoxy to install a structural liner within a section of corroded 30-in diameter steel pipe. It is envisioned that other trenchless methods that provide a structural liner will be investigated. These technologies should provide the logistical advantage of rehabilitating large pipelines in difficult-to-access locations and minimize above-ground disruption to traffic, streets, and private property.

On a long-term perspective, PWD stands to gain most on leakage reduction from improved policy regarding customer service line leaks. The current policy, which places the entire burden to arrange leak repairs on the customer, is known to be ineffective as average repair time runs approximately 4 weeks. Many utilities have reduced this average repair time to 2 to 3 days, saving much lost water from known, pinpointed leaks. Changes in policy require consensus-building and diplomacy, thus time will be needed to enact a meaningful improvement in this regard.

PWD can also benefit from refinements in its work order tracking for leakage events. PWD is currently piloting a new work order management system and this is envisioned to present the opportunity to improve the accuracy and reliability of leakage tracking. The PWD will continue to improve its overall capabilities to stem leakage in its operation of perhaps the oldest water distribution system in the United States.

A.1.6 Addressing Apparent Losses

The joint efforts of the PWD and WRB have been successful in creating the largest water utility AMR system in the United States and one of the first Revenue Protection Programs in a water utility. The city is also poised to take another major leap forward in its accountability with the startup of a new computerized customer billing system in 2008.

Philadelphia's FY2006 water audit indicates that the city's apparent loss volume of 7905 million gal (21.7 mgd) is just over one-third of its real loss volume of 22,464 million gal (61.5 mgd). Remarkably however, apparent losses exert an impact of $30.8 million annually on the city due to lost revenue potential, compared to the real loss impact of $5.1 million largely as excess production costs. This stark difference occurs since apparent losses are valued at the retail cost charged to customers, which is a much higher unit cost than the variable production costs used to valuate real losses. The variable production cost and retail costs used by the PWD and WRB in the city's water audit are given in Table A.1.1. Since apparent losses represent service rendered without revenue recovered, these losses are usually highly cost-effective to recover since they occur in the fundamental operations of customer metering, meter reading, accounting, and billing.

Prior to 1997, the city was greatly hampered in establishing a reliable evaluation of its apparent losses. Although the customer population was essentially fully metered, the city's customer billed consumption existed with a suspected large degree of error since manual meter reading efforts were often unsuccessful. Customer meters exist within customer buildings for most accounts in Philadelphia. Modern lifestyles result in many residential properties being absent of any inhabitants during weekday business hours when manual meter reading was attempted. By the mid-1990s the WRB's meter reading success rate was in the mid 60%s and residential reads only the mid-30%s. With quarterly meter reading cycles in use with a monthly billing interval, only 1 out of every 7 water bills issued was based upon an actual customer meter reading. In addition to compromising the accuracy of customer water consumption data, estimated water bills resulted in frequent billing adjustments and high customer call/complaint volume.

Between 1997 and 1999 the city and its contractors successfully installed the largest water utility AMR system in the United States. With continued integration of AMR into all account types, over 487,000 AMR units are now deployed in the PWD service area. Reading is accomplished in mobile fashion, remotely via radio transmission to vans patrolling regular meter reading routes. The project included the use of city plumbing contractors who installed replacement customer meters manufactured by Badger Meter and meter reading devices and services from Itron. The initial 2-year installation phase and short-term subsequent period following installation witnessed significant billing adjustment activity since AMR was installed in many properties that had not been visited in years. Billed consumption actually dropped slightly during the initial transition to AMR due to this heightened adjustment activity. As the AMR system implementation has progressed to over 98% of customer accounts, its high read accuracy rate has created great confidence in customer billing data. The AMR system also includes tamper detection capabilities that have greatly helped to thwart unauthorized consumption by customers.

While employing AMR the PWD and WRB reorganized its metering and meter reading groups since manual meter reading has been greatly reduced to only a small group of hard-to-access properties. A "revenue protection" mission was added to the metering group, which now focuses on customer investigations as well as meter replacement and repair. Considerable resources are now devoted each year to investigate a large number of suspect accounts. Such accounts include chronic zero consumption accounts and nonbilled accounts. The latter represent customer accounts that have had billing suspended for one of a number of administrative reasons. The population of nonbilled accounts grew during the 1990s without close monitoring. Many of these supposed nonwater using accounts have been found to be actually consuming water, but without meter reading or billing. Accounts encountering multiple cycles of zero consumption have been investigated and up to 45% of these annual investigations were confirmed to be meter tampering by customers that accounted for the zero consumption readings. In the course of conducting its many investigations, the Revenue Protection Program has identified a number of gaps in the permitting, accounting and billing data handling procedures of the city which have since been corrected. These gaps allowed many accounts to remain improperly in unbilled status when they had actually returned to water-consuming status.

Lax accounting at the city's municipally owned buildings has also been a problem. Sometimes believed to be downplayed in importance since these accounts do not generate net revenue to the city, many municipal buildings have gone without water meters, meter maintenance, or meter reading. Many properties have escaped accounting in the city's billing system altogether, avoiding tracking of any water consumption. The city's largest water treatment plant and largest water consumer of more than 2 mgd was unmonitored for many years; but metering and billing have been established for this account. Several other plant and pumping facilities were found to lack meters, accounts, or both. The Revenue Protection Program is ensuring that billing accounts, water meters, and automatic meter reading exist for all customers. AMR routes and billing procedures are being improved to ensure that all accounts are monitored effectively. During its first 8 years of operation the city's Revenue Protection Program has recovered revenue totaling over $14 million. The success of this program has led to the recommendation to expand its resources and scope of work.

The PWD has also achieved success in one unusual source of lost water that has plagued older urban centers in the United States: fire hydrant abuse. With aboveground fire hydrants and large inner city populations, hydrants have often been opened illegally as a means of heat relief during hot summer periods. In addition to high water loss, these dangerous events have worked to draw down distribution system pressures below safe levels to fight fires and protect against backflow. The PWD has achieved success in checking this phenomenon by installing center compression locks (CCLs) on most of its fire hydrants. The device requires that a special adapter be used to open the hydrant by compressing an internal coil. The adapter must stay on the hydrant to keep it open. The adapter is removed to close the hydrant. Although some individuals have found ways to defeat the CCL, they can only open one hydrant at a time and usually oblige the PWD by closing the hydrant (removing their makeshift adapter) when finished. This results in much less lost water than the pre-CCL era when a single illegal wrench could be used inscrutably to open numerous hydrants, which usually remained running at length before PWD personnel arrived to close them.

The PWD and WRB look forward to the full implementation of a new customer billing system in 2008. The billing software package—Basis2, by Prophecy International Holdings, Ltd.—is being implemented via a city contract with Oracle, and is much anticipated to provide greatly improved functionality over the dated system that has existed. PWD utilized its consulting contract with WSO to conduct billing system data mining and analysis in the current system to reveal the extent of negative billing adjustments and other billing aberrations on the annual volume of billed consumption. This work, carried out on billing data from the fiscal year 2003 to 2006 period, pointed to a number of shortcomings in the structure of the dated billing system. Once the Basis2 billing system is integrated into normal operation, it is envisioned to provide capabilities that avoid many of the shortcomings of the former billing system. Implementation of the Basis2 customer billing system has been a long-term endeavor. It is anticipated to offer the PWD and WRB the same leap in billing capabilities that the AMR system provided in meter reading starting in 1999.

Reducing apparent losses is attractive since it offers high economic payback. In this way it "creates" previously uncaptured sources of funding and allows utilities to delay rate increases by equitably spreading costs among all users. The City of Philadelphia has achieved considerable progress in reducing apparent losses but, with over $30 million of such nonrevenue water still existing, much work remains.

A.1.7   Advancing Water Loss Control in the Water Industry

PWD has been active in a number of water industry trade organizations and partnerships with regulatory agencies. PWD personnel serve in a number of leadership roles in many of these relationships and have had an influential effect on the development of new policy and regulation in water resources management. PWD had a direct role in the execution of two recent AWWA Research Foundation projects on water loss and assisted in the development of the Free Water Audit Software published by the American Water Works Association. PWD has served as an advisor to several regulatory efforts including those in the state of Texas that now require water utilities to submit a period water audit. Similar efforts are envisioned in several states such as California, New Mexico, and Georgia. PWD will continue to be an active participant in the advancement of new technology and methods in its operations, as well as the water industry outreach needed to effect true change in our industry.

A.1.8   Philadelphia's Water Loss Future

The Philadelphia Water Department and Water Revenue Bureau have achieved national recognition as a leader in implementation of water loss control programs in its water supply system, and as a leader in promoting water efficiency to the water industry at large. Much success has been achieved in a 15-year effort to improve water accountability; but much work remains. The major focus areas for the city include

- Implementing the Basis2 Customer Billing System starting in 2008
- Pursuing fixed network Automatic Meter Reading capability in District Metered Area 5 as a pilot for the next generation of AMR in the city
- Expanding the Revenue Protection Program
- Improving work order tracking for leakage occurrences
- Continue to monitor DMA5 and implement further leakage control refinements to determine the lowest achievable levels of leakage possible for the PWD distribution system
- Pursuing development of an improved policy regarding the repair of customer service connection leaks

The PWD is moving on several fronts to address the above recommendations. In addition to major opportunities to make long-term reductions in leakage, the city will continue with its efforts to reduce apparent losses and recoup lost revenue. The combined savings of recovered real and apparent losses since 2000 are believed to be much greater than the cost of the effort expended. This is confirming the notion that water loss recovery is a cost-effective undertaking. Philadelphia has a long history of taking progressive action to better its level of service to its customers. Its work on water loss control is an important chapter in this history and one that stands to influence a greater understanding of water loss in North America and the need to control it.

Case Study A.2: Real Loss Reduction—Halifax Water, Halifax, Nova Scotia, Canada

Carl D. Yates, M.A.Sc., P.E. *General Manager, Halifax Water*

Graham MacDonald, *Halifax Water*

Tom Gorman, *Halifax Water*

A.2.1   Abstract

Halifax Water was the first utility in North America to adopt the IWA/AWWA methodology for real loss reduction in its distribution systems. By March 31, 2006, Halifax Water has reduced leakage in the Dartmouth system by 16 million L/d with a corresponding plant output reduction from 59 to 43 million L/d. In addition, Halifax Water tackled leakage within the Halifax distribution system and reduced system input by an additional 18 million L/d. The total leakage reduction of 34 million L/d represents annual savings of $550,000. In addition to direct savings, the customers of Halifax Water see increased public health protection (a leaking system has more potential for contamination) and reduced service disruption and property damage as leaks are now found in a proactive manner.

A.2.2   Background

In 1996, the Halifax Regional Water Commission (HRWC) was formed as part of the amalgamation of four municipal units to make up the Halifax Regional Municipality (HRM). The amalgamation brought immediate challenges and opportunities as the utility dealt with the pressing need to construct a new water treatment plant and transmission main in Dartmouth. The $60 million project was completed in 1998, on time and on budget. With the completion of this project, Halifax Water embarked on a continuous improvement program under the vision of becoming a world class utility. A priority that emerged for the utility was to reduce aggravated leakage in the distribution system. This was particularly important in the Dartmouth system where losses were in the order of 35% and the new plant produced the highest cost water in the region, predominantly due to the requirement to boost the water from the plant. A reduction in leakage would see immediate reduction in plant costs and deferral of capital costs associated with future upgrades to increase plant capacity. A cross departmental team was created to determine the best practice for water loss control. The investigation initially focused on North American efforts where the water profession was centered on the reduction of "unaccounted-for water" which was also the traditional approach followed by Halifax Water. Since this approach had obvious shortcomings, Halifax Water expanded its search and discovered an emerging methodology being promoted by the Water Loss Task Force of the International Water Association (IWA) which included a representative from the American Water Works Association (AWWA). The IWA/AWWA approach was holistic in nature but required a paradigm shift to implement. It was based on the concept of "accountability." Halifax Water put the methodology into action in 1999 and formally adopted it as a best practice in April, 2000.

A.2.3   Innovation and Excellence

The IWA/AWWA methodology for real loss reduction is all about accountability and an integrated approach to water loss control. The IWA/AWWA standard water balance and corresponding strategies were adopted by Halifax Water which required a change in thinking. It started with a ban on the term "unaccounted-for water" and a recognition that the standard water balance had a place for everything and everything in its place.

Four key strategies support the IWA/AWWA methodology, namely, active leak detection, pressure management, speed, and quality of repairs and asset management.

At Halifax Water, active leak detection encompasses noise mapping surveys of the system twice a year using acoustic equipment and digital noise correlation to supplement acoustic methods to pinpoint leaks. Leak detection activities are also supported by a Supervisory Control and Data Acquisition (SCADA) system which is utilized for flow trend analysis within each District Metered Area (DMA) of the distribution system. In this manner, leak crews can be sent to zones of the distribution system immediately when trends indicate active leakage.

Pressure management has been actively pursued to ensure pressure within the distribution system is optimized for customer service and kept at levels to minimize leakage. There are clear correlations between pressure and leakage, as identified in the concepts of fixed and variable area discharges paths (FAVAD) and component analysis of burst and background leakage estimates (BABE). Halifax Water has also explored the more advanced applications of pressure management whereby the pressure in the distribution system is intentionally reduced in the nighttime when water usage normally drops off with a corresponding pressure and leakage increase. Halifax Water has had initial success with flow modulated pressure control as part of AWWARF Project No. 2928 ("Leakage Management Technologies").

Speed and quality of repairs are centered on the reduction of leakage run times. Accordingly, speed of repairs in this context does not solely mean the actual repair of the leak itself. There are three components that make up the leakage run time: the awareness time of the leak, the location time for pinpointing, and the actual repair time. In some utilities where leak surveys are only carried out once every 2 years, the average leak will have been active for 1 year. Even a small service leak can add up over a 1 year period.

Asset management is more of a long-term strategy, but an important one. Funds should be set aside to replace or rehabilitate aging and leak prone mains on a regular basis. The HRWC has a proactive main renewal program with funding through dedicated depreciation reserves and capital from operating revenue. The establishment of depreciation as an operating expense is by itself being recognized as a best practice and in all likelihood will be incorporated with the implementation of Bill 175 in Ontario. In addition to pipes, another important asset to install and maintain is meters. HRWC has

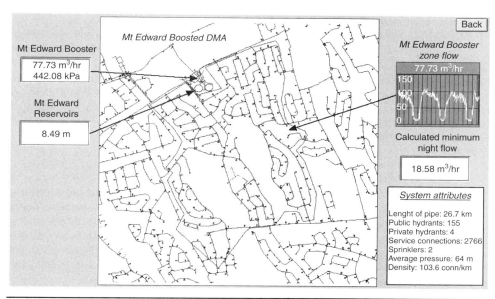

FIGURE A.2.1   Mount Edward DMA, Dartmouth, Nova Scotia. (*Source:* HRWC)

universal metering for monitoring customer usage and a fleet of master meters within the DMAs of the distribution system.

All of these strategies make up a holistic approach to water loss reduction but it is worthy to comment on the importance of DMAs and SCADA. Halifax Water has over 65 DMAs and a robust SCADA system. These tools are used in tandem for night flow analysis for leakage assessment and to determine best achievable benchmarks in system flows. A typical DMA (see Fig. A.2.1] incorporates a zone in the distribution system with a maximum pipe length of 30 km or approximately 2500 customer connections. Some zones can be smaller if there is a discreet elevation boundary or the zone is boosted. If DMAs are not established, finding a leak is like finding a needle in a haystack. The basic purpose of the DMA establishment is to break up the haystack into smaller ones and use the SCADA system to tell you which one has the needle (leak).

Night flow analysis is important to determine how "low you can go" with real loss reductions. Technical staff calculates the nighttime use of residential customers, measures the exceptional commercial/industry usage, and compares it with flows recorded through the SCADA system to determine the active leakage in the system. Efforts can then be zeroed in on zones where active leakage intervention will give the biggest return, that is, "bang for the buck." The utility's investment in leakage control can be measured in terms of recaptured water and corresponding value of the water. This economic assessment should influence a utility's decision to either increase or reduce leak detection activities in a particular zone of the distribution system.

In accordance with the IWA/AWWA methodology, the overall assessment to measure performance is the infrastructure leakage index (ILI). The ILI is the ratio of real system losses to the unavoidable system losses. Real losses are derived from the IWA/AWWA standard water balance, a calculated volume, and unavoidable losses are derived from an established empirical database. Unavoidable losses are related to the length of piping in the public system, the density of service connections, and normal system operating pressure. It is logical that a system with higher service connection densities and higher water pressure is assigned higher unavoidable losses. The benefit of using the ILI as a performance indicator is that utilities can measure themselves against any other utility in the world. The old way of comparison based on "unaccounted-for water" was inconsistent and subjective without a standard approach and terminology. A new way has emerged. In 2003, AWWA and the Canadian NRC InfraGuide recognized the IWA/AWWA methodology as best practice, 3 years after it was formally adopted by Halifax Water.

A.2.4 Implementation, Results, and Lessons Learned

Adoption of the IWA/AWWA methodology for water loss control was carried out with Halifax Water's vision of becoming a world class utility. To start the initiative, a steering committee was formed with representation from all departments, namely, distribution system operations, engineering, plant operations, and finance and customer service. Interdepartmental cooperation can sometimes be a double-edged sword and many initiatives can get stalled due to the extra coordination required. When cross department initiatives go well, however, they can produce breakthrough results. Such is the case with the water accountability venture put forward by Halifax Water.

With the operations department playing a leadership role, and the support of senior management, staff conducted an international search to find the best practice for water loss control. This search took them to water professionals working with the Water Loss Task Force of the IWA. The Water Loss Task Force was given a mandate to develop a world class methodology and strategies for leakage reduction. In 2000, the task force completed the project with the standard water balance and strategies as we know them today.

In 1999, HRWC hired an international expert associated with the IWA/AWWA methodology to ensure staff understood the loss reduction strategies and documentation of inputs to the standard water balance. Over 50 employees of HRWC were exposed to the methodology with operations staff receiving advanced training with a standing order for annual workshops to keep abreast of leading edge applications. The engineering department played a strong supporting role to operations with the development of drawings for regular noise mapping of the distribution system. In addition, engineering used the corporate geographic information system (GIS) to assist with DMA design. Several areas of the distribution system were transformed to incorporate DMA principles.

Halifax Water has also embarked on a project to monitor flows to large (high consumption) customers in real time through the SCADA system in support of water loss strategies. This was a mutually beneficial installation as the customer knows when they have aggravated leakage, and the utility doesn't send crews out to look for false leaks in the distribution system. Halifax Water notifies the customer of large increases in flow and the customer hires a work crew to find and fix a leak if one is identified.

The success of the real loss reduction program is well documented. The performance of the program is measured by the reduction in ILI, which fell from 9.0 in 1998 to 3.0 as of March 31, 2006 (see Fig. A.2.2]. The ILI is reported on a quarterly basis as a rolling annual measurement. The total real losses recovered by Halifax Water amount to 34 million L/d with system inputs reduced from 168 to 134 million L/d which represents annual savings of $550,000.

Although it is recognized that an ILI of 1.0 is attainable from a theoretical viewpoint, many utilities have challenged themselves to demonstrate economic viability. In

HRWC Regions results	ILI 1997/98	ILI 1999/00 *	ILI 2000/01	ILI 2001/02	ILI 2002/03	ILI 2003/04	ILI 2004/05	ILI 2005/06
Central	NA	1.6	1.2	1.0	1.0	1.5	1.1	1.0
East	NA	4.4	4.5	2.9	3.1	2.4	2.4	2.0
West	NA	11.7	11.7	11.5	9.2	7.3	6.9	5.3
Corporate	9.0	6.4	6.3	5.5	4.7	4.0	3.8	3.0

* Formal adoption of IWA/AWWA methodology.

Figure A.2.2   Regional ILI performance results. (*Source:* HRWC)

other words, a utility should not spend more than a dollar to save a dollar. Halifax Water is no different and based on research carried out by AWWARF; the utility is very close to the economic level of leakage.

In addition to direct economic benefits associated with leakage reduction in the distribution system, other direct and indirect benefits are realized. A reduction in system inputs allows for the deferral of capital investment if plant capacity needs to be increased to match future demand. Since the production and distribution of drinking water is energy intensive, other indirect benefits include reduction of greenhouse gases. When it comes to promoting water conservation, it is also easier to get buy in from customers to reduce if a utility can demonstrate it is doing everything that it can to control leakage.

There are also good service and social reasons to reduce water leakage proactively. Since the vast majority of leaks are found early using the IWA/AWWA methodology, they can be repaired under controlled conditions to minimize service disruption and property damage to adjacent properties. Adoption of the IWA/AWWA methodology can also help minimize the liability of the water utility from damage claims as it demonstrates a commitment to best practice in water loss control.

Last but not least, it should be recognized that water utilities are in the public health protection business. A distribution system with aggravated leakage is much more prone to contamination, in recognition that water and sewer pipes often share a common trench.

A.2.5   Project Sustainability and Policy Framework

The water accountability program of Halifax Water directly supports its strategic plan and the sustainability goals of its parent organization, the Halifax Regional Municipality (HRM). Halifax Water has utilized a balanced corporate scorecard to measure the performance of its strategic plan which places an emphasis on stewardship of the environment and infrastructure. One of the key scorecard performance indicators to measure success and establish objectives is the ILI which is the key benchmark associated with the IWA/AWWA methodology.

HRM has established sustainability goals with the development of its own corporate scorecard. One of the themes of HRM's scorecard is preservation of the environment with ties to the Halifax Water scorecard through the ILI measurement. The adoption of the IWA/AWWA methodology by Halifax Water also directly supports HRM's objective to reduce greenhouse gases since reduced water system inputs mean there are less chemicals and energy used at water treatment plants.

Reducing leakage in the distribution system is like doing the laundry; it is never done. In this regard, Halifax Water is committed to the IWA/AWWA methodology for the long term and expects to make further inroads in water loss reduction. The goal of Halifax Water is to get to its economic level of leakage, which correlates to an ILI of approximately 2.5. This represents a further leakage reduction of 2 million L/d within the distribution network.

The IWA/AWWA methodology for water loss reduction is expected to continue indefinitely at Halifax Water since all departments have bought in and breakthrough results have already been attained. These breakthrough results reflect an integrated approach to a significant problem and have strengthened interdepartmental relationships. The holistic approach of the IWA/AWWA methodology to water loss reduction is like the multiple barrier approach to maintain water quality, which is also paramount to Halifax Water.

Halifax Water has received national and international recognition for its water accountability. In June, 2005 Halifax Water was awarded the Sustainable Community Award in the water category through the Federation of Canadian Municipalities for its approach to water loss control. In September, 2005, Halifax Water hosted the IWA Leakage 2005 conference and has participated in research with AWWARF and the National Research Council of Canada.

Case Study A.3: Water Loss Control Program—Metro Water Services, Nashville, Tennessee

Leanne B. Scott, P.E., *METRO Water Services,*
1600 Second Avenue North, Nashville, TN 37208

Paul V. Johnson, P.E., *WSO, 102 Space Park South Drive,*
Nashville, TN 37211

A.3.1 Background

Beginning in 2002 (FY 2001/2002), Metro Water Services (MWS) of Nashville, Tennessee, contracted with WSO to perform a series of three water audits for Fiscal Year 2001/2002, 2002/2003, and 2003/2004. Prior to this, MWS had attempted to control losses through various leak detection programs, both in-house and contractor supplied, with varying degrees of success. Several discussions about possibly privatizing MWS led to a decision to significantly improve the control and management of water losses. In 2002, MWS decided that the standardized IWA/AWWA water audit methodologies provided them with a different tool to accurately assess and control their water losses in place of the methodologies they had been using. The attractive difference in the IWA/AWWA methodologies was that it not only allowed MWS to analyze their system but enabled them to design and implement the most appropriate intervention strategy against water loss for their system.

The data for the MWS distribution system is as follows:

MWS—Metropolitan Nashville and Davidson County, Davidson County, Tennessee

Average daily system input, mgd   92

	Year 1	Year 2	Year 3
UARL (MGD)	2.877	2.878	3.620
CARL (MGD)	15.058	18.785	22.792
ILI	5.23	6.53	6.30
NRW*	26.3%	32.6%	28%

% of system input by volume
(*Source*: MWS)

TABLE A.3.1   Comparison of Water Loss Performance Indicators

Omohundro and K.R. Harrington Water Treatment Plants, 100% surface water

Miles of Main:	2888
Residential Services:	157,006
Commercial Services:	14,621
Fire Hydrants:	19,511
Valves:	60,040
Reservoirs:	44
Total Reservoir capacity, million gal:	93.5
Average System Pressure, psi:	74.8
Marginal Cost of water, $/1000 gal:	$0.277
Average Retail Cost, $/1000 gal:	$6.39
Population:	500,000+

MWS utilizes the Cumberland River for source water. With the two water treatment plants and the current source, capacity and supply are not huge problems for MWS. Even with the current 3 year drought being experienced by Nashville, water supply has not been a significant problem. The main driver behind the water audits and subsequent leak detection programs were to improve the operating efficiency of the MWS distribution system.

The IWA/AWWA water audits undertaken were extensive and to a high level of detail. Validation of key water audit components was undertaken during the consecutive audits improving the overall confidence related to the audit results. Table A.3.1 provides some of the performance indicators determined during the three consecutive audits.

In the next section, we will go into more detail about the audits, the results, the reasons why the performance indicators changed from Year 1 to Year 3 and the decision to implement the leakage detection program.

A.3.2   Water Audit

The first water audit was very thorough and concentrated on producing a water balance and developing a first look at the distribution system. The first audit highlighted that several key components of the water audit (notably the source input volume) had

2003	C.L. Real Losses	±68.4%
2004	C.L. Real Losses	±9.1%
2005	C.L. Real Losses	±4.9%

(*Source:* MWS)

TABLE A.3.2   Improvement of Confidence
Level Related to Calculated Real Loss Volume

a low level of confidence due to the difficulty in accurately testing the source meters. These confidence limits needed to be raised to achieve the desired confidence in the audit results. Even with the confidence level being not as good as desired, there was clearly a business case for designing and implementing a leak detection program. The potential impact of the low confidence levels in the audit was tested through sensitivity analysis to make sure the decision to start a leak detection program would still be economically justifiable once the confidence levels in the audit results were improved. The real losses were significant enough that a leak detection program was called for to start reducing those losses.

The second water audit was done to follow up the initial audit and to improve the reliability (confidence limits) of the initial audit through testing of the master meters or system input meters. An important component of the second audit was the determination of the economic level of leakage (ELL) for the MWS distribution system. Field measurements, including setting up sample District Measurement Areas (DMAs) for minimum night flow (MNF) measurements and analysis, were done to verify the audit results. Furthermore, a real loss component analysis was developed to further refine the determination of the real losses in the system.

The third audit focused on further improvement of the reliability of the data used for analysis of the apparent losses through a vigorous analysis of the small meters. This analysis focused on the data derived from over 1500 domestic meters that were tested for accuracy and showed that, overall, the domestic meters were in better condition than expected with the exception of one small group of meters that was much worse than average and that stopped meters were accounting for more of the apparent losses than the rest of the domestic meter inaccuracy combined.

From looking at the Table A.3.1, it would appear that the performance indicators (PIs) for the system deteriorated over the 3 years the audits were performed. That is not the case. In reality the reliability of the data improved, as shown below, to the point where the case for performing a leak detection program is much more defensible than it would have been with only the original data.

It was much easier to defend the decision based on the real losses being between 7910.4 million gal and 8725.6 million gal (Year 3 Real Losses ±4.9%) rather than real losses being between 2166.6 million gal and 11,546.4 million gal (Year 1 Real Losses ±68.4%).

A.3.3   Development of an Intervention Strategy

The recommendation from the first water audit was to perform a leak detection program on the MWS distribution system, even though the confidence limits for the Real Losses were not as good as desired. Results from Year 2 and Year 3 reinforced the correctness of the decision made from Year 1. Various options were reviewed including

- In-house versus consultant
- Noise loggers versus manual sweeps
- Annual detection on the whole system versus half of the system

To perform the leak detection program in-house would require expenditures for equipment and training, in addition to the overhead cost of hiring more personnel, while hiring a consultant would put that burden on the consultant. Noise loggers versus manual sweeps came down to a decision to go with qualified people rather than depending entirely on equipment and the decision to do a comprehensive investigation rather than just a "quick, overall" check.

While it was determined that leak detection on the whole system on an annual basis was optimal in reducing real losses, the economics of the situation required a slightly scaled back version of the program. The leak detection program was designed to cover approximately half of the distribution system each year and was also designed to be a dynamic program to be able to change with the needs of the program as it progressed. The initial term of the leak detection program was to cover 5 years, three of which had been completed at the time of the writing of this case study.

A.3.4   Intervention against Real Losses

The MWS Leak Detection Program was set up to be a combination of temporary DMA measurements and leak detection to reduce the recoverable leakage in the DMAs. Nashville had the advantage of having approximately 73 districts already designed from former Pitot zone measurements. These districts were defined either by natural boundaries or by valve closures with designated inlets and storage where required. To measure the DMAs was simply a matter of measuring the flows at the inlet and ensuring the boundary valves were closed.

In Year 1 of the Leak Detection Program, MWS prioritized the districts in the distribution system based on past experience with locating and repairing leakage. Fifty DMAs were measured with 31 DMAs surveyed for leakage during Year 1 and 13 DMAs carried over and surveyed for leakage detection in Year 2 because of a lack of time to check them during Year 1. Six DMAs were determined to have levels of recoverable leakage low enough to not be economical to survey for leakage. Approximately 878.32 mi of main were surveyed in Year 1 (including 65.20 mi of transmission mains) with approximately 260 leaks located including 30 leaks of 50+ gpm. Total leakage located in Year 1 was estimated at 4367 gpm or 6.29 mgd. The survey method used for leak detection was to check all services for leakage as well as the distribution system around the services. The survey started with a general sweep where the services were sounded and then a follow-up investigation where leak indications were heard to pinpoint leaks.

Initial plans were to measure the DMAs, locate the leaks, have the leaks repaired, and then remeasure the DMAs to determine how much of the recoverable leakage had been removed. Leak repair scheduling problems within MWS did not allow this plan to work, so the remeasure of the DMAs was dropped in favor of moving ahead to areas that initially had been planned for later in the project.

Year 2 of the Leak Detection Program included the measuring of 48 DMAs, 23 that were measured for the first time and 25 DMAs that were repeated from Year 1. There were 45 DMAs surveyed for leakage during Year 2 (including the 13 DMAs carried over from Year 1) with 16 DMAs not surveyed for leakage due to low recoverable leakage levels. Approximately 1352 mi of main (including 50.70 mi of transmission mains) were checked for leakage in Year 2. During Year 2361 leaks were located with 28 of those leaks being larger than 50 gpm. A total of 3666 gpm or 5.28 mgd in leakage was located during Year 2 of the Leak Detection Program.

Year 3 of the leak detection program included the measurement of 51 DMAs and surveying 36 of those DMAs for leakage, with 15 DMAs not being surveyed for leaks. Total nonrevenue water located during Year 3 amounted to 8341 gpm or 12.01 mgd from 622 leaks in 1302.18 mi of main surveyed (no transmission mains we surveyed in Year 3). During Year 3, there were 34 leaks found larger than 50 gpm.

Year 3 had some interesting differences when compared to Year 1 and Year 2. Year 3 had over 300 more meter leaks than Year 1 or Year 2, mainly attributable to three DMAs that only shortly before the DMA measurements were switched to AMR meters, many of which were found to be leaking. The other outstanding item is the 8341 gpm of nonrevenue water located in Year 3 which included a 6-in blow-off (approximately 4000 gpm) on an 18-in main that had been opened. While this was not technically a leak, it definitely was nonrevenue water (NRW) that would not have been located as quickly if MWS/WSO had not been performing DMA measurements.

The temporary DMAs are a powerful tool to prioritize the leak detection efforts in those areas where the biggest benefit is derived from the effort and money spent on leak detection.

A.3.5   Conclusions and Lessons Learned

MWS/WSO has learned that leak detection for the MWS system has to be an on-going program due to the reoccurrence of leakage in the system. Some DMAs have been surveyed three times in this program with essentially the same amount of leakage being found each time. In addition, the same number of large main breaks is found each year indicating that large leaks are reoccurring after existing leaks have been repaired. This reoccurrence of leakage can be attributed to a number of causes: changes in temperature, high pressure in the system, large areas of rock throughout the distribution system, and cycles of pressure due to pump operations in the system.

DMA measurements have clearly proven to be a good tool for identifying and prioritizing areas for leak detection. The blow off was found open while trying to measure the DMA in which it was located.

One aspect of DMA measurements that is currently being refined is the accounting for exceptional night users (ENU) in the DMA. The MWS system was not originally designed to facilitate identifying these ENUs in each DMA. WSO has been working to refine the identification of these users and improve the information available from the DMA measurements. The ENUs are being identified by reviewing the entire large meter database from WMS and placing the large meters in their respective DMAs so that the ENUs can be monitored when the DMA is measured.

Case Study A.4: Italian Case Study in Applying the IWA WLTF Approach: Results Obtained

Marco Fantozzi Eng.* Gussago (BS), *Italy*

A.4.1   Abstract

In Italy, nonrevenue water (NRW) levels range from 15 to 60% of total system input volumes. Passive control is the approach most widely used for leakage control. The water systems often lack the necessary maintenance and rehabilitation to provide an adequate service to the customers. Therefore, minimization of losses in the network is a key requirement in Italy. The paper describes ongoing initiatives in Italy to promote the application of international best practice and measurements in water loss management. Specifically, to help improve the management of water losses in the Italian water Industry, a Water Loss Group has been created. The paper also includes case study material from the author's own experience and from that of colleagues in the Italian water industry. Speaking of the future in Italy regarding water loss management following 18 months of activities of the Italian Water Loss Group, we can say that now many Italian utilities have the practical methods and the tools for achieving this important goal.

A.4.2   Introduction

In Italy, nonrevenue water (NRW) levels range from 15 to 60% of total system input volumes, the average being 42% (ISTAT 2003). Some European countries—notably the United Kingdom and Malta—have fully sectorized distribution networks, with continuous night flow measurements, and frequent interventions to locate unreported leaks. In Italy however, the majority of water utilities only repair "reported" leaks, and do not practice any regular form of active leakage control or pressure management, except perhaps as an emergency response during droughts.

In an effort to stem these losses, the Decree no 99/97 (DISPOSIZIONI IN MATERIA DI RISORSE IDRICHE: Decreto Ministero Lavori Pubblici n°99 del 8.1.1997) regarding water balance calculations was issued on January 8, 1997. According to the Decree, Italian water utilities are required to calculate the water balance for all their networks and to report on water loss from each network.

Italian Decree 99/97 introduced some important recommendations regarding pressure and flow measurement. However, it lost the opportunity to give the Italian utilities a practical tool for developing a strategy for management of NRW based on a better understanding of the reasons for NRW, and the factors which influence its components. In Italian Decree 99/97 water losses, as well as NRW and leakage, are still quoted as a percent by volume of system input (or water production). This indicator is unreliable for benchmarking the operational management of real losses as it is so strongly influenced by consumption, and changes in consumption. Over 25 years ago the UK National

---

* Marco.fantozzi@email.it, www. studiomarcofantozzi.it, www.leakssuite.com.

Water Council (Report 26) had identified this problem, as did the German DVGW in 1986. More recently, the American Water Works Association, as well as national organizations in a number of countries, and the World Bank Institute, are no longer recommending the use of percentages for this purpose.

Following the publication of the IWA "Best Practice" Performance Indicators,[1] it is now recognized that liters per service connection per day, and infrastructure leakage index (ILI)— the ratio of current annual real losses to unavoidable annual real losses— are the preferred means of comparing leakage management performance in most systems. The large range of connection densities (per kilometer of mains) experienced in Italy makes m^3/km of mains/day unsuitable for this purpose, according to the IWA International Report on Water Loss Management and Techniques.[1]

A.4.3   Ongoing Initiatives in Italy

Minimization of losses in the network is a key requirement in Italy, as water loss levels are very high for a developed country. In an effort to better manage water loss from water networks, Italian regulators are looking at new legislative measures to make water utilities report their water loss. With these moves underway, there is an urgent need for water managers to gather information and to use tools for implementing such requirements.

Therefore, to promote the application of international best practice and measurements in water loss management, and, more generally, to help improve the management of water losses in the Italian water industry, two organizations created a Water Loss Group—GOA—an acronym which stands for "Gruppo Ottimizzazione Acquedotti." They are: Fondazione AMGA—a member-supported, nonprofit organization that sponsors research to enable water utilities, public health agencies, and other professionals to provide safe and affordable drinking water to consumers— and Feder Utility, the organization representing 400 water and gas utilities which supply water to around 36 million people in Italy. The activities of the group, which has already gathered more than 80 members from Italian utilities, universities, and water institutions, began officially on 25 October 2004 in Genoa (Italy) during the FederUtility Workshop "Towards More Effective Management of Water Losses in Distribution Systems." The Water Loss Group is a vehicle for:

- Increasing water utility awareness of the importance and economic benefits of improved management of pressure to reduce new burst frequencies and leak flow rates

- Acting as a national center for promoting International Water Association (IWA) specialist information to the Italian water industry

- Disseminating the practical approach developed by the IWA Water Loss Task Force to a wide number of potential end-users and to obtain their feedback

- Communicating available methodologies and innovative techniques for efficient water loss management, allowing end users to make contact with each other and exchange ideas and experiences

The crucial issue was the general acceptance of the approach. Moving forward, Fondazione AMGA has initiated a series of training workshops to further extend the application of the methodology in Italy. So far more than 180 technicians from utilities and regulatory bodies from all over Italy have been trained in the practical application of the methodology.

The Italian water industry itself views the IWA methodology in a very positive light. The important thing is that most advanced utilities are already on board and that, on the regulatory side, Emilia Romagna region has already issued a new Guideline on Water Balance Calculation, which introduces several concepts and key performance indicators from the IWA practical approach. The ILI— the performance indicator which measures how effectively real losses are being managed at current operating pressure— has been given particular attention.

Many utilities have started using the IWA water balance and key IWA performance indicators, using a specialized software developed to take into account the requirements of the Italian Decree no 99/97. This allows export of data from the Italian water balance to build the IWA water balance. This software, based on PIFastCalcs, has been made available through the Italian Water Loss Group. PIFastCalcs is based on the standard water balance methodology and IWA Performance Indicators recommended by the IWA. The performance indicators are calculated and compared with the values from an international data set used in the "AQUA" December 1999 paper,[2] and an initial European data set. PIFastCalcs is part of the LEAKS (Leakage Evaluation and Assessment Know-How Software) suite of softwares using best practice methods promoted by IWA Water Losses Task Forces. Further information is available at www.studiomarcofantozzi.it or at www.leakssuite.com .

Most Italian utilities do not have an active leakage control (ALC) program and budget, and are unaware of the extent to which the annual volume of their real (physical) losses could be significantly reduced by limiting the average run times of unreported leaks. After being trained on how night flow measurements can assist in the timing of individual ALC interventions, some Italian utilities have successfully applied a new methodology to predict economic frequency of ALC intervention for their systems, and to calculate an annual budget for economic ALC (excluding repair costs) and the economic level of unreported real losses.[4]

Italian utilities have also begun to understand that effective management of distribution system pressures is the foundation of any successful and economic policy for leakage management. One example of this is the pressure management scheme in Torino, reported at the October 2004 Fondazione AMGA Genoa Workshop. In this case, the installation of a well-placed booster station resulted in a 10% reduction in night pressures (and average pressures) over a major part of the city, and has resulted in a sustained reduction of around 50% in annual repair costs, as well as a reduction in real losses. The presentation of this scheme at the October 2004 wrkshop, coupled with an explanation of the evolving theories of pressure—burst frequency relationships, and international examples of burst reduction by pressure management—stimulated three other utilities to successfully attempt pressure management schemes. These were reported at the April 2005 Genoa Workshop. As a result of such Italian success stories, supported by the expertise and commitment of the Water Loss Task Force, and the efforts of the Italian Group, it is hoped that real progress in reducing water losses, which are clearly far too high at present, will be achieved in Italy.

Good flow metering also helps the utility operator to understand NRW. Customer metering is common in Italy but, as in most countries, not enough attention is given by utilities to ensure that the meters have an acceptable performance over their lifespan. Apart from a few advanced cases, the replacement policy is not actually based on economic aspects, and, in general, utilities do not consider that meter performance decreases over time.

A.4.4   Some Italian Case Studies

The following case studies clearly demonstrate that through the practical application of advanced methodologies, a significant improvement in the efficiency of Italian distribution systems is not only feasible, but that Italian experiences can also provide examples to encourage other countries to improve their performance.

DMA Management at Enia, Reggio Emilia (Italy)

The strategy selected by Enia Reggio, Emilia (a multiservice company at the forefront in Italy in water loss management operating in Emilia Romagna region) in order to address and reduce water leakage within their water distribution systems was to implement District Metered Areas (DMAs) and Pressure Management Areas (PMAs) (see Fig. A.4.1).

The leakage reduction program has been implemented so far in 3524 km of the network, representing 75% of the total length of the network (4700 km), where the distribution systems were divided into around 100 DMAs. DMA creation also allowed a more efficient pressure management with a 20% reduction of average system pressure from 50 m to 40 m. This methodology is relatively new in Italy, but is recognized being both appropriate and effective.

To enable efficient control of recoverable losses, DMAs are being used both to identify and reduce recoverable leakage in the short term and then to monitor and control leakage in an ongoing manner.

A sensitive flow measurement device is permanently installed onto the inlet pipes to each DMA and flow and pressure profiles are recorded using data loggers. These

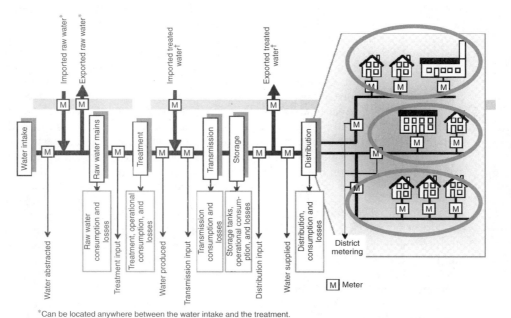

*Can be located anywhere between the water intake and the treatment.
†Can be located anywhere downstream treatment.

Figure A.4.1   District Metered Areas. (*Source*: Italian Case Study in Applying the IWA WLTF Approach: Results Obtained)

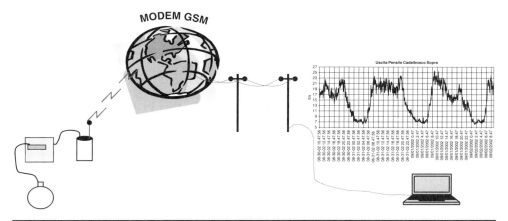

Figure A.4.2   Monitoring system. (*Source*: Italian Case Study in Applying the IWA WLTF Approach: Results Obtained)

profiles are transmitted via GSM to a personal computer in the Enia control room (see Figs. A.4.2, A.4.3, and A.4. 4) and allow real time monitoring of each DMA.

For each DMA, minimum night flow (MNF) profiles are analyzed, in conjunction with pressure profiles recorded by other pressure loggers strategically placed inside the DMA at the average zone point (AZP) and at the critical point (CP), to identify where an intervention with active leakage control is economically justified. This methodology allows Enia engineers to prioritize areas of high leakage and to quantify the rate of rise

Figure A.4.3   Measurement points.(*Source*: Italian Case Study in Applying the IWA WLTF Approach: Results Obtained)

Figure A.4.4 Measurement points. (*Source*: Italian Case Study in Applying the IWA WLTF Approach: Results Obtained)

of unreported leaks to be used in payback calculation and in calculation of intervention frequency with active leakage control.

After high leakage areas are identified and leakage volume is quantified, the individual leaks are located by acoustic detectors (leak noise correlators, geophones, and noise loggers). Once the DMA has been cleared of detectable leaks, a pressure-dependent baseline flow is determined and the area is monitored to identify when leakage starts to develop again.

Enia selected the engineering firm of Marco Fantozzi to implement a methodology to analyze minimum night flow (MNF) profiles and compare real losses calculated from night flows and water balance, using a special software (named StiperzEnia) developed by Allan Lambert and translated into Italian. In Fig. A.4.5 you can see Real Losses calculated from Night Flows and Water Balance, Best Estimate, and comparison with DMA specific unavoidable annual real losses (UARL) for a single DMA managed by Enia. The assessed reduction in night real losses, from 93 to 45 m^3/d is clearly evident.

At Enia, there is now an automated process that determines the average flow rate for each DMA between 3 and 4 a.m. Each morning it is possible for each DMA to compare the average night flow rates with established benchmarks and calculate the difference between the benchmark and the most recent night flows. DMAs with high night flows can then be analyzed in detail to reveal burst time and flow rate are quickly identified.

StiperzEnia software allows Enia to better estimate Real Losses in each DMA, improving overall management of DMAs and economic intervention calculation with active leakage control.

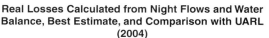

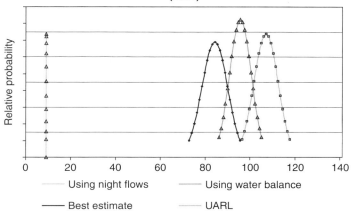

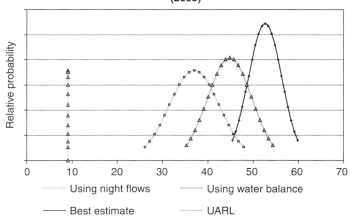

FIGURE A.4.5   StiperzEnia Software comparing real losses calculated from night flows and water balance for a DMA in 2004 and 2005. (*Source:* Italian Case Study in Applying the IWA WLTF Approach: Results Obtained)

The DMA program, in the last four years (from 2001 to 2005) has obtained a 13% reduction of the per capita daily inflow and a 28% reduction in the number of repairs, mainly due to pressure reduction, as shown in Figs. A.4.6 and A.4.7.

Economic Frequency of Active Leakage Control at a Small network in Northern Italy

In the remainder of the paper, a recent example applying a simple method for calculating these parameters on a system-specific basis is presented, based on sonic leak detection by regular survey (sounding fittings, noise loggers, or similar techniques).

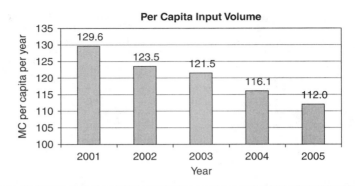

FIGURE A.4.6   Reduction in per capita input volume.

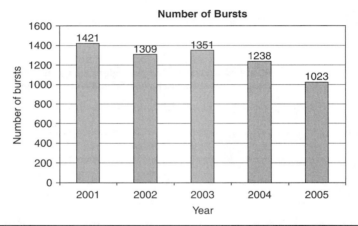

FIGURE A.4.7   Reduction in number of bursts. (*Source*: Italian Case Study in Applying the IWA WLTF Approach: Results Obtained)

This approach, with enhancements such as 95% confidence limits for all parameters and predictions, has now been incorporated into software for calculations of various economic leakage levels. Versions of this software (ALCCalcs) are also now available in Australia, Canada, Croatia, Cyprus, UK, and the USA.

A stated key objective of the present Water Losses Task Force is 'to develop a quick and practical method for calculating economic intervention (for active leakage control to locate unreported leaks and bursts), and short-run economic leakage level (SRELL).

Clearly, there is little point in attempting to calculate, or to achieve, an economic level of real losses for a particular system, unless the utility commits to undertake (to an appropriate extent) all four components of real losses management (speed and quality of repairs, pressure management, active leakage control, and rehabilitation). Pending the development of a method for calculating economic leakage levels, a practical approach successfully used by utilities such as Malta Water Services Corporation, and Halifax Regional Water Council (Canada) has been to identify and implement a mixture of initiatives within the four components that individually have the highest benefit: cost ratio, or shortest payback period. When no further economically viable initiatives can be identified, it can be reasonably assumed that an economic leakage level has been

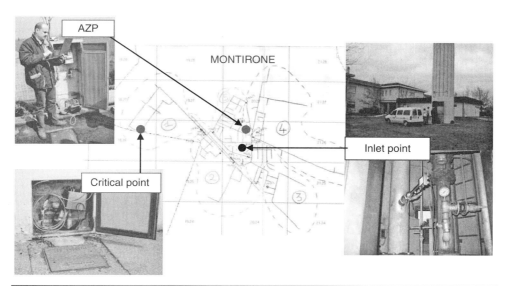

FIGURE A.4.8   Montirone network and monitoring points. (*Source*: Italian Case Study in Applying the IWA WLTF Approach: Results Obtained)

achieved—although it must be recognised that the economic leakage level will change with time.

Using assumptions similar to economic stock control theory, it can be shown[6] that the economic frequency of intervention occurs when the cost of a "full system" intervention (excluding repair costs) equals the value of the unreported leakage volume. Thus the economic period between interventions can be calculated accordingly.[7]

The case study here described is a small system in Northern Italy with 1300 service connections, 23.2 km of mains, and with no permanent inflow metering.[6]

Figure A.4.8 shows Montirone network and chosen monitoring points where flow meters and pressure gauges with data loggers and GPRS have been placed. Components used include

- One battery-powered electromagnetic converter, with internal data logger and GPRS for the acquisition of flow and pressure at the inlet point.

- Two battery-powered electromagnetic converter, with internal data logger and GPRS for the acquisition of pressure at the average zone point (AZP) and at the critical point(CP).

- The software WIZCalcs for the economic management of hidden leaks. The software uses night flow measurements collected by the field instruments to decide when it is economic to do a leak detection exercise. WIZCalcs uses the calculation method described in the paper: *Recent Advances in Ccalculating Economic Intervention Frequency for Active Leakage Control, and Implications for Calculation of Economic Leakage Levels* (presented at IWA International Conference on Water Economics, Statistics, and Finance in 2005).

Figure A.4.9 shows the links between the different components of the monitoring system applied in Montirone. Data collected by field instruments are recorded by the

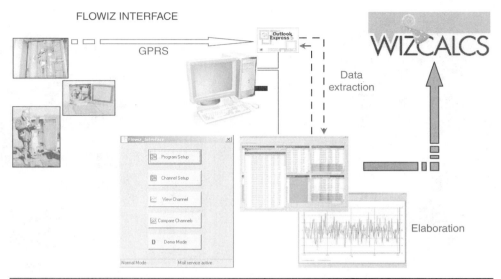

FIGURE A.4.9 Monitoring system applied in Montirone. (*Source*: Italian Case Study in Applying the IWA WLTF Approach: Results Obtained)

loggers and transmitted daily via email to a personal computer. Data are extracted and key information are then used by the software WIZCalcs to evaluate the flow and pressure values inside the network.

Figure A.4.10 shows best achieved minimum night flow after an active leakage control intervention done in 2003 and recent minimum night flow in April 2006. It is possible to see that, without any further active leakage control since 2003, the night flow in the distribution system gradually increased with time, because of "unreported" leaks and bursts, even though all "reported" leaks and bursts have been promptly repaired.

The actual night flow is checked against estimates of customer night use and background leakage, to calculate recoverable losses. The average rate of rise that occurred is system-specific and irregular, being influenced by several local factors; but the average rate of rise has been assessed from periodic night flow measurements at times of year when industrial and irrigational use at night is considered to be minimal.

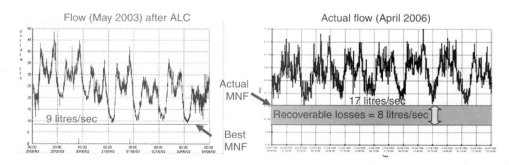

FIGURE A.4.10 Comparison of night flows. (*Source*: Italian Case Study in Applying the IWA WLTF Approach: Results Obtained)

Data entry	Calculated values	Data from another worksheet		
Utility	Enter licensee's name when issuing	Country	Italy	Conf. limits±
System	Montirone	Currency	Euro	
Length of mains	23.2	km		1.0%
Number of service connections	1300			2.0%
Natural rate of rise of unreported leakage RR	232	m^3/day in a year		20.0%
	178.8	Litres/connection/day per year		
	10.0	m^3/km mains/day/year		
This is categorized as being	Very high			
Variable cost of water CV	0.114	Euro/m^3		10.0%
Full system intervention cost CI	5000	Euro		5.0%
Economic intervention every	12	Months		2
Last intervation was	**34**	**months ago, so an intervention is overdue**		
Annual budget for intervention	4.9	Thousand Euro		0.0
Economic unreported leakage	43	Thousand m^3/year		7
	91	Litres/service connection/day		15
	5.09	m^3/km of mains/day		0.83

Figure A.4.11 WIZCalcs Software applied in Montirone. (*Source:* Italian Case Study in Applying the IWA WLTF Approach: Results Obtained)

Figure A.4.11 shows results from WIZCalcs. The method applied, presented in this paper, requires only three parameters: average rate of rise of unreported leakage, variable cost of water, and cost of intervention to determine economic intervention frequency with active leakage control, annual budget for intervention, and economic volume of unreported leakage.

As in Montirone calculated natural rate of rise of unreported leaks is very high and the previous intervention with active leakage control was 34 months before, WIZCalcs quickly shows that an intervention is overdue in April 2006. The annual budget for intervention and economic volume of unreported leakage have also been calculated and reported in Fig. A.4.11. Repairs of unreported leaks found during a recent active leakage control intervention, following this analysis, are actually ongoing and should allow the target minimum night flow to be achieved.

A.4.5 Conclusions

Conclusions of this paper are as follows:

- Over the last decade, a technically robust set of methods and concepts have been developed for assessment and management of nonrevenue water and its components.
- These methods, adopted in an ever-increasing number of countries, are being successfully promoted internationally by members of the IWA Water Losses Task Force (WLTF).
- European utilities (Malta Water Services Corporation, Lemesos Utility in Cyprus, and the like.) which rapidly accepted the methodologies have achieved impressive results. Some European regulators are also becoming positively interested.

- Recently, the most extensive upsurge in interest in the methodologies in Europe has been in Italy, where intensive training in collaboration with the Italian Water Association took place and some utilities started to apply the methodology.

- The Italian case studies outlined in this paper have the aims to better estimate real losses in DMAs, improve overall management of DMAs, and calculate economic intervention frequency for active leakage control.

A.4.6   Acknowledgments

To Allan Lambert who has greatly contributed to the development of these methodologies and to their dissemination in Italy. Particular thanks to Nicola Bazzurro, Francesco Calza (Enia), and to the many members of the Italian Water Loss Group for their support and participation.

References

1. Alegre, H., W. Hirner, and J. Baptista, et al."Performance Indicators for Water Supply Services." Manuals of Best Practice: IWA Publishing , ISBN 1 900222 272.

2. Lambert A., T.G. Brown, and M. Takizawa, et al."A Review of Performance Indicators for Real Losses from Water Supply Systems." AQUA, Dec 1999. ISSN 0003-7214.

3. Alegre, H., W. Hirner, and J. Baptista, (in press). "Performance Indicators for Water Supply Services." Manuals of Best Practice 2d ed:. IWA Publishing , ISBN 1843390515.

4. Lambert A . "Water Science and Technology:Water Supply" *International Report on Water Loss Management and Techniques*, vol 2, no. 4,2002, pp. 1–20.

5. WIZCalcs. International software to calculate IWA Water Balance and Performance Indicators with 95% confidence limits. Contact ILMSS Ltd and Studio Fantozzi for details.

6. StiperzEnia,(2006). International software to compare Real Losses calculated from Night Flows and Water Balance with 95% confidence limits. Contact ILMSS Ltd and Studio Fantozzi for details.

7. Lambert A., Fantozzi, M. "Recent advances in calculating Economic Intervention Frequency for Active Leakage Control, and implications for calculation of Economic Leakage Levels," IWA International Conference on Water Economics, Statistics and Finance, Rethymno (Greece), 2005.

8. Calza F.. *StiPerZEnia: Experiences in in-depth management of DMAs in ENIA Group*, Leakage 2006 Conference, Ferrara (Italy). Available: www.leakage.it, 2006.

9. Guazzoni R. *Leakage control by use of Flowiz electromagnetic flow meter*, Leakage 2006 Conference, Ferrara (Italy). Available: www.leakage.it, 2006.

Case Study A.5: The Water Loss Control Program in São Paulo, Brazil

Francisco Paracampos, *SABESP*

Paulo MassatoYoshimoto, *SABESP*

A.5.1   Introduction

The water loss control program in Sao Paulo is ongoing—this case serves to update the São Paulo case study featured in the first edition of the *Water Loss Control Manual* .

The metropolitan region of São Paulo has 19 million inhabitants settled in 800 km². The landscape is hilly, varying from 730 to 850 m above sea level. The São Paulo Water & Sewer Co., SABESP supplies water sanitation services through a distribution network of 29,500 km of mains, with 3.6 million connections for their customers, and bulk sales to six municipalities. The water system is fully metered and consumers have individual building storage tanks.

The average water production of 65.5 m³/s has been stable for the last 3 years, despite a growth of one hundred thousand new connections yearly. The year 2006 figure for total water losses (a rolling average for 12 months) is ±502 L per connection per day.

A.5.2   The Water Losses Control Program

The key strategy was to disaggregate the water distribution system into smaller components, typically a sector or smaller subsectors. For each of the zones a careful analysis was carried out following the IWA/AWWA standard methodologies. The results of these detailed water loss assessments for each zone were then used to define the most appropriate intervention methods for each zone. Several teams were simultaneously allocated to develop the most suitable intervention strategies. Advanced statistical tools and analysis were used to quantify the level of apparent losses and to develop apparent loss intervention strategies. For real losses long-and short-term intervention strategies were developed.

A.5.3   Key Actions

1. Desegregation of system components, for analyses, prior to implementation of field interventions
2. Optimization of average pressure in the distribution network
3. An intensive leak detection survey program
4. Renewing of the weakest point of the infrastructure—the service connections
5. Improved maintenance of customer meters, especially large ones (keeping them in optimal working condition)
6. Reinforcement of antifraud actions—reduction of unauthorized consumption
7. Improvement in the commercial database information

Desegregation of System Components

São Paulo has 120 sectors—all of them were ranked based on IWA water balance approach. Those which presented a poor performance were analyzed in depth, considering real and apparent losses, with in-depth evaluation of water loss components and field test of N1 and infrastructure condition factor (ICF), in order to have the best possible diagnosis.

The prioritization of intervention and the selection of the most appropriate intervention tools were based on the analysis and field tests mentioned above. A specific intervention strategy was developed for each area.

Optimization of Average Pressure in the Distribution Network

The water distribution network of São Paulo works on a gravity principle, and much of the system is made up of old piping (nearly 30% of pipes are older than 40 years). The city has experienced steady growth for decades, although recently a lower growth rate has been observed. However, 100,000 new service connections are still made annually.

The following picture summarizes the system, the distribution network usually has high velocity in old pipes, commonly undersized for the actual demand, and so significant head losses occur during peak hours. Hence, an advanced pressure control strategy was compulsory to cope with such scenarios.

Another characteristic related to the São Paulo water supply system is the low overall distribution storage capacity (1,500,000 m³), although there is a positive impact as the customers have domestic roof tanks.

The pressure reduction program was essentially done with pressure reducing valves (PRVs) and rezoning works in some sectors. The key approach was to start the implementation of PRVs in large areas, regardless of the total head to be reduced in each area. SABESP has found that even a small amount of pressure reduction over a large area will provide excellent results, on both volume and frequency of new leaks reduction. Undoubtedly, such a view drove SABESP to obtain the most significant savings with the overall program. It is worth remembering that a traditional concept for the application of PRVs is to seek higher-pressure reduction in a smaller area. This was adopted in São Paulo when very critical points were addressed.

The evaluated savings for the 954 installed PRVs are 3.1 m³/s.

Leak Detection Program

The program is designed to survey the entire distribution network each year. Critical parts of the system are investigated twice or ever three times, depending on the results achieved in the general survey.

Renewing the Infrastructure

The weakest part of São Paulo distribution network is the old service connections. In the old part of São Paulo, 100,000 new connections have been changed each year, as well as rehabilitation works in another 1% of the total existing mains.

Results from this action using new materials (HDPE) are a very low frequency of failures in previously problematic areas as can be seen in Fig. A.5.1.

Efficient Metering Performance

In the search for a good metering performance, 450,000 meters have been renewed each year which allowed 2.2 m³/month recovery per residential meter. Larger meters are continuously resized, according to the change of customer's consumption patterns and a target of 2 years was set as an average lifetime for large meters.

Reinforcement of Antifraud Actions

Some strategies for gathering better data have proved successful like using a long historical commercial data (5 years), matched with standard parameters data for typical

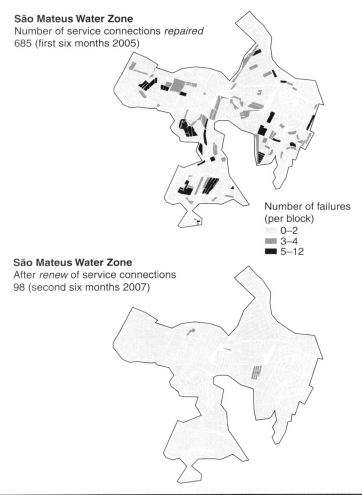

São Mateus Water Zone
Number of service connections *repaired*
685 (first six months 2005)

Number of failures
(per block)
0–2
3–4
5–12

São Mateus Water Zone
After *renew* of service connections
98 (second six months 2007)

FIGURE A.5.1   Change in service connection repairs after Intense Service Connection Replacement Program. (*Source*: SABESP.)

activities, and additional information collected from several sources. All this information is combined in order to define suspicious locations for further detailed inspections.

Also, an intensive training course to employees and contractors combined with the field training was conducted beside the utilization of modern equipments like micro cameras and sounding equipments. All that brought an overall recovered volume of 3,800.000 m³ this year.

Reinforcement of the Commercial Database

Special attention was given to the commercial database in the last 3 years. Relevant points to be confirmed monthly, apart from volume consumed for billing purpose, are the number of inhabitants at the address, major usage of water (residential, commercial, kind of commercial activity, hours of use per day, and so on), and eventually any anomalous situation at the meter set. Also, statistical evaluation of consumption is made on a routine basis.

A.5.4   Summary

SABESP is striving to reduce water losses using a continuous and standardized approach of assessment, validation, intervention, and reassessment. The target for total water loss in 2012 is 250 L per connection per day or about half of the current volume. Key changes in the intervention process since the last report in the first edition of the manual are the large number of service connections, which are being replaced, and the savings generated from this.

Case Study A.6: Proper Meter Sizing for Increased Accountability and Revenues

John P. Sullivan, Jr., P.E. *Chief Engineer, Boston Water and Sewer Commission*

Elisa M. Speranza, *Special Project Manager, Boston Water and Sewer Commission*

A.6.1   Background

Each year in the water industry, billions of gallons of water are "lost." The American Water Works Association Research Foundation (AWWARF) estimates revenue losses due to "unaccounted-for water" range from \$158,000,000 to \$800,000,000 nationwide. The problem of unaccounted-for water, which has been the subject of dozens of studies, reports, and books, can basically be summarized as follows:

1. All of the water purchased does not reach its intended destination.
2. The retailer is never paid for some of the water which does reach its intended destination.

The Boston Water and Sewer Commission (BWSC, the commission) provides retail water and sewer services to over one million people who live and work in the city of Boston. The BWSC is the largest customer of the Massachusetts Water Resources Authority (MWRA), a regional authority that provides wholesale water and sewage treatment to 60 communities.

Because the commission purchases about 40% of the total water sold by the MWRA, when Boston's water distribution system is losing water, an artificially high demand in efforts, including leak detection and repair, are aimed at reducing the amount of water lost between the time the commission purchases it and the time it is sold to the customers. Through these efforts, the commission makes an expensive contribution toward the goal of avoiding expensive and environmentally damaging water augmentation projects in the future.

Just a few years ago, the diversion of the Connecticut River to supplement Greater Boston's water supply seemed like a certainty. Through aggressive "demand management" programs, demand on MWRA water system has been reduced from 317.2 million gallons per day (mgd) in 1976 to 290 mgd in 1990. The latter figure is well below the system "safe yield" of 300 mgd. As a result, the Connecticut River diversion project has been placed on hold indefinitely.

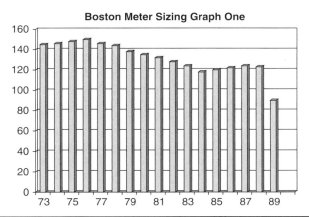

Boston Meter Sizing Graph One

FIGURE A.6.1   From 1976 to 1990, water consumption in Boston dropped by 26%, from 150 to 110.2 mgd. (*Source:* Boston Water and Sewer Commission.)

The second problem, unbilled water usage, is a potential untapped revenue source for the commission. In these times of fiscal austerity, the BWSC's ratepayers must be assured that all customers are paying their fair share, and that the commission is maximizing its income to meet the rising costs of providing water and sewer services.

As shown in Fig. A.6.1, from 1976 to 1990, water consumption in Boston dropped by 26%, from 150 to 110.2 mgd. Between 1998 and 1990, the Boston Water and Sewer Commission brought water consumption down by 9.3% from the 1988 figure of 121.5 to 110.2 mgd. The dramatic drop in water usage can be credited to the commission's aggressive leak detection, repair, and other water conservation programs. Unaccounted-for water—the difference between the amount of water purchased from the MWRA and the amount billed to BWSC customers—dropped by 18%, from 33 to 27% of total.

A.6.2   Unaccounted-for Water Task Force

While the consumption decrease in Boston is significant, the city's unaccounted-for water percentage is still unacceptably high. According to industry studies, unaccounted-for water values of 20 to 30% are not uncommon for older systems, particularly those in the northeast. However, the commission believes a concerted, agency-wide focus can significantly reduce unbilled water, even in an older urban system.

In response to the challenge of accounting for more of the water the commission purchases, the Executive Director formed an Unaccounted-for Water Task Force, in March 1990 to conduct a comprehensive review of the source of unaccounted-for water, and to investigate potential strategies to address this issue. The task force was unusual in that it included staff from various departments which had not necessarily dealt with the question of unbilled water in the past, including field services, meter installation, billing, water operations, planning, and engineering services.

Unaccounted-for water had been previously reviewed as part of BWSC's Water Distribution Study conducted by Camp, Dresser, and McKee on behalf of the commission in May 1987. Based on 1985 data, the consultants determined that there was a gap of about 32% between what the commission purchased from the MWRA and what it billed its customers. The study acknowledged the downward trend in unbilled water since the commission's creation in 1977, and described various efforts—particularly leak detection and repair—which resulted in a reduction of water purchased and an increase in water billing.

The study also identified various reasons for unbilled water, including: metering and billing problems, unmetered consumer use, unmetered public use, unavoidable leakage, and potentially recoverable leakage. Therefore, a percentage of what was termed "unaccounted-for water" was, in fact, being used for legitimate purposes, but was not being identified as such, nor was it necessarily being billed.

At that time, potentially recoverable leakage represented the largest share of unaccounted-for water at 18.5 mgd—about 49% of the 38.1 mgd total unaccounted-for water and 16% of the total average daily water purchase of 119 mgd. The task force estimated that potentially recoverable leakage still represents almost half of all unaccounted-for water.

In December 1990, the task force issued its first report. Many specific recommendations were made, including suggested revisions to metering, billing, and record-keeping practices, new efforts to improve water accountability, and the continuation of successful programs such as leak detection. The task force's recommendations in the metering area dealt with proper sizing, reading, slippage, repair and replacement of meters. This paper focuses on one of these issues, proper meter sizing.

A.6.3   Past Metering Practices

The BSWC has over 86,000 meters in service, about 10% of which are larger than 1 ½ in. Past metering practices generally required that meter size be determined by the size of the supply pipe. Thus a 1 in meter was installed on a 1 in service pipe, a 2 in meter on a 2 in pipe, and so on. The pipe size was determined using applicable plumbing codes, taking into account total required volumes and maximum allowable pressure drop through the pipe.

These extremely conservative calculations, made by developers, often resulted in the installation of meters, which were larger than was needed. It should be noted that many older fixtures used more water than their modern counterparts, so usage assumptions were, perhaps correctly, higher.

Because the cost of water was so low, the city was not generally concerned about missing water at low flow rates through oversized meters. The major concern was to guarantee no additional pressure loss through the meter. On pipe sizes 3 in and above, the common practice was to install a compound meter, a complicated mechanical device capable of recording low, moderate, and high flow. The meter consisted of a small 5/8 to 3 in metering device and a larger 3 to 8 in turbine type meter, which would work together to record the total flow.

The compound meters actually worked quite well. Unfortunately, however, preventive maintenance programs were inadequate and the compound meters fell into disrepair. By 1974, most of the larger compound meters were partially or totally malfunctioning. Based on historical repair records and a cost comparison of turbine to compound meters, the city decided to replace the 3 in and above meters with new, state-of-the-art turbine meters. The newer turbine meters were far less complicated and easier to maintain, but they could not accurately register flows less than five gallons per minute (gpm) for 3 in meters, 10 gpm for 4 in meters, and 20 gpm for 6 in meters. It was generally assumed that most of the flow in a building serviced by these larger meters would fall into the meter's range. The amount of water used at lower flow rates was unknown and was not a factor in determining meter type and size.

In 1976, the combined water and sewer rate was about $1.02/1000 gal. Since then, the combined water and sewer rate has increased by 429% to $5.40/1000 gal in 1991. In

1985, when the MWRA took over the ailing metropolitan water and sewer system, rates began to rise sharply. Massive capital projects such as the $7 billion Boston Harbor cleanup and a proposed water filtration plant will drive rates up even further over the next 10 years. By the year 2000, the BWSC predicts that the average family in Boston will be paying $14.40/1000 gal—over $1000 in annual water and sewer bills. Consequently, the cost of water has gone from being an insignificant factor in meter sizing to being an extremely important consideration.

In 1988, the commission began to investigate the possibility of downsizing meters. All new accounts generally have been required to install a meter, which is one size smaller than the nominal pipe size, and developers are required to submit forecasted water demands. Although the commission recognizes that compound meters could accomplish the goal of accounting for water at all rates of flow, experience, and judgment dictate that, in most cases, simpler devices would better serve the commission's needs. Because there was no available methodology for evaluating whether existing meters were properly sized, the BWSC developed a pilot project to address this issue.

A.6.4 Project Approach

The problem confronting the commission in implementing the task force's recommendations was how to account for more water used at low rates of flow, without violating the customer's high-end flow requirements. The project approach developed was based on the theory that (a) water lost at low flow rates was significant enough to warrant a major effort to recover it and (b) people do not use water at previously assumed rates.

Project Team

The commission formed a meter downsizing project team as a subcommittee of its Unaccounted-for Water Task Force. The team consisted of staff from the engineering services, field services, meter installation, and meter reading divisions.

Since August 1990, the project team has met every Monday morning to coordinate its efforts. At these meetings, staff brings cases to the table from various sources and agrees on the proper meter size for a particular customer. Water requirements are analyzed, potential problems are discussed, and proper meter sizes are assigned based on the operating ranges of different sized meters.

Among the factors taken into consideration was that 2 in disc meters, in contrast to the older turbine meters, can accurately register as low as 2 gpm and will record flow as low as gpm with 95% accuracy. The trade-off in using the smaller meters is a limit to the operating range (a 2 in meter's maximum is 160 gpm vs. 350 gpm with a 3 in meter) and the added pressure loss (10 psi loss at 160 gpm with a 2 in meter and 1 psi at 160 gpm with a 3 in meter).

Data Bases

The first task, which faced the project team, was to develop lists from which the meter testing, investigation, and installation crews could work. The project team has focused primarily on meters over 1½ in. Although they represent only 10% of the meters in service, large meters account for roughly 63% of the water the commission sells.

The first priority was to evaluate recently changed large meters. These meters would be easier to downsize because fittings would be new, control valves were functioning, and the commission would have had recent contact with the customer. In addition, accurate consumption information would be more readily available.

A second list involved the generation of a database of large users with apparently too low average daily use (ADU) records. The MIS department was asked to generate a list of meters over 3 in with ADUs of 0 to 300 ft³/d.

A third category of meters investigated was derived from special cases which were brought to the attention of staff through various other sources, such as customer services, construction, or the routine large meter testing and change-out program.

Account Investigations

The project team has assigned special crews to conduct field investigations and collect detailed information on water usage, including number and type of fixtures, type of building, number of stories and units, and whether the building has central air conditioning and a pump. A current meter reading is taken, and measurements are made from flange to flange. The investigating crew is instructed to gather all the information it can when visiting the site, in order to avoid duplication of effort, and to provide enough background material for the project team to determine the proper meter size.

Flow Testing

The project team decided that more accurate rate-of-flow measurements were needed to make better-informed decisions about meter sizing. When determinations regarding proper meter sizing cannot be made using fixture unit evaluations or other methods, flow search equipment is used.

Because the computer technology available to analyze rate-of-flow information is relatively new, the commission sought out manufacturers of equipment, which would meet its needs. Two manufacturers F.S. Brainard Co. and Schlumberger Industries responded by supplying, respectively, the Meter Master and the Flow Search. During the commission's pilot project, both companies have fine-tuned their equipment and related software packages using input from Boston meter tests.

Both the Meter Master and Flow Search work on a similar principle—magnetic pulses, which vary from meter to meter, are emitted from the spin of the turbine or disc in the meter. Depending on the meter size, the pulses reflect different volumes of water, which are defined in the related software. The data is imported to a computer, where the software translates the pulses into total rates of flow over various time intervals. The Meter Master, which is designed to work with most makes of meters, uses a sensor placed on various locations on the meter, wherever the signal is the strongest.

The Flow Search was designed for use with Neptune Meters. The sensor is placed directly on top of the meter, after the register is removed, and picks up pulses from there. The register head then fits back on top of the sensor to continue recording consumption.

It should be noted that the BWSC ruins flow tests during expected periods of peak flow, usually for 3-day periods, sometimes during the week and sometimes over a weekend. Meter sizing decisions are often based on a combination of flow test data and best engineering judgment because testing over longer periods of times is not always possible. So far, out of over 400 large meters downsized, only one has been upsized again, due to pressure problems.

The commission has recently purchased three Meter Masters and three Flow Searchers to conduct flow testing on large meters, which are candidates for downsizing. Six more Flow Searchers are on order. Rate-of-flow testing equipment is also used to track changes in consumption.

Follow-Up

In order to track the progress of the program and to obtain estimates of water recovered, a specialized database was established. The database records account number, address, old meter sizes, work orders, new meter size and number, and former average daily use. The commission then takes meter readings 30, 60, and 90 days following the installation of the smaller meter, and keeps a running total of the change in ADU recorded.

At first, due to the condition of the customer database and the prevalence of estimated readings, it was difficult to obtain accurate estimates of actual usage. The project team decided to track both the change in water consumed and the changed in billed consumption by deriving a "true" actual former usage number from the last available actual reads. When compared with the estimated usage, the net changes in billed consumption and actual consumption were surprisingly quite close.

Unusually large gains or losses in consumption are investigated in order to ensure the data has been correctly entered, and to explain any aberrations in consumption.

The commission plans to read accounts 1 year after downsizing to obtain a truer picture of change in recorded usage, accounting for seasonal consumption and other factors.

A.6.5   Results

In analyzing "candidates" for downsizing, the project team attempted to identify trends and generalizations among various categories of customers. While downsizing is not always appropriate, and does not always result in an increase in recorded consumption, the team has identified many accounts where smaller meters would likely have an immediate impact on unaccounted-for water.

The following case studies, in public housing, apartments, schools, commercial and institutional buildings, and municipal property, are representative of some of the emerging trends the team has identified, which have provided guidance for subsequent decisions

Public Housing

The Boston Housing Authority (BHA) is the BWSC's largest customer, bringing in over $6 million in revenue annually. Because the BHA represents such a significant portion of the BWSC's customer base, the Unaccounted-for Water Task Force decided to focus attention on a representative sample of the BHA's accounts.

In examining water consumption trends at various public housing developments, the project team took several factors into account.

- Most daily housing apartments do not have dishwashers, central air conditioning, or in-unit laundry facilities, which would drive up water consumption.

- Most developments are no more than four stories high, which would obviate most pressure considerations.

- Many developments have undergone recent renovations, which would likely include the installation of newer, water-saving fixtures.

Most meters at BHA developments were 3 and 4 in. Based on the factors above, and on earlier flow measurements, the Unaccounted-for Water Task Force recommended that most meters at public housing developments could be downsized to 2 in. The downsizing project team has implemented that recommendation, which has resulted in significant increases in water accounted-for at several developments.

The 2 in meters have delivered sufficient high-end flows to buildings with 50 to 124 units, and have picked up thousands of gallons of water which previously slipped by the larger meters at low rates of flow.

For example, at the Bunker Hill housing development, a Meter Master rate-of-flow recording system was used to monitor flows in one 119-unit building in the development in which a new 2 in meter was installed. Unfortunately, an analysis of the impact of downsizing at Bunker Hill was skewed by several aberrations in the commission's billing account system. If the "problem" accounts are removed from the analysis, the ADU of 15 accounts jumped from 109,118 to 127,220 gal per day (gpd) after downsizing, a 17% net increase of 18,102 gal a day.

At the BHA's Fidelia Way housing development, the commission downsized two 4 in turbine meters, with an operating range of 10 to 450 gpm, to two 2 in displacement meters, with a normal operating range of 2½ to 160 gpm. Rate-of-flow test data showed that the flow through one of the 4 in turbine meters ran below the minimum flow rate nearly all night. After downsizing, test data showed an increase in registered water from midnight to 6 a.m. of 13,039 to 18,477 gal, or a 42% increase. It also showed an increase from 21,866 to 28,667 gal from 10 p.m. to 6 a.m., or a 31% increase.

Apartments/Condominiums

Market rate and luxury apartments were studied as a separate category from public housing for several reasons, including

- Different lifestyles of market rate tenants and condominium dwellers would indicate different rates of water consumption, particularly where fixtures such as dishwashers, washing machines, and central air conditioning are present.

- Many multiple-unit apartments and condominiums are in high rise buildings where water is pumped to upper floors; the presence of a pump eliminates most water pressure concerns, which should always be taken into account in making meter sizing decisions.

- When apartments have been recently renovated, new plumbing code requirements and the availability of water-saving fixtures may influence water usage.

At the Foundry, a newly-renovated condominium complex in South Boston, a 2 in meter was installed in place of an existing 3 in turbine meter. From June 1988 through March 1990 the average daily use was recorded at 1.728 gal. After the smaller meter was installed, the recorded average daily use jumped 33% to 2304 gal.

At 65 Commonwealth Avenue, a 16-unit condominium complex, a 4 in meter was downsized to 2 in. The ADU before downsizing was 419 gpd. After the meter was changed, the ADU jumped to 3456 gpd—an increase of 74%.

Down the road at 12 Commonwealth Avenue, the 3 in meter at a 57 unit apartment building was downsized to 1½ in, resulting in an increase of 3%, from 6006 to 6193 gpd.

Schools

The project team has analyzed fixture units and potential water demand at several public and private schools in the city and believes most meters in schools are oversized.

For example, at St. John's parochial school, an old 3 in meter was feeding 16 sinks, 24 toilets, and 5 urinals. The account was using approximately 1668 gpd. A 3 in meter was found supplying 25 toilets, 25 urinals, and one shower at the Beethoven School in West Roxbury, with an ADU of 785 gpd. At Boston High School, a 4 in meter was supplying 80 toilets, 24 sinks, and one shower, with a pumped system. That school uses approximately 1242 gpd. All three schools are scheduled for downsizing to 1½ in meters based on an analysis of the fixture units and data previously collected in buildings with similar fixture units.

At the end of the 1991 school year, the commission plans to downsize approximately 120 meters at Boston public school buildings, mostly installing 1½ in displacement meters in place of the 3, 4, and 6 in turbines currently in service.

Commercial

Downsizing is also appropriate for many commercial buildings, although careful attention must be paid to the type of business and potential process-related fluctuations in water use.

Subsequent to flow testing at 109 Lincoln Street, which houses offices and a garage, a 4 in meter was downsized to 2 in, with a resulting 42% increase in the ADU from 3104 to 4421 gpd. After downsizing, tests showed the maximum flow at 33 gpm, well within the range of the 2 in meter, which probably could have been downsized even further to a 1 in meter.

At an office building at 40 Court Street, a 3 in turbine meter was downsized to 1½ in following a flow test which showed a maximum flow of 35 gpm.

According to the project team's analysis, commercial laundry facilities most often require larger size meters. Although large amounts of water are not used on a constant basis, flow measurements have shown peaks in consumption, which require the higher maximum flow through a large meter.

Institutional

Many of Boston's largest water units are institutions such as universities, hospitals, and museums. Therefore, the project team decided to focus attention on institutional users as a separate category. Within user categories, and even within accounts, water use can vary widely. For example, university dormitories will have different consumption patterns than classroom buildings and therefore could require different meter sizes.

Northeastern University facilities personnel were reluctant to allow the commission to downsize meters. Flow testing, however, showed that many of the meters feeding the university could safely be downsized. At the Ryder Hall, 139 Forsyth Street, flow testing showed a maximum flow through a 3 in meter of 25 gpm, with a minimum of 0 after downsizing, the new 1½ meter picks up flows of around 2 gpm which were previously missed.

At another Northeastern building at 370 Huntington Avenue, the maximum flow through a 3 in meter of about 21 gpm, with a minimum of 0 this meter was downsized to 1½ in, and flow measurement shows a ½ gpm flow running all night. Thirty days after downsizing these two Northeastern meters, the ADU has increased 146%, from 9993 to 24,624 gpd. At Wentworth Institute, 550 Huntington Avenue, a 3 in turbine was downsized to 2 in disc, resulting in a 126% increase from 3231 to 7286 gpd.

Flow measurement at the Museum of Fine Arts indicated that the average flow rate was 25 gpm and a minimum flow of 7 gpm. The project team therefore concluded that the 4 in meter should be downsized to 2 in to capture low flow. In addition, a large constant night flow was detected.

At the Young Men's Christian Union, 48 Boylston Street, a 3 in meter was flow tested at a maximum flow of 43 gpm, with a minimum of 0. After the meter was downsized to 1½ in there was a 20% increase in recorded consumption from 6440 to 7719 gpd.

Municipal

The city of Boston is the BWSC's second largest customer. Almost all city facilities are metered, and represent a broad spectrum of user categories, from municipal office buildings to fire and police stations to parks and other recreational facilities.

AT the city's largest recreational facility, the James Michael Curley Recreational Center (known as the L Street Bathhouse) flow testing indicated that the 6 in meter could be safely downsized to 2 in. Since that meter was changed, the ADU has increased by 35% from 9784 to 13,240 gpd. In addition, a constant flow was measured at night, indicating a leak at the premises. This figure also shows that the normal operating minimum of a 6 in turbine meter was far too high to pick up the water being used at low rates of flow.

Flow testing at the Boston Fire Department building at 125 High Street, showed a maximum flow of only 20 gpm through a 3 in meter with usage under 4 gpm most of the time this meter is scheduled to be downsized to 1 in.

A 4 in turbine meter at the Curtis Hall Municipal Building was downsized to 1½ in following a flow test showing a maximum flow of 8 gpm and a minimum flow of 0. A close-up view of the low flow testing after the meter change reveals a constant minimum flow.

A.6.6   Conclusions

Downsizing Works

Downsizing has been successful for a wide variety of the commission's accounts. Although results have varied from case to case, as of May 17, 1991 the commission has recovered over 57,474 ft^3/d (429,905 gal p d/156,915,320 gal per year) of water by downsizing over 400 meters 1 ½ in and larger. When multiplied by the BWSC's current water and sewer rates, the downsizing effort could generate over $700,000 annually to offset the commission's rate revenue requirements, in addition to cutting unaccounted-for water.

In some cases, the value of downsizing meters is immediately apparent, such as has been shown by the analysis of most public housing and schools. In all cases, previously held assumptions about meter sizing should be questioned.

To illustrate this point, the project team analyzed several cases to determine what the meter size should have been, using standard fixture unit assumptions, versus the actual rates of flow measured with flow testing equipment. At 216 Tremont Street, a nine-story office building yielded a fixture unit value of 1838, which would indicate a meter size of 3 in. Flow measurement indicated this meter could be downsized to 1½ in.

A 12 story office building at 40 Court Street had a 3 in meter, as was dictated by the fixture value of 3256. Flow testing accurately predicted that a 1½ in meter would be sufficient.

In some cases, even though the fixture value indicated a 1½ in meter would be appropriate, 3 in and even 4 in meters were installed. Such was the case at 12 and 65 Commonwealth Avenue, both of which have recently had meters downsized to 1½ in based on flow measurements.

The commission recognizes that the meters being replaced may be performing slightly below AWWA standards, which may contribute to the increase in registered consumption after the new meter is installed. However, data from the BWSC's large meter testing program indicated that most large meters are within accuracy standards, so the commission does not consider this a significant factor in measuring downsizing results.

Factors Affecting Downsizing Decisions

Several outside factors not previously considered have significant impact on water consumption and rates of use.

Pumps   When analyzing the proper meter size for a building over five stories, it is essential to know whether the building uses a pump to deliver the water to upper floors. If the water is not pumped, an undersized meter may have an adverse impact on water pressure. A smaller meter may be used if the building is equipped with a pump.

Air Conditioning   Air conditioning make-up water is another factor, which should be taken into consideration. Central air conditioning can use from 3 to 10 gal of water per minute to make up for evaporation, depending on the size of the unit. Flow testing and water consumption evaluations performed during the winter will not be accurate during the summer months, when air conditioning increases water usage.

Flushometers   A third factor to take into account is the amount of water required by "Flushometer" type fixtures. Depending on the type of fixture, Flushometers can use water at a rate of approximately 35 gpm in a 15-second burst. It is also important that sufficient pressure be maintained so the fixtures will reset properly.

Space Limitations   In some cases space limitations prevent smaller meters from being installed. For example, at a Boston University building at 632 Beacon Street, the pipe leading to a 3 in meter must be replaced and a new flange installed before a properly sized 2 in meter can be installed. At the commission's headquarters at 425 Summer Street, a similar situation exists, making downsizing a complicated and time-consuming endeavor.

Water Conservation Devices   As mentioned previously, where water conservation devices such as low flow showerheads, faucet aerators, toilet dams, and low-flush toilets have been installed, previous estimates of water usage should be reconsidered when making meter-sizing decisions.

Additional Benefits of Downsizing

Capital Costs   An obvious additional benefit to meter downsizing is the reduction of capital costs for large meter replacement.

Leak Detection   During the course of flow investigations, many leaks have been found at customers' premises. As mentioned, flow search at the city's James Michael Curley Bathhouse revealed a constant night flow of 3 gpm. The commission received a letter of thanks from another customer, the Roxbury Boys and Girls Club, for discovering a 6-g p m leak as a result of a flow search investigation. Although discovery of these leaks and the sharing of data can offset some of the additional revenue generated by meter downsizing, it fosters positive customer relations.

Water Conservation The commission has also determined that smaller meters act as flow restrictors by increasing headloss. Therefore, downsizing can actually promote water conservation without the installation of new water-saving fixtures.

Data Collection Another important benefit of the program has been the opportunity to clean the customer database through detailed meter investigations. Meters, which had not been read for long periods of time, have been located and are now regularly read. Illegal connections have been discovered and remedied, preventing water theft. In general, the program has enabled the commission to gain more knowledge about its system and about customer water consumption, both of which have contributed to a significant reduction in unaccounted-for water, and an increase in revenues.

References

1. American Water Works Association Research Foundation. *Water and Revenue Losses: Unaccounted-for Water*, AWWA, December 1987.
2. Male, J.N., R.R Noss, and I. C. Moore. *Identifying and Reducing Losses in Water Distribution Systems*. Park Ridge, NJ: Noyes Publications, 1985.
3. Camp, Dresser and McKee. Boston Water and Sewer Commission, *Water Distribution Study*, May 1987.
4. Schlumberger Industries. *Water Division Product Catalog*, 1989.
5. Hensley, J. (ed.). *Cooling Tower Fundamentals*: Marley Cooling Tower Company, Kansas City, Missouri.

Case Study A.7: Apparent Loss Control Program—Bulk Water Sales, Halifax, Nova Scotia, Canada

Carl D. Yates, M.A.Sc., P.Eng., *General Manager, Halifax Water*

Graham MacDonald, *Halifax Water*

Tom Gorman, *Halifax Water*

A.7.1 Abstract

In 2002, as part of its comprehensive water loss control program, Halifax Water replaced its old "designated hydrant" program for bulk water sales with a new bulk water delivery program using six automated bulk water fill stations that incorporated prepaid smart cards and readers. The introduction of the fully metered bulk water fill stations provided greater control, measurement, and accounting of bulk water sales. The switch away from hydrants to the automated bulk water fill stations also resulted in improved water quality, a significant reduction in the theft of water, and enhanced levels of protection and security for the water system infrastructure.

A.7.2 Background

Halifax Water is a municipally owned utility, which serves the Halifax Regional Municipality in Nova Scotia, Canada. The population served in the urban core is approximately

325,000. There are approximately 77,000 customer connections served by two main water treatment plants, with approximately 1250 km of water main in the distribution system.

In 1999, Halifax Water initiated a comprehensive, utility-wide water loss control program in order to reduce real and apparent water losses in the system. Halifax Water adopted the IWA methodology as best practice in April, 2000.

A key component of the loss control program was the apparent losses associated with the existing bulk water sales program. At that time, Halifax Water operated a designated hydrant program to manage the various requests for bulk water from water haulers. In the Halifax region, bulk water is generally used for construction work, street cleaning, domestic consumption, filling pools, and lawn maintenance/landscaping. Bulk water haulers, whose tankage and trucks were inspected and approved by Halifax Water, were permitted to connect their hoses to certain designated hydrants throughout the distribution system.

Under the old designated hydrant program, bulk water haulers were required to track and report the times and estimate the volume of water they were taking. The water haulers were responsible for reporting and paying for the bulk water on a monthly basis. The honor system of self-reporting did not provide a particularly accurate or timely means of measuring the real consumption.

Additionally, bulk water haulers did not always go to the nearest designated hydrant. Haulers would, on occasion, connect to "more convenient" hydrants. When this happened in residential areas where the mains were typically smaller, the large draws on the system would disrupt flow patterns and frequently cause dirty water complaints. This would result in Halifax Water having to investigate the complaints and flush the lines where required. The frequent connections to hydrants throughout the region by a wide range of businesses and users invariably led to damaged hydrants and occasionally, to main breaks caused by rapid opening and closing of hydrants.

Part of Halifax Water's loss reduction program included the sectorization of the distribution system through the establishment of District Metered Areas (DMAs) using enhanced flow metering and SCADA capabilities. The DMAs allowed for significantly improved metering accuracy and real-time system monitoring and data recording within the distribution system. As more DMAs (with the improved metering capabilities) were established throughout the distribution system, it was evident that the actual volume of bulk water being drawn from the system was more than what was being recorded through the honor system of self-reporting used by the water haulers. The new DMAs also confirmed that water was being taken from areas where no designated hydrants existed (i.e., theft of water).

A.7.3   Bulk Water Fill Stations

The old program was seen as a problematic, money-losing service that had to be provided to the haulers. From a variety of drivers (water loss prevention, revenue recovery, dirty water events, and system security), an improved bulk water sales/delivery program was needed. In 2001, Halifax Water initiated an investigation to determine what had been done by other water utilities to address the various issues associated with bulk water sales.

In 2002, as a result of the industry review, Halifax Water ultimately chose to install six bulk water fill stations around the urban core to replace the old designated hydrant program. The fully-automated bulk water fill stations include prepayment "smart card" technology to manage the accounts and usage. The volume of water taken could be pre-selected by the user, similar to typical gas station setups. The bulk water fill stations are available 24 hours a day, year round, which was an improvement over the old program because the designated hydrants had certain day-time and seasonal restrictions. The new stations represented a significant value-added service to the bulk water customers.

The cost of this enhanced service for the bulk water fill stations was funded through a new water rate for bulk water sales. The bulk water rate is based on the base water rate and a return on the utility's capital investment for the cost to deliver the enhanced service. Cost recovery and revenue projections for the new rate were based on the self-reported bulk water volumes taken under the old honor system.

The automated bulk fill station program (with convenient features such as 24/7 access, prepaid cards, and preselection of volumes) was generally well received by the water haulers. The recorded sales through the automated bulk water stations in the first year confirmed that the volumes taken were significantly higher than what had been reported under the old program. In 2001, the last full year of data prior to the switch-over, the total estimated volume of bulk water taken was 84,000 m^3, which generated a gross revenue of \$36,287.83 including permit and vehicle inspection fees. In 2003, the first full year after the transition, the measured total volume of bulk water sold was 99,778 m^3. The bulk water sales volume has been trending upward in the 5 years since the stations were introduced. The volume of bulk water sold during the 2006/2007 fiscal year was 120,616 m^3 and generated a gross revenue of \$212,977.60. Since the implementation of the new program the sale of bulk water has consistently generated revenue for the utility.

As noted previously, there were significant security concerns over the wide range of users that had access to hydrants under the old system. A small number of users were reluctant to break the habit of connecting to convenient hydrants and use the bulk fill stations. As part of the switch-over to the bulk water fill stations, Halifax Water revised its rules and regulations to make it illegal for any person except the fire department or a water utility employee to connect to a fire hydrant. Halifax Water also raised public awareness regarding the security issues associated with water system infrastructure and encouraged the public to report any unauthorized use of hydrants. Halifax Water was successful in prosecuting and fining a number of people who illegally connected to hydrants after the new program was implemented.

The implementation of the bulk water station program in Halifax is an ongoing success story. Combined with revisions to the rules and regulations under which Halifax Water operate, the program resulted in a sustained reduction in apparent losses and nonrevenue water, and a corresponding increase in revenue water. Authorized bulk water haulers are now accurately metered, and random unauthorized hydrant usage has been eliminated with the added benefit of a more secure water distribution system.

Case Study A.8: Tracer Gas Testing of Conduits and Closed Systems: Procedures and Methodology

Dave Southern, *Technical Service Operations,*
Hetek Limited, London, Ontario, Canada

A.8.1   Introduction

The tracer gas process of locating leaks in buried conduits from the ground surface involves inserting a known amount of tracer gas, helium gas, or helium/air mixture into the pipeline or closed system after the existence of leakage has first been determined by a hydrostatic or other test. The tracer gas leak location process is most often utilized in the commissioning of new pipelines that have failed hydrostatic pressure testing, thereby indicating the presence of a leak in the system. A failure to hold static pressure at a predetermined pressure for a predetermined time is considered a hydrostatic test failure.

The helium type tracer gas procedure includes a number of logistical and safety requirements and is generally utilized after other traditional methods of leak detection have been employed without success. Preparation for the test requires dewatering the pipeline, placing test holes over the pipeline or system, at predetermined intervals, securing a supply of helium and conducting background specific gravity tests for gases in the test holes. All other methods of leak detection should be explored, other than excavation, prior to attempting the helium tracer gas procedure.

A.8.2   Equipment and Materials

Gasophon—General Description

The Gasophon is an instrument that was developed as a self-contained portable leak detector to locate leaks in buried conduits from the ground surface, using helium as a tracer gas. The procedure is carried out by a portable, highly sensitive battery-operated gas detector. The instrument can be utilized to detect a wide variety of gases with specific gravities different from atmospheric air. The instrument has the ability to locate gases of many types, light or heavy, flammable or nonflammable. The expansion of the scale through three measuring ranges enables even minor gas concentrations to be detected and differentiated. The instrument does not indicate the tracer gas percentage of volume, and consequently, cannot be considered a quantitative instrument. The heavier or lighter the gas being detected in relation to atmospheric air, the more sensitive the instrument is to that gas. Helium is a desirable gas for locating leaks in pipeline or closed systems because it is a nontoxic, inert gas, which is readily available in pressurized cylinders. It is the second lightest gas, having a specific gravity of 0.137, and because of this, is easily detected in small concentrations by the Gasophon from ½ in test holes placed in the ground over the buried pipeline systems.

Principle of Operation

A simplified explanation of the operation of the Gasophon is as follows. The velocity of sound in air is approximately 330 m/s. With gases, the velocity of sound will be greater or less than air depending upon whether the gas is heavier or lighter than air. The instrument consists primarily of two tubes, one for sample measurement and the other containing atmospheric air as standard for comparison. Each tube is fitted with a sonic transmitter at one end and a receiver at the other. An internal pump draws the sample to be analyzed through the measurement tube. The difference in the velocity of sound in the two tubes is directly proportional to the specific gravity and amount of tracer gas contained in the sample. The velocity difference is measured electronically and is displayed as a meter deflection. The heavier or lighter the gas, the more sensitive the instrument.

Plunger Bar   The plunger bar is a manual impact tool designed to drive a ½ in diameter rod to a depth of up to 3 ft.

Helium Injection Gauges and Manifold   The regulator is a standard gas regulator such as an L-Tec Trimline Model R-76-150-580 with gauges 0 to 5000 psi on the inlet and 0 to 100 psi on the outlet side. The ½ in diameter outlet hose goes to the inlet of the manifold. The manifold has a ½ in diameter inlet and a standard Chicago fitting for the compressor inlet. At the other end of the manifold is another Chicago fitting which can be removed if the manifold is to be attached directly to the injection point instead of being attached to the compressor outlet.

Compressor   Any compressor with an outlet in the range of 175 ft³/min such as Ingersoll-Rand 175 can be used provided that it has Chicago fittings on the outlet and has no oiler.

Pneumatic Rock Drill   Any 30 to 60 lb air operated rock drill with a 1 in chisel bit is suitable.

Helium Gas Cylinders   Industrial grade helium is readily available from welding suppliers. Standard welding trade associations should be consulted to obtain further information.

A.8.3   Prerequisites

The success rate for the type of test is relatively high, however, there is no guarantee that every leak can be located. In order to avoid problems and assume that the operator has a reasonable probability of success, the following background information is required.

The Type, Diameter, and Length of the Pipe or System to be Tested

The type of pipe (PVC, Ductile Iron, other) is necessary as different pipe materials may have different test pressure limits set by the manufacturer. Also, the pipeline or system may have to be located if as-built drawings are not available. If the pipeline is PVC, does it have a tracer wire? The diameter and length of the pipeline or system are needed to calculate the volumetric capacity, which is necessary to calculate the amount of helium required. Some common pipeline capacities are listed below.

Results of any Hydrostatic Pressure Testing

The test requirements for residential water mains are that the mains hold a pressure of 150 psi for a period of 1 hour to be accepted. There is also a provision for allowable loss based on AWWA Standard C600-82. In the event of a test failure, it is important to know the amount of water required to raise the pressure back to 150 psi, as this will indicate the relative size of the leak or leaks. Whether or not the pressure drops to that of an adjacent tied-in main, to another pressure, or to atmosphere pressure during the test, may indicate that a tie-in valve is not sealing tightly in the closed position or, if a joint leak is reseating itself at a certain pressure. If this occurs, the pressure during the helium test must be greater than the lowest pressure during the hydrostatic test, to ensure leakage of the tracer gas from the joint. At the same time, care should be taken not to pressurize the pipe above design limits.

The Depth of the Main and Type of Fill

This information is used to determine the appropriate venting time and spread pattern into the soil atmosphere of the tracer gas from the leak location. If the fill material is not uniform along the entire length of the main, test hole spacing may have to be adjusted. If native backfill is used, it is possible that organics may be present which could release methane gas into the soil atmosphere during the decomposition process. Since methane gas is significantly lighter than air, it will register on the device as a light gas similar to helium. For this reason, test holes are tested and a base line established prior to helium injection.

As-Built Drawings of the Depth and Alignment of the Main

Plans should be provided, if possible, to determine whether the conduit is a straight run of pipe or if mechanical fittings, such as elbows or tees are present. If so, more than one blow-off point may be necessary. Any change of depth in the main which produces a low area where water could sit in the pipe must be known, because if this water cannot be removed by purging or use of a pig, the pipe should be excavated and exposed at the low point and cut or drilled to release the water. Failure to do this may result in a "no leak" result if the leak happens to be in this section. In order for the gas (helium) to escape, the water must first be pushed out of the leak. This can sometimes take a considerable length of time, particularly with larger diameter mains.

Location of Other Underground Utilities

Plans of other utilities should be obtained prior to placement of test holes in order to identify sites away from known utilities. It can be very hazardous and embarrassing to inadvertently drill a hole in one pipeline, such as a gas main, while trying to find a leak in another. It can also be very expensive.

Access to an Air Compressor

For large diameter mains, or long runs of pipe, a compressor is required to blow the main down and during injection, to provide air as a carrier for the tracer gas. Also, in the event that some or the entire main is under asphalt, concrete, or a heavy frost cap, the test holes may have to be placed using a pneumatic rock drill which also requires use of an air compressor.

Personnel Requirements

The water utility and/or construction contractor must provide sufficient personnel to place the test holes, to open and close blow-off valves, and to provide traffic control where required.

Weather Conditions

Tracer gas testing is somewhat dependent upon weather conditions, particularly rain. Rain, which is heavy enough to saturate the ground and/or fill in the test holes, makes testing impractical. A heavy frost cap, which extends below test hole depth, may cause tracer gas to spread under the cap and not be detectable at the test hole directly above the leak.

A.8.4   Field Operations

Based on the information received, the operator will proceed with field operations in the following order:

Preparation of the Water Main for Helium or Helium/Air Injection

1. The main should be emptied prior to the arrival of the operator. Air valves at the high elevations and drain or blow-off valves at the low elevations of the pipeline should be opened, and water allowed to drain out and air allowed to fill the pipeline. This includes opening the valve at the end of any branches if it is not a straight run of pipe.

   Caution: From the time that the conduit or closed system is dewatered or emptied and put under pneumatic, rather than hydraulic, pressure, it is important to understand that the system is longer dealing with a liquid system, but a high pressure gas system, with all the additional hazards that this involves.

   Since liquids are not compressible, an event such as an explosive decompression of the pipe, or a joint/fitting failure while under hydraulic pressure, even at the hydrostatic test pressure of 150 psi, would only result in the loss of a small amount of water in a short time with little risk of injury. However, since gases are compressible, the result of an explosive decompression under pneumatic pressure is very different. Enough force may be generated to blow the pipe apart and send many razor sharp fragments of the main in all directions. In one such incident, a piece of 8 in diameter water main was removed from a tree 50 ft away from the excavation. All main line valve caps and curb box tops should be removed to provide a venting point for the helium.

2. The compressor is then connected to the fitting at the high end of the main, making sure that the Chicago fittings are wired together. The blow-off valve(s) are closed, and the main is pressurized with air to a reasonable pressure. With the compressor still running, the blow-off valves are then opened beginning with the blow off closest to the compressor, and the main is depressurized. This will help to purge the main of any residual water left after draining. This procedure can be repeated as many times as necessary to ensure that as much water as possible is removed from the main.

Test Hole Placement and Marking

The purpose of placing test holes in the ground as opposed to surface sampling is for the following reasons:

Unlike natural gas pipelines, which are relatively shallow, other conduits are generally at depths of 5 ft or greater, which means that the helium exiting from the leak is going to be diluted to a much greater extent as it vents through the soil atmosphere.

Most natural gas pipeline leaks have been venting gas into the soil for an extended period of time, as opposed to the relatively short venting time of a helium leak.

The placing of a test hole creates a miniature well in the ground ½ in in diameter and 18 in deep to which the helium can migrate without being diluted by atmospheric air.

Advance gas testing of the test holes, prior to helium injection, provides a base line background for comparative purposes for subsequent tests.

1. Location of all underground utilities should be noted before test holes are placed. Where the conduit or system to be tested is under fill material with no hard surfaces, the standard method is to place ½ in diameter plunger bar holes

at uniform intervals and depths along the conduit and at known fitting locations prior to inserting the tracer gas. Bar hole spacing is determined by depth of main and type of cover from available information. Generally, a bar hole interval of about 10 ft is most practical; however, if joint and fitting locations are known, this can be expanded. Conversely, if the fill material is granular, such as sand or gravel, the bar hole spacing may be shortened. The depth of the bar holes should be about 18 in, except in frost conditions, when it is necessary to go below the frost cap.

2. Where the main to be tested is under concrete, asphalt, or frost, all of the above apply, except that the holes are placed with a compressor and rock drill. It would also be prudent prior to drilling to arrange for filling of these holes after testing.

3. After placement, the test holes are numbered with paint. Care must be taken at this point to spray the marking paint adjacent to, and not near the test hole, as the solvents in the paint can affect the test. This marking identifies the test hole if records are kept, and makes each test hole easy to locate should a subsequent test run be required.

Testing the Existing Soil Atmosphere prior to Injection

The soil atmosphere in each test hole is analyzed to detect lighter than air readings on the "10" scale prior to injection of helium. The "10" scale is the most sensitive setting. This is done to predetermine the density of the soil atmosphere relative to atmospheric air in the ground cover over the buried conduit system. A Lexan test probe approximately 3 ft long by ¼ in diameter with side holes in the bottom 6 in is ideal. It is reasonably strong and clear. If any water is sucked up by the equipment from a test hole, it is readily visible before it enters the equipment. This pretest procedure is necessary to detect and either eliminate or compensate for the following conditions:

1. *Neutral reading*: no needle movement. This is the ideal situation where the specific gravity of the soil atmosphere is the same as atmospheric air. This generally occurs in new installations where clean-engineered fill is placed over the conduit. No remedial work is required in this case and helium injection can proceed without site complications.

2. *Upscale deflection*: lighter than air reading. Since the instrument is designed to detect any gas with a specific gravity difference to air, upscale readings prior to helium injection indicate the presence of other lighter than air gases in the soil atmosphere. The most common gases are
   a. *Naturally occurring methane gas*: Methane gas is sometimes present when native material containing organics is used for backfill over the main. This gas should be removed from the soil atmosphere by purging if possible. However, if this is not practical, and the readings are relatively small (full scale on 10 or less), it may be possible to use the 100 scale by enriching the helium/air mix being injected into the main.
   b. *Natural (pipeline) gas*: This gas may be in the test area if there is a leak in an adjacent natural gas distribution pipeline. Natural gas (pipeline gas) consists largely of methane and displays similar specific gravities as naturally occurring methane (CH_4). Unlike naturally occurring methane gas, pipeline

gas is introduced into the soil atmosphere under pressure, and can travel through the soil a considerable distance. When natural gas leakage is confirmed suing CGI and ethane identifier, the gas company should be notified immediately in the interest of public safety. Helium testing is not recommended until the gas leak is repaired and the natural gas has been purged from the soil atmosphere.

3. *Downscale deflection*: heavier than air reading. Since the instrument is also designed to detect gases with a higher specific gravity than air, downscale readings prior to injection of helium indicate the presence of heavier than air gases in the soil atmosphere. The most common contaminants are

a. *Heavy hydrocarbons* (petroleum products): These products are generally found in the soil around or adjacent to fuel storage and distribution areas. After confirmation with a CGI and charcoal filter, the appropriate authorities (fuel safety branch) should be notified immediately. Helium testing is not recommended until remedial work, such as removal or treatment of the contaminated soil is completed.

b. *Carbon dioxide*: This is generally present when clean fill, native or engineered, is used over the conduit system. It is generally caused by the aerobic decomposition of organics in the fill. The presence of carbon dioxide, or other heavier than air gases in the soil atmosphere over the conduit being tested, is important when using lighter than air tracer gas. The heavy gas and the light tracer gas may mix, resulting in a neutral gravity and a corresponding readout on the instrument (no reading). Carbon dioxide can be filtered out by passing the sample through a filtering medium of calcium oxide or the helium/air ratio can be increased to compensate.

c. *Water vapor*: Helium testing can proceed in this case, provided that there is not enough water vapor present to condense inside the equipment. A water trap or hydrophobic filter placed in the inlet sample line can usually reduce or eliminate this problem.

Injection of the Helium/Air Mixture into the Water Main

1. Connect the air compressor to the injection point. If the injection point is a fire hydrant, or a service line with a curb valve, the hydrant or curb valve must remain open for the duration of the test. Closing either of these after injection of helium will allow the gas to escape through the hydrant barrel or the curb valve through the self-draining mechanism. The conduit pressure is allowed to return to zero by shutting off the compressor feed valve on the manifold and opening the blow-off(s). The gauges are attached to the helium bottles, and the helium feed line is connected to the manifold. It is recommended at this stage to inject a certain amount of 100% helium into the main prior to opening the delivery valve of the compressor. This is accomplished by opening the tank valve on one of the helium bottles, and bleeding off about 500 lb of the 2500 lb pressure in the tank. This slug of helium will mix with the air already present in the conduit and will ensure that a detectable amount of helium will appear at the blow-off(s).

2. The mixed gases are then injected into the system. It is important to differentiate between "delivery pressure" and "maximum output pressure" of the compressor.

    a. *Maximum output pressure*: Most compressors, which have a delivery volume of 175 ft3/min, have a "maximum output pressure" of 100 to 120 psi before they automatically shut down to an idle. This pressure can be determined by running the compressor, shutting the outlet valve and noting the pressure at which the compressor goes idle. For our purposes, this pressure reading is not important unless the situation calls for a test pressure on the conduit above the maximum output pressure of the compressor. If the maximum output pressure is 110 psi and for some reason the main has to be pressurized to 120 psi, the compressor outlet valve can be closed, and the additional pressure can be made up by feeding helium from the tanks by turning the regulator up to 120 psi.

    b. *Delivery pressure*: A compressor running unrestricted with the outlet valve open generally has a delivery pressure of about 30 psi. When this compressor is attached to a main with the blow-off(s) open, it is still not working against a great deal of backpressure. The amount of backpressure depends upon the diameter of the conduit, the length of the run, and how much of a restriction is caused by the blow-off(s). When the blow-off(s) are closed, the pressure in the conduit begins to rise, as does the delivery pressure of the compressor, when equilibrium is reached at the maximum output pressure, the compressor will go to idle.

3. Compressed air from the compressor delivery point and the tracer gas are fed into the main at a ratio of approximately one unit of tracer gas to 10 units of air. This is accomplished by noting the delivery pressure of the compressor, which will remain constant as long as the blow-off(s) remain open and adjusting the regulator on the helium gauges attached to the tanks to read 5 psi above this pressure. The delivery pressure can be determined at any time during the injection process by shutting off the helium feed valve.

4. The blow-off points at the ends of the system are not closed until a sample taken with the Gasophon indicates the presence of the tracer gas. Testing should be done in sequence starting with the blow off nearest to the injection point. Each blow-off point is closed when the proper concentration of tracer gas is indicated. The time required for the helium to reach the blow-off point will, of course, vary directly according to the total volume of the conduit and the delivery volume from the compressor and helium tanks.

5. When all blow-offs are closed, the delivery pressure of the compressor will start to rise as backpressure is built up in the conduit system. By constantly monitoring this delivery pressure (as discussed before) and adjusting the regulator on the helium tank upward to maintain 5 psi higher difference, a readily detectable concentration (±10%) of helium will be injected into the conduit system.

6. When the appropriate test pressure is reached, the compressor delivery valve, and the helium tank valve are closed. The test pressure can be observed on the helium injection gauge. To ensure that there is a detectable spread of the tracer gas into the soil atmosphere prior to testing the bar holes, a time period of at least one hour should be allowed after the conduit is pressurized. The only exception to this procedure would occur if a significant pressure loss (2 to 5 psi depending on the total volume of gas in the main) were observed before the 1-hour interval has passed.

Testing for Tracer Gas in the Soil Atmosphere

Gas emerging from the bar holes are tested with the Gasophon in order starting with the injection point and working toward the blow-off(s). A sample of the soil atmosphere is drawn into the Gasophon and analyzed. The result of this analysis is noted and compared with the initial test at the same location. If no readings show up on the initial test, the procedure can be repeated at regular intervals until the tracer gas is detected. When repeating the testing, be certain to always start at the injection point and work downstream. This equalizes the time that the gas has been in each section of the conduit. This is particularly important when dealing with long runs of pipe involving numerous bar holes. Assuming a quarter mile section of conduit with test holes spaced at approximately every 10 ft, we are dealing with at least 132 bar holes. If the operator finishes the run at the blow-off and then works his way back to the injection point, a leak, which is near the injection point may have introduced so much helium into the soil that pinpointing becomes difficult due to the spread.

Pinpointing Helium Locations

"Pinpointing" is defined as the process to determine the exact location of the excavation needed to effect leakage repair. Whether or not the process is successful is determined by a number of factors. Helium is a nontoxic, noncombustible, and inert gas having a specific gravity approximately one-tenth that of air. Because the size of the molecule is so small, it has the ability to leak in detectable quantities from even the smallest of failures. It spreads from a leak by displacing the natural soil atmosphere that normally occupies the space between soil particles. The shape and size of the leakage pattern is determined largely by the resistance of the soil atmosphere to gas venting from the leak. Because helium is considerably lighter than the soil atmosphere, it tends to rise rapidly through the soil to the surface. However, if restricted, the leaking helium will seek the path of least resistance in developing a spread pattern. Some factors influencing a helium-spread pattern are as follows:

1. *Line pressure*: Generally the higher the pressure in the main the faster the helium vents to the surface.

2. *Leak sizes*: The volume of gas entering the soil in a given time period is directly influenced by the size of the leak. In general, the larger the leak the greater the spread pattern.

3. *Depth of cover*: If all other factors are equal, an increase in the depth of cover will result in a larger spread pattern at the surface and an increase in venting time to the surface. Leak patterns are normally in the shape of an inverted cone.

4. Type of cover
   a. *Light soils*: Engineered fill (sand and gravel) and light, porous soils offer little resistance to the flow of gas. The spread pattern, if unrestricted by frost or a hard surface, is generally a small circular pattern with very little lateral spread. It may be necessary to space the bar holes somewhat closer together under these conditions. Also, particularly in sand, surface sampling with a bellows-type probe may be successful if there is a pressure drop in the main indicating leakage. The helium may be venting between the bar holes.
   b. *Medium soils*: Because loamy soils are less porous, there is more resistance to the flow of gas. Therefore, a somewhat larger pattern normally occurs at the surface. A bar hole spacing of 10 ft. intervals should be sufficient.

    c. *Dense soils*: Heavy clay soils greatly restrict the flow of gas and the spread pattern can be large. There is also the possibility that the helium may not appear in the test hole directly above the leak or in any adjacent test holes because the natural sealing qualities of the clay can force the gas to follow cracks, fissures, and voids in the soil and vent elsewhere. Extreme care must be taken when locating leakage in this type of material. Additional bar holes may be necessary.

5. *High groundwater level*: Displacement of the soil atmosphere by a high water table which covers the conduit causes resistance to the flow of gas. However, this is generally not a problem. In some cases, where the groundwater level is at or near the bottom of the test holes when the conduit is pressurized with air prior to injecting the helium bubbles will appear indicating a leak. This area should be noted and the bubbles can be tested for helium after injection. It is best to proceed with the helium test in any event since it is never certain that the leak, which is bubbling, is the only leak in the conduit.

6. *Other underground utilities*: Other underground utilities in the area of the test may act as well and collect helium, particularly in the case of dense clay soils. Such things as natural gas curb boxes, communication system and hydro risers, and sanitary, storm, and communication system manholes create wells in the ground to which the helium can migrate. It is a good policy to locate and test these collection points as well as the test holes. Curb boxes and main-line valve boxes on water mains can act as collection points. Care must be taken when pinpointing helium leaks on water main under test. For example, a helium reading at a water curb box may indicate two situations, particularly if the curb box is on the water-main side of the street. The curb box itself may be leaking, or helium may be migrating along the service trench from the main connection or an adjacent point. This can sometimes be confirmed by placing a bar hole next to the curb box. Since the curb valve is directly under the curb box a leak on this valve should vent directly up. Consequently, readings in the adjacent bar hole should be zero or very minor.

7. *Frost*: A layer of frost over the water main has a pavement-like effect and causes a larger spread pattern. Frost penetration is dependent upon soil moisture content, soil type, temperature, and the insulation properties of the surface. Snow cover and sod act as insulators and retard the penetration of frost. Pavement has poor insulating qualities and allows deeper penetration. Variations in soil moisture and surface insulation cause irregular frost penetration. The frozen areas restrict venting while the nonfrozen areas permit venting. As is the case with dense soils (clay), the helium will sometimes vent through cracks and fissures in the frost cap rather than into the test holes. Any frost pattern, which is deeper than the test holes, makes the helium test impossible as the helium will spread under the frost cap and not be detectable in the test holes.

When the tracer gas is detected, the maximum point of concentration is determined by comparing the individual readings in each test hole quantitatively using the Gasophon. If two or more bar holes show a similar concentration or if the initial bar hole spacing is greater than the size of the desired excavation needed to effect the repair, additional bar holes can be placed between the existing ones to determine the maximum concentration of the tracer gas in the soil atmosphere. Where two bar holes give the same reading, the Gasophon pump can be used to purge each hole. The time required for each reading to go to zero will indicate which hole is closer to the leak. It often happens that one of the test holes can be purged while the other cannot.

There are several tests that can be conducted to pinpoint the leak after the initial site is excavated. Never assume that there is only one leak present in the system under test. It is important that approximately 1 hour after the first leak has been pinpointed, another test of all bar holes and venting points be conducted before leaving the site. This will normally ensure that there are no more leaks present. In some cases, it may be prudent to repeat the helium test process after the initial repair is complete.

Excavation, Leak Location, and Repair

Before excavation of the helium indication begins, there are two operations, which must be performed. Firstly, the helium injection manifold and gauges must be removed. Before the regulator is put away, turn the regulator adjustment knob counterclockwise until it is loose. This will relieve the pressure on the spring and diaphragm. Secondly, all pressure remaining in the conduit should be relieved for safety reasons prior to repair.

It is possible that the helium location may be excavated and the source of helium is not found in the excavation. Since there will be residual helium remaining in the soil on each side of the excavation, helium readings taken in bar holes placed horizontally each way may indicate the direction of the leak. Once exposed, the leak location can be reconfirmed with the Gasophon. After repairs it is proper to recharge the conduit and conduct a hydrostatic test to confirm integrity.

Case Study A.9: Water Main Leakage Detection by Means of Ground Penetrating Radar

Rodney Briar

A.9.1   History

As with much innovative technology, ground-penetrating GPR was initially developed by the U.S. military during the Vietnam War to assist in finding the enemy in their underground passages and bunkers. The technology eventually found its way into the commercial field and, initially, Geophysical Survey Systems Inc. developed ground-penetrating GPR (GPR) into a viable tool for shallow penetration up to about 16 ft. and named their product "subsurface interface GPR." This name relates to the fact that GPR is able to produce an image of what is below the ground, by reflecting GPR frequency waves, emitted by a transmitter, from any interface in the ground, such as earth/water, earth/rock, rock/air, and so on, back to the receiving antenna. Usually this antenna is built into the same box as the transmitter, and drawn over the ground, producing data which can be processed and converted into a vertical cross-section or slice of ground below where the transmitter/receiver, henceforth referred to as the "antenna," has been drawn.

Dr. Hylton White, a South African physicist, was working in the United States in the 1980s and became involved with GPR and returned to South Africa to continue his involvement, working first for the South African Chamber of Mines and then a company who obtained the sales rights for Geophysical Survey Systems Inc in South Africa, during 1990. At this same time, a large contract for the replacement of water mains in the Central Business District (CBD) of Johannesburg was underway. Covering half the water mains in the CBD, or around 170 city blocks, this contract was widely publicized in the press, and attracted the attention of Dr. White. The rest, as the saying goes, is history.

The water mains in Johannesburg were principally steel and particularly the distribution pipes of 6-inch diameter and below had been laid under the footpaths. For ease of laying the precast concrete slabs forming the footpath, they had been laid on a fine sand locally available, residue from the processing of gold, which apart from still containing a tiny fraction of gold, now being processed out of the remaining stock piles of the "mine sand," also contained residual acids from the processing.

Johannesburg, South Africa is known as the thunderstorm capital of the world. During heavy summer thunderstorms, acids are slowly leached out of the "mine sand" into the subsoils and corrode the steel water mains buried underground.

All the city blocks have buildings with basements several floors deep and water leakage into the basements was a huge problem. Conventional leakage detection methods were being used to trace the leaks but, with a plethora of other municipal utility infrastructure also underground and impacted by the action of the "mine sand" acids, the success rate in finding leaks was lower than normal with listening methods. Leaks rarely surface in Johannesburg as the city is built on a high rocky ridge, with excess water entering into the ground, disappearing and finding its way to the older, abandoned, small mine workings close by. Due to interference with much ambient noise in the city streets, the only time conventional leak detection work could be conducted was during very restricted hours of 1 to 5 a.m. The presence of many noises can confuse leak correlators, particularly when there are long distances between access points for soundings. The longest straight length without a tee being 230 ft. and then another tee across the street at a further 50 ft. Sources of extraneous sounds include airplanes since Johannesburg is under the flight path into the international airport. Noisy electrical cables and transformers close to both internal and external water pipes also generate sound interference. These problems hinder the successful acoustic detection of leaks in many major cities.

GPR is only affected by one external influence, which is high voltage overhead cables, which do not exist in city centers, and even then they only leave a particular pattern on the T.V. screen if on-site detection is being used, or the computer monitor if additional later processing is being used to enhance the data.

A.9.2   The Growth of the Use of GPR

The introduction of GPR meant that reported leaks into basements could be searched for at any time of day or night. If a leak was not found in the adjacent small distribution pipe, the large transmission main or mains in the roadway could also be checked. This checking of the transmission mains could not be done by listening methods because the size of the mains restricted the effective length over which correlators could work and the access points for listening were already much further apart than on the smaller distribution pipes.

The use of GPR slowly grew and even spread to suburban conditions where "nuisance water," otherwise water causing a nuisance, was present and could not be traced to an adjacent leak by methods already employed by trained operators of the local municipal authority. GPR was the last resort. In one notable case, a leak sent water into the basement of the Johannesburg Stock Exchange Building. This leak had defied all attempts at detection of its source for over a year. GPR was finally called in and the mains within the immediate area were scanned without conclusive findings. A scan around the building revealed that water was making its way to the basement wall from an adjacent telephone cable trench. This water was traced by GPR to a leak within the Central Business District over 5000 ft. away. The water was traveling in the cable trench under the ducts, where it is always difficult to compact the earth during the laying of any pipe, and exiting at a low point in the cable duct run into the Stock Exchange Building.

The successes of GPR were such that, in 1993, the Johannesburg Municipality was beginning to think about trying to reduce the leakage losses across the entire system. During the late 1980s leakage losses were roughly 35% of water system input and it was desired to reduce leakage to a more acceptable figure. A full-scale leakage detection contract was envisaged. In order to choose the methods for this attack on the leakage situation, head to head trials between GPR and an expert correlator team from a European correlator manufacturer were held in a suburb of Johannesburg, with the goal of finding leaks on distribution pipes. Each of the two methods found exactly 11 leaks, and it was decided to draw up a contract for leakage detection on 375 miles of water pipes in Johannesburg. A further test was done, using three different correlators and it was found that the maximum length and diameter of pipe over which the correlators of the time available in South Africa could give reliable results was 985 ft. and 12-inches, respectively, on steel pipe, and lesser capabilities on fiber cement and PVC pipes. The contract was let using correlators on pipes up the 6-in diameter and GPR on pipes above that size.

The writer was the contract manager and it was a huge success with correlators finding 30 leaks in 280 miles of pipe tested and GPR finding five leaks in 93 miles of pipe scanned. Although the success rate of the GPR appears to be inferior to correlators, the transmission mains are usually constructed to a higher standard than distribution and with better supervision during that construction. There are also no consumer connections, which accounted for half the distribution leaks found by the correlator.

With each leak found being assessed by the municipality in terms of the volume of water lost, the contract was also a financial success with a payback ratio over the value of water being lost over 1 year being 5.5:1.

The 1980s were a time of drought in South Africa, but the 1990s proved to be the opposite. In addition to plentiful rainfall and the impending commissioning of additional water supply from the Lesotho Highlands Water Scheme in 1997, the will of the municipality to reduce the leakage losses weakened and the GPR was put to other uses where there was a need to "see" into the ground.

However, the acoustic finding of the source of nuisance water and suspected leaks as opposed to GPR scanning continued. Several smaller contracts were let for scanning with the GPR for leaks in Johannesburg and Cape Town for lengths up to 62 miles at a time. As time passed, an alternative method had been developed for finding leaks on distribution pipes which involved merely listening with listening sticks at available access points on the distribution piping and correlating where a leak noise was heard. This increased the price differential between acoustic methods and GPR, and effectively restricted GPR to transmission mains as opposed to distribution pipes. However, whenever there was a difficult leak to find, or when several noises confused the geophones and correlators, GPR was always called in. Additionally, GPR was used to confirm the presence of leaks found by other methods where the signal might have been doubtful and the cost of excavation was high, such as in a street in the Central Business District.

One of the smaller contracts of about 44 miles was carried out for the South African Bureau of Standards. Agency officials were very impressed that GPR was used in place of listening sticks, which were being used on the distribution piping in the township of Soweto. This area is particularly difficult to track leaks due to meters being inside the properties, the confined nature of the area increasing transient noise, the inherent dangers of night listening, and the practice of all rubbish being dumped at street corners on top of the valves. The South African Bureau of Standards was so impressed with the ground penetrating method of leakage detection that the method was included in the next edition of their Code of Practice on Water Loss Control, S.A.B.S. 0306 (1999). By the end of the 1990s, GPR was well established as a leakage detection tool in South Africa, and was still used in a number of other successful applications.

A.9.3   The Evolution of GPR

Despite the good rains of the 1990s and the additional water from the Lesotho Highlands Water Scheme, the change in the political scene in South Africa saw a tremendous effort made to provide treated, distributed water to many millions of people who previously had to fetch water from local boreholes and wells. The construction of piped water distribution networks has brought relief to the previously disadvantaged peoples of South Africa. However, since there is little skill and money available to maintain the domestic side of the new systems, and water is often left running where the piping terminates at an outside standpipe, there is tremendous waste and a new load on water resources to the extent that the Lesotho Highlands Scheme will be augmented and extended, and municipalities now have to report and account to national government for nonrevenue water, or losses from pipelines.

This pressure on the municipalities has forced them once again to consider leakage detection seriously, and the result in Johannesburg alone is a 2100 mile leakage detection contract, launched in the year 2000. The contract was planned to cover about 50% of the water mains in the city. Of the 2100 miles mentioned, 500 miles are transmission mains, for which ground-penetrating GPR was chosen. Due to the current pressure on water resources, there are other relatively large municipalities in South Africa which will require leakage detection throughout the early years of the new millennium.

A.9.4   The Method

The GPR unit which has been used for leakage detection in South Africa since 1990 is a model S.I.R. 3 sold by Geophysical Survey Systems Inc. of New Hampshire with a Model 39 visual display interface, which allows instant color monitoring of the scan results instead of the standard paper roll black-and-white printout. A high quality cassette tape recorder is used to record interesting data for later processing and selective printout of results. Clients like to see a "picture" of a leak! The antenna is drawn along the ground and is connected to the GPR unit by a 100 ft.–long cable. The whole system is carried in a pickup truck. A minimum of three persons are required to scan for leaks, a driver who, due to the walking pace at which the antenna is drawn, can also watch the moving image on the screen for signs of leakage, a person to handle the antenna, and a third person to look after the cable. The whole system runs at 12 V and is powered from a 12 V, 50 to 100-A/hr vehicle battery, which will last for a day before recharging is necessary. It is not connected directly to the vehicle battery due to the voltage fluctuations during charging.

More modern and compact systems allow for one man to be able to carry the GPR unit, look at a small monitor, and draw the antenna. But in South Africa, for several reasons (security, the parlous exchange rate to the dollar, and so on), investment in a new system has been resisted until a continuous year's work ahead can be seen. The present GPR unit has been very reliable, only once needing outside help in 10 years when it was sent to Allied Associates Geophysical in the United Kingdom for repair.

The GPR unit has an almost infinite combination of color palettes which can be used, all of which are useful under different circumstances and tasks. Once one has used the GPR with some success one tends to use the same palette again. In our case a background of black and gray was used. until a strong reflector to the GPR shows up, one of which is water, but pieces of steel, concrete, and rock are also strong reflectors. Strong reflectors appear as white shapes, which are very visible on the monitor to a driver piloting a vehicle along the road.

It is necessary to eliminate images which are similar to leaks but are not actual leak images. A digital geophone is used in these cases to listen above the potential leakage site. If a modern digital geophone is used, it is usually possible to ascertain whether a leak is present or not. The readout gives nine comparable readings which can be used to construct a "curve" of results, which indicate a leak if maximum noise is coincident with the GPR image. The geophone also listens for constant noises typically produced by a leak, as opposed to variable transient noises which are not leaks and have been historically problematic for geophone operators for years.

It is not possible in this overview to give detailed instructions to set up GPR for leakage detection. Because there are many different makes and models available, the user must refer to product specific literature. The advantages and disadvantages of the use of GPR should be mentioned, however. The reader can sense leakage detection must be approached using a combination of techniques in order to ensure reliable results. Because of the nature of the leak detection companies (most of them have an agency for some form of leak detection equipment), the predominant form of attack is single-technique. This approach is now becoming discredited in all branches of engineering geophysics, and the philosophy is spreading to leakage detection. However, the two-pronged attack is taking longer to manifest itself in leakage detection because most leakage detection is carried out by small or one-man companies with few financial resources to have a stock of different methods in their armory.

The Pros and Cons of GPR Leakage Detection

GPR is to all intents and purposes unaffected by the typical interferences that hinder acoustic methods. Scans or investigations can be carried out almost anywhere quickly and the results can be assessed on the site either during scanning, or immediately after scanning by rerunning the recorded data. Alternatively, data can be assessed back at base on a larger, clearer monitor, before or after processing to enhance the images. The only geographical terrain where GPR is difficult to use is one where the antenna cannot easily be drawn smoothly across the ground, that is, through very long grass, shrubbery, boulder strewn areas, and steep cross falls. Modern hand portable GPR assists in avoiding this problem. However, if there are great distances of adverse terrain to cover during the execution of a leakage detection exercise on a major cross-country pipe route, it is possible to use aerial means, such as a helicopter to carry the equipment, and conduct the survey. Careful cost–benefit justification should be proven prior to the undertaking of such means of leakage detection.

The combination of GPR and geophone is powerful, but using listening sticks and correlators on the smaller sized pipes is cheaper and just as effective. GPR comes into its own when the pipe sizes get above 6 inch, the pipe materials change from steel to fiber cement or PVC, or access points are available at great distances of more than 650 ft. In effect, when the conditions make the use of listening methods either unreliable or too expensive, then GPR should be used. In addition, there are circumstances outside the above when GPR needs to be brought into play. The presence of certain noise generators within a distribution system, or close thereto can make the use of listening methods impossible. In line pressure reduction valves, large district water meters, and all forms of electrical substations close to the pipes can transmit high-noise levels which can either drown out leak noises to a geophone, and/or confuse a correlator, as do overhead electrical equipment mounted on steel poles. It is then necessary to check with GPR for reliable results.

Sometimes unusual fittings or configurations in the distribution piping, such as a series of severe bends, will cause a "leak noise," and create a false positive pinpointing. Excavating a pipe only to find no leak (a "dry" hole) is a common inefficiency and can be quite disruptive in the middle of a busy city street. Under those circumstances, a scan with GPR will confirm whether it is worth excavating or whether the listening methods have picked up a false signal. Confirmation with two methods is always better, particularly when the alternative is the needless generation of a traffic disruption, or worse. Consider a situation where a complaint is received by the manager of a water network that there is a leak on a water main which is getting into the adjacent ducts of the local telephone system and causing faults and cable damage, and listening methods do not produce results. It is possible to employ GPR to ascertain whether groundwater or over-irrigation, for example, is the culprit, by scanning the ducts and the possibly innocent main to determine the source of the water. While it is not strictly related to leakage detection, GPR can also assist managers of water networks in the location of underground pipes, and chambers which have been "lost" because of inadequate drawings, or construction work and road improvements which may have, unknowingly or knowingly, been carried out over valve chambers and other access points to mains. Unfettered access to chambers assisted by GPR is a useful tool in the fight against elusive leakage.

The Future of GPR in Water Main Leakage Detection

As with all electronic devices to date, it is certain that GPR will become more portable, faster, specialized, and easier to use.

Taking the above four adjectives in order, portability is already high, but will probably be improved by manufacturers producing GPR which can be carried above ground in probably the same manner as a briefcase, with the images being observed on one eyepiece of a pair of goggles similar to the virtual reality viewers, the whole being run for a day by a small pack of lithium-ion batteries.

Speed of leakage detection could increase to the point already reached by GPR when used for carrying out road condition surveys, up to 30 miles/hr, with the antennae rear mounted on the vehicle just above road level. The unmetalled surfaces above the distribution systems could be covered in the same way by quad bikes.

Most of the GPRs produced today are normally for general purpose applications, i.e. anywhere a need to "see" into the ground is of advantage. The future GPRs will be produced for specific applications. A major simplification of the setup and controls could be made if only one function, such as leakage detection were envisaged. For instance, mine worthy, flameproof GPR is now the norm down mines, but the data recording capacity is not up to that needed for other purposes. Production of GPR for specialized purposes will make it easier to use by the specialists in those fields.

Case Study A.10: Severn Trent—Leakage Management Process

Martin Kane

A.10.1   Overview

This case study describes how Severn Trent, Plc (ST), a large water and waste water utility in the United Kingdom, harnessed new technology as part of its leakage strategy and achieved a step change in leakage performance. Critical to ST's success has been the deployment of an intelligent leak noise logger (Permalog) which has allowed the end-to-end process of leakage detection and repair to be reengineered. Leak noise logger technologies in general, or the Permalog system in specific, are not a panacea for leakage control. But it is the way that this technology is deployed and the overall leakage process within which it sits that delivers real business benefit.

A.10.2   Severn Trent

The United Kingdom operation of ST provides water services to eight million people across an area of 8000 mi^2. The watershed area includes the United Kingdom's second city of Birmingham and 10 other major industrial cities in Central England that hold a large proportion of the engineering and industrial base. Leakage is given a special focus within the company's distribution system operations, with all activity and management reporting under a Leakage Process Manager.

A.10.3   Background

Severn Trent was originally formed in 1974 as part of a major reorganization of the water industry in the United Kingdom. Leakage levels gradually rose from then until the early 1980s under a generally passive approach to detection. Leakage was not one of the more visible targets for the water industry and government was generally relaxed about the figures. In 1983, the 3 year average leakage level was 574 ML/d or 30% of water supplied. From this time on an active leakage control policy was followed and by 1986 ST had established an extensive coverage of district meter areas (DMAs), with 2200 DMAs now in play. The impact of DMAs and pressure management initiatives was immediate with leakage levels reduced to 525 ML/d in 1989. Following the United Kingdom water industry privatization in 1989, leakage levels rose across the industry. By 1994, ST's leakage level had reached 665 ML/d, or 32% of water supplied.

In 1995, the United Kingdom experienced a drought, with a number of water companies having to issue restrictions to customers to conserve supplies. The media embarked on a campaign to raise the profile of leakage and water resources as national issues. The company declared a program of water supply mains reinforcements and source development to ensure that we were capable of providing sufficient water under "worst case" scenarios. In parallel with this activity, the company committed to reducing leakage by 50% by the year 2000. The leakage target agreed with the government at that time was 342 ML/d to be achieved by March 2000.

A.10.4   Severn Trent Leakage Management Strategy

The leakage management strategy addressed the following issues:

- Accurate measurement of the company's district meter areas (DMAs)
- Implementation of valve control procedures to track all strategic and tactical valve operations to ensure DMA integrity
- Pressure management and optimization of existing PRVs
- Meter reading/repair frequencies set and achieved
- IT systems enhanced to manage leakage data and provide effective targeting of the detection effort
- Introduction of response time targets to fix leaks once detected
- Investment in the latest leakage detection equipment, including leak noise correlators and the "Aqualog" model of leak noise loggers
- Development of an in-house leak detection training facility with all appropriate staff trained and assessed in the latest methods and best practice

The new process showed significant reductions brought about in the first 3 years from 665 to 399 ML/d, albeit rather expensive, as the initiative was labor intensive. While the cost of fixing leaks remains fairly static, the cost of detection follows an exponential curve as the lower the leakage level, the greater the cost of reducing leaks.

The Economic Level of Leakage

Given the mounting criticism from the United Kingdom government, national media, and consumer groups on the levels of leakage being reported by the industry, the debate turned to what the economic level of leakage for any water company should be. Each water company undertook this work for its own operation with the outputs differing to reflect the diverse nature of the mains systems, supply headroom (the difference between available water resource allocations and the supply being utilized), and cost of water among other factors.

Leakage targets were then defined which, while meeting government and regulatory requirements, resulted in the lowest cost to the company of supply and demand over a 30 year planning horizon.

The outcome of the work was a least cost curve, which sets out a target for ST of 330 ML/d for year 2002/2003, equating to a leakage rate of about 18% of total supply. The curve shows that lower levels of leakage rapidly become more significantly expensive, driven by the increased costs of detection.

The study also concluded that the company's approach to leakage control represented water industry best practice. However, in the light of the United Kingdom water regulator's (Office of Water Services, or OFWAT) determination of a 14% reduction in income from 1 April, 2000, the cost of achieving 342 ML/d during the period 1999/2000 together with the predicted costs of making any further in-roads into the target of 330 ML/d were becoming prohibitive, using traditional leak detection methods.

Traditional Leakage Detection

Over the past 20 years, leak localization (finding the general area in which leakage is located) has been carried out in the United Kingdom using one of three methods:

1. "Stop tap bashing," or listening at service connection curbstops, still accounted for the vast majority of leak localization activity although it is a slow and highly repetitive task.

2. "Step testing" is still used to localize significant mains bursts, although it is less popular nowadays due to the necessity for night work and the potential for water quality problems.

3. "Noise logging" has been introduced effectively in recent years with an overall improvement in cost efficiency.

All these methods are labor-based and therefore incur significant operational expenditure. In the current climate, water companies face the challenges of achieving and maintaining lower leakage levels at lower cost, as well as improving customer service by maintaining continuity of supply, reducing the overall water lost through leakage, and responding faster to incidents when they occur.

The existing noise logging devices in the market, such as "Aqualog," showed the potential for the technology to pinpoint where to look. However the units were quite "intelligent" and required programming each time they were used. They were also expensive. Retrieval of data required a visit with a laptop PC to download the readings. These were then taken back to base for analysis of what leaks might exist and where. In short, they did not produce the reduction in process time needed to match future cost reduction requirements.

A.10.5 Permalog

In late 1998, ST entered discussions with Palmer Environmental about an opportunity to become involved in the development of an exciting new concept called Permalog, which addressed this fundamental issue. As the name suggests, this is a development of the Aqualog.

In order to overcome the costs of data retrieval, the new device had a built-in processor. It would still listen for noise in the pipe, but would be able to track a stable pattern and alarm when noises were heard that were outside the anticipated profile.

Rather than downloading data from the logger, the area in which loggers have been deployed is patrolled with a remote hand-held receiver, known as a "Patroller." This is usually deployed from a moving vehicle that drives round the area under investigation and automatically receives, decodes, and analyzes transmissions from the loggers, communicating its findings to the operator.

Each logger transmits to the Patroller whether it has a "noise" or not. Clearly, any two adjacent loggers having a noise indicate that a leak exists on the main between those points. Once a potential leak has been exposed, a two-man team will go and pinpoint the exact location using standard correlation techniques.

Early 1999: Permalog Pilot Trials

During the first half of 1999, ST piloted the Permalog system. The aims were to prove the performance of the new technology in a realistic environment and to demonstrate the operational and economic benefits that could be obtained. Several DMAs, representative of ST's distribution network, were selected as suitable locations.

One of these DMAs was Castle Donington, located in the east of ST's region. A total of 173 Permalog units were deployed across the whole DMA. Immediately following deployment, an initial Permalog patrol was carried out which identified many areas of interest for follow-up, and 13 leaks, including a significant main break (burst), were confirmed by correlation and/or surface sounding. Once these had been repaired a second patrol identified a further five leaks. Each patrol took only 2 to 3 hours to complete.

July 1999: Full Scale Trials in Worcestershire
In July 1999, ST purchased 14,500 Permalogs in order to carry out full-scale deployment in four of its nine operational areas. 10,000 of these loggers were assigned to just one of the Company's operational areas, Worcestershire

- Area = 1981 km^2
- Population = 616,900 people
- Properties = 250,000
- Average daily consumption = 170,000 m^3/d
- Length of water mains = 4610 km
- Service reservoirs = 59
- District meter areas = 300

Over the course of 8 months, ST developed world-leading leakage processes in the application of this new technology, enabling it to reduce leakage levels in Worcestershire from 27 to 15% of water supplied. This equates to water saving of 21,000 m^3/d.

The process has now moved on to cover Birmingham, a city of over one million people, where an innovative "lift-and-shift" strategy, using Permalogs, has delivered significant reductions in leakage.

This "lift-and-shift" strategy has further built on the process management innovations developed for Worcestershire, provided for enhanced cooperation between distribution network controllers and field service staff, and allowed for more directed reward schemes for field staff. The process has been closely coordinated with pressure reduction and optimization schemes and is now being rolled out across the whole of the company's distribution operation.

Benefits of Technology
Permalog challenges the traditional concepts of leakage detection. Traditional methods relied upon targeting an area and then deploying any number of field teams to sound between fittings to find out if leaks existed or not. As leakage levels fell the cost of both maintaining the position and bettering it became proportionally higher.

New technology has allowed the end-to-end process of leakage detection and repair to be totally reengineered. The shift from manual sounding to a reliable IT-based alternative that sits as a constant monitor on the network enables leaks breaking out on the network to be detected much earlier. Repairs can be scheduled to follow the detailed detection phase, as it is known, with more certainty that leaks will be found.

As total leakage is a function of the number of leaks breaking out and the time they run before detection and repair, the optimized detection process using Permalog can generate significant benefits in leakage management. Once areas have stabilized, the deployment density can be reduced and/or the patrolling frequency reduced, depending on the desires of the water utility, available funding, and the targets expected to meet. The benefits from Permalog lie not just with the application of the technology, but also in the ability to significantly reengineer the whole leak detection and repair process.

Achievements to Date
Leakage levels in ST have nearly halved since the introduction of its Leakage Control Strategy in 1996. Leakage in 2000 stood at 340 ML/d compared with 665 ML/d in 1994.

The United Kingdom Regulator (OFWAT) set mandatory industry targets for 2001 to 2002, and ST's target was set at 333 ML/d.

As of the year 2000, the company had the second lowest unit leakage level in the United Kingdom water industry. More significantly, ST held the lead position in the "virtuous quadrant" of low leakage and low water consumption by its unmeasured domestic customers.

Demand has correspondingly reduced from a level of 2100 ML/d in 1995 to a current level of 1880 ML/d. This reduction obviously has significant implications for the company's resource strategy. The additional costs of active leakage control are currently estimated to be in the order of £16 million per year. Clearly, there has to be an economic level of leakage below which further investment cannot be justified.

OFWAT assessed ST's own calculation of the economic level of leakage as robust and recommended that the company should be measured against its progress toward this level. They do require, however, that the economic level of leakage be recalculated on a biannual basis. Given the early achievements of the Permalog technology, it is likely that the company's success in improving pressure reduction, Permalog and other leakage detection and repair techniques will become widespread, leading to an ongoing downward direction for the economic level of leakage for the water industry.

A.10.6   Conclusion

Harnessing new technology such as Permalog has enabled ST to make a step change in leakage performance and maintains its industry leading position. It must be emphasized, however, that new technology is not in itself a panacea in the war against leakage. At the end of the day, Permalog is simply one strategic tool that sits firmly alongside a number of tried and tested methods and practices. It is the strategic leakage management process within which it is used that delivers the real benefits.

Case Study A.11: Conservation Project Saves $24 Million for Utilities

Tim Brown, *Heath Consultants*

One of the largest energy and water conservation projects carried out in the United States saved $24.4 million per year at a cost of $2.7 million for 278 water utility companies in the state of Tennessee as of January 1991. This project achieved a benefit cost ratio of 9.5:1, representing a payback period of just 38 days. The average system savings was $91,398 per year.

The Tennessee Department of Economic and Community Development, Energy Division, provided the funding for, and implemented the water accountability project. More than 400 water utilities were eligible for participation in the program.

The Tennessee Energy and Water Conservation Program was submitted to the State of Tennessee and was nominated by State Governor McWherter for national award consideration by the U.S. Department of Energy. Government officials, scientists, engineers, and others then evaluated the project. Upon completion of the evaluations, the program was honored with a National Award.

The Tennessee Association of Utility District oversaw and administered the project on behalf of the State of Tennessee, and the water system audits, meter accuracy testing, and leak detection/pinpointing surveys were performed by a consultant. The project was divided in two phases:

- Phase I: to identify energy and water loss and to make recommendations for corrective action
- Phase II: to conduct a leak detection/pinpointing survey of the distribution system.

Phase II actions were initiated when the benefit-to-cost ratio determined that this activity was economically justified.

A.11.1   A Description of the Program

In January 1988 Heath Consultants contracted with the Energy Division of Tennessee's Department of Economic and Community Development to conduct the two-phase program to identify energy and water loss and to make recommendations for corrective action.

Phase I of the program included a detailed audit of the water produced and purchased, operational costs, electric consumption (pumping costs), and the daily operations of the utility. All testable master and commercial/industrial water meters 2-inch and larger were tested to determine their accuracy, since inaccurate meters figure significantly in determining the water system's product accountability.

The purpose of Phase I was to accurately determine the amount of nonrevenue water, and the cost of that water based on the cost to produce and/or purchase and distribute it. This figure is known as the avoidable cost. The total avoidable energy loss (BTUs) and total dollar loss to the utility due to nonrevenue water was used to determine the benefit-to-cost ratio for corrective action.

Before the program began, there was confidence that the program would save a great deal of energy, water, and money, but it appears the magnitude of savings was grossly underestimated!

When Phase I was complete, a total of 119 audits had been compiled, yielding a cumulative total of $9,010,224 of avoidable cost within 1 year at a cost to the State of Tennessee of only $409,132. This represented a payback period of 16.6 days. Included in the avoidable cost was 72,698,052,000 avoidable BTUs representing $1,496,860 of energy savings. In addition, a total of $342,909 of avoidable revenue loss per year was identified mostly due to inaccurate meters.

It was also reported that a total of 77 systems had completed the leakage detection phase, pinpointing a total of 4,175,118,600 gal/year of water loss due to system leakage. This represents $4,793,863 of lost water per year. The cost of pinpointing this lost water was only $511,944 with a payback period of just 39 days. The system needed only to excavate the pinpointed leakage locations and repair the leakage to realize their savings.

The innovation, transferability, energy savings, and economic impact encompassed by this program made it a winner. Innovation was demonstrated in the program's ability to assist water systems throughout Tennessee to identify and correct deficiencies in order to operate more efficiently. Utilization of funds from the state "oil overcharge fund" to finance this extensive program was a breakthrough. Many of the managers and operators of the system had never been exposed to such in-depth study of virtually all aspects of their water system operations. An objective "third party" review and a full explanation of all activities that were being performed throughout the program, coupled with the reduction of energy and water loss identification, made this a very popular program.

Transferability comes into play when experts "transfer" their knowledge to system operators and their own in-house staff. For example, utility operators observed the meter testing process and learned the how and why of meter accuracy testing, and the importance of controlling losses. Having observed this procedure, some operators elected to conduct meter accuracy tests in-house, utilizing existing personnel.

When a leakage control survey is found to be necessary, the consultant often requests that a utility company employee accompany the leakage technician while the survey is in progress so that the utility employee can be trained to detect and pinpoint the source of leakage in the distribution system. The first thing the operator learns is that the vast majority of leaks, for a variety of reasons, will not come to the surface. Many water operators have now learned that they cannot wait for leaks in the system to emerge in order to find them; they must go out and find out where the hidden leaks exist. This knowledge alone goes a long way toward controlling leakage in the system. When the leakage control survey is completed, each utility is left with a program to control their leakage in the future and to respond to leakage complaints with a logical and systematic plan.

Energy savings generated by the program were very easy to demonstrate. The program identified the amount of energy loss, which can be saved by corrective action. During this project, most of the loss was due to system leakage, which was pinpointed and repaired, resulting in immediate energy savings due to the decrease in required pumping volumes.

The obvious economic impact of this program was derived from savings, which were realized by the individual water systems. Many water utilities became able to operate on a more economically sound basis. Many utilities looked to upgrade their systems to operate more efficiently to supply higher quality water to the consumer at a reasonable cost, with the end saving realized by the consumer. Many systems capitalize on their potential return on investment for this type of service. Future budgets within these water systems should include such services to allow the systems to be maintained and operated under sound economic practices. This will conserve energy and drinking water, which is beneficial to the industrial and residential growth of every society.

Case Study A.12: Water Temperature Predicts Maintenance Peaks*

Scott Potter

The Louisville Water Co., a Kentucky utility chartered in 1854 as a municipal corporation, is a nationally recognized utility with demonstrated technical competence in all areas of water utility management. LWC is a member of the Partnership for Safe Water and one of the first utilities to be evaluated by the AWWA QualServe program. As such, LWC is proactive in dealing with legislation and regulations under the Safe Drinking Water Act and continuously maintains a rigorous research program to effectively deal with possible future requirements of state and federal regulations. As part of its proactive program, LWC replaces or rehabilitates 45 mi of water main each year—approximately 1.5% of the system—for an annual capital expenditure of $10 million.

Evaluating main breaks is an important part of the replacement and rehabilitation program. When looking at the entire transmission and distribution system, one factor stood out more than others as a contributor to main breaks—finished water temperature (FWT). The age of the cast iron pipes varies throughout the LWC distribution system from older than 130 years to brand new. The pressure also varies significantly, from a minimum of 40 to 100 psi, as do the soil conditions, from clay to sand. These variations do not appear to affect the number of breaks throughout the system as much as

*Reprinted from Opflow, vol. 26, no. 7 (July 2000), by permission. Copyright © 2000, American Water Works Association.

FWT and, except for the temperature of and drought effect on the soil, were not considered in the following discussion.

A.12.1   Verifying Operational Observations

LWC decided to examine closely FWT because several people within the operational group used it as an informal indicator for probable increased break activity. Experienced operations staff knew that as finished water temperatures dropped toward 40°F (4.4°C), break activity would increase. A detailed survey of temperature trend data confirmed this informal observation.

LWC experiences dramatic water temperature changes because raw water temperatures from the Ohio River vary from 33°F (0.5°C) to 85°F (29.4°C) over the course of a year. One riverbank infiltration well has a slight moderating influence on FWT in LWC's elevated service area, but the breaks appear consistently throughout the system, and the FWT discussed here represents the main plant's discharge temperature. Analysis demonstrates that extreme temperatures (either low or high) produce above-normal break activity.

Data collected from December 25, 1998, through March 8, 2000, demonstrates a strong correlation between the FWT and the propensity for main breaks. When the FWT reached 39°F (3.9°C) on two separate occasions, the number of main breaks increased dramatically, and when the FWT approached 90°F (32.2°C), main break activity increased as well (Fig. A.12.1).

The first interval of extremely low FWT was during the first 20 days of January 1999 when the water temperature was below 39°F (3.9°C). Workers from Local 1683 of the American Federation of State, County and Municipal Employees repaired 163 main breaks over this 3-week period, for an average of 7.76 breaks repaired per day. On January 21, the finished water temperature reached 40°F (4.4°C), and continued to

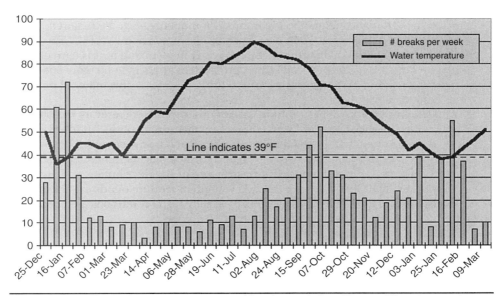

FIGURE A.12.1   Finished water temperature and break correlation. (*Source:* Water temperature predicts maintenance peaks.)

increase to 44°F (6.6°C) through the next month, and main break activity leveled off. It is important to note that the FWT never fell below 40°F (4.4°C) after January 21. From January 24 through March 13, 56 main breaks were repaired—an average of 1.17 per day. This represents an 85% decrease from the early January main break rate.

On March 18, the FWT began to increase for the spring and summer period. Break activity also began to increase. Figure A.12.1 shows, while not exactly parallel, that summer break activity also increases with higher FWT.

The colder the FWT, however, the stronger the correlation in increased break activity. The second survey period when the finished water temperature fell below 39°F (3.9°C) was between January 21, 2000, and February 11, 2000. During that interval, the union repaired 134 main breaks, averaging 6.38 break repairs per day. The peak, 99 breaks from December 23, 1999, through January 14, is presumed to be caused by a combination of rapidly declining finished water temperature and persistent drought.

While the FWT did not drop as low as 39°F (3.9°C) the rapid decrease in temperature, increase in water density, and severely dehydrated soil conditions caused by drought generated break activity almost equal to that when the FWT actually reached 39°F (3.9°C).

A.12.2   Year-Long Analysis

LWC experienced a total of 967 main breaks in 1999, an average of 2.65 breaks per day. This was a record year for the company: the two periods when FWT fell below 39°F (3.9°C), coupled with the prolonged drought (August 2 through November 9, 1999), contributed to the extraordinary number of breaks (the shifting, cracked soil conditions, and high water demand during the drought are also considered to be factors in the breaks during those periods). The number of breaks per day when the FWT was 39°F or lower was 70% higher than the number of breaks per day when averaged over the entire year.

The data strongly supports the conclusion that a FWT of 39°F (3.9°C) or lower will result in a dramatic increase in the number of main breaks to be repaired. The reasons for this phenomenon have not been specifically researched by LWC. There is general consensus within LWC that the density of water maximizing in this temperature range plays a large role. The cast iron within our system appears to be more susceptible to a rapid decrease in FWT: a rapid transition to 39°F (3.9°C) in this material produces even higher break activity.

Data from the LWC Distribution Operations ground temperature measurement system, which provides constant soil temperature measurements at 1-ft intervals, from 1 to 6 ft, demonstrated that the soil temperature at 3 ft and below never fell below 45°F (7.2°C) over the 1999 to 2000 winter season. This indicates that there may be a slight heating effect on water within the buried infrastructure of the distribution system at temperatures below 39°F (3.9°C). Also, if finished water temperature trends are at extremely high levels, break activity may increase, too, especially if soil conditions are poor.

A.12.3   Conclusions

Finished water temperature is a great advance warning system. LWC Distribution Operations uses this information for advance planning and the identification of the need to initiate the winter emergency plan. For instance, if long-range weather forecasts indicate extreme low temperatures over a sustained period and FWT is dropping quickly toward or is already below 39°F (3.9°C), it is reasonable to assume that the unusual break activity is going to begin to persist. This may require contractual assistance in main break repair, notification to authorities of the possibility of longer-than-normal repair completion rates, and other activities.

LWC is also gathering data to identify the effects, if any, of mixing the demonstration Riverbank Infiltration Well discharge water with water from our normal Ohio River source on finished water temperature. An unexplored possible benefit to the Riverbank Infiltration Well water is that water's temperature stability when compared with Ohio River water. Other research, including the continuing examination of soil temperatures, is also planned. For, in observing and analyzing the patterns that contribute to a problem such as main breaks, LWC can continue to be proactive in its efforts to supply safe drinking water to its customers.

Case Study A.13: Santana Zone—SABESP Sao Paulo—A Successful Case Study for Water Loss and Energy Reduction

Mario Alba

Milene Aguiar

The Santana supply zone is located in the northern part of Sao Paulo supplied by the Cantareira treatment facility. The average flow into the zone is 700 L/s which supplies a population of 174,000 through approximately 44,000 connections and 320 km of mains. Table A.13.1 shows the key system statistics.

The supply zone has two pressure zones within it: the high zone which is supplied by a booster and the low zone which is supplied by the transmission system directly. Figure A.13.1 shows some shots from the zone.

The Santana booster station has a reservoir which is partially below ground with a capacity of 12,000 m^3 and is supplied by excess pressure from the transmission main from the low zone. Transmission main pressure and the booster inlet pressure are controlled by SCADA. The booster station has five pumps which were controlled by a pressure sensor in the water tower. Figure A.13.2 shows the old hydraulic system prior to this project.

System	Cantareira
Flow	700 L/s
Active connections	44,000
Population	174,000
Mains length	320 km
Residential properties	68.514
Commercial properties	6.310
Industries	601
Public buildings	83

Source: Mario Alba and Milene Aguiar.

TABLE A.13.1   Santana Supply Zone Key System Statistics

FIGURE A.13.1   Pictures of Santana supply zone. (*Source*: Mario Alba and Milene Aguiar.)

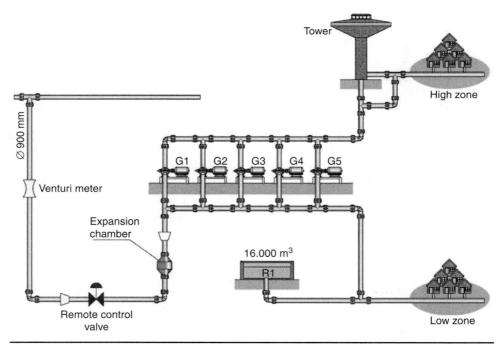

FIGURE A.13.2   Old hydraulic system prior to this project. (*Source*: Mario Alba and Milene Aguiar.)

With this configuration the inlet pressure to the station was reduced by a valve and reducer. The reduction in pressure from 25 to 5 meters head (m) was necessary in order to ensure supply to the low zone; however, this meant that it was necessary to boost pressure through the station to double the pressure necessary as can be seen in Fig. A.13.3.

Figure A.13.3   Changes to pressure. (*Source*: Mario Alba and Milene Aguiar.)

Before	After
2 200 HP pumps + 3 100 HP which were activated from tower level.	4 100 HP (3 functioning and 1 reserve), with VFD and main pressure sensor

Source: Mario Alba and Milene Aguiar.

Table A.13.2   Characteristics of Old and New Pumps

A proposal was made to separate the mains which fed the high and low zones so that it was not necessary to boost the pressure to double the required pressure.

A pressure study showed that during the night it was not necessary for the booster station to operate. The pumps were changed out for lower power pumps as shown in Table A.13.2.

The hydraulic configuration after the project was completed and can be seen in Fig. A.13. 4.

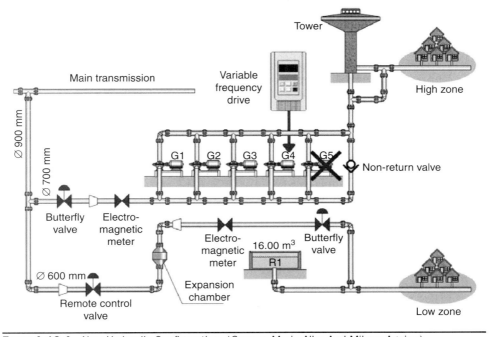

Figure A.13.4   New Hydraulic Configuration. (*Source*: Mario Alba And Milene Aguiar.)

Before	After	Advantages
Poor quality source metering with one 900 mm venturi	Good quality source metering with 2500 mm lectromagnetic meters	Improved accountability
Booster station with 2 200 HP pumps and 3 100 HP pumps	Booster station with 4 100 HP pumps	Equally sized modern pumps with improved performance and ease of maintenance
Fixed speed pumps	Variable speed drives	Optimized operation
Peak demand of 480 kW and off peak demand of 500 kW	Peak demand of 180 kW and off peak demand of 270 kW	Energy conservation
Station operated with 4 duty pumps and 1 reserve pump	Now operates with 3 duty pumps and 1 reserve pump	Renergy conservation
High zone fed by booster pumps	High zone fed by booster pumps during the day and by gravity at night	Reduction in leak volumes due to reduced pressure at night

Source: Mario Alba and Milene Aguiar.

TABLE A.13.3   Benefits of New Hydraulic Configuration

The new configuration brought many benefits which can be seen in Table A.13.3.

A.13.1   Results

As a result of the installation of variable frequency drives (VFD) on the pumps the outlet pressure became a function of the supply main and not the level in the tower which adequately supplied the necessary pressure to the system as required.

The VFD reduced pumping during peak load hours while also guaranteeing the necessary pressure at the critical points in the system which are monitored by SCADA.

Another benefit was reduced nighttime pressures as previously pressure went up significantly in the system when there was little consumption and headloss was low. As well as energy reduction the project also brought about a reduction in leakage volumes and new leakage frequency.

The results are shown as follows:

1. Energy conservation: 100,000 kWh/month
2. A reduction in leakage volume of 283 L/conn/day (850 to 570 L/conn/day)

The night pressure before and after the project can be seen in Fig.A 13.5.

Figure A.13.6 shows the reduction in supply volume after the project was completed and Fig. A.13. 7 shows the reduction in minimum night flow (MNF).

A simple cost benefit analysis shows the following results:

Total cost: R1,000,000

Monthly energy savings: R28,600

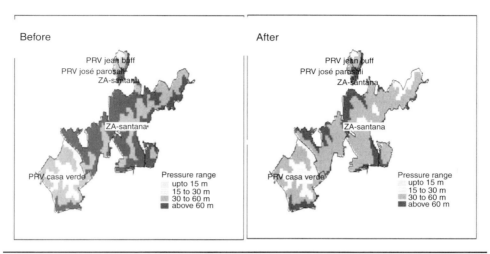

FIGURE A.13.5   Night pressure before and after project implementation. (*Source*: Mario Alba and Milene Aguiar.)

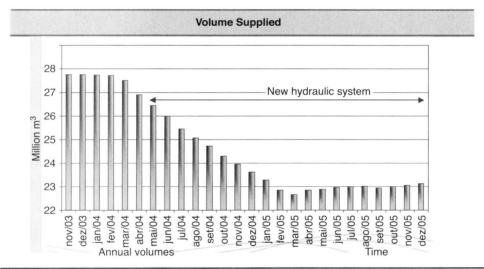

FIGURE A.13.6.   Reduction in supply volume before and after project completion. (*Source*: Mario Alba and Milene Aguiar.)

Monthly savings on leakage volume: R253,333

Total savings: R281,933

Project payback: 4 months

The project payback was calculated only on reduction in energy use and leakage volumes. However, it is likely that the frequency of new breaks in the system also reduced adding even more benefit to the project—SABESP is currently studying the effects of reduced pressure and reduced break frequency and will report on this when data is available.

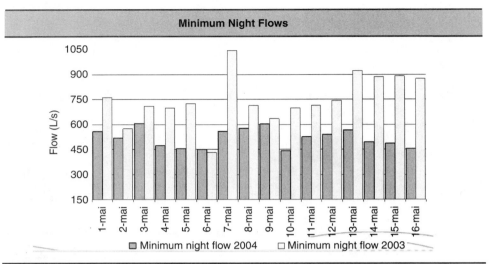

Minimum Night Flows

☐ Minimum night flow 2004   ☐ Minimum night flow 2003

FIGURE A.13.7   Reduction in minimum night flows before and after project completion. (*Source:* Mario Alba and Milene Aguiar.)

Case Study A.14: Advanced Water Pressure Management in the Berea–Alexander Park Supply District[*]

Allen Young

A.14.1   Background

The Eastern Local Council of the Greater Johannesburg Metropolitan Council (the GJMC) calculated in 1999 that 18.6% of the volume of all water acquired from its bulk water supplier, was being lost in the storage and distribution system, over and above an allowance of 12% for expected losses.

No data is available on what proportion of the unaccounted-for water may be due to leakage from pipes; however, a rough indication based on comparison of observed night flows in the Berea–Alexander Park district, suggest that as much as 50% of this might be due to leakage.

In November 1998 the concept of advanced pressure management using electronic controllers fitted to pressure reducing valves (PRVs) to achieve modulation of pressures, was presented to engineers of the GJMC by Mr. Julian Thornton of the Brazilian company BBL/Restor. Case studies based on his experience in applied pressure management in Sao Paulo, Brazil were presented, which provided the motivation for the council to include pressure management as one of its strategic initiatives, to reduce unaccounted-for water. The ambit of the GJMC project was to locate existing pressure controlled districts that would benefit from modulated pressure control, or alternatively to identify suitable districts for introduction of pressure management. The latter approach proved to be the most fruitful, and the Berea-Alexander Park supply district

---

* By permission of the Greater Johannesburg Metropolitan Council.

was selected as one of the districts that presented good potential for successful implementation of advanced pressure management.

A.14.2   Selection Criteria for the District

The criteria that favored selection of this district were

- Its large size and potential isolation from neighboring supply areas
- Lack of known low water pressure problems
- Adequate static pressures and topography that lends itself to an overall reduction of the hydraulic grade across the district
- Suitable positions for PRVs that would allow adequate working head for pressure control

A.14.3   Description of the District and Condition of the Pipes

The district covers an area of some 1370 hectares and is a predominantly residential area with a range of lot sizes from 0.06 to 0.15 hectares. The residential area has commercial centers consisting mainly of clustered shops and small shopping complexes, with the Bruma office park and hotels on the eastern side of the district. The area houses four sports clubs and the Kensington golf course, which are potentially large users of water for irrigation.

Other vital statistics of the supply area are set out in Table A.14.1.

The district is supplied with water from two linked reservoirs situated on the western side of the district. These reservoirs are constructed with approximately the same top water levels, and supply a maximum static pressure in the lower parts of the district

---

- Number of consumer connections: 8577
- Population: 30,230
- Type of consumers
  - Residential: 95%
  - Commercial: 5%
- Mains lengths
  - Primary mains: 200 to 750 mm in diameter, 34 km
  - Secondary mains: 20 to 160 mm in diameter, 136 km
- Pipe materials, proportion of total length, and average age
  - Steel, 71%, 45 years
  - HDPE, 10%, 13 years
  - UPVC, 18%, 14 years
  - Fibre cement, 1%, 52 years

Source: GJMC.

TABLE A.14.1   Statistics of Berea–Alexander Park Supply District of 12 bars, while a minimum static pressure of 4.9 bars is provided at the highest point in the district. There was no existing pressure reducing valve sites within the zone.

The older supply mains in this zone are of rolled steel with a 6 mm wall-thickness and caulked spigot and socket joints. Pipes are coated internally and externally with bitumen. Although external bitumen coatings on older pipes are generally intact, internal linings exhibited loss of binding with entrapped pockets of water between the lining and pipe wall with resultant pockets of advanced corrosion under the lining. Spalling of the bitumen lining was also observed. Exposed caulked joints were found to be weeping which together with the general internal condition of the pipes, indicated that leakage from the older primary mains, which comprise about 15% of the distribution system, is a likely occurrence.

Little is known of the condition of the smaller diameter piping and consumer connections, but the age of the older sections of the district (>45 years) indicates that corroded house connections would be a cause of leakage.

A.14.4   Preinstallation Investigation and Initial Pressure Management Plan

Following a desktop study in which the district was identified as a possible candidate for pressure management, pre-installation investigations were undertaken in order to develop a pressure management plan for the district. Field investigations included the following:

* Gathering of infrastructure and demographic data for the district
* Field inspection of proposed PRV sites and consumer types in the critical high areas
* Checking of normally closed valves that isolate the district from adjacent supply areas
* Measurement of flows and pressures at the feed points, and logging of pressures at critical high points and other selected points in the reticulation. (Use was made of portable electromagnetic and turbine insertion meters as an economical method to measure transient flows.)
* Analysis of data and the estimation of leakage reduction for a proposed diurnal modulated pressure profile using a statistical model
* Estimation of costs for installation of PRVs and performance of a cost–to-benefit analysis to test the viability of the proposal

The pressure management plan was to install two new PRV stations on the two reservoir feeds into the district. Pressures would be modulated to give an average reduction of 1 to 2 bar throughout the zone with a maximum reduction of pressure of 2 bar during off peak times based on a target pressure at the critical high point of 3 bar.

Use would be made of the Technolog Autowat PRV control equipment to modulate pressures. Theoretical pressure modulation profiles for the two PRVs were designed as a starting point. The interaction between the two PRVs is a function of the head loss in the reticulation and would require observation in the field. The pressure control profiles would have to be set empirically in the field once the effect of installation of PRVs on the pressures and flows at the feed points and critical points had been observed. It was anticipated that one of the new PRVs (the one at a lower elevation) would be set at a fixed outlet and would probably remain closed except during peak draw off periods. Pressure modulation in the zone would then take place by controlling only the other PRV.

The practical location of pressure control stations took the following into account:

- The PRV should have the maximum upstream working head possible taking into account the expected head loss through the PRV.

- The points at which the least number of PRVs would be required to effect control on the district.

- The ease with which the site would fit into the existing pipe layout.

- The ease of access to the future station.

- The environmental acceptability of the site.

Field measurements of flow and pressure were carried out at the proposed PRV sites and at the critical high point and average pressure points in the district. Portable electromagnetic insertion flow meters were used to obtain temporary measurements of the flows.

The field data was used in a statistical software model that estimated the effect of modulating pressures on reduction of background and burst leakage in the pipe system, taking into account estimated reduction in pressure related consumption.

The field data was further used to size PRVs and meters for the proposed pressure control stations.

A calculation was done of expected saving of water. The forecast monetary saving through reduction of leakage using the statistical model was R795,816 per annum. A cost-to-benefit calculation yielded a payback period of some 8 months confirming the economic viability of the pressure management plan, and it was decided to proceed with implementation of the pressure management plan.

A.14.5 Final Design

Each PRV station comprised a 250 mm Claval PRV with a pot strainer mounted upstream and a Meinecke meter positioned 5 diameters downstream of the PRV. A 250 mm bypass was constructed around the meter and PRV to facilitate future maintenance.

Reinforced concrete chambers were constructed, one of which was partially positioned under a roadway. The chamber was enlarged and partially repositioned to create access from the sidewalk—an important consideration for safety and ease of access for regular data downloads and checking of equipment.

A.14.6 Results

After commissioning of the PRVs they were both set to maximum fixed outlets. It was found that the PRV at Montague Street remained shut (inlet pressure 7.5bar/outlet pressure 6 bar) while the PRV in Berea Road continued to feed the zone without a problem (inlet pressure 5 bar/outlet pressure 4 bar).

Night flows were observed to have diminished from 350 to 110 m^3/hr.

The outlet pressure of the PRV in Berea Road (black line) was seen to drop from 3.2 bar during the daytime to 2.2 bar at night. The pressure at the critical high point became more even and reduced only by 0.5 bar on average. (This may indicate a smaller uncontrolled feed into this area that is able to sustain pressures at high point of the zone, which will require further investigation.) The pressure at the average zone pressure point was reduced on average by 1 bar and showed less fluctuation, indicating that leakage is being reduced throughout the district and there is less stress on the pipe network.

In addition to reduction in night flows, there was a clear reduction in consumption during peak periods. Part of this was due to reduction in leakage and part due to reduced pressure related consumption, for example, garden sprinkling.

Total reduction in consumption (leakage and usage) was calculated. The minimum reduction in leakage was estimated for the period 21:00 to 05:00 when normal consumption is minimal. These figures are shown in Table A.14.2 and Table A.14.3, respectively.

The total cost for the two sites including professional and construction costs was R850,000. A realistic payback period of 6 to 9 months was therefore achieved.

Volume: 2259 m^3
Period: 23.75 hour
Savings per day: 2283 m^3
Savings per annum: 833,141 m^3
Rand value: R1,749,595

Source: GJMC.

TABLE A.14.2   Total Reduction in Consumption

Volume: 1110 m^3
Period: 6 hours
Savings per day: 1110 m^3
Savings per annum: 405,223 m^3
Rand value: R850,968

Source: GJMC.

TABLE A.14.3   Nighttime Reduction in Leakage from 21:00 to 05:00

Case Study A.15: Case Studies in Applying the IWA WLTF Approach in the West Balkan Region: Pressure Management

Jurica Kovac*, *IMGD Ltd., A. Georgijevica, Croatia*

A.15.1   Abstract

The purpose of this paper is to present the results obtained so far in promotion and implementation of the IWA WLTF (International Water Association—Water Loss Task Force) approach in solving problems regarding losses in water distribution networks in the region of Western Balkan.

The situation in the region regarding losses is serious (NRW is in average above 50%) and it is necessary for all water utilities to consider implementation of plans and programs for proper quantification of losses and creation of water losses reduction strategies.

One of the most important steps in this program is the selection of an appropriate methodology. In the past, before the IWA WLTF approach, reliable benchmarking and evaluation of options was not possible because of the many different approaches used for calculations of water balance and performance indicators. Also, very often, these previous approaches were unsuccessful, and were associated with high costs, little sustainable reduction in losses, and low motivation to continue. In our region the losses are still presented in terms of percent of nonrevenue water (NRW). Some individual utilities are now starting to use IWA terminology (or similar), but usually with some exceptions and modifications that sometimes produces more confusion.

Our intention is to present our experience in implementation of the IWA terminology and WLTF approach, to encourage others to follow. To help everyone with an interest in water losses problems to "get started," we have translated a simple international software for calculation of the IWA Water Balance and basic performance indicators (CheckCalcs). The software is free of charge and can be an excellent first tool for quantification of losses, and for a first realistic benchmarking between water utilities. CheckCalcs helps in understanding where we really are, and priorities as to how to proceed (presentation of main measures needed and simple calculation of benefits regarding pressure reduction in the system). Our goals are to start with implementation of the IWA WLTF approach by individual water utilities, and to promote acceptance nationally. This should result in a better understanding and faster improvements, and at the end saving of water that is so important for all of us.

Presentation of Results Regarding Analyses of Real Losses in Water Distribution Systems from the Region

All our water utilities are public companies, owned by municipalities or towns. This means there are a large number of utilities, quite small with weak financial strength and lack of qualified and trained staff (for example, the Croatia population is 4.3 million

---

* jurica.kovac@imgd.hr.

with 116 public water utilities). Also, the problem of losses in distribution system was for a long time considered less important than increasing the coverage of population with safe drinking water. Very often the same utility is responsible also for the sewers, and sometimes also for some other communal activities like waste collecting, maintenance of parks, cemeteries, and the like.

In the last couple of years many large water utilities have invested in equipment for leakage detection and pipeline inspection (ground microphones, leak correlators, mobile flow, and pressure meters). But very often they had not developed loss reduction programs based on pressure management, or active leakage control for awareness and location of unreported leaks. Some midsize and small utilities received some equipment through donations or by other kinds of international help (for example, Bosnia and Herzegovina, Serbia, Monte Negro, Croatia); but in most cases equipment was purchased without proper selection and at the end often without proper (or without any) staff training.

Knowledge regarding district measuring areas (DMAs) is getting more accepted in the region but is still not used enough. The reason is that old systems were developed with many interconnections for emergency supply and water quality objectives. However, the utilities with lowest losses are using system zoning with installed control flow meters.

Installations for pressure control in the systems are rare, perhaps because we have lacked knowledge regarding the influence on pressure on leak flow rates and burst frequencies. We have cases where pressure reduction valves (PRVs) are installed because of very high pressures, but without proper maintenance they malfunction, resulting in higher losses and frequent bursts.

We have also, more recently, some positive examples where utilities with lowest losses are implementing PRVs or other solutions in pressure control.

We must also underline a serious problem in our water utilities: the lack of qualified, trained, and motivated staff. Sometimes the problem is technicians who are responsible for the leakage detection and pipeline inspections (untrained or underpaid). More often, managers do not understand importance of managing losses, have lack of knowledge of practical effective methods, or are simply too occupied with other obligations; this sometimes results in the incorrect conclusion that losses can be effectively reduced only by replacing old pipelines.

From our experience it is most often the case that utilities have the staff necessary for successful implementation of losses reduction program, but the staff are not adequately managed.

From the beginning of 2005, IMGD started to use the IWA terminology in calculations of all the components of the water balance, including real Losses. Also, other concepts promoted by the IWA WLTF are now becoming part of our activities (BABE and FAVAD concepts, active leakage control, pressure management, and the like).

An important evolution was the introduction of the performance indicator ILI (infrastructure leakage index), which is the ratio of CARL (current annual real losses) to UARL (unavoidable annual real losses). This was a major step forward for our water utilities considering that for the first time we could assess unavoidable annual real losses on a system-specific basis, taking account of local characteristics (main length, number of service connections, meter location, pressure).

In Croatia (see Fig A.15.1), it has been traditional to consider NRW losses of less than 25% as being a good performance, without allowing for different system characteristics (current average NRW is 40% for 2005).

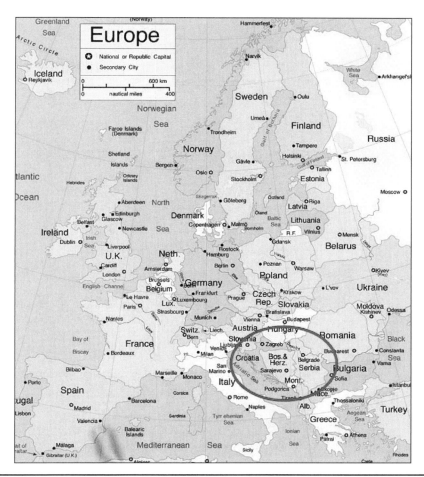

Figure A.15.1   Western Balkan region.

In the table below, %NRW and ILI are calculated for 12 Croatian systems and one from Bosnia and Herzegovina (Table A.15.1). Note: some received data from some users are based on approximate data and some errors are possible (unbilled authorized consumption, unauthorized consumption, average pressure) but we assume that in all cases the confidence limit is acceptable for initial comparisons of this kind. In the future with more experience regarding the new methodology, the accuracy will be better.

When these data are compared with international data sets where we have mean ILI 438[1] it is evident that situation in our water distribution systems regarding real losses is similar to the world scale.

It is also important to emphasize that %NRW is not adequate for assessing performance in managing real losses (Fig. A.15.3.). For example, in systems 3 and 9, the %NRW is similar (39% and 38%), but the ILI provides more meaningful performance information for real losses management. Because each system has different specific characteristics and different unavoidable annual real losses, we can see from the ILIs that real losses management in system 3 (ILI = 2, 7) is twice as good as in system 9 (ILI = 5, 8).

Distribution system	Pipelines length (km)	Number of service connections	NRW (% WS)	CARL (% WS)	CARL (L/conn/d)	UARL (L.conn/d)	Average Pressure (m)	ILI
1	142	6310	33	31	111	73	55	1,5
2	1500	42000	27	25	168	99	60	1,7
3	259	4834	39	35	259	96	45	2,7
4	991	30375	42	39	277	82	50	3,4
5	1500	23000	54	50	451	122	60	3,7
6	338	9000	37	33	290	82	60	3,7
7	713	33073	24	19	302	73	65	3,7
8	550	21700	41	35	230	55	40	4,2
9	435	12000	38	34	464	80	50	5,8
10	265	13995	52	46	346	47	40	7,4
11	97	4535	53	49	486	73	45	7,5
12	117	9184	49	43	345	40	35	8,7
13	769	42308	70	65	1069	63	50	1,7

Source: Jurica Kovac, IMGD Ltd

TABLE A.15.1 Comparison of System Characteristics and Performance Indicators for 13 Systems

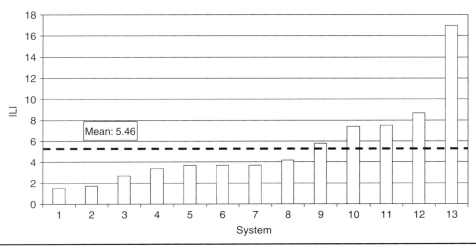

FIGURE A.15.2   ILIs for 13 Systems in Croatia and Bosnia Herzegovina. (*Source:* Jurica Kovac, IMGD Ltd)

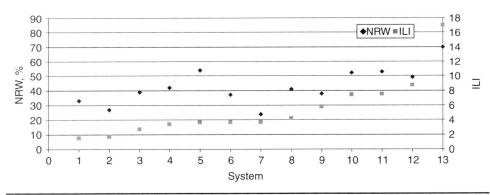

FIGURE A.15.3   13 Systems in Croatia and Bosnia and Herzegovina NRW and ILI. (*Source*: Jurica Kovac, IMGD Ltd)

The following case studies are used to demonstrate how implementation of activities like zoning, pressure control, and the like also supported by IWA WLTF approach, can be effective. We will present them briefly with the most important steps undertaken and the results obtained.

Case study 1: Pilot Project Zagreb, Croatia

Reduction of Leakage through Pressure Control—Development and Result Obtained   The water distribution system in Zagreb city, the capital of Croatia, is one of the largest in our region (over 2900 km of pipelines and more than 100,000 connections, serving a population of approximately 800,000). In October 2005 we have started a pilot project regarding pressure control for leakage (losses) reduction. The selected zone (see Figs. A.15.4, A.15.5, and A.15.7.) is a residential area with multistorey buildings

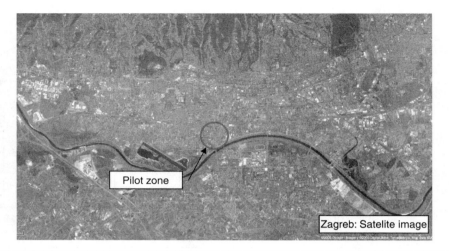

FIGURE A.15.4   Location of pilot zone. (*Source*: Jurica Kovac, IMGD Ltd.)

FIGURE A.15.5   Pressure control chamber. (*Source:* Jurica Kovac, IMGD Ltd.)

(averaging 10 floors), high pressures, and a suspected high level of leakage. The zone has 13.5 km of cast iron mains, and 653 service connections (cast iron, galvanized iron, and PEHD).

The first step was initial measurement of flow and pressure within the zone (after all boundary valves had been closed and checked). IMGD selected the location, and specified all details regarding chambers, for installation of pressure reduction valves and all other equipment (PRV DN250, Woltmann type flowmeter, valve controller, and remote GSM monitoring—Fig. A.15.5).

Also we have established three selected locations for pressure monitoring (with GSM data transfer) inside of the zone (Fig. A.15.6).

Implementation of the project had the following outcomes (with following initial data minimum flow: 44 L/s (160 m^3/hr) and initial inlet pressure: 6.50 bar (day) up to 7.10 bar (night)).

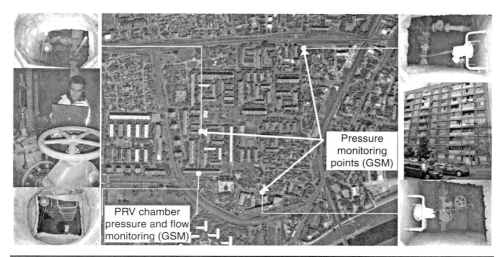

FIGURE A.15.6   Location of pressure monitoring points and PRV chamber. (*Source:* Jurica Kovac, IMGD Ltd.)

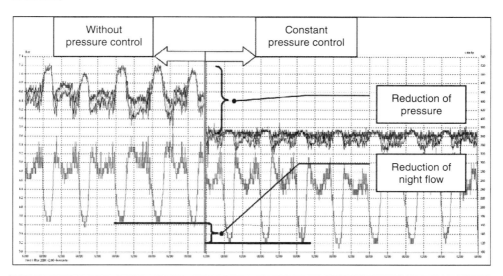

FIGURE A.15.7   First step of regulation: fixed outlet pressure (5.70 bar). (*Source:* Jurica Kovac, IMGD Ltd.)

1st step of regulation: fixed outlet pressure (5.70 bar)—Fig.A.15.7. .Night flow reduced by 24% (total 24 hour inflow reduced by 11%).

Second step: outlet pressure varies with flow (day pressure 5.70 bar; night pressure down to 4.80 bar)— Fig. A.15.8.

Night flow reduced by 39% (total 24 hour inflow reduced by 14%)

Total 24 hour inflow reduced from 6300 m^3 to 5400 m^3 (900 m^3/d savings)

Detailed estimation is underway—using FAVAD method and calculating infrastructure condition factor (ICF).

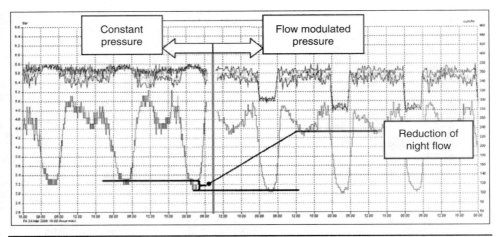

FIGURE A.15.8 Second. step: outlet pressure varies with flow (day pressure 5.70 bar; night pressure down to 4.80 bar). (*Source:* Jurica Kovac, IMGD Ltd.)

This reduction in pressure had no influence on consumer's standard of service for water supply.

Case study 2: Project Gračanica, Bosnia and Herzegovina

Pressure : Burst Frequencies Relationship, Development and Results Obtained The gravitational water distribution system in the town of Gračanica (see Figure A.15.9), north Bosnia and Herzegovina has 70 km of mains and 4500 service connections, mainly private

FIGURE A.15.9 The town of Gračanica. (*Source:* Jurica Kovac, IMGD Ltd.)

Zone	1	2	3	4	5	6
Name	Grad north	Grad center	Grad south	Čiriš north	Čiriš south	Mejdanić

Figure A.15.10   Name of zones. (*Source:* Jurica Kovac, IMGD Ltd.)

houses with two floors, and a population of approximately 15000, and has for a long time experienced water shortage, especially in summer time. In the first half of 2005 we analyzed the system and concluded that pressure control is most favorable regarding short-time benefits. The key objective was to reduce current leakage, but we also wished to explore the relationship between pressure reduction and burst frequency.

The first step was initial measurements of flow and pressure and separation of the system into six zones. The system was already separated into three areas based on ground elevation.

Separation of the system in six zones was made in order to implement flow and pressure control in more detail [introduction of district measuring areas (DMAs), and especially to separate central area (Grad) into three smaller zones] (Figs. A.15.10 and A.15.11).

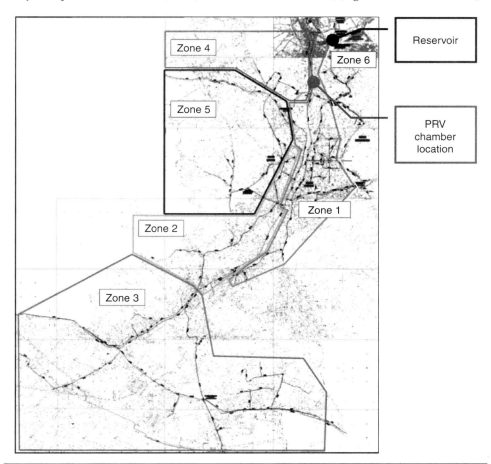

Figure A.15.11   Location of zones and their boundaries. (*Source*: Jurica Kovac, IMGD Ltd.)

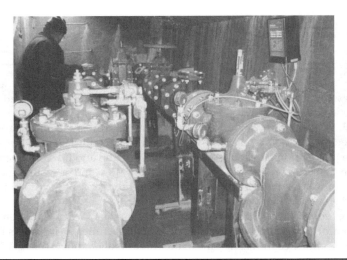

Figure A.15.12   Pressure control chamber. (*Source:* Jurica Kovac, IMGD Ltd.)

Pressure control was implemented in the area Grad (in our new zoning this area covers zones 1, 2, 3, see Fig. A.15.11.) with pressure reduction by 20%. IMGD selected the location, and specified all details regarding chambers, for installation of pressure reduction valves and all other equipment (two PRVs DN150, Woltmann type flowmeter, valve controller, and remote GSM monitoring—Fig. A.15.12).

Pressure before implementation of control and reduction was in the range between 4.80 and 5.30 bar (average 5.00 bar). Reduction of pressure and control was tested in two steps; first step with constant pressure at PRV outlet of 4.00 bar, and second step with pressure modulated by PRV controller according to current flow registered by flowmeter inside of the chamber (see Fig. A.15.13).

Figure A.15.14 clearly demonstrates the existence of a pressure: burst frequency relationship. With reduction and control of pressure, the number of bursts is dramatically reduced. *Note: presented bursts in Fig. A.15.14 are for the whole distribution system but pressure control was implemented for zones 1,2,3. Determination of results only for zones 1, 2, 3 is currently under way.*

Accomplished results for a 20% reduction of inlet pressure in area Grad (Zones 1, 2, 3); for complete system: Mains bursts reduced by 59%; service connection burst reduced by 72% (percent reductions based on PressCalcs software calculation comparing bursts rate 638 days before pressure control and 272 days with pressure control).

Another important outcome of pressure control was reduction of losses (leakage). Daily inflow was reduced by 12% (average savings 450 m^3/day)—for the complete system.

Figure A.15.15 presents data received by remote monitoring via GSM, showing how pressure control (blue line) is modulated by current flow (red line). This mode of pressure control secures adequate pressure according to current demand (for example in case of fire fighting, the system recognizes the rise in flow and automatically increases the pressure). This mode of operation can be used to ensure that all consumers will always have enough pressure. It is also important to have remote monitoring of modulated systems, because new leaks and bursts will also produce a rise of flow and a rise

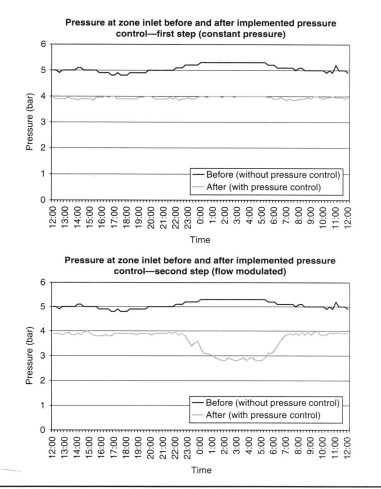

FIGURE A.15.13 Zone inlet pressure before and after implementation of pressure control. (*Source:* Jurica Kovac, IMGD Ltd.)

in inlet pressure, and such events may not be noticed if they do not generate customer complaints of low pressure or no water.

Pressure Management Projects Implementation in the Region In the last year pressure management became more recognized as efficient solution regarding water leakage reduction and bursts frequency reduction. Couple of projects are under way led by IMGD. On the map (Fig. A.15. 16) are presented locations where projects are in preparation or in implementation phase.

Promotional Activities in the Region Regarding IWA WLTF Approach

Our case studies and many others from around the world are good examples of benefits that can be achieved, and we hope that others will follow our way. In most cases, implementation of the IWA WLTF approach is also cost effective in the short term, which is one more argument to start as soon as possible.

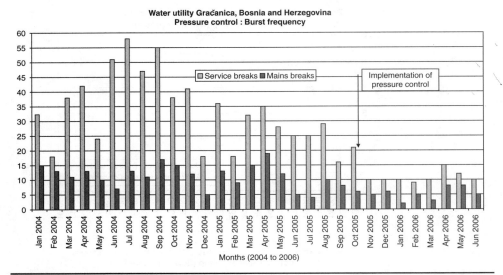

FIGURE A.15.14 Burst frequency before and after implementation of pressure control. (*Source:* Jurica Kovac, IMGD Ltd.)

The first important step regarding this approach is familiarization with the basics of the IWA WLTF methodology and terminology. For this purpose, different computer softwares have been developed. The free CheckCalcs was developed by ILMSS Ltd.—Allan Lambert—as part of the LEAKS software suite[2] (see Fig. A.15.17). With this software a water utility can quickly and easily calculate basic indicators according to both

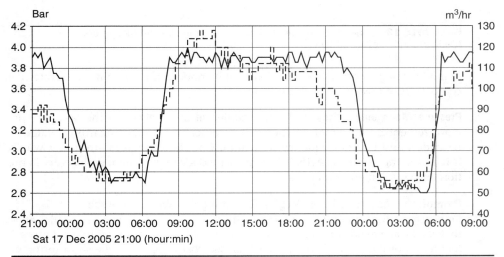

FIGURE A.15.15 Flow modulated pressure control. (*Source:* Jurica Kovac, IMGD Ltd.)

FIGURE A.15.16 Location of pressure management projects (in preparation or in implementation phase). (*Source:* Jurica Kovac, IMGD Ltd.)

old (%NRW) and new methodology (CARL, UARL, ILI) and benchmark their own performance with others from around the world or in the region. Also this software uses evaluation (ranking) recommended by the World Bank Institute. The software explains all basic terms and gives explanations how to proceed further with more advanced softwares in the LEAKS Suite.

Our goal is to help everyone interested in IWA WLTF approach. CheckCalcs software is already translated into the Croatian language but other language versions of the software for the region are also underway. CheckCalcs is available free of charge from IMGD (requiring only user registration) and because it is in Microsoft Excel it can be widely used.

Besides promotion through free software IMGD will undertake other steps in our region.

First is cooperation with the government agency Croatian Waters on promoting the IWA WLTF approach in Croatia. Our goal is to integrate this approach at a national level and to improve the traditional existing approach which uses %NRW as the main performance indicator.

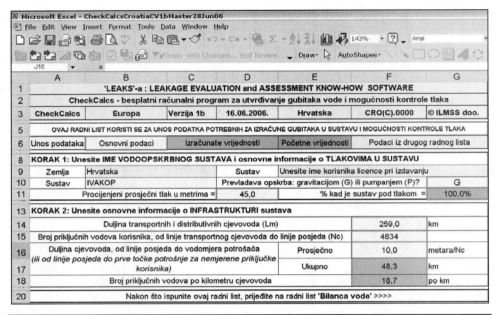

FIGURE A.15.17 CheckCalcs. (*Source*: Jurica Kovac, IMGD Ltd.)

The second step is already undertaken and consists of the transfer of knowledge through our services for water utilities.

The third step involves promotion in other countries in the region through attendances at conferences and seminars. Also we have started close cooperation with UNDP (United Nations Development Program) office in Croatia regarding water loss management.

In the beginning this program will start with implementation in Croatia. UNDP representatives are hoping that after Croatia this approach will be recognized and used in wider region.

References

1. Lambert, A. and R. Meckenzie. "Best Practice performance indicators: practical approach." *Water 21*. 43–45, 2004.
2. Lambert, A. CheckCalcs, Western Balkan version, software manual, Leaks Suite of Softwares, 2006.
3. Kovac, J. "Case Studies in Applying the IWA WLTF Approach in the West Balkan Region: Results Obtained," International conference on Water loss management, telemetry and SCADA systems in water distribution systems, Skopje, Macedonia, 2006.

Case Study A.16: Ramallah,Palestine, Case Study for Reducing Leakage from Al Jalazon Refugee Camp Water Network

Nidal Khalil, *Jerusalem Water Undertaking*

A.16.1   Background

Jerusalem Water Undertaking (JWU) is located in the central part of the West Bank, 16 km north of Jerusalem. JWU provides drinking water to most of the population centers in Ramallah and Al Bireh Governourate.

The Governourate includes one major urban area, the twin cities of Ramallah/Al Bireh, and about 100 villages, municipalities, and refugee camps.

The region that obtains its water services from the JWU is a densely populated area of approximately 205,000 people in 1999. It consists of the two largest municipal areas of Ramallah and Al Bireh cities, which form the political, economical, and cultural heart of the Governourate. And four other towns, Betunia, Beit Hanina, Bir Zeit Dier Dibwan, and Silwad which fall within 15 km radius to the northeast and west of the heart of the district. It also includes some 40 villages and four refugee camps.

JWU is a self-sufficient autonomous, nonprofit, national utility established in 1966. JWU, governed by a board of directors, has a well-deserved reputation for efficiency, and responsiveness, to community needs. JWU has the responsibility for the planning, design, maintenance, and the overall management of the water supply schemes in its service area.

A.16.2   Introduction

Al Jalazon refugee camp was established after the Arab–Israeli war of 1948. It is situated on approximately 0.85 km^2 of land located 6 km to the north of Ramallah city and is inhabited by 6400 inhabitants. Until 1980 the camp lacked a water distribution network and a decent sewage collection system.

In 1980 a water network was installed in the camp. It was laid using galvanized pipes for diameters 2 in and less and steel pipes with internal cement lining and external asphalt coating for diameters greater than 2 in. The bad condition of the open sewage collection system resulted in increasing the corrosivity of the soil and accelerated the deterioration of the network.

A.16.3   Definition of the Problem

The crowded conditions in the camp, the large variations in the topography, the large inlet pressure, and the condition of the sewage collection system all contributed to the following:

1. Over 50% of the network is subjected to pressures ranging from 10 to 16 bars.

2. Over 60% of the galvanized network and house connections are badly corroded.

3. A large difference in the average daily water billed (322 m^3) and what is registered at the inlet connection (520 m^3) indicated the existence of a serious problem, which needed to be solved.

4. Large number reported of bursts due to corroded pipes (20 in 1998) compared to the length of the network (6.56 km).

A.16.4 Investigations Made

Using the BABE software, JWU conducted an investigation for the amount of leakage from the network. It consisted of measuring the inflow and pressure at the inlet connection and the pressure at the point of AZNP (average zone night pressure) and target point over 24 hours. All the data was incorporated into BABE software and results for the expected amount of leakage, usage, and total inflow were obtained as shown in Table A.16.1. Figure A.16.1 shows the relation between pressure and flow before pressure reduction for the different flows (use, losses, and total flow).

In addition to that JWU investigated the samples of the pipes obtained from the reported bursts caused by corrosion or other causes and found that most of the samples showed that the network is badly corroded and needs replacement. Figure A.16.2 shows a picture of a sample of old pipe against a sample of new pipe.

A.16.5 Results of Investigation

The results of the investigation showed that the difference between the total amount of water, which entered the system, and the estimated use was 103.2 m^3. This amount is associated mainly with leakage from the network and was purely theoretical.

In order to deal with the problem, the pressure at the inlet was reduced (using pressure control valve) by 23 m and another set of readings were taken at the same points for the pressure and flow. This reduction resulted in reducing the leakage from 103.2 to 85.8 m^3/d thus realizing a saving of 17.4 m^3/d while maintaining the same amount of usage. Table A.16.2 and Fig. A.16.3 show the results of this reduction.

A.16.6 Solutions Proposed

In light of the water savings achieved at a reasonably low reduction of pressure at the inlet connection and without actual pressure management of the system, it was proposed to do the following:

1. To install a permanent pressure control valve at the inlet connection in order to reduce the inlet pressure to 130 m (otherwise water will not reach the target point). The total cost of the proposed system was US$6200.00.

2. To sector the camp into two zones in order to reduce the difference in elevation between low points and high points.

3. To replace the water network taking into consideration the existing sewage collection system. A new design of all the system was made and the estimated cost of replacing the network was US$336,336.

A.16.7 Financing of Solutions

JWU being a nonprofit public utility does not have the necessary funds to finance large replacement projects. It depends on external donors to finance such projects. The Al Jalazon replacement network project (among other projects) was proposed to a number of donors; unfortunately so far we have not been successful in mustering the necessary funds.

As for installing the pressure control system the cost was reasonable and within the capacity of JWU.

Sectoring the network was not possible due to the many interconnections within the network.

Time (Hours)	Average Hourly Pressure at: (m)	Average Inflows (m)	Period Inflow Point (m)	AZP Point (L/s)	Target Measured (m³/hr)	Losses (m³/hr)	Use (m³/hr)
00 to 01	158.0	135.0	108.0	3.9	14.0	5.46	8.54
01 to 02	160.0	139.0	111.0	2.8	10.0	5.63	4.38
02 to 03	162.0	139.0	110.0	2.2	8.0	5.63	2.38
03 to 04	162.0	139.0	111.0	2.2	8.0	5.63	2.38
04 to 05	163.0	139.0	111.0	1.9	7.0	5.63	1.38
05 to 06	157.0	130.0	100.0	2.5	9.0	5.26	3.74
06 to 07	145.0	124.0	92.0	3.6	13.0	5.02	7.98
07 to 08	131.0	102.0	75.0	4.4	16.0	4.13	11.87
08 to 09	115.0	86.0	58.0	6.4	23.0	3.48	19.52
09 to 10	112.0	81.0	51.0	7.5	27.0	3.28	23.72
10 to 11	108.0	78.0	44.0	11.4	41.0	3.16	37.84
11 to 12	110.0	80.0	48.0	9.7	35.0	3.24	31.76
12 to 13	116.0	88.0	50.0	8.6	31.0	3.56	27.44
13 to 14	120.0	85.0	55.0	8.3	30.0	3.44	26.56
14 to 15	125.0	95.0	65.0	8.6	31.0	3.84	27.16
15 to 16	133.0	102.0	73.0	6.9	25.0	4.13	20.87
16 to 17	126.0	98.0	68.0	7.5	27.0	3.97	23.03
17 to 18	115.0	82.0	55.0	7.5	27.0	3.32	23.68
18 to 19	104.0	75.0	46.0	6.4	23.0	3.04	19.96
19 to 20	115.0	82.0	50.0	8.6	31.0	3.32	27.68
20 to 21	130.0	102.0	61.0	7.5	27.0	4.13	22.87
21 to 22	140.0	115.0	85.0	6.7	24.0	4.65	19.35
22 to 23	149.0	124.0	95.0	4.7	17.0	5.02	11.98
23 to 24	155.0	129.0	100.0	4.4	16.0	5.22	10.78
Averages	133.79	106.21	75.92	6.02	21.67	4.30	17.37
Maximum	163.00	139.00	111.00	11.39	41.00	5.63	37.84
Minimum	104.00	75.00	44.00	1.94	7.00	3.04	1.38
	Daily Totals in m³/d 520.0	103.2	416.8				
	% Losses	19.8 %					

Source: Jerusalem Water Undertaking.

TABLE A.16.1   Pressure and Flow before Control

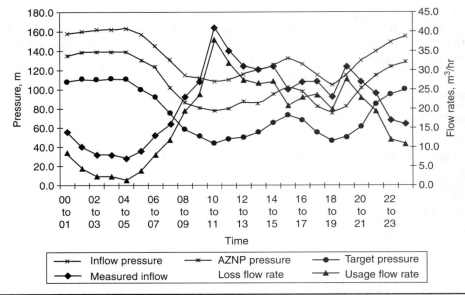

FIGURE A.16.1 Pressure and flow before control. (*Source*: Jerusalem Water Undertaking.)

FIGURE A.16.2 Corroded piping. (*Source:* Jerusalem Water Undertaking.)

A.16.8 Solutions Implemented

The pressure control system was implemented and is functional as planned.

Some replacement work was done for the network. No solution was made for the open sewage system.

A.16.9 Results of Implementation and Loss Reduction

The following is the result of installing the pressure control system at the inlet connection:

Saving in water in 1 year: $17.4 \times 365 = 6351$ m^3

Marginal cost of 1 m^3 of water: $0.684

Time (Hours)	Average Hourly Pressure at (m)	Average Inflows (m)	Period inflow Point (m)	AZP Point (L/sec)	Target Measured (m³/hr)	Losses (m³/hr)	Use (m³/hr)
00 to 01	135.0	115.0	84.0	3.9	14.0	5.44	8.56
01 to 02	135.0	115.0	87.0	2.8	10.0	5.44	4.56
02 to 03	142.0	120.0	91.0	1.9	7.0	5.67	1.33
03 to 04	140.0	119.0	89.0	2.2	8.0	5.63	2.38
04 to 05	139.0	119.0	90.0	1.4	5.0	5.63	-0.63
05 to 06	131.0	106.0	79.0	1.9	7.0	5.01	1.99
06 to 07	124.0	90.0	64.0	3.1	11.0	4.25	6.75
07 to 08	103.0	70.0	46.0	4.7	17.0	3.31	13.69
08 to 09	90.0	60.0	32.0	5.3	19.0	2.84	16.16
09 to 10	81.0	50.0	20.0	6.7	24.0	2.36	21.64
10 to 11	75.0	40.0	7.0	9.4	34.0	1.89	32.11
11 to 12	85.0	42.0	14.0	8.3	30.0	1.99	28.01
12 to 13	87.0	45.0	20.0	10.0	36.0	2.13	33.87
13 to 14	95.0	57.0	27.0	8.6	31.0	2.69	28.31
14 to 15	92.0	55.0	28.0	8.1	29.0	2.60	26.40
15 to 16	100.0	70.0	41.0	8.1	29.0	3.31	25.69
16 to 17	90.0	54.0	33.0	6.7	24.0	2.55	21.45
17 to 18	87.0	51.0	25.0	6.7	24.0	2.41	21.59
18 to 19	76.0	49.0	17.0	7.2	26.0	2.32	23.68
19 to 20	86.0	49.0	16.0	6.1	22.0	2.32	19.68
20 to 21	95.0	68.0	35.0	8.3	30.0	3.21	26.79
21 to 22	111.0	80.0	51.0	6.4	23.0	3.78	19.22
22 to 23	117.0	91.0	63.0	5.8	21.0	4.30	16.70
23 to 24	125.0	100.0	72.0	5.8	21.0	4.73	16.27
Averages	105.88	75.63	47.13	5.81	20.92	3.57	17.34
Maximum	142.00	119.00	91.00	10.00	36.00	5.67	33.87
Minimum	75.00	40.00	7.00	1.39	5.00	1.89	−0.63
	Daily Totals in m³/d 502.0	85.8	416.2				
	Losses %	17.1%					

Source: Jerusalem Water Undertaking.

TABLE A.16.2 Pressure and Flow after Control

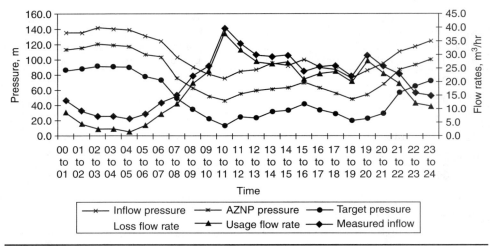

FIGURE A.16.3   Pressure and flow after control. (*Source:* Jerusalem Water Undertaking.)

Total value of water saved per year: 6351 × 0.684= $4344

Costs of material and labor: $6200

Payback period: 1.4 years

This saving will be realized without doing major replacement work on the network.

Case Study A.17: Ductile Iron Pipe in Stray Current Environments

R W Bonds, P.E., *Ductile Iron Pipe Research Association,*
Birmingham, Alabama

A.17.1   Introduction

Stray currents pertaining to underground pipelines are direct currents flowing through the earth from a source not related to the pipeline being affected. When these stray direct currents accumulate on a metallic pipeline or structure, they can induce electrolytic corrosion of the metal or alloy. Sources of stray current include cathodic protection systems, direct power trains or street cars, arc-welding equipment, direct current transmission systems, and electrical grounding systems.

To cause corrosion, stray currents must flow onto the pipeline in one area, travel along the pipeline to some other area or areas where they then leave the pipe (with resulting corrosion) to reenter the earth and complete the circuit to their ultimate destination. The amount of metal lost from corrosion is directly proportional to the amount of current discharged from the affected pipeline.[1]

Fortunately, in most cases, corrosion currents on pipelines are only thousandths of an ampere (milliamps). With galvanic corrosion, current discharge is distributed over wide areas, dramatically decreasing the localized rate of corrosion. Stray current corrosion, on the other hand, is restricted to a few small points of discharge and, in some cases, penetration can occur in a relatively short time.

Considering the amount of buried iron pipe in service in the United States, stray current corrosion problems for electrically discontinuous gray iron and ductile iron pipe are very infrequent. When encountered, however, there are two main techniques for controlling stray current electrolysis on underground pipelines. One technique involves insulating or shielding the pipeline from the stray current source; the other involves draining the collected current by either electrically bonding the pipeline to the negative side of the stray current source or installing grounding cell(s).[2]

Inquiries to the Ductile Iron Pipe Research Association (DIPRA) show that, of the different sources of stray current previously mentioned, impressed current cathodic protection systems on both bare and polyethylene encased iron pipe. The cause, investigation, and mitigation of this source of stray current on iron pipe are the focus of this paper.

A.17.2   Ductile Iron Pipe Is Electrically Discontinuous

Ductile iron pipe is manufactured in 18- and 20-ft lengths and employs a rubber-gasketed jointing system. Although several types of joints are available for ductile iron pipe, the push-on joint and, to a lesser degree, the mechanical joint, are the most prevalent.

These rubber-gasketed joints offer electrical resistance that can vary from a fraction of an ohm to several ohms, which is sufficient for ductile iron pipelines to be considered electrically discontinuous. A ductile iron pipeline thus comprises 18- to-20 ft long conductors that are electrically independent of each other. Because the joints are electrically discontinuous, the pipeline exhibits increased longitudinal resistance and does not readily attract stray direct current. Any accumulation which is typically insignificant is limited to short electrical units.

Joint resistance has been measured at numerous test sites as well as in operating water systems. Forty five joints were tested at a DIPRA stray current test site in an operating system in New Braunfels, Texas. In 830 ft of 12-in diameter push-on-joint ductile iron pipe, nine joints were found to be shorted. Such shorts sometimes result from metal to metal contact between the spigot end and bell socket due to the joint being deflected to its maximum. Due to oxidation of the contact surfaces, however, shorted joints can develop sufficient resistance over time to be considered electrically discontinuous with regard to stray currents.

The ability of electrically discontinuous ductile iron pipe to deter stray current was demonstrated in an operating system in Kansas City, Missouri, where a 16-in ductile iron pipeline was installed approximately 100 ft from an impressed current anode bed. A 481-ft section of the pipeline was installed so that researchers could bond all the joints or only every other joint. When current measurements were made on this section of pipeline, it collected more than 5½ times the current when all the joints were bonded than when every other joint was bonded.

The effect of joint bonding on stray current accumulation has also been demonstrated in the laboratory. The pipe was installed so that researchers could test combinations of bonded joints, unbonded joints, polyethylene-encased pipe, and bare pipe. It was found that pipe with bonded joints, collected three times more current than pipe with unbonded joints (Fig. A.17.1). Also, when exposed to the same environment, the bare pipe collected more than 1100 times the current collected by the pipe encased in 8-mil polyethylene.[3]

A.17.3   Cathodic Protection Systems

Cathodic protection, which is a system of corrosion prevention that turns the entire pipeline into the cathode of a corrosion cell, is used extensively on steel pipelines in the

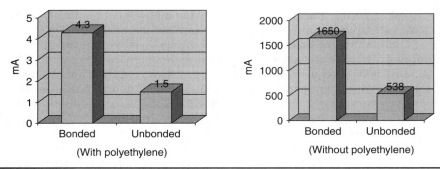

Figure A.17.1   Effects of joint bonding—laboratory installation rectifier output 8 A. (*Source:* Ductile Iron Pipe Research Association.)

oil and gas industries. The two types of cathodic protection systems are galvanic and impressed current.

Galvanic cathodic protection systems utilize galvanic anodes, also called sacrificial anodes that are electronically more active than the structure to be protected. These anodes are installed relatively close to the structure, and current is generated by metallically connecting the structure to the anodes. Current is discharged from the anodes through the electrolyte (soil in most cases) and onto the structure to be protected. This system establishes a dissimilar metallic corrosion cell strong enough to counteract normally existing corrosion currents (Fig. A.17.2). Galvanic cathodic protection systems normally consist of highly localized currents, which are low in magnitude. Therefore, they are generally not a concern of stray current for other underground structures.[4]

Stray current corrosion damage is most commonly associated with impressed current cathodic protection systems utilizing a rectifier and anode bed. The rectifier converts alternating current to direct current, which is then impressed in the cathodic

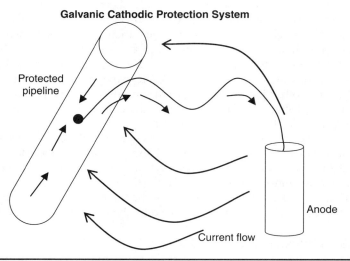

Figure A.17.2   Galvanic cathodic protection system. (*Source:* Ductile Iron Pipe Research Association.)

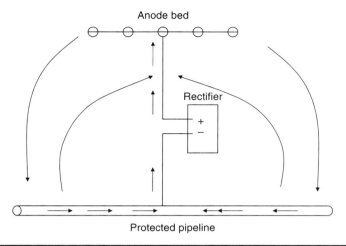

FIGURE A.17.3   Impressed current cathodic protection system. (*Source:* Ductile Iron Pipe Research Association.)

protection circuit through the anode bed. The rectifier's output can be less than 10 V or more than 100 V, and less than 10 A to several hundred amperes. The impressed current discharge from the ground bed travels through the earth to the pipeline it is designed to protect and returns to the rectifier by a metallic connection (Fig. A.17.3). Unlike galvanic cathodic protection systems, one impressed current ground bed normally protects miles of pipeline.

A.17.4   Ductile Iron Pipelines in Close Proximity to Impressed Current Anode Beds

Whether an impressed current cathodic protection system might create a problem on a ductile iron pipeline system depends largely on the impressed voltage on the anode bed and its proximity to the ductile iron pipeline. In general, the greater the distance between the anode bed and the ductile iron pipeline, the less the possibility of stray current interference.

If a ductile iron pipeline is in close proximity to an impressed current cathodic protection anode bed, a potential stray current problem might exist. Around the anode bed (the area of influence), the current density in the soil is high, and the positive earth potentials might force the ductile iron pipeline to pick up current at points within the area of influence. For this current to complete its electrical circuit and return to the negative terminal of the rectifier, it must leave the ductile iron pipeline at one or more locations, resulting in stray current corrosion.

Figure A.17.4 shows a ductile iron pipeline passing close to the impressed current ground bed and then crossing the protected pipeline at a more remote location. Here, if the current density is high enough, current is picked up by the ductile iron pipeline in the vicinity of the anode bed. The current then travels down the ductile iron pipeline, jumping the joints, toward the crossing. It then leaves the ductile iron pipeline and is picked up by the protected pipeline to complete its electrical circuit and return to the negative terminal of the rectifier. At the locations where the current leaves the ductile

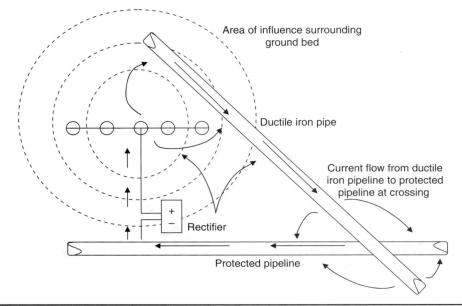

Area of influence surrounding
ground bed

Ductile iron pipe

Current flow from ductile
iron pipeline to protected
pipeline at crossing

Rectifier

Protected pipeline

FIGURE A.17.4   Stray current from a cathodic protection installation. (*Source:* Ductile Iron Pipe Research Association.)

iron pipeline, usually in the vicinity of the crossing and/or in areas of low soil resistivity, stray current corrosion results.

Figure A.17.5 shows a ductile iron pipeline paralleling a cathodically protected pipeline and passing close to its impressed current anode bed. Again, if the current density is high enough, the ductile iron pipeline may pick up current in the vicinity of the anode bed, after which the current flows along the ductile iron pipeline in both directions and leaves to return to the protected pipeline in more remote areas. This may result in current discharging from the ductile iron pipeline in many areas, usually in low soil resistivity areas, rather than concentrated at the crossing as in the previous example.

Normally, electrically discontinuous ductile iron pipeline will not pick up stray current unless it comes close to an anode bed where the current density is high.

A.17.5   Pipeline Crossings Remote to Impressed Current Anode Beds

Usually, a stray current problem will not exist where a ductile iron pipeline crosses a cathodically protected pipeline whose anode bed is not in the general vicinity. A potential gradient area surrounds a cathodically protected pipeline due to current flowing to the pipeline from remote earth. The intensity of the area of influence around a protected pipeline is a function of the amount of current flowing to the pipeline per unit area. If a foreign pipeline crosses a cathodically protected pipeline and passes through this potential gradient, it tends to become positive with respect to the adjacent earth. Theoretically, the voltage difference between pipe and earth can force the foreign pipeline to pick up cathodic protection current in remote sections and discharge it to the protected pipeline at the crossing, causing stray current corrosion on the foreign pipeline (Fig. A.17.6). Because the intensity of the potential gradient around the protected pipeline is small negligible for well coated pipelines—and because ductile iron pipelines are electrically

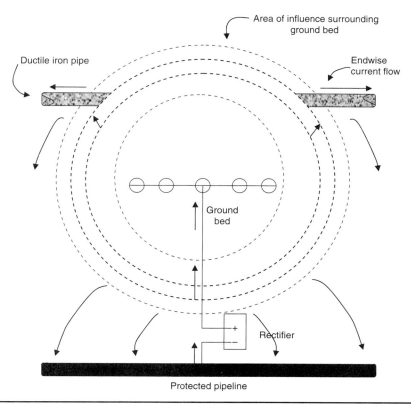

FIGURE A.17.5 Ductile iron pipeline paralleling a cathodically protected pipeline and passing close to its impressed current anode bed. (*Source:* Ductile Iron Pipe Research Association.)

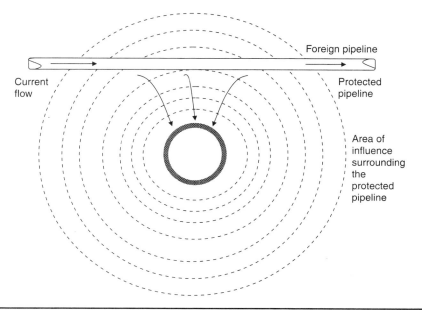

FIGURE A.17.6 Foreign pipeline passing by protected pipeline. (*Source:* Ductile Iron Pipe Research Association.)

discontinuous—stray current corrosion is rarely a problem for ductile iron pipe systems crossing cathodically protected pipelines if the impressed current anode bed is remote. At these locations, the ductile iron pipeline can be encased with polyethylene per ANSI/ AWWA C105/A21.5 for 20 ft on either side of the crossing for precautionary purposes.

A.17.6   Investigation of the Pipeline Route Prior to Installation

It is important to inspect the pipeline route during the design phase for possible stray current sources. If stray current problems are suspected, mitigation measures can be designed into the system, the pipeline can be rerouted, or the anode bed can be relocated.

If during the visual inspection, an impressed current cathodic protection rectified anode bed is encountered in the general vicinity of the proposed pipeline, one method of investigating the possibility of potential stray current problems is to measure the potential difference in the soil along the proposed pipeline route in the area of the anode bed. This can be done by conducting a surface potential gradient survey using two matched half-cell electrodes (usually copper–copper sulfate half-cells) in conjunction with a high resistance voltmeter. When the half-cells are spaced several feet apart in contact with the earth and in series with the high resistance voltmeter, earth current can be detected by recording any potential difference. The potential gradient in the soil, which is linearly proportional to the current density, can be evaluated by dividing the recorded potential difference by the distance separating the two matched half-cells.

When conducting a surface potential gradient survey, one half-cell can be designated as "stationary" and placed directly above the proposed pipe alignment while the other half-cell is designated as "roving" (Fig. A.17.7). Potential difference readings are then recorded as the roving half-cell is moved in intervals along the proposed route. A graph of potential versus distance along the proposed pipeline can then be constructed. Normally, depending on the geometry of the ground bed, cathodically protected pipeline and foreign pipeline locations, the highest current density will be found closest to the anode bed. Usually, the higher the current density the greater the possibility of encountering a stray current corrosion problem on the proposed pipeline.

The installation of a ductile iron pipeline typically will not appreciably change the potential profile. This allows the engineer to make recommendations based on the surface potential gradient survey conducted prior to pipeline installation. Figures A.17.8 and A.17.9 are surface potential gradient survey graphs of stray current test sites located in

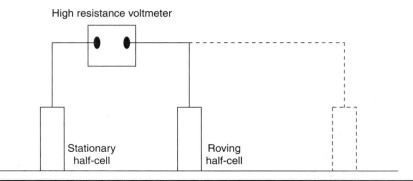

FIGURE A.17.7   Surface potential gradient survey. (*Source:* Ductile Iron Pipe Research Association.)

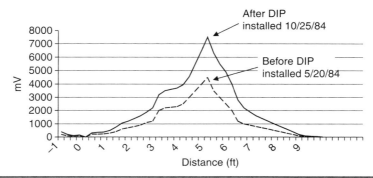

Figure A.17.8 Potential profile comparison: New Braunfels, Texas May 20 and October 25, 1984. (*Source:* Ductile Iron Pipe Research Association.)

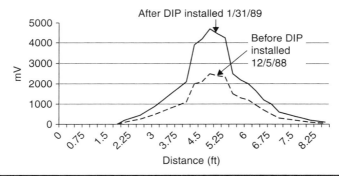

Figure A.17.9 Potential profile comparison: San Antonio, Texas, December 5, 1998 and January 31, 1999. (*Source:* Ductile Iron Pipe Research Association.)

New Braunfels, Texas, and in San Antonio, Texas, respectively, which compare the current density profile before and after installation of the ductile iron pipeline. As can be seen, there is very little difference in the current densities of the two profiles regarding their slope and their boundaries—a fact evidenced in numerous other installations and test sites.

Pipeline installations can vary by geometry, soil resistivity, water table, pipe sizes, pipeline coating, rectifier output, and the like. Yet by knowing the potential gradient prior to installation, the engineer can predict—using conservative values—whether the proposed pipeline will be subjected to stray current corrosion.

A.17.7   Mitigation of Stray Current

Electrical currents in the earth follow paths of least resistance. Therefore, the greater the electrical resistance a foreign pipeline has, the less it is susceptible to stray currents. Ductile iron pipelines offer electrical resistance at a minimum of every 18 to 20 ft due to their rubber-gasketed joint systems. This, in itself, is a big deterrent to stray current accumulation. The effect of joint electrical discontinuity can be greatly enhanced by encasing the pipe in loose dielectric polyethylene encasement in accordance with ANSI/AWWA C105/A21.5.

The electrical discontinuity of ductile iron pipelines and the shielding effect of polyethylene are effective deterrents to stray current accumulation and are all that is required in the vast majority of stray current environments. This would include any crossing of cathodically protected pipelines and/or where the ductile iron pipeline parallels a cathodically protected pipeline. At these locations the potential gradient is created by the protective current flowing to the protected pipeline and is normally small.

There are isolated incidents where electrical discontinuous joints and polyethylene encasement would not be adequate to protect the pipe, for example, the ductile iron pipeline passing through, or very close to, an impressed current cathodic protection anode bed. When this is encountered, consideration should be given to rerouting the pipeline or relocating the anode bed. If neither of these options is feasible, the potential area of high density stray current should be defined (this can be accomplished by concluding a surface potential gradient survey), the ductile iron pipe in this area should be electrically bonded together and electrically isolated from adjacent pipe, polyethylene encasement should be installed in accordance with ANSI/AWWA C105/21.5 through the defined area and extended for a minimum of 40 ft on either side of the said area, and appropriate test leads and "current drain" should be installed. A typical installation is shown in Fig. A.17.10.

In the defined area, the ductile iron pipe most probably will collect stray current. This area needs to be electrically isolated from adjacent piping that will not be collecting stray current. One method of achieving this is installing insulating couplings. Bonding of joints in this area ensures that corrosion will not occur at the joints.

Polyethylene encasement of the pipe in the defined area dramatically reduces the amount of collected stray current. This helps to contain the area of influence and reduces the power consumption of the cathodic protection system. The polyethylene encasement extending on either side of the said area shields the pipe from collecting stray current.

Test leads for monitoring are normally installed on each side of the insulators and in the location of the crossing, if one exists. By having test leads on each side of the insulators, their effective electrical isolation can be ascertained. The test leads on the insides of the insulators can also be used to check whether the bonded section is, in effect, electrically continuous.

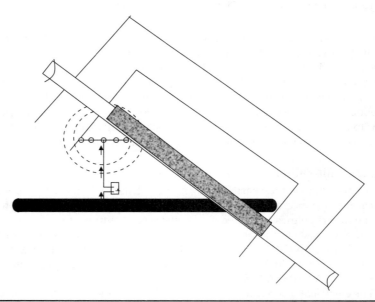

FIGURE A.17.10 Typical installation. (*Source:* Ductile Iron Pipe Research Association.)

The collected current then will need to be effectively drained back to the cathodic protection system. This can be accomplished by installing a resistance bond from the affected area of the ductile iron pipeline to the protected pipeline or to the negative terminal of the rectifier. Resistance can then be regulated to achieve a desired potential on the ductile iron pipeline and reduce the current consumption from the cathodic protection system. Another method of draining the collected current is the design and installation of grounding cells. These grounding cells normally consist of anodes located in areas of current discharge.

A.17.8 Conclusions

DIPRA has conducted numerous investigations in major operating water systems where ductile iron pipelines crossed cathodically protected gas and petroleum pipelines. These investigations involved rectifiers and anodes located in the immediate vicinity (within several hundred feet of the crossing), as well as those located at remote distances.

When the anode beds were remote to the crossings, all investigations indicated that the amount of influence on the ductile iron pipe was negligible and would not be considered detrimental to the expected life of the system. In installations where the anode bed was located in the immediate vicinity, the findings were influenced by factors such as rectifier output, soil resistivity, diameter of the respective pipelines, condition of the coating on the protected line, and the like. Despite these variables, several observations confirmed the findings of laboratory tests. The most significant was the efficacy of rubber-gasketed joints and polyethylene encasement in deterring stray current from ductile iron pipelines.

Throughout the United States, thousands of ductile iron and gray iron pipelines cross cathodically protected pipelines. Yet very few actual failures from stray current interference have been reported. This is additional strong evidence that stray current corrosion will seldom be a significant problem for electrically discontinuous ductile iron pipelines. The bonding of joints and the use of galvanic anodes or drainage bonds may well be a solution to stray current interference in high current density areas, but these systems must be carefully maintained and monitored. If the anode-grounding cell becomes depleted or the drainage connection broken, the bonded ductile iron pipeline will be more vulnerable to stray current damage than if the pipe had been installed without joint bonds. Therefore, such measures should be taken only where stray current interference is inevitable. In most cases, passive protective measures such as polyethylene encasement are more desirable.

References

1. Peabody, A.W. *Control of Pipeline Corrosion*, National Association of Corrosion Engineers, Houston, Texas, 1967.
2. Wagner, E.F. "Loose Plastic Film Wrap as Cast-Iron Protection," Presented September 17, 1963, at AWWA North-Central Section Meeting, *Journal AWWA*. 56(3):361–368, March 1964.
3. Stroud, T.F. "Corrosion Control Measures for Ductile Iron Pipe," National Association of Corrosion Engineers Conference, 1989.
4. Smith, W. H. *Corrosion Management in Water Supply Systems.* New York: Van Nostrand Reinhold, 1989.

Case Study A.18: Leakage—How Low Can You Go? Cheadle Water Works Project—a Unique Opportunity to Minimize Leakage

Ian Elliott, *Director of Engineering, Severn Trent Water Ltd.*

John Foster, *Principal Engineer, Severn Trent Water Ltd.*

The case study covers rehabilitation of water distribution infrastructure to a small town of circa 8000 people in the United Kingdom.

The work involved the replacement of 32 km of distribution mains with diameters ranging from 50 to 350 mm with new MDPE pipe utilizing a number of no dig techniques. All customer services were renewed including in many cases the customer owned service pipes. Radio read meters were installed in all properties.

The result will be a unique opportunity to determine "lowest practical leakage level" and gain a detailed understanding of degradation of systems integrity with time.

A.18.1   Background

Severn Trent Water in the United Kingdom

- Severn Trent Water is a leading provider of water supply and waste water services in the United Kingdom.
- Severn Trent Water Limited is part of the Severn Trent Plc group and has a market capitalization of $6.5 billion.
- We provide water and sewage services to eight million people across the heart of Britain and to communities in 15 states of the United States of America.
- Since privatization in 1989 more than $6.4 billion has been spent on mains upgrading, replacing the distribution system and our services.
- Severn Trent Water has the lowest average water service charges in England and Wales.
- Severn Trent Water has the best overall quality in the United Kingdom. We achieve 99.9% compliance with U.K. and European drinking water standards—the most stringent in the world.
- The level of treatment provided by our sewage treatment works is the highest in the country.

Water Charges

U.K. domestic water supplies have historically been charged for through a "rateable value" system based on the value of the property irrespective of the number of occupants or water usage.

The privatization of the water industry in 1989 introduced compulsory metering of all new properties to ensure realistic charges based on water usage. Generally existing properties remain unaffected but customers are able to opt for a meter to be installed and pay charges based on usage if they choose to do so.

Severn Trent Water (STW) introduced compulsory metering of high usage properties, for example, those with swimming pools, garden sprinkler systems and provided a free meter installation scheme for those customers who wished to switch to a different method of charging.

Service Pipes

Pipes linking properties to the mains network have usually had a shared ownership, the water company being responsible for the pipe up to the property boundary, whilst thereafter the pipe becomes the sole responsibility of the owner. This has raised concerns in two specific areas—leakage on the customer side and potential problems where properties have been fed via a shared pipe.

A.18.2   Cheadle Water Works Company (CWW)

The private CWW Company had been established in the early nineteenth century and fed the small market town of Cheadle, Staffordshire in the North Midlands of England. The town is located entirely within the Severn Trent Water supply area and consisted of approximately 3800 properties of varying age with a mixture of rural properties on the town fringes and major industrial user in JCB.

In 1997 average demand was recorded as 2.5 to 3 ML/d when accepted per capita demand figures suggested this should be of the order of 1.25 ML/d.

The company's assets were in very poor condition and consisted of

- Twenty kilometers of very old unlined iron mains
- Two very old leaking service reservoirs (circa 1830 and 1935)

The limited resources of the Cheadle Company and restrictions of the old distribution system had already resulted in the new development on the fringes of town being supplied with water by STW.

The CWW was unable to meet current water demand effectively and did not have sufficient financial resources to fund investment required to meet increasing demands on the fringes of the town for new development areas. Massive rate increases to begin the process of improving its assets were being considered, but instead an approach was made to STW to take over the company for a nominal sum.

Severn Trent recognized that a significant program of rectification was required and set up a Project Board to manage the assimilation of CWW.

A.18.3   Project Organization

A Project Board was established early in 1997 and included representatives from all disciplines to be involved. The overall task of the Project Board was to coordinate the activities of the individual aspects of the project, which included mains and service renewal, metering, borehole, and reservoir. Project Engineers for each one of these activities regularly reported to the Project Board to allow an overall coordination program to be controlled.

The overall project was tacked in three stages:

- Establish the true factual situation at takeover.
- Gather data and implement short-term solutions to safeguard supplies and improve levels of service.
- Define and implement a longer term strategy for the implementation of the old Cheadle system into STW.

Problems

The Company had suffered from a serious lack of investment for many years, which had resulted in the following problems:

- A continuous gardening watering ban which had been in place for some years
- Variable water quality
- Inconsistent and unfair charging policy
- Inadequate distribution network
- Leakage estimated to be of the order of 50% of the water into supply
- Inadequate pressures in some key areas
- Few operable valves and no method of zoning the distribution system
- No network meters to monitor flows
- Only marginal chlorination with inadequate safety measures
- Leaking unstable service reservoirs with inadequate security measures
- Inoperable stop taps or in many cases none at all

Data and Information

Investigations into the mains network were complicated by the fact that there was no adequate record plans. The only mains layout plan dated from the 1920s.

Sample sections of the most critical mains were taken and confirmed the system to be undersized and in very poor condition showing significant problems with leakage and encrustation.

No guarantee could be placed on any of the existing pipes and it was concluded that it would be impractical and uneconomical to try and repair the system piecemeal. The decision was taken to renew the complete distribution network.

Network design was commissioned and a Stoner model was established to determine the main distribution system required. This work was completed in November 1997 and indicated that a minimum of 27 km of new mains was required.

Initial Priorities

The immediate concern was to secure supplies into the area and enable the garden-watering ban to be lifted. This was affected by providing a short new link main into a neighboring STW supply zone and upgrading an existing booster. Work was completed in the summer of 1997.

The Cheadle reservoirs were then taken out of normal service and the whole town was fed from the adjacent system bypassing the old works. This had a dramatic effect on the water lost through leakage and meant that the long-standing garden-watering ban could be lifted.

With the short-term objectives secured the process of developing the project in more detail commenced.

A.18.4   The Cheadle Project

Outline Proposals

The base project included the complete replacement of the distribution system and the construction of a new reservoir together with the refurbishment of the borehole source and provision of a treatment system.

In addition, the company decided to extend the project by offering to provide every property with a new separate service pipe and meter, where practical. This would create a discrete area in which every property would be supplied through new pipework and a meter, which could be remotely read by computer from the office base at Leicester some 60 mi away.

This would provide the company with valuable information with respect to

- Leakage levels in a newly refurbished distribution system
- Leakage detection and localization
- Water usage patterns
- A complete new distribution system, which could be monitored over a period of years to provide information on developing leakage patterns to assist with defining an economic level of leakage

Construction and installation work was broken down into three main elements:

1. Mains and service pipes
2. Reservoir and borehole reconstruction
3. Meter and radio read installation

Mains and Services

The work was to include the replacement of the whole of the mains network and individual service pipes to provide separate pipes to each individual property where possible.

Timescale was a major factor in tendering the contract on a design and construct basis—reducing the time period for design and construction by approximately 3 to 4 months.

The tenderers were provided with the network design indicating pipe sizes and lengths with the documents stating that the tenderers had the choice of construction method. It was also implicit within the tender that the contractor would be responsible for specific customer care aspects including service pipe surveys and the normal warning procedures for disruptions to supplies.

The company's intention was also to replace as many service pipes as possible including private side services up to the property wall. This work could only be done with the consent and agreement of the owners and because of the lack of any record information involved a complete survey of all the properties to identify pipe runs.

During the survey customers were asked if they would wish to have their pipes replaced—initial indications showed a potential take up of almost 90%, though this was later complicated by the issues surrounding common services, on which owners were given certain conditions which had to be met before work could proceed.

Teamwork

The unique nature of the project and the high impact on both individual customers and the town as a whole resulted in an almost unique teamwork approach to the design and construction process.

The mains and services project team included representatives from the following groups:

- *Engineering*: overall project management, contractual detail and customer interface, quality and specification compliance, supervision of construction
- *Contractor*: surveys of the service pipes, design details, construction methods, and management and customer liaison
- *Operations*: confirmation of final design details and operational advice on the existing network
- *Customer relations*: additional contact point for general customer issues and support with specific problem areas
- *Highways Authority*: continued liaison for planning of roadworks and advice on quality issues
- *Marketing*: general public relations and media contact

Construction Methods

The restrictions within the town meant that the most cost-effective methods would be low dig techniques (conditions permitting). The final choice of method was determined by the main contractor D.J. Ryan.

The following methods were used, (see Fig. A.18.1):

- Pipe sizes ranged from 25 mm for service pipes to 350 mm distribution mains. All pipes used were MDPE.
- The contract has used a very open teamwork approach. Close liaison took place at all times with the local highways authority which included regular progress and program meetings to review the effects on traffic and standards of construction.

Specific Problems Addressed

- Traffic management
- Density of construction operations
- Ground conditions
- Maintenance of supplies during works
- Vulnerability of the existing system

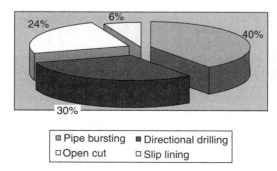

FIGURE A.18.1   Construction methods. (*Source:* Severn Trent Water Ltd.)

- Quality problems during the rehab work due to high velocities
- Resolution of customer issues
- Customer options

The unique nature of the project and the resulting cooperation required from domestic customers resulted in a hierarchy of options being developed.

Customers were given the following options:

- Meter fitted either internally or externally (within the boundary stop tap box)
- Service pipes could be replaced up to the property wall by agreement with the customer
- Payment for water could be either by metered rates or through the rateable value charging system at the customer's request

This provided customers with an extensive range of choices, which would enable them to find an option most suited to their own circumstances.

Though all properties would eventually be fed through a meter, the ideal scenario of all properties having their own meter and separate service pipe was not achievable because of problems with access to certain properties and the issue of common service pipes.

Reservoir and Boreholes

The existing reservoirs were known to be in very poor condition; only 25% of the total capacity was available for use because of structural and leakage problems.

The site itself was in very difficult location, but in spite of this, the limitations on other water sources within the area resulted in the decision to refurbish the boreholes and provide new treatment and storage facilities rather than to abandon the site.

The main construction problems revolved around the restricted nature of the site, which was located on a hill at the center of the town and included a very difficult access route, which was narrow and winding running very close to very old properties.

Two well/boreholes existed on the site; these were up to 70 m deep. A closed-circuit television (CCTV) survey found them to be in good overall condition and samples showed good quality water. Test pumping proved the number one borehole capable of yielding up to 2.5 ML/d.

Construction work commenced on site during September 1998.

Meters

Current STW Metering Policy   STW are now promoting internal metering because of the customer benefits, but as a result need to identify a cost-effective method of remote reading because of potential access difficulties.

Current STW policy is to fit Fusion System Equity water meters with remote read by a Talisman Touch Pad read system.

Metering on the Cheadle Project

The Cheadle area offered the unique opportunity of a high-density meter population within the context of a pipework system, which will be almost completely new. The new network giving improved customer service could in addition provide an opportunity to offer further customer benefits and to give the company a chance to obtain very precise information on system leakage and general usage.

The objectives were identified as follows:

- To monitor all inlet and outlets within the Cheadle DMA system
- To enable automated calculation of leakage based on the integrated sum of inlet and outlet totals
- To enable ad hoc readings of individual or groups of meters and to provide customers with up-to-date information on their own water usage
- To enable automatic collection of nighttime readings when usage is at its lowest to determine unusual flow patterns
- To support possible future automation of readings for billing purposes

Radio Read Systems Design

The system was designed to serve meters, which could be installed either internally within the property or externally in the boundary stop tap box. System Equity electronic meters were used for internal fits and Kent encoded meters for boundary box locations.

Each of the meters is linked to a Genesis Meter Module (GMM) radio transponder supplied by Itron via a waterproof wired connection developed by Fusion meters for use with all utility meters and remote devices.

The town area was broken down into a network of 15 radio areas each of which included repeaters, slaves, and master receivers. These were sited on street light standards and telegraph poles with the relevant permissions. Signals were then relayed via Vodafone wide area link to a remote system computer. The system has been developed to read and collect information from every meter within the network at 15-minute intervals between 1 and 3 a.m. This includes six main distribution meters and the consumption meters, which will number in excess of 3500.

The network design included designation of overlapping areas. The siting of repeaters, slaves, and masters being crucial to allow areas to overlap and enable rerouting capability for the maximum system dependability given the topographical nature of the area.

Difficulties Encountered

Logistics and Timings: The operational complexity was heightened by the need to follow closely behind the pipeline rehabilitation work. The need for operational flexibility and traffic management requirements resulted in a varying program, which made planning difficult.

High Water Levels Groundwater levels in the lower areas of the town began to rise as abstraction from the borehole ceased.

Customers Issues Customers were given the choice of internal or external meter installation. Initial survey figures indicated that the preference would provide an 80/20 split. As data from the first installations built up, a ration of 50:50 was observed. However, as time progressed, the ratio was raised to 70:30 as word spread that internal installations were performed with great attention to customer preferences.

A.18.5   Progress to July 1999

Mains and Services

- Over 32 km of mains have been replaced and most of the old mains had been abandoned.
- Over 3150 "Company" service pipes have been replaced and some 2200 "Customer" service pipes had been renewed.
- The mains and service works were substantially completed by the end of March.

Reservoirs and Borehole

- The new reservoir has been constructed and final testing is in progress.
- The new borehole head works has been constructed and the new disinfecting and control equipment is being commissioned.
- Completion is expected by the end of August 1999.

Meters and Radio Read Installation

- About 3537 properties have been fitted with meters.
- The meters at over 2600 properties can now be read remotely on either an ad-hoc basis or automatically and as individuals or in groups.
- Up to nine readings per meter can be taken during the nighttime period of 1 and 3 a.m.
- Installation and commissioning of the AMR system continues and is expected to be completed by the end of August.

See Table A.18.1 and Fig. A.18.2.

Item	Major contractors	Forecast final cost ($million)
Secure supplies		0.29
Mains and services	D J Ryan & Sons	6.72
Reservoirs and borehole	Mowlem Construction	2.05
Meter and radio read installation	Kennedy Iron Ltd	2.25
Other		0.16
Total		11.47

Source: Severn Trent Water Ltd.

TABLE A.18.1   Costs

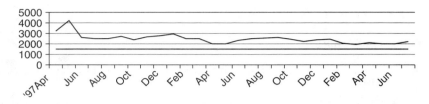

FIGURE A.18.2 Water into supply. (*Source:* Severn Trent Water Ltd.)

A.18.6 Scheme Benefits

Customer Benefits
The major benefits were identified as follows:

- Vastly improved supplies to modern day treatment standards with a secure source
- Removal of the garden-watering ban
- The option to monitor water usage (and therefore potential costs) directly through a meter without any commitment to switch from the rateable charging system
- The opportunity for leakage on customer pipes to be detected
- A telephone contact to give the opportunity for customers to discuss their individual circumstances and options to switch to lowest charging method
- The possibility of individual meter readings remotely to verify reported leakage

Severn Trent Water
Major benefits from the system can be described as follows:

- Almost 4000 new customers with completely new assets and a secure borehole source
- Significant reduction in water lost through leakage
- Future provision of accurate usage and leakage calculations
- Data to enable the monitoring of leakage changes and water usage over time (including an accurate assessment of private side leakage)
- Information on daily and seasonal variations
- A potential long-term assessment of the new pipe network and changes over time
- An assessment of the effect of high meter concentrations on general water usage indication of "lowest practical" leakage level

A.18.7 Summary and Conclusions

- STW set out to renew infrastructure and establish a complete metered system to monitor water usage and minimize leakage using a radio read system—the project should be fully operational by the end of August to provide online leakage detection and a new model system.

- It is the first known project of its kind, combining the provision of an almost totally new system, including the replacement of customer owned pipes, with all outlets metered and measured by an intelligent system.

- The work has been carried out within a very short timescale and overcome some significant problems during the process.

- Initial benefits of the improved service to customers are already apparent, supplies are more reliable and quality more consistent, and the garden watering restriction has been lifted.

- The longer term benefits from the leakage and water usage information cannot yet be quantified but data gathered should provide benefits to both STW and the water industry as a whole.

A.18.8   Acknowledgments
The authors wish to thank Severn Trent Water for giving permission to publish this paper.

Case Study A. 19: Performance-Based Nonrevenue Water Reduction Contracts*

R. Liemberger, †*Liemberger & Partners, Klagenfurt, Austria*

W.D. Kingdom, ‡*The World Bank, Washington, D.C.*

P. Marin, §*The World Bank, Washington, D.C.*

The Challenge of Reducing Non-Revenue Water (NRW) in Developing Countries— How the Private Sector Can Help: A Look at Performance-Based Service Contracting, WSS Sector Board Discussion Paper #8, World Bank, 2006, by William D. Kingdom, Roland Liemberger, and Philippe Marin.

A.19.1   Introduction
One of the major challenges facing water utilities in the developing world is the high level of water losses either through physical losses (leakage) or theft of water from the system, or from water users not being properly billed. This difference between the amount of water put into the distribution system and the amount of water billed to consumers is known as "nonrevenue water" (NRW).

The total cost to water utilities caused by NRW worldwide can be conservatively estimated at $15 billion/year. More than a third of that occurs in the developing world, where

---

*This paper is excerpted from: The challenge of reducing Non-Revenue Water (NRW) in Developing Countries—How to private sector can help: A look at Performance-Based service contracting, WSS sector board discussion paper #8, world bank, 2006, by William D. Kingdom, Roland Liemberger, and Phillippe Marin.
†roland@liemberger.cc.
‡wkingdom@worldbank.org.
§pmarin@worldbank.org.

> ## Box A.19.1
>
> **The three components of nonrevenue water**
>
> **Physical (real) losses** consist of leakage from the system and overflows at the utility's storage tanks. They are caused by poor operations and maintenance, inadequate leakage control, and poor quality of underground assets.
>
> **Commercial (apparent) losses** are caused by customer meter under-registration and data handling errors, as well as thefts of water in various forms.
>
> **Unbilled authorized consumption** includes water used by the utility for operational purposes, water used for fire fighting, and water provided free to certain consumer groups.

about 45 millions m³ are lost daily through water leakage in the distribution networks—enough to serve nearly two hundred million people. Similarly, close to 30 millions m³ are delivered every day to customers but are not invoiced because of factors like pilferage, employees' corruption, and poor metering. These challenges seriously affect the financial viability of water utilities through lost revenues, lost water resources, and increased operational cost, thus reducing their capacity to fund necessary expansions of service, especially for the poor.

A high NRW level normally indicates a poorly run water utility that lacks the governance, autonomy, accountability, and the technical and managerial skills necessary to provide reliable service. The private sector, through well-managed performance-based service contracting, can help water utilities with the technical and managerial skills to carry an effective NRW reduction programs.

The Case for Reducing NRW

Research by international institutions is helping us understand the true magnitude of the losses from NRW, since utilities responsible for the losses have proven either unwilling or unable to provide such information. The World Bank database on water utility performance, known as IB-Net (www.ib-net.org), includes data from more than 900 utilities in 44 developing countries. The average figure for NRW level in developing countries' utilities covered by IB-Net is around 35% (Table A.19.1)—representing a value of $5.8 billion (Table A.19.2).

- What are the sources and costs of NRW? The principal components are leaks and unbilled consumption.
- *Water leaks*: Every year 33 billion cubic metres of treated water physically leaks from urban water supply systems around the world, while 16 billion cubic metres are delivered to customers for zero revenue. Half of these losses are in developing countries, where public utilities are starving for additional revenues to finance expansion of services, and where most connected customers suffer from intermittent supply and poor water quality. It is estimated that US$15 billion is lost every year by water utilities around the world, more than a third of that by water utilities in developing countries. The scale of the problem is obvious and cannot be ignored.
- *Commercial losses:* The value of water lost every year in developing countries through commercial losses—water actually delivered but not invoiced—is

	Supplied Population, Millions (2002)	System Input (l/capita/d)	NRW as Share of System Input (%)	Estimates of NRW				
				Ratio (%)		Volume (billion m³/year)		
				Physical Losses	Commercial Losses	Physical Losses	Commercial Losses	Total NRW
Developed countries	744.8	300	15	80	20	9.8	2.4	12.2
Eurasia (CIS)	178.0	500	30	70	30	6.8	2.9	9.7
Developing countries	837.2*	250	35	60	40	16.1	10.6	26.7
				TOTAL	**32.7**	**15.9**	**48.6**	

Source: World Health Organization, IB-Net, and authors' estimates.
*Based on a total population having access to safe water supply of 1902.7 million people, with 44 % of these receiving water through individual household connections.

TABLE A.19.1   Estimates of Worldwide NRW Volumes (billions of cubic metres/year)

	Marginal Cost Of Water (US$/m³)	Average Tariff (US$/m³)	Cost of Physical Losses	Lost Revenue due to Commercial Losses	Total Cost of NRW
			Estimated Value (US$ billions/year)		
Developed countries	0.30	1.00	2.9	2.4	5.3
Eurasia (CIS)	0.30	0.50	2.0	1.5	3.5
Developing countries	0.20	0.25	3.2	2.6	5.8
		TOTAL	**8.1**	**6.5**	**14.6**

Source: Authors' calculations.

TABLE A.19.2   Estimated Value of NRW and its Components

estimated at US$ 2.6 billion. This is about a quarter of the total yearly investment in potable water infrastructure in the developing world. It is also more than the World Bank, the biggest water financier among international financial institutions, lends every year in aggregate for water projects in developing countries.

Although more analysis is needed, it is already clear that a sizeable portion of this commercial loss is likely to come from fraudulent activities and corruption—such as illegal connections, fraudulent meter reading, or meter tampering. This should be the cause of concern for both developing countries' governments and the donor community alike. The benefits of reducing NRW are clear (see Box A.19.2).

Box A.19.2

Clear benefits from reducing NRW

Reducing NRW to just half the current level in the developing world would deliver the following benefits:

- Every year 8 billion m³ of treated water would be available to service customers.

- Ninety million more people could gain access to water supply, without increasing demand on endangered water resources.

- Water utilities would gain access to an additional US$2.9 billion in self-generated cash flow, equivalent to more than a quarter of the amount currently being invested in water infrastructure in the developing world, and this without affecting in any manner the debt capacity of those countries.

- Fairness among users would be promoted by acting against illegal connections and those who engage in corrupt meters reading practices.

- Consumers would have improved service from more efficient and sustainable utilities.

- New business opportunities would be created for NRW reduction activities, with thousands of jobs created to support labor-intensive leakage reduction activities.

A.19.2   Why Utilities Struggle with NRW—and How the Private Sector Can Help

NRW reduction is not a simple matter to implement, which explains why so many water utilities fail to address it effectively. New technical approaches have to be adopted and effective arrangements established in the managerial and institutional environment—often requiring attention to some fundamental challenges in the utility.

Not understanding the magnitude, sources, and cost of NRW is one of the main reasons for insufficient NRW reduction efforts around the world. Only by quantifying NRW and its components, calculating appropriate performance indicators, and turning volumes of lost water into monetary values, can the NRW situation be properly understood and required action taken. Other issues concern the inherent weaknesses of water utilities in developing countries. Utilities in developing countries

- Often operate under a weak governance and financial framework, with utility managers having to face multiple political and economic constraints

- Must provide service to customers on a daily basis using deteriorated infrastructure

- Often lack the proper incentives and the specialized management and technical expertise necessary to carry out an effective NRW program

- Operate under an inadequate incentive framework

Because the water utilities in developing countries typically lack the capacity, incentives or governance to put in place NRW reduction programs, they need external assistance.

Potential for Private Sector Involvement

A potential source of assistance is the private sector, where involvement can take many forms, ranging from long-term public-private partnerships (PPP) to service contracts or subcontracting of certain tasks. Depending on the option chosen, the private sector can bring

- New technology and the know-how to use it efficiently
- Better incentives for project performance
- Creative solutions for the design and implementation of the program
- Qualified human resources
- Flexibility for field work (e.g. night crews)
- Investment, under certain conditions.

The key message, one too often overlooked, is that NRW must not be considered in a vacuum, but within the broader context of utility reform. The design of any NRW program needs to take into account the incentives open to the managers and staff of the program, as well as the other parties involved. Any program should ensure, as far as possible, that incentives are properly aligned with the objective of developing an efficient and effective utility that meets the needs of its consumers.

The paper excerpted here deals with *performance-based service contracting* (PBSC), a relatively new and flexible approach to the NRW challenge. Under PBSC, a private firm is contracted to implement an NRW reduction program. It is paid for services delivered and provided incentives to meet contractually enforced operational performance measures. With the proper balance of government oversight and private sector initiative, PBSC can provide an enabling environment and incentives conducive to reducing NRW, with immediate operational and financial benefits. But it is not a substitute for carrying out the broader institutional reforms necessary to promote the sustainability of the sector.

In practice, the applicability of PBSC to an NRW reduction program depends on the level of risk that the private sector is willing or able to take. Although PBSC is a relatively new concept for the water sector in the developing world, it is increasingly contemplated in other sectors as a way to improve efficiency and accountability of contracts with private providers. This is the first full study of large NRW reduction performance-based service contracts, and it considers key issues in contract design, management practices, outsourcing options, technical assistance, risk management, and other lessons learned.

Case Studies A.19.3: Reducing Lost Water and Increasing Revenue

To date only a small number of large contracts have been let, and little information has yet become publicly available. However, the authors were able to study four significant and diverse projects. In Selangor, Malaysia, a large contract for reducing physical and commercial losses has been in place since 1998 between the water utility (at that time state owned) serving Kuala Lumpur and its surroundings, and a consortium led by a Malaysian company. In Thailand, the Metropolitan Waterworks Authority (MWA) that supplies Bangkok outsourced physical loss reduction to private contractors from 2000 to 2004. In Brazil, SABESP, the water utility that serves the São Paulo Metropolitan

Region, experimented with different contractual approaches with the private sector for reducing commercial losses. And in Ireland, the Water Division of the Dublin City Council contracted in 1997 an international private operator to implement a two-year contract for reducing physical losses.

The following six key factors were used to evaluate these contracts:

- *Scoping:* What is the role of the private contractor? What are the NRW reduction targets?
- *Incentives:* How is the performance-based element of the contract structured?
- *Flexibility:* To what extent does the contract allow the private sector to be creative in the design and implementation of the NRW reduction activities?
- *Performance indicators and measurement:* How is NRW reduction measured?
- *Procurement/selection:* How was the private contractor selected?
- *Sustainability:* What happened after the performance-based service contract was completed? Does the contract include specific clauses to ensure transfer of know-how to the public utility?

Selangor, Malaysia: the Largest NRW Reduction Contract to Date

In 1997 the population of the Malaysian State of Selangor (and the Federal Territory of Kuala Lumpur) experienced a serious water crisis caused by the El Niño weather phenomenon. The crisis provided the trigger for the government to start dealing with the high level of NRW that had affected the water utility for many years. An estimated 40% of the water produced was not invoiced, with leakage estimated at 25 %, or around half a million m^3/d. Halving the amount of physical losses would provide sufficient water to serve the equivalent of 1.5 million people and thereby avert the water shortage in Kuala Lumpur.

Faced with this crisis, the State Waterworks Department accepted an unsolicited proposal from a consortium led by a local firm, in joint venture with an international operator. The contractor committed to reduce NRW by a specified amount agreed in advance, in a given time. The contractor had full responsibility for designing and implementing the NRW reduction activities with its own staff, in exchange for an agreed lump sum payment.

The incentives for achieving the targets included (1) penalties for noncompliance of up to 5% of the total lump sum, and (2) a performance guarantee of 10% of the contract value. The lump sum included all necessary activities like establishment of district metered areas (DMAs), pressure management, leak detection and repair, identification of illegal connections, and customer meter replacement as well as the supply of all equipment and materials. The contractor was free to select the zones within the network in which to conduct its NRW reduction activities.

Phase 1 of the contract demonstrated that the concept worked: a private firm can be contracted to efficiently reduce NRW level to specific targets, provided it has the flexibility to conduct the NRW activities and the payment arrangement covers all necessary work and materials. One of the technical innovations in this case was the universal use of pressure-reducing valves (even in very low-pressure situations) not only to reduce leakage through the reduction of excessive pressures but to also protect the already repaired DMAs from upstream pressure fluctuations. The performance of Phase 1 actually exceeded the target (18,540 m^3/d), achieving savings of 20,898 m^3/d (approximately equally between commercial and physical losses). Twenty-nine DMAs were established with average savings of 400 cubic metres per day in each DMA and around 15,000 meters were replaced. The cost to the State Waterworks Department was equivalent to US$ 215 per m^3/d.

The Phase 2 contract had a number of shortcomings but was significant in its size—the contractor was committing to an ambitious target of around 200,000 m^3/d NRW reduction, something that had never been done under a PPP arrangement.

The long-term sustainability of the project is not clear. The Phase 1 contract included training of the client's staff. Training on its own, however, proved insufficient for the client to maintain the improvements, and the Phase 1 zones were handed back to the contractor to operate in Phase 2. Obviously, any NRW strategy must address the issue of what to do once the contract ends.

Bangkok: Plugging Leaks

Water services in Bangkok are provided by a public utility, the Metropolitan Water-works Authority (MWA). Like most water utilities operating in the megacities of south-east Asia, MWA has been struggling for years to cope with demand from a fast-growing population. Major investments were made to increase production capacity, with production raised from 1.7 to 3 millions m^3/d between 1980 and 1990. It seemed that NRW was also reduced from 50% in 1980 down to about 30% in 1990. However, the reduction in percentage terms was mainly the result of the substantial increase in production capacity. Despite significant efforts, the volume of NRW remained stable during this period, at a high of about 900,000 m^3/d.

During the 1990s, as the system's supply swelled from 3.0 to 4.5 million m^3/d, NRW rose dramatically, both in percentage and in volumetric terms, reaching a peak in 1997 (1.9 million m^3/d, or 42%), presumably caused by supply improvements and pressure increases. System input was then again reduced to below 4 million m^3/d, and NRW consequently decreased and stabilized in 1999, albeit at a rather high level of 1.5 million m^3/d.

Subsequent efforts have resulted in NRW reduction by 200,000 m^3/d (to 1.3 million m^3/d, or 30 %) even as the system input increased to 4.2 million m^3/d. A significant part of the reduction in NRW can be traced to performance contracts, which the MWA decided to award to private contractors in 2000. The objectives of these contracts were to reduce physical losses in three of the fourteen service branches of Bangkok (each representing around 100,000 customers). The duration of the contracts was four years. They were competitively bid, but only two companies were prequalified and submitted proposals. Both received contracts.

The design of these contracts was significantly different from the case of Selangor. There was no fixed target for leakage reduction, and payment was based in part on the actual water savings achieved through leakage reduction. While each contractor was free to carry out leakage reduction activities (such as detection, pipe repairs, main replacement, installation of hydraulic equipment) as they saw fit, this was done through the use of local firms based on reimbursables (on a cost-plus basis). Instead of a lump sum payment, as used in Selangor, the remuneration of the contractor comprised three elements: (1) a performance-based management fee to cover overhead, profits, and for-eign specialist staff, (2) a fixed fee covering essentially the cost of local labour, and (3), the largest part of the project's cost, reimbursables for all outsourced services, work, and materials performed in the field.

In terms of technical performance, the contracts can be considered a success—but the cost efficiency of the three contracts varied widely (US$ 246 vs. 518 per m³/d water loss reduction). Physical losses in these three areas were reduced by 165,000 m³/day. To give some sense of perspective, the amount of water saved is equivalent to the volume needed to serve an additional half-million inhabitants.

It is interesting to compare the three Bangkok contracts to the Selangor contract.

> *Advantages compared with Selangor*: There were neither arbitrary targets nor lump sum remuneration, but instead a true performance-based element, based on the actual volume of NRW saved. In addition, the fact that two different contractors were in place simultaneously allowed for some useful benchmarking.

> *Disadvantages compared with Selangor:* The high proportion of reimbursables transfers a substantial amount of risk from the private to the public partner. Basic activities, such as leak detection, should have been included in the performance fee.

In terms of sustainability, it does not seem that the contractors put proper control and management systems in place, which the MWA staff could then continue to use. However, MWA is aware of the problem and has recently tendered a project for advanced network monitoring, DMA establishment, and so on.

Sao Paulo: Payments and Collections

SABESP, the utility that serves the São Paulo Metropolitan Region, is one of the largest public water utilities in the world (supplied population: 25 million). It has put in place a proactive approach to water loss reduction with the help of the local private sector. Leakage reduction is routinely carried out by a series of leak detection contractors that are paid per length of distribution network surveyed. Some 40% of the 26,000 km network is surveyed every year.

However, customer metering and billing, including identifying and replacing under-registering meters, had been traditionally left to in-house crews. In 2004, it was estimated that SABESP was incurring daily revenue losses equivalent to one million cubic metres per day. Faced with this situation, SABESP decided to experiment with performance-based arrangements with the private sector. One of the contracts discussed below dealt with the reduction of bad debts (which is not, strictly speaking, part of NRW but has a similar negative impact on the utility's financial equilibrium); the other focused on meter replacement.

The concept of the first contract was simply to contract local private firms to negotiate unpaid invoices and collect the agreed amount. The scope of the contracts was limited to domestic and commercial customers; SABESP still dealt directly with public institutions. Several contracts were awarded covering all of SABESP's branches. The initial contracts started in 1999 for a two-year term. The contractors were remunerated by retaining a percentage of the debt collected. That percentage was bid on by the contractors; the winning bid in each branch offered the lowest percentage figure.

The São Paulo Metropolitan Region is the industrial heartland of Brazil and industrial and large commercial customers and large condominium buildings account for a major portion of SABESP's revenues. In fact, 28 percent of total billed metered consumption and 34% of all revenues come from just 2% of SABESP customers. Because meters were suspected of under-registering true levels of consumption, the strategy of the second contract was to upgrade and optimize the metering system.

SABESP came up with an innovative solution to this problem by tendering a series of turnkey contracts for meter replacement. The project target was to replace the meters of 27,000 large accounts identified by SABESP. Five 36-month contracts were put in place, and the contractor was responsible for the analysis, engineering and design, supply and installation of the new meters. There was no upfront payment, and the contractor had to prefinance the entire investment. The contractor was entitled to a payment based on the average increase in billed volume through a complex formula.

The concept of performance payments, rather than just paying for supply and installation, was chosen because resizing and flow profiling of the meters were the most critical activities in the contract. Given the high daily consumption of the large customers concerned, proper calibration could significantly increase metered flows and billing. By linking payments to the improved billed volumes, SABESP ensured that the contractor would focus on these critical issues.

The results of the contract were remarkable. The total volume of metered consumption increased by some 45 million m³ over the contract's three years, while revenues increased by US$72 million. Of this, US$18 million was paid to the contractors, with a net benefit to SABESP of US$54 million.

In terms of sustainability, the contracts for reduction of bad debt have now become standard practice for SABESP. With new, properly sized, commercial customer meters installed it should now be easy for SABESP to maintain the accuracy of these meters and thus maintain the higher billed volumes from this customer category.

Dublin: Upgrading a Very Old System

In January 1994 the city of Dublin had to deal with a severe water shortage caused by decades of underinvestment in the distribution network, combined with the absence of systematic leakage control, which had allowed physical water losses to reach very high levels. Several areas of Dublin had only intermittent water supply.

The first reaction was to ask for funds to build new treatment plants and expand existing ones. However funding was not made available because of the high level of leakage. A comprehensive study then identified, for the first time, the volume of water being lost: every day some 175 million L of water, more than 40% of the existing treatment capacity, was estimated to be leaking away from the distribution network. The European Commission was approached, and the request for cofinancing of the planned Dublin Region Water Conservation Project was approved, with a focus on reducing physical water losses.

The project target was very ambitious: to reduce leakage over a two-year period from 40% to 20% (in volumetric terms from 175,000 to 87,000 m³/d). Given the aggressive nature of the reduction program, there was no alternative but to engage an experienced contractor to assist the city.

In November 1996 eight consortia were invited to submit bids. The contract was of limited duration—only two years—and focused on physical loss reduction. The contractor was responsible for establishing DMAs throughout the network, locating and repairing leaks, installing pressure-reducing valves, rehabilitating parts of the network, and training the Dublin water utility staff. It was designed essentially as a target cost contract expressed in monetary terms. It included a bonus-and-penalties mechanism to provide some incentive for performance, based on a complex methodology combining actual project expenses with the marginal cost of physical losses.

The contract was won by a UK water utility on a quality/cost basis. Significant details were left to be resolved during contract negotiations. The contractor's remuneration in the winning bid covered a management fee, technical labour, and all leak-detection equipment. This did not include the cost of leak repairs, repair materials, or network rehabilitation, which were carried out through local subcontractors and covered separately as reimbursables under what were known as "compensation events." The contract's accomplishments were significant.

Establishment of a total of 500 small DMAs (less than 1000 connections each), covering the whole distribution network. Some 15,000 leaks were repaired and about 20 km of mains replaced. Total leakage was reduced from 175,000 m^3/d to about 125,000 m^3/d, and although the 20% leakage target was not achieved, the project was considered a success. (There was broad consensus that the original 20% target was not realistic given the short duration of the contract). The savings made were sufficient to end the water crisis.

In terms of sustainability, training and capacity building were components of the contract and were taken seriously by both parties. Substantial transfer of technology took place in practice, and the Dublin water utility now controls leakage as a regular part of its day-to-day operations.

A.19.4 Lessons Learned

It is not feasible to eliminate all NRW in a water utility, but reducing by half the current level of losses in developing countries is a realistic target. Figures of such magnitude, even though based on estimates, should be enough to capture the attention of donors and developing country governments. In practice, good paybacks are possible with well-designed NRW reduction programs; therefore, if nothing else, NRW reduction makes business sense, although each opportunity has to be assessed in terms of its particular cost-benefit ratio.

Performance-based service contracts appear a viable way of reducing NRW losses. However, successful project implementation requires two essential and related elements: preparing good contracts and setting realistic baselines.

The case studies show various levels of quality in contract preparation, baseline setting, and—as a consequence—project effectiveness. Contract design must be clear about what the utility expects from the contractor and how it envisions success. All NRW reduction contracts should include basic guidelines concerning risk transfer, an indicator for leakage, and provisions for effective oversight by utility managers. Contracts should set viable targets and allow for flexibility in responding to challenges and opportunities.

To be successful, however, the study shows that good preparatory work is required. The starting point is to develop a strategy based on a sound baseline assessment of the sources and magnitudes of the NRW. Such a strategy needs to consider both the short and long terms (for example, the achievement of short-term reductions vs. how to maintain lower levels of NRW over the long term). It is during strategy development that opportunities for teaming with the private sector can be identified. Once those opportunities are known, policy makers must create an incentive framework that will encourage the private sector to deliver reductions in the most cost-effective manner, allocating risk appropriately between the parties.

References

Baietti, A., W. Kingdom, and M. Van Ginneken, "Characteristics of Well-Performing Public Water Utilities," Water Supply and Sanitation Working Notes 9. Washington, DC: The World Bank, February 2006.

Brocklehurst, C. and J. Janssens, "Innovative Contracts, Sound Relationships: Urban Water Sector Reform in Senegal," Water Supply and Sanitation Sector Board Discussion Paper 1. Washington, DC: The World Bank, January 2004.

Brook, P., and S. Smith, "Contracting for Public Services: Output-Based Aid and its Applications," Private Sector Advisory Services. Washington, DC: The World Bank, http://rru.worldbank.org/Features/OBABook.aspx.

Kingdom, B., Liemberger, R., and Marin, P. "The Challenge of Reducing Non-Revenue Water (NRW) in Developing Countries—How the Private Sector Can Help: A Look at Performance-Based Service Contracting", WSS Sector Board Discussion Paper No. 8. Washington, DC: The World Bank, 2006, http://siteresources.worldbank.org/INTWSS/Resources/WSS8fin4.pdf

Liemberger, R. "Competitive Tendering of Performance Based NRW Reduction Projects," IWA Conference on Efficient Management of Urban Water Supply, Tenerife, April 2003.

Liemberger, R. "Outsourcing of Water Loss Reduction Activities—the Malaysia Experience" New Orleans: AWWA ACE June 2002.

Liemberger, R. "Performance Target Based NRW Reduction Contracts—A new Concept Successfully Implemented in Southeast Asia. Berlin: IWA 2nd World Water Congress, October 2001.

Marin, P. "Output-Based Aid (OBA): Possible Applications for the Design of Water Concessions," Private Sector Advisory Services. The World Bank. http://rru.worldbank.org/Documents/PapersLinks/OBA%20Water%20Concessions%20PhM.pdf

Ringskog, K. M-E Hammond, and A. Locussol. "The Impact from Management and Lease/Affermage Contracts," PPIAF, 2006.

APPENDIX B

Equipment & Techniques
Flow Metering, Pressure Measurement, Pressure Control, and Leak Detection

Reinhard Sturm

Julian Thornton

George Kunkel, P.E.

B.1   Introduction

When undertaking hands-on evaluation of a water system or facility's losses it is vitally important to use good, accurate valid data taken from the field. To do this, we can either use portable, temporary equipment; or we install new, permanent equipment (or rehabilitate old) alongside data loggers or recorders to collect time-based trends. These allow us to analyze the actual dynamic situation. Once the current volume of water losses has been accurately identified (by the audit as mentioned in Chap.7) then detection equipment in the case of real losses is used to pinpoint individual problem areas.

> **T**here is no substitute for good, accurate, and accountable field data.

This appendix outlines some of the equipment types and methodologies in today's market. Additional and complimentary information can also be found in Chaps. 16 through 19 which deal more with field intervention methodologies, although some overlap is present between App. B and these chapters.

B.2   Portable Equipment

Many different types of flow measurement equipment are available on the market today and, as with leak-detection equipment, the operator must ensure that whichever equipment he decides upon, he can be trained and supported locally to use and maintain the equipment in good working order. The operator must be confident in the accuracy of the equipment he is using and in its reliability in different situations. To use test equipment where the operator is unsure about its credibility will only lead to confusion. Often the reason for measuring the flow is to try to identify why something is happening; why a master meter is overreading; why or if there is leakage, and the like. This in itself is often difficult enough with test equipment, which is familiar to the operator, without throwing in the added uncertainty.

The discussion, which follows, is designed to highlight some of the methodologies used in field measurements, not intended to promote one methodology over another or indeed one manufacturer over another.

B.2.1   Portable Insertion Flow Meters

One of the most common portable flow meters is the insertion type meter. Insertion meters come in various types from many different manufacturers. The most common operating principles found in insertion flow meters are electromagnetic probes, turbine probes, and differential pressure probes. Each type of equipment has its strong and weak points; however, with good technical support, calibration, and maintenance most of the instruments on the market today do the job.

Some of the most common insertion meters are

- The pitot rod (differential pressure probes) (see Fig. B.1).
- The turbine insertion meter (see Fig. B.2).
- The electromagnetic insertion meter (see Fig. B.3). The most common forms of these meters are actually singlepoint velocity meters. The measurement point is selected either as a measured average point in the pipe or often the centerline is used with a calculated factor to take into account both the blockage of the meter and the difference between centerline velocity and average velocity. Once the average velocity has been either calculated or recorded it is then multiplied by the effective area of the flowing pipe to give flow.

Some of the above-mentioned meters, in particular the differential pressure (DP) meter and some electromagnetic insertion flow meters have an averaging function built in. They measure the velocity at predetermined points, perform an automatic average, and give out flow units. If the idea of the portable equipment is to make measurements on varying pipe diameters then it is more likely that a single point meter will be used as this allows the user to work in virtually any pipe diameter without having to order varying rod sizes.

In order to calculate either the average velocity point or the factor between the theoretic average and the centerline velocity, it is necessary to perform a velocity profile. The velocity profile is a series of velocity measurements taken across the diameter of the pipe sometimes in both 90° and 180° degree axis depending on the accuracy required, see ISO 7145–1982 E for more details on this methodology. See Fig. B.4 for an example diagram, Fig. B.5 for velocity profiling, and Fig. B.6 for a sample spreadsheet used for calculating this task. As with many waterworks tasks the use of a simple spreadsheet can save time and effort in reducing the need for repeatable calculations. The manufacturer coefficients should be programmed into the spreadsheet for each different type of meter used. Some manufacturers provide profiling programs with their equipment as part of the package.

FIGURE B.1 Pitot rod installation. (*Source:* BBL Ltda.)

FIGURE B.2 Turbine insertion meter and data logger. (*Source*: Reinhard Sturm.)

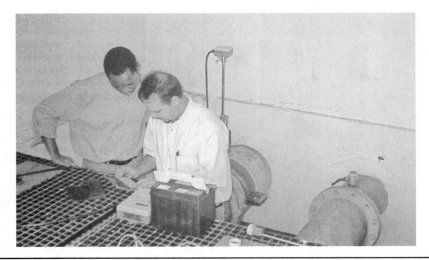

FIGURE B.3   Electromagnetic insertion meter installation. (*Source:* Restor Ltda.)

Alternately many meter manufacturers supply approximate points or coefficients for uses on varying pipe sizes see Table B.1 for an example. It can be seen that as the pipe size increases the coefficient increases. This is a function of the reduced effect of blockage on the pipe area by the meter.

The decision of when to profile and when to use approximate industry standard figures really comes down to the accuracy the operator is expecting from the meter. This

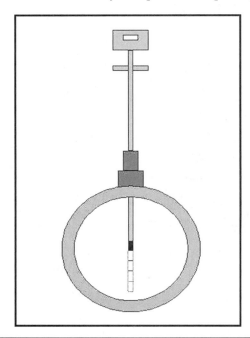

FIGURE B.4   Velocity profiling. (*Source:* Julian Thornton.)

Figure B.5   Velocity profiling using a turbine insertion meter. (*Source:* Reinhard Sturm.)

will be very task specific. For example, if the operator is testing a master meter, which is supposed to be accurate to ±0.5%, then he must at a very minimum perform a velocity profile. However, if the operator is looking for orders of magnitude of variation in flow to ascertain which zones have night flow for a leakage ranking exercise or using the meter to size a valve which is quite forgiving in flow pattern, then obviously it may not be deemed necessary to spend the time and effort on a velocity profile but rather use the industry standard figures. (These sorts of decisions should be noted in the comments column of the audit sheet and the data collection sheet.)

Hot Tapping

Insertion meters are usually fitted to the pipe through a hot tap or tee connection on the pipe, see Fig. B.1. Fitting a hot tap is a relatively simple procedure and is usually done under pressure using a tapping machine, see Fig. B.7. Care should be taken to ensure that the tap is of sufficient diameter that the rod of the meter can pass through the valve without damage. For example, if the meter rod external diameter is 1 in or 25 mm then care must be taken to select a valve or tap which has at least a 1 in or 25 mm internal bore.

Note that some valves are not identified by their internal diameter. Consideration must also be given to the pipe material.

Hot taps are usually tapped directly into the pipe if the pipe has a metallic wall; however some utilities prefer to use a weldolet see Fig.B.8, or a tapping sleeve see Fig. B.9. In all cases where the pipe is not metallic a tapping sleeve should be used. In addition, care should be taken to ensure that the cutting bit is sharp so as not to crack the pipe (especially in the case of AC or cement pipes).

Installation Process

The first step, and one of the most important, is to select a suitable metering location, away from other fittings and disturbances in the flow, such as gate valves, meters, PRVs, and the like.

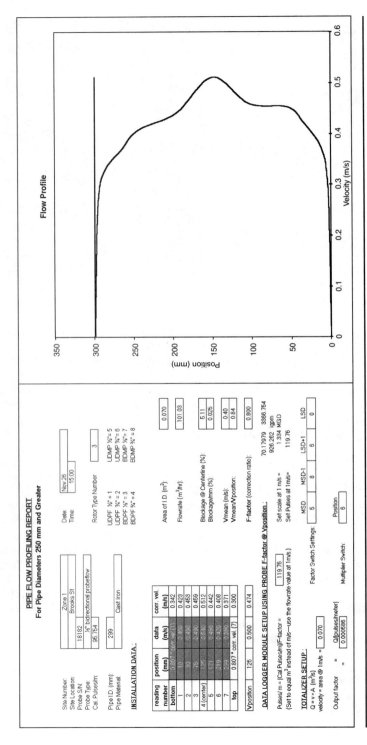

FIGURE B.6 Velocity profiling spread sheet. (*Source:* Julian Thornton.)

Pipe ID (mm)	Centerline
150	0.658
200	0.753
250	0.798
300	0.823
400	0.847
600	0.863
800	0.867
1000	0.869

Source: Julian Thornton.

TABLE B.1   Coefficients for Velocity Profiling (*Source*: Quadrina data).

FIGURE B.7   Under pressure tapping. (*Source:* Restor Ltda.)

Most manufacturers will indicate the number of upstream and downstream diameters of pipe needed for accurate measurement. If this information is not available a good rule of thumb is 30 diameters from an upstream disturbance and 20 from downstream disturbances (see Fig. B.10). Measurements can be made closer to disturbances but may result in either unstable velocity recordings or unstable velocity profiles, which will result in error. (If this was the only option then this would be noted on our measurement sheet mentioned in Chap. 5 along with an estimated error and would become part of the audit trail).

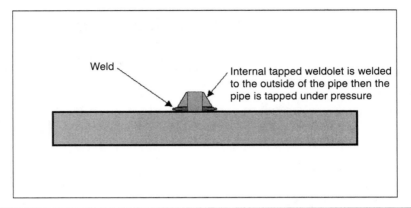

FIGURE B.8 Weldolet can be used on metallic pipe. (*Source*: Julian Thornton.)

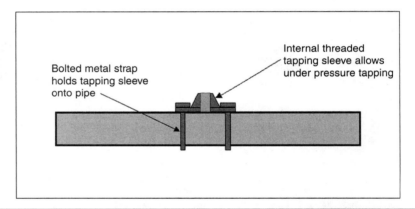

FIGURE B.9 Tapping sleeve. (*Source:* Julian Thornton.)

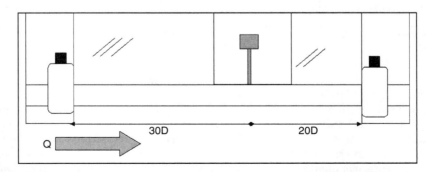

FIGURE B.10 Measurement should be made away from disturbances. (*Source*: Julian Thornton.)

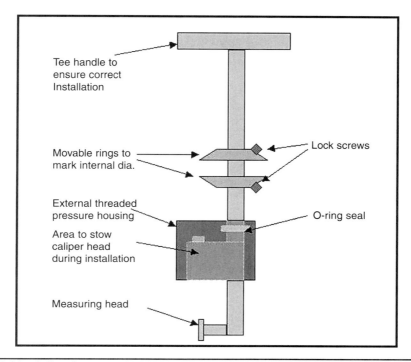

Figure B.11   Caliper for measuring internal pipe diameter. (*Source:* Julian Thornton.)

Once the hot tap has been installed the exact pipe internal diameter must be measured to ensure accuracy. Figure B.11 shows a caliper, which can be used for this process.

Profiling

After the pipe diameter has been measured a decision must be made as to whether or not to profile the velocity. Singlepoint insertion flow meters estimate the flow through a pipe by measuring the velocity in one point of the pipe. Therefore any distortion of the velocity profile will have significant effects on the accuracy of flow measured. Hence, the authors do strongly recommend that a velocity profile is undertaken every time a singlepoint insertion meter is installed. Only the results from a velocity profile measurement will provide the operator with all the necessary information to achieve measurement results of desired accuracy.

If a velocity profile is to be undertaken, it should be done during stable flow conditions. Depending upon the pipe diameter, the operator will select a number of positions; install the meter taking care to tighten the pressure fittings before opening the valve. Then the operator will measure and record the velocity at the predetermined points and enter them into the program or spreadsheet. It is a good idea to take several readings at each point, as even during stable conditions velocities tend to vary. For additional accuracy three profiles should be taken and an average used. The program or spreadsheet will give either a factored value to be used at centerline, or will predict the point at which average velocity can be found. Once the profile has been taken or the decision not to profile, the operator must decide whether to fit the meter at center point

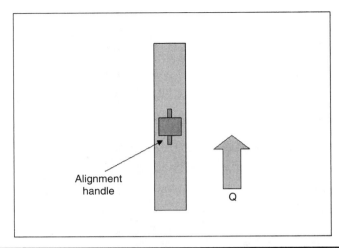

Alignment
handle

Q

FIGURE B.12 Proper alignment is important for accurate measurement. (*Source*: Julian Thornton.)

and factor the data as discussed above or to locate the average point and use raw data. Usually the operator will fit the meter at the center point when low velocities occur, to allow the meter to work within its minimum velocity limits.

Positioning and Fitting the Data Logger

Once the metering location has been attained the operator should secure the locking nut ensuring that the meter is positioned correctly along the axis of the pipe, see Fig. B.12. Most meters are quite sensitive to this positioning and measurement errors can occur if care is not taken.

At this point the operator must fit the data logger if one is to be used. Key things to remember when using a single point velocity probe to measure flows are

- Always record all data during setup. Clearly mark any assumptions that were made.
- The accuracy stated by the manufacturer is the accuracy and repeatability of the probe. This figure does not allow for human error during the setup process.
- An incorrect diameter measurement will seriously affect flow accuracy.
- Incorrect positioning of the probe within the velocity profile will affect the accuracy of the flow calculation.
- Incorrect factoring of a centerline velocity will also seriously affect the accuracy of a flow calculation
- Incorrect axial positioning will affect the accuracy
- When installing the meter be sure to hold the meter head to ensure it doesn't inflict bodily damage when opening the valve to insert the meter
- Be sure to withdraw the meter completely from the pipe before closing off the valve when a metering exercise is complete. Failure to do so will seriously damage the probe and also require a total system shutdown to remove the damaged probe.
- Ensure that the rod length and the available space in a chamber are compatible.

- Identify battery life and monitoring period required. Ensure that backup batteries can be fitted if required.

- Always store raw and manipulated data in case of error during setup.

The inexperienced operator may feel that insertion metering is a lot of work with many potential areas for error. However, with care and a little practice insertion meters can give very good performance. While they are intrusive and an entry point does have to be provided which can sometimes be a stumbling block, this methodology does have the benefit of being able to accurately measure the internal diameter of the pipe. This is not always the case with other nonintrusive methodologies and can be critical. Insertion flowmeters are nowadays widely used by water companies and consultants to undertake on-site verification of critical flowmeters (e.g., system input meters, export meters, zone meters, and the like.) and to provide temporary flow measurements for various purposes (e.g., calibration of hydraulic models, leakage measurements in discrete supply areas , and the like.). As for every piece of equipment there are advantages and disadvantages of singlepoint insertion flow meters some of them are listed below.

Advantages

- Relative low cost.

- Insertion meters can be used on different diameter pipes (if long enough they are relatively independent from pipe diameter).

- Do not require a shutdown in supply.

- Most insertion probes do not require power supply.

- They cause small headlosses.

Disadvantages

- Accuracy is very sensitive to changes in flow profile.

- Require ample straight length of pipe out, and downstream without disturbances to flow profile.

- Operator must be proficient and experienced in use of flowmeter.

- Require a hot tap to be installed.

B.2.2   Portable Ultrasonic Meters

Portable ultrasonic meters have been around for about 25 years and have recently become very sophisticated and accurate. Some operators don't like ultrasonic meters; however, with the right care during installation they can provide very accurate information. In some cases the operator may just want to have an idea about the flow variance, and in this situation ultrasonic meters are perfect as they are completely nonintrusive and do not require a hot tap or entry point to the pipe.

There are various types of ultrasonic meter on the market; however, the most common fall into two categories:

- Ultrasonic Doppler

- Ultrasonic transit time (sometimes referred to as time of flight)

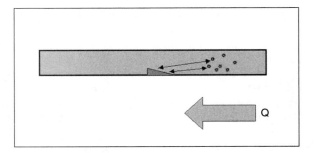

Figure B.13   Doppler effect sensors. (*Source:* Julian Thornton.)

Ultrasonic Doppler meters are normally used to measure liquids, which have either particles or entrained air. The Doppler principle works on a reflective basis when an ultrasonic signal is reflected by suspended particles or gas bubbles in motion as seen Fig. B.13. Ultrasonic Doppler utilizes the physical phenomenon of a sound wave that changes frequency when it is reflected by moving discontinuities in a flowing liquid. The ultrasonic Doppler flowmeter transmits an ultrasonic sound into a pipe with flowing liquids, and the discontinuities reflect the ultrasonic wave with a slightly different frequency that is directly proportional to the rate of flow of the liquid. For this reason Doppler meters are usually not used in clean water applications.

Ultrasonic transit time meters work by sending and receiving signals from one sensor to another as seen in Figs. B.14 and B.15. Figure B.16 shows a typical installation in reflex mode where the sensors are mounted to the same side of the pipe. In this case the

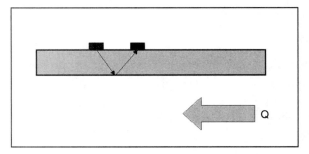

Figure B.14   Transit time reflex pattern. (*Source:* Julian Thornton.)

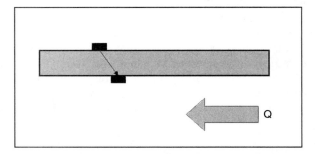

Figure B.15   Transit time measurement. (*Source*: Julian Thornton.)

FIGURE B.16   Clamp on ultrasonic meter. (*Source*: Reinhard Sturm.)

signal bounces off the pipe wall and back to the second sensor. Figure B.15 shows an alternative method of installation where the signal is transmitted directly at the second sensor.

The time it takes for the signal to travel from one sensor to the other sensor changes as the velocity in the pipe changes. This is stated in Faraday's law.

Pipe materials affect the transmission of the signal. Most experienced users of ultrasonic meters will know that sometimes it can be hard to get a good signal on old corroded cast iron pipe, as the corrosion tends to deflect the ultrasonic beam.

As the pipe material and diameter change so will the angle of the beam as it travels through the pipe material. It is therefore very important to know the pipe material prior to programming the unit, as this will dictate the sensor separation. If an incorrect separation is used then the signal received by the second sensor may be weak or nonexistent. It is also important to measure the pipe wall thickness, which is hard to do even with an ultrasonic thickness gauge as the pipe may be lined or corroded; however, best estimations and measurements will need to be made.

> **I**t is often difficult to measure flow accurately on old corroded pipes.

Installing the Meter

As with the insertion meter a location should be chosen with ample straight length of pipe before and after the meter away from other pipe fittings.

The pipe wall material should be identified and where possible internal and external measurements made. If not ultrasonic thickness gauges can be used in some cases to measure the pipe wall thickness. In many cases these gauges do not measure layers of

> **T**he 30-diameter, 20-diameter rule also works well for ultrasonic meters if no other rule is specified.

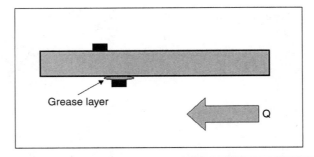

Grease layer

Q

Figure B.17   Grease helps signal pass through pipe wall. (*Source*: Julian Thornton.)

corrosion or pipe linings, these have to be estimated. It is a good idea to have a set of pipe tables with the equipment, so that in the case where internal measurements cannot be made at least a good estimate from a pipe table can be entered.

The outside of the pipe should be cleaned. Corrosion or grease and dirt will affect the strength of the signal and therefore the effectiveness of the meter at this point.

The pipe information should be programmed into the unit according to the manufacturer instructions and the unit will in most cases tell the operator to install the sensor in either reflex or direct mode.

Before installation the pipe should be clearly marked for sensor installation. It is important that the sensors are positioned exactly and well aligned otherwise a poor signal may result.

Prior to applying the sensors to the pipe wall conductive grease should be applied to the sensor to allow it to bond better to the pipe wall (see Fig. B.17). The grease helps the signal transmit directly into the pipe and not reflect the outside of the pipe. Care should be taken not to over grease the sensors as excessive grease on the pipe in particular on small pipe diameters can create a situation where the signal passes directly from one sensor to the other. (On smaller pipes the sensors tend to be placed closer together).

Once the sensors are in location it is a good idea to check the signal strength, by gently moving the sensor around the location. Once the highest signal strength is located the sensors should be secured to the pipe wall most manufacturers supply suitable strapping or magnets for this purpose. In the worst case scenario duct tape works quite well.

Flow Measurement

Once the sensors are securely strapped in place then flow measurement can begin. At this point it is often recommended to undertake a zero flow calibration if possible. This allows an in-place calibration of the unit. Flow should be closed off momentarily and the unit informed that the flow is zero. After this the flow may be carefully turned back on again. When turning flow on and off in a system, take care to do it very slowly as careless operation of valves can create water hammer, which can create additional leakage and damage to the system.

At this point a data logger may be fitted to record flow and time ratios. Most units nowadays will come equipped with an internal data logger and some may even have real time graphs in LCD displays.

Ultrasonic Transit Time Meter Readings

Key things to remember when using an ultrasonic transit time meter to measure flow are

- Bad data in bad data out. Incorrect pipe material or wall thickness data will result in error.
- Disturbances to the flow will cause errors or erratic readings.
- Old corroded pipes are hard to measure with these devices although not impossible.
- Always carry a tube of grease.
- Identify battery life and required monitoring period. Ensure that backup batteries are available and can be fitted if necessary.
- Always carry pipe tables.

As with the insertion meter, an inexperienced operator may feel insecure about the number of potential errors. However, experienced operators will confirm that ultrasonic meters can be effectively used in most situations as long as the limitations at each particular installation point are respected. As for every piece of equipment there are advantages and disadvantages of portable ultrasonic flowmeters, some of them are listed below.

Advantages

- Relative low cost.
- Can be used on different diameter pipes.
- Do not require a shut down in supply.
- They cause no head losses.
- Most units combine flowmeter and data logger.

Disadvantages

- Accuracy is very sensitive to changes in flow profile.
- Require ample straight length of pipe out, and downstream without disturbances to flow profile.
- Operator must be proficient and experienced in use of flow meter.
- Require power supply for long-term measurements.

Selecting Portable Equipment

Some points to consider when selecting portable equipment

- The two types of portable equipment discussed are very flexible and with certain limitations can be used to gain a good idea of flow or velocity in a pipeline where no existing permanent equipment is available.
- Both types of equipment are available in battery-operated or main power formats. Some equipment can run for longer than others, can run on internal batteries, and some can have backup batteries fitted with more ease than others can.
- Some project situations require bidirectional flow recording and most of the equipments available on the market today can be equipped to do this; however, care should be taken to perform this function if required.

- The equipment can be used in many situations to check the accuracy of permanent metering equipment and will do so reasonably accurately as long as the data input into the setup criteria is correct. However, when checking the accuracy of permanent metering equipment a calibrated volumetric test is preferable although not always possible.
- Local support is invaluable.

Typical Applications
Some of the typical applications for this type of equipment are

- Zone flow analysis
- Leak volume monitoring
- Comparative accuracy testing of permanent metering equipment
- C-factor testing
- Demand analysis
- Hydraulic model data collection
- Valve sizing
- Pump testing

There are of course other project-specific applications for which this equipment can be used.

B.2.3   Portable Hydrant Meters
This type of meter is used to check fire flows, C factors, and, in general, when the operator needs to put a known flow onto the system, maybe to calibrate an existing meter. Portable hydrant meters come in two main types, turbine and differential pressure. Figure B.18 shows a turbine type meter in action.

Differential Pressure Hydrant Meter
The differential meter is simple to use and is normally either handheld or screws on to the hydrant port. The unit is fitted to the port and the flow slowly turned on. If the handheld device is to be used the sensing area should be held steadily in the flow

Figure B.18   Turbine hydrant meter. (*Source*: Julian Thornton.)

stream. The screw on version is automatically positioned correctly. Then it is a simple matter of reading the gauge pressure and relating the flowing pressure to a volumetric flow, usually by means of a chart, which will be provided by the manufacturer. Normally this type of meter has no moving parts or electronic components and is therefore robust, easy to use, and easy to maintain.

When testing areas with old mains, in particular unlined old cast iron hydrants to be tested should be flushed well before and after testing. This ensures that any debris in the line does not enter and damage the meter and that there are no dirty water complaints after the testing.

Care should be taken when using any hydrant meter to ensure that no damage is caused to the surrounding area. The force of the flowing water from the hydrant is often sufficient to dig large holes in grass verges and often dig up and damages asphalt road or pavement coverings.

To avoid damage to any surrounding area the use of a hydrant diffuser is recommended. The hydrant diffuser is a cone with various baffles inside, usually made of strong mesh. The diffuser either thread straight onto the hydrant in the case of the handheld differential type meter or onto the downstream end of the turbine type meter. Diffusers are simple to manufacturer and can be made by most sheet metal shops. The number of baffles required will vary with the potential flow and pressure from the hydrant.

Checklist

- Ensure the area around the hydrant to be tested is barricaded off with the necessary cones and signs to warn traffic and pedestrians as to the testing, as there will be significant discharge of water

- If the flows and pressures are potentially high protect the ground where the water will flow with plastic sheeting to reduce impact and damage

- Use a flow diffuser where potential damage or high flows may occur

- If using a handheld differential pressure meter be sure to hold the unit in the center of the flow

- Operate hydrants slowly to reduce negative hydraulic impact

- Ensure that there is sufficient drainage to take away the water, there will be a significant amount of discharge

- When testing in areas with basements ensure that water cannot back up into basements

- Flush hydrants well before installing meter to ensure that damage doesn't occur through rust and corrosion passing through the meter

- Flush hydrants well after testing to ensure that there are no dirty water complaints

B 2.4   Flow Loggers

Flow loggers or recorders are a special type of data logger, which can take a pulse directly from the magnetic drive of virtually any meter, convert it into an electronic signal, and record rates of flow (see Fig. B.19).

Flow loggers are used for

- Comparative accuracy tests
- Demand analysis

FIGURE B.19 Meter-master flow logger. (*Source*: F.S. Brainard & Co.)

- Rate of flow recording from volumetric meters
- Leak detection by zone flow analysis
- Hydraulic model data collection
- And various other projects, which require flow data collection

The flow loggers are relatively easy to set up, by following manufacturer instructions and inputting data such as flow units, pulse significance, and desired recording time.

The secret to good flow logging lies in the positioning of the sensor to pick up optimum pulse strength when using a flow logger as shown in Fig. B.19. Obviously missed pulses equal missed or inaccurate flow recording. Fig. B.20 shows a chart with various examples of sensor location for different meter types.

Some meters have a protective ring around the meter head magnets. The idea of this is to stop theft of water by placing a magnet around or near the drive magnets to interfere with the normal operation of the meter. This of course also interferes with the ability of the flow recorder to pick up the pulses emitted from the magnets. High sensitivity units are available which can pick up pulses even in these situations although these are usually a speciality item. If using high sensitivity units special care should be taken to ensure that all pulses picked up are actually from the meter rotation and are not caused by external influences. If the operator feels that external influences could be interfering with good recordings this can often be resolved by wrapping the unit in aluminum cooking foil. The foil has the effect of blocking the external signals and allowing the logger to pick up only the meter pulses.

Checklist

- Identify meter types to be logged and obtain pulse factor from the meter manufacturers.
- Ensure that a good healthy pulse signal can be obtained from the manufacturer's recommended sensor location. If not attempt to manually locate best spot.

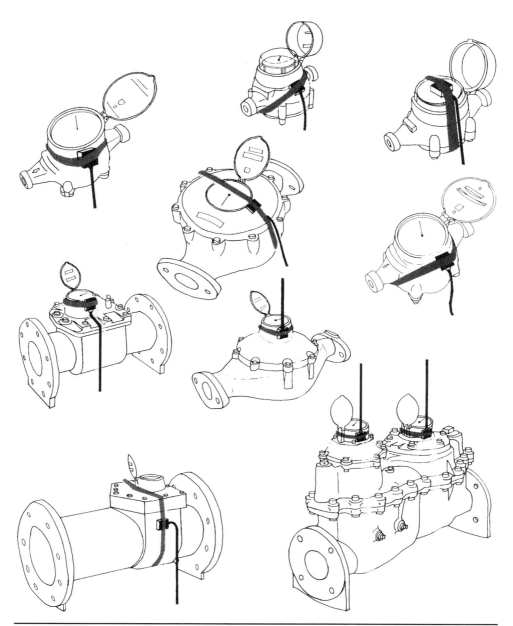

FIGURE B.20   Sensor locations for different meter types. (*Source*: F.S. Brainard & Co.)

- Check pulse significance by recording a sample of flow and volume while on site and comparing this to the volume that passes through the meter according to the meter register. Perform a volume and time calculation to estimate approximate flow rates and compare it to the flow loggers calculated flow.

- If external signals are picked up to attempt to shield the unit with aluminium foil.
- Ensure that both raw pulse files and calculated flow rate files are saved. In the case of an error, the raw data files can be used to recalculate the proper flow rate.
- If the meters to be tested have security magnetic shielding be sure to use a high sensitivity unit.

B.2.5   Flow Charts

In some cases a utility will have older style flow charts. There are various types of flow chart in existence. Some take a differential pressure from a DP meter and some take a 4 to 20 mA signal, which represents flow rate. Flow charts are relatively easy to set up by following the individual manufacturer's instructions; however, in all cases care should be taken not to allow water to be spilt on the chart and not to allow pen ink to dry up.

Some charts will have a mechanical clock and some an electronic drive to allow flow rate to be recorded over time. In the case of the mechanical drive units care should be taken to select either a 30 day, 7 day, or 24 hour clock. Charts should not be confused between the various options of maximum flow rate and time period.

Checklist

- Select correct clock mechanism for recording period required.
- Select correct chart for maximum flow rate required and clock mechanism installed.
- Ensure pen is clear and recharged with ink.
- Do not allow water to be spilt on the chart.
- Ensure that the mechanism is free to rotate and not jammed.
- Be sure to change the chart at the end of the recording period.
- Clearly record the date and nature of the test on the chart after removing for analysis.
- Ensure that the pen arm is properly calibrated at zero flow. Often the pen arm gets bent when changing charts.

B 2.6   Step Testers

In Chap. 16 we have discussed step testing as a means of isolating sections of leaky main, identifying volume of loss, and ranking and pinpointing repair programs.

Step testers have been around for many years. Older model step testers work, on a swing gate principle. The gate swings further when flow is higher which in turn moves a pen arm on a chart. The chart has a clock, which moves the chart in a circular motion in function to time, and a graph is recorded. As the flow drops in response to a shut in leak the graph will display the difference in flow.

It is important to ensure local manufacturer support when considering the first time purchase of high-tech equipment.

Nowadays most operators use a simple data logger and flow meter. Some step-testing data loggers have a function, which automatically analyzes the leak volume, and some have a radio function, which allows the operator to work in a remote vehicle.

B.3   Permanent Equipment

Previous sections of this appendix have discussed various types of temporary equipment which is often used during field-testing exercises. Now we will discuss some permanent options. Before undertaking fieldwork where measurements must be taken it is a good idea to inspect any permanent meters, which may be used for measuring purposes. Meters should be tested to ensure proper accuracy and should be sized correctly for the flows in the field.

In addition to proper calibration and sizing, it is important to understand what type of meter is being used, its principles of operation, and how it transmits data. This is necessary so that a compatible recording or logging device or mode can be selected to collect the data, or in the absence of these options manual reads can be scheduled. Section B.3.1 shows some of the industry standard types of meter and shows some figures, installation recommendations, and flow-limitation charts.

Each manufacturer will have their own specifications for their meters, and therefore it is necessary to have individual data sheets for different manufacturers' equipment. However, in the United States, for example, all manufacturers will manufacture their meters to be equal or better than the AWWA specifications. In other areas of the world manufacturers will perform equal to or better than the ISO specifications, which vary from those of the AWWA. AWWA and ISO specifications are readily available from the respective organizations and can usually be ordered over the web.

B.3.1   Meter Types

- Turbine and turbo meters (see Figs. B.21 and B.22 and Table B.2)
- Propeller meters (see Fig. B.23 and Table B.3)
- Compound meters (see Figs. B.24 and B.25 and Table B.4)
- Piston residential meters (see Fig. B.26)
- Velocity residential meters (see Fig. B.27)
- Magnetic meters (see Fig. B.28)
- Ultrasonic meters
- Differential pressure meters
- Vortex shedding meters

Each meter type has its benefits and its negative points and it is not the intention of this book to promote one methodology or another. It is important, however, that the operator become familiar with the meter types in the area in which they will work.

Data Transmission

Most of the meters mentioned above will transmit data in one of several ways. The most common are

- Manual read and indexed stored info
- Analog output of 4 to 2 mA (see Fig. B.28)

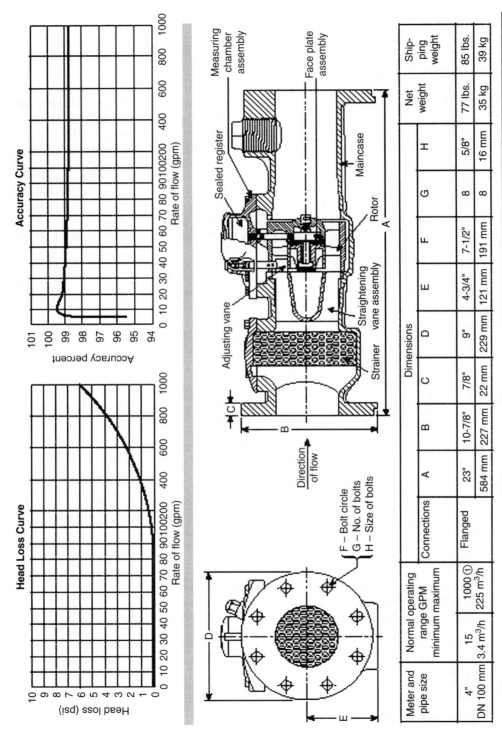

Accuracy Curve

Head Loss Curve

Meter and pipe size	Normal operating range GPM minimum maximum		Connections	Dimensions									Net weight	Ship-ping weight
				A	B	C	D	E	F	G	H			
4"	15	1000 ①	Flanged	23"	10-7/8"	7/8"	9"	4-3/4"	7-1/2"	8	5/8"	77 lbs.	85 lbs.
DN 100 mm	3.4 m³/h	225 m³/h		584 mm	227 mm	22 mm	229 mm	121 mm	191 mm	8	16 mm	35 kg	39 kg

F – Bolt circle
G – No. of bolts
H – Size of bolts

Direction of flow

Measuring chamber assembly

Sealed register

Adjusting vane

Face plate assembly

Maincase

Rotor

Straightening vane assembly

Strainer

Figure B.21 Turbine and Turbo meter specifications. [*Source*: Invensys Metering Systems (formerly Sensus Technologies).]

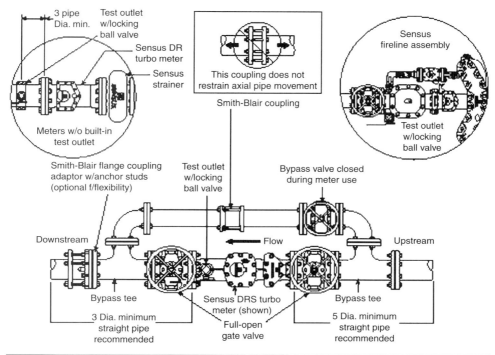

FIGURE B.22 Sample turbo meter installation. [*Source*: Invensys Metering Systems (formerly Sensus Technologies).]

- Pulse output (see Fig. B.29)
- Frequency output (see Fig. B.30)

Output Correlation

In general we might suggest that the following types of meters fall into the three categories of output:

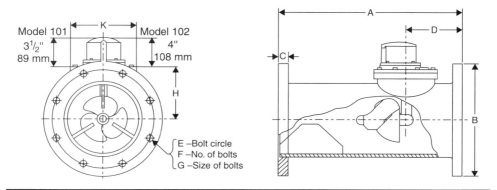

FIGURE B.23 Propeller meter. [(*Source:* Invensys Metering Systems (formerly Sensus Technologies).]

Size	Model	Main Case	Normal Flow Limits (100.0% # 1.5%) Gpm	Extended Flow Unit (Intermittent Flows)	Low Flow (95%) Gpm	Head Loss @ Maximum Flow	End Connections
1-1/2 in	W-120DR	Bronze	4-120	160	3	13.5	1-1/2 in size two boll oral, AWWA 125-pound class
1-1/2 in	W-1200 RS	Bronze	4-120	160	3	15.4	1-1/2 in size, two boll oral, AWWA 123-pound class
2 in	W-160 DR	Bronze	4-160	200	3	5.6	2 in size, boll slot oral, AWWA 125 pound class optional 2-11-1/2 in MPI, internal threads
3 in	125WDR	Cast Aluminum	10-350	400	10	8	2-1/2-7-1/2 in NSI (National Standard Fire Hose Coupling Thread) furnished unless otherwise specified
3 in	W-350DR	Bronze	5-350	450	4	5.0	3 in size, round ANSI 125-lb class
4 in	W-1000DR	Bronze	15-1000	1250	10	3.6	4 in size, round ANSI 125-lb class
4 in	W-1000DRFS	Bronze	15-1000	1250	10	6.3	4 in size, round ANSI 125-lb class
6 in	W-20000DRS	Bronze	30-2000	2500	20	6.2	6 in size, round, ANSI 125-lb class
6 in	W-2000DRSL	Bronze	30-2000	2500	20	3.0	6 in size, flat lace, 125-lb class
6 in	W-2000DRFS	Bronze	30-2000	2500	20	6.7	6 in size, round ANSI 125-lb class
8 in	W-3500DR	Bronze	35-3500	4400	30	8.3	8 in size, round ANSI 125-lb class
8 in	W-3500DRFS	Bronze	35-3500	4400	30	8.5	8 in size, round ANSI 125-lb class
10 in	W-5500DRFS	Bronze	55-5500	7000	35	6.1	10 in size, round ANSI 125-lb class
10 in	W-5500DRFS	Bronze	55-5500	7000	35	6.0	10 in size, round ANSI 125-lb class
16 in	W-10,000DR	Cast Iron	250-10,000	12,500	200	5.3	16 in size, round ANSI 125-lb class

Source: Invensys Metering Systems (formerly Sensus Technologies).

TABLE B.2 Flow Data for Turbo Meters

Meter Size	Low Flow gpm (m³/h)	Normal gpm (m³/h)	Dimensions									Shipping Weight Pounds (kg)
			A	B	C	D	E	F	G	H	K	
3 in	80	100-250 GPM	16 in	7-1/2 in	¾ in	6-1/2 in	6 in	4	5/8 in	3-3/8 in	5 in	70 lbs
DN80 mm	18.2 m3/h	23-57 m3/h	406 mm	190 mm	19 mm	165 mm	152 mm		16 mm	86 mm	127 mm	32 kg
4 in	82	125-500	18 in	8 in	5/5 in	7-1/2 in	1-1/2 in	8	5/8 in	3-7/8 in	7-1/2 in	85 lbs
DN 100 mm	18.6	28-114	457 mm	228 mm	16 mm	190 mm	190 mm		16 mm	99 mm	190 mm	39 kg
6 in	160	220-1200	22 in	11 in	11/16 in	9 in	9-1/2 in	8	¾ in	5 in	9 in	115 lbs
DN 150 mm	36.3	30-273	550 mm	279 mm	17 mm	229 mm	241 mm		19 mm	127 mm	229 mm	52 kg
8 in	100	250-1650	24 in	13-1/2 in	11/16 in	9 in	11-3/4 in	8	¾ in	6 in	9 in	150 lbs
DN 200 mm	43.2	57-375	610 mm	343 mm	17 mm	229 mm	208 mm		19 mm	152 mm	229 mm	68 kg
10 in	260	330-2500	26 in	16 in	11/16 in	10 in	14-1/4 in	12	7/8 in	7-3/6 in	11 in	200 lbs
DN 250 mm	50.0	75-568	660 mm	406 mm	17 mm	254 mm	362 mm		22 mm	187 mm	279 mm	91 kg
12 in	275	350-3500	28 in	19 in	13/16 in	10 in	17 in	12	7/8 in	8-3/8 in	11 in	290 lbs
DN 300 mm	62.4	80-795	711 mm	483 mm	21 mm	254 mm	432 mm		22 mm	213 mm	279 mm	132 kg

Source: Invensys Metering Systems (formerly Sensus Technologies).

TABLE B.3 Flow Data for Propeller Meters (*Continued*)

Meter Size	Low Flow gpm (m³/h)	Normal gpm (m³/h)	Dimensions									Shipping Weight Pounds (kg)
			A	B	C	D	E	F	G	H	K	
14 in	350	450-4500	42 in	21 in	1-3/8 in	12 in	18-3/4 in	12	1 in	9-1/4 in	13-1/2 in	450 lbs
DN 350 mm	79.5	102-1022	1067 mm	533 mm	35 mm	305 mm	476 mm		25 mm	235 mm	343 mm	204 kg
16 in	450	550-5500	48 in	23-1/2 in	1-7/16 in	12 in	21-1/4	16	1 in	10-1/4 in	13-1/2 in	550 lbs
DN 400 mm	102.2	125-1249	1210 mm	597 mm	37 mm	305 mm	504 mm		25 mm	260 mm	343 mm	249 kg
18 in	550	752-7250	54 in	25 in	1-9/16 in	15 in	22-3/4 in	16	1-1/8 in	11-5/8 in	13-1/2 in	620 lbs
DN 450 mm	124.9	165.1647	1372 mm	635 mm	40 mm	381 mm	570 mm		29 mm	295 mm	343 mm	281 kg
20 in	700	850-9000	60 in	27-1/2 in	1-11/16 in	15 in	25 in	20	1-1/8 in	12-5/8 in	13-1/2 in	820 lbs
DN 500 mm	150.0	193-2044	1524 mm	699 mm	43 mm	381mm	635 mm		29 mm	321 mm	343 mm	372 kg
24 in	1000	1300-13000	72 in	32 in	1-7/8 in	18 in	29-1/4 in	20	1-1/4 in	12-5/8 in	13-1/2 in	1000 lbs
DN 600 mm	227.1	259-2592	1829 mm	813 mm	48 mm	457 mm	740 mm		32 mm	321 mm	343 mm	454 kg
30 in	1600	2100-18500	84 in	38-3/4 in	2-1/8 in	18 in	36 in	28	1-1/4 in	12-5/8 in	13-1/2 in	1150 lbs
DN 750 mm	363.4	477-4224	2123 mm	984 mm	54 mm	457 mm	914 mm		32 mm	321 mm	343 mm	522 kg
36 in	2400	3000-24000	96 in	46 in	2-5/8 in	20 in	42-3/4 in	32	1-1/2 in	12-5/8 in	13-1/2 in	1350 lbs
DN 900 mm	545.0	681-5450	2436 mm	1168 mm	67 mm	508 mm	1086 mm		38 mm	321 mm	343 mm	613 kg

Source: Invensys Metering Systems (formerly Sensus Technologies).

TABLE B.3 Flow Data for Propeller Meters (*Continued*)

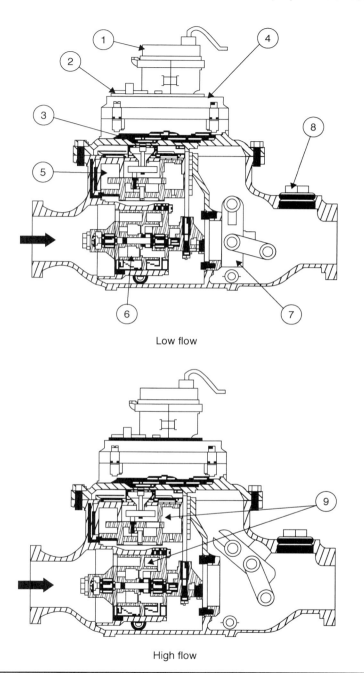

Low flow

High flow

FIGURE B.24   Compound meter. [*Source*: Invensys Metering Systems (formerly Sensus Technologies).]

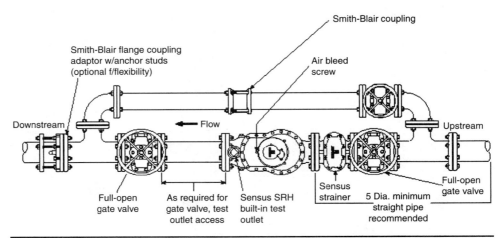

FIGURE B.25   Sample compound meter installation. [*Source:* Invensys Metering Systems (formerly Sensus Technologies).]

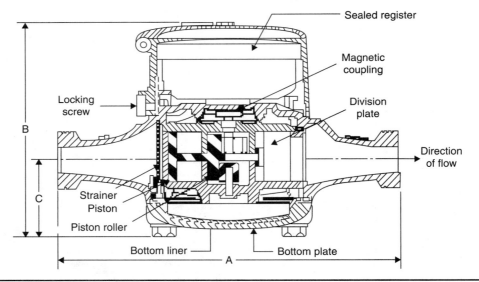

FIGURE B.26   Piston residential meter. [*Source:* Invensys Metering Systems (formerly Sensus Technologies).]

1. Manual read option: All of the above meter types

2. Four-to-twenty milliampere option: This is usually reserved for the electronic type meters such as the magnetic and ultrasonic type meters, the electronic vortex shedding meters and differential pressure meters, which have electronic converters.

3. Pulse and frequency output option: Pulse output is usually available from most of the electronic meters such as the magnetic and ultrasonic meters. It is also often available for the turbine, propeller meters, and residential meters (PD and velocity jet type) although sometimes this option needs to be requested for later.

Nominal Size/ Model	(GPM) Normal Operating Range	Low Flow (GPM) Accuracy @ 95%	AWWA Maximum Continuous Flow (gpm)	Minimum Intermittent Flow (gpm)	Minimum Accuracy Crossover	Head loss @ Maximum Intermittent Flow
2* SRH	2–160	¼	80	160	95%	5.0
3* SRH	4–320	½	160	320	95%	5.3
4* SRH	6–500	¾	250	500	95%	3.2
5* SRH	10–1000	1–1/2	500	1000	95%	13.0
8* Manifold	16–1600	2	800	1600	95%	13.2

Source: Invensys Metering Systems (formerly Sensus Technologies).

TABLE B.4 Flow Data for Compound Meters

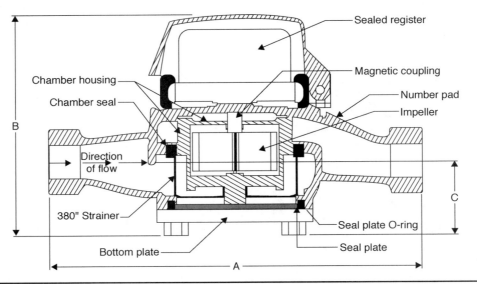

FIGURE B.27 Velocity residential meter. [*Source:* Invensys Metering Systems (formerly Sensus Technologies).]

FIGURE B.28 Magnetic meter. (*Source:* Chris Bold, Invensys Metering RSA.)

Obviously before we can select equipment to record flows from permanently installed equipment, we need to understand the type of output the meter has, if it has one. Then we need to identify the correct recording equipment for the job.

This is easier when the permanent equipment is to be installed as part of the project as the meter and the recording device can be selected ahead of time ensuring compatibility.

B.3.2   Metering Types/Characteristics

As we will see in the following sections meters can be fitted with various types of pulse output, giving faster or slower responses to changes in flow.

The operator must learn to properly select the right type of output for the job at hand, always ensuring that there are sufficient pulses to accurately record the flow or volume, while ensuring that there are not too many pulses to either confuse the equipment or use up the data storage allocation in the logger before the test period is complete.

> **A**lways check that the output is compatible with the recording device and the ranges are similar.

Electronic Meters

Electronic meters usually allow the selection of the output pulse value. This is done via either mechanical switches or through programming the meter with a computer or handheld device. It is relatively simple to do in most cases, with minimal training from the equipment supplier or by reading the operational manual.

Mechanical Meters

Mechanical meters have different types of pulse output. Normally they are either

- Reed switch sensors tend to give a lower pulse output than optical switches. The reed switch functions in conjunction with the drive magnets in the meter. See Fig. B.29(*a*) and (*b*) for different configurations, which might be encountered. Others are also available.

FIGURE **B.29a**   Pulse output.

FIGURE B.29b   Pulse output.

- Optical switch usually functions in conjunction with a special dial. See Fig. B.30(*a*) and (*b*) for different configurations, which might be encountered. Other configurations are also available.

If the metering site is going to be underground in a chamber, it is often a good idea to ensure that the cables and meter heads are waterproof, (usually NEMA 6 rating in the

FIGURE B.30a   Frequency output.

FIGURE B.30b   Frequency output.

United States and IP68 in United Kingdom, other areas of the world will have other classi-fication ratings), as this tends to be an area where readings can fail. Most manufacturers will have waterproof dials and sensors available, but it may be an option not the norm.

B.4   Output Readings

B.4.1   Understanding Pulse Recording

The preferred method for recording flows in the field with data loggers is a pulse out-put. The reason for this is that the logger counts a determined number of pulses over the complete time period. Each pulse equals a determined volume of water passing through the meter. Accuracy is high and accountable if the recording equipment is set up correctly. To set up the equipment properly the operator needs to understand several things about the way in which they will record the pulse.

> **Q.**   What is the difference between pulse and frequency?
>
> **A.**   A frequency is a fast pulse.

Some manufacturers' equipment will have a frequency output stated in hertz. One hertz equals one pulse per second. So a frequency output is also a pulse output although most operators think of a pulse as a fairly slow occurrence, for example, 10 pulses per minute would be considered a pulse output. Where 10 pulses per second would prob-ably in most cases is stated as a frequency output of 10 Hz.

Most data loggers can record pulse and frequency and can count it. In most cases, however, they have a limit as to what they can count. Above that the logger may revert to a sampling system where the logger opens a window of time and counts the number of pulses recorded then shuts down and opens up after another predetermined period of time. If this is the case the operator stands the chance of loosing valuable changes in flow rate between the time windows when the logger is actually recording. The effect will be a severely averaged flow profile, which for some applications as mentioned above is not a problem; however, for others such as meter sizing it can create a problem.

If frequency sampling is to be used it is recommended that the operator calculate the average frequency and ensure that the time window is sufficient to allow 10 pulses or more to be recorded. The effect of a missed pulse in the time window would then give an error of 10% or less. If a better resolution is required then the operator should open the time window until the possibility of one lost pulse has a significance of less then the allowable error. See "Pulse Count Recording" for an example.

Pulse Count Recording

By far the best way of recording flow, especially in situations where portions of a flow profile may be used for volumetric analysis is to use a pulse count option.

Once the operator has decided to use a pulse count option as opposed to frequency sampling or analog sampling then he or she must decide on one of two modes of storing the data

- Averaging mode
- Event mode

Pulse Counting Using Averaging Mode

Most data recorders or loggers allow the user to define which type of data storage they will use.

The averaging option operates by counting pulses over a certain period and then storing the average value at defined intervals. In this way the logger is not storing every single pulse (which might be a significant amount depending on the flow rate and the output unit). By not storing every single pulse the operator can prolong the memory of the logger to allow longer storage periods between downloads. The averaging option is useful for recording flow profiles where the important factor is the change in flow profile as in as zone flow measurement for leakage detection.

When selecting an averaging mode it is important that the operator realize that the results will be an averaged accumulation of many data points. When this is understood they can decide what time window to use. It is important that the time window realistically reflects the type of resolution the operator is expecting to get from the data.

For example, if a meter has a very slow pulse output lets say one pulse per 5 minutes and the operator sets a time window of 10 minutes then the logger may only record one or two pulses within that time frame. Obviously that would leave quite a lot of room for error (one in two) in the assumed flow rate depending on if the second pulse falls just inside or outside of the time window limit (see Fig. B.31). A better time window for this pulse would be every hour, which would mean that the logger would record 12 readings within the time window (see Fig. B.32). In this case the potential for error would be 1 in 12.

In many cases the operator would want a faster response than one reading or value every hour so in this case the operator would change the type of pulse output.

Event Recording

Event recording means that the logger or recorder reads and stores every pulse and records the time between each pulse to infer flow rate. This is the most accurate way of

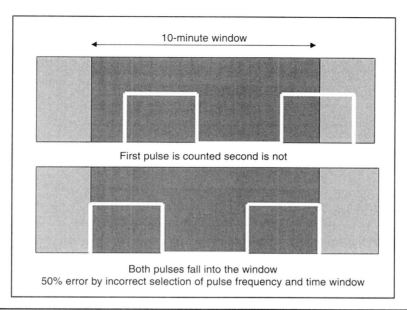

FIGURE B.31   Fifty percent potential error caused by incorrect selection of pulse frequency and time window. (*Source:* Julian Thornton.)

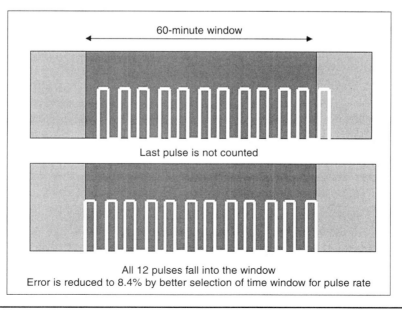

FIGURE B.32   Error is reduced to 8.4% by better selection of time pulse frequency and time window. (*Source*: Julian Thornton.)

> **E**vent recording is the best means of data collection for direct pressure demands.

recording a flow profile and should be used when sizing meters, for example, as this method shows the true peaks and spikes of usage.

However, care should be taken when using this method so as not to fill the memory of the logger or recorder prior to the end of the test period.

State Recording

State recording involves the logging of the status usually of a switch. The switch is either on or off. This type of logging is particularly useful for recording the status of pumps over a period of time. It may be that the operator wishes to optimize the pumping routines but is unsure as to when each pump kicks on and off. By recording the pump state in a bank of pumps the operator can fine-tune the sequence and in many cases improve energy consumption efficiency.

This may seem like a lot of information for the inexperienced operator; however, the meter and logger manufacturers are usually very keen to help out and will in many cases assist the operator with their application in the field. It is therefore recommended that inexperienced operators use equipment with local support options.

B.4.2   Understanding Analog Values

As discussed above many meters have a 4- to 20-mA output. In most cases this will be a scalable output where 4 mA equals zero flow and 20 mA equals maximum flow. In the case of bidirectional outputs, some will have a second channel and some will divide the output, for example, 4 to 8 mA would be available for negative flows and 8 to 20 mA would be available for positive flows. The individual manufacturer instructions will identify methodology.

In most cases the data logger will wake up at a predetermined interval say every 5 minutes, take a sample of the value and store it and go back to sleep. This method works well again for tasks, which do not need every single peak in flow. It is possible to record continuously and register every peak using this methodology; however, in the case of most portable data loggers this would use up the available memory very quickly and therefore in most cases is not an option. This type of analog recording is often used in SCADA systems where the data is sampled very fast say every second or tenth of a second and then transmitted to a central base. In this way the data is stored in a large computer.

Rotating Store or Store till Full

Most data loggers and recording devices have these two modes of data storage

- The rotating store mode functions in a cycle. Memory is used from the front to the back. As the memory begins to become full the earliest data is erased allowing the latest data to always be on file. This type of memory allocation is used in long-term field logging exercises. Care must be taken to ensure that the number of data points to be recorded per time interval and the whole memory are enough, that when the unit starts to overwrite earlier data the historic data bank is of a long enough period for the analysis required.

- Store till full mode is usually used when a specific event or number of event over defined time period are to be recorded. This mode does not cycle like the one above but rather stores data until the memory is full then stops recording. Care should be taken in this mode that the number of pulses and time allocation are correctly related to allow all of the vents in a test to be recorded before the logger turns off.

B.4.3 Logging Summary

The best way of recording data in the field with recorders and loggers is very site specific and depends on the resolution of data required. The operator will soon learn to match the best combinations of recording device with the best combination of output options available for his metering devices. Until the operator learns these skills it is best to rely on skilled local support to avoid unnecessary loss of data and costly rerecording.

Checklist

- Check output types
- Check recorder options
- When using pulse select suitable pulse generator for field application
- Identify correct number of pulses per unit of time to ensure maximum resolution
- Choose averaging method when recording data over long periods of time where lots of pulses may interfere with the recorder memory
- Choose event logging where real demand profiles are required as in the case of meter correct sizing
- When using a frequency select suitable sample rate and time window
- When using analog recording remember to set offset to 4 mA if using 4 to 20 mA output
- Select suitable sample rate
- Identify either store till full mode or rotating store depending on the type of test
- Ensure that outputs and input cables are waterproof where required

B.5 Calibration, Testing, Dead Weight Tests

This section discusses calibrating flow-measuring equipment, comparative flow versus volumetric testing, and dead weight tests.

Any portable or permanent metering equipment needs to be tested on a periodic basis. The time between tests very much depends on the economic factor for permanent meters, (a combination of time, volume and water quality, environmental conditions, and the like) and depends on the type of use and transportation, and the like for portable equipment.

> **A**ll permanent and temporary equipment should be periodically tested to ensure accuracy.

Portable equipment should be tested before and after any major field data collection exercise and sometimes periodically in between. Either portable or permanent equipment is usually tested in one of two ways:

FIGURE B.33   Volumetric meter testing. (*Source:* Chris Bold, Invensys Metering RSA.)

- By performing comparative flows against a calibrated high-resolution permanent meter in a test rig.
- By performing volumetric tests against a calibrated tank volume or weight (see Fig. B.33).

Sometimes permanent meters can be field calibrated by either removing the metering chamber and substituting it for a recently calibrated one and comparing before and after flows to a hydrant or other controlled flow or volume source such as a reservoir or tank with known volume. Alternatively some meter manufacturers supply filters or in-line strainers which are the same size as the meter body. (Use of in-line filters or a strainer is recommended in any case to protect the internal parts of the meter and preserve the good working life of the meter). The strainer can be temporarily removed and a calibrated meter chamber inserted. The two meters can then be both manually read for volume comparison or data logged for flow time profile comparison. Once the testing is complete the reference meter chamber goes back to the shop for retest and calibration to ensure that the reference is indeed valid. In-the-field testing is by far the best means of testing equipment, as it is not always the equipment that fails and causes error but incorrect installation that causes error. By field testing the equipment can be tested in situ. If an error is found in the field but not when the equipment is retested on a calibrated flow rig then the error is obviously in the field in either temporary or permanent installation and can be rectified before further work is undertaken.

Checklist

- Always test volumetrically where possible.
- Always test in the field where possible.
- Ensure that the reference meter chamber has been recently calibrated in the case of comparison.

- Ensure that portable equipment is not damaged in transit.
- Ensure that portable equipment is calibrated frequently at least before and after each major project.
- Ensure that permanent equipment is calibrated according to a predetermined, modeled economic frequency.

B.5.1   Flow Meter Testing Summary

Flow measuring and data logging equipment whether permanent or temporary is often a big investment for a utility, operator, or contractor and should be well maintained and calibrated to ensure continued accountable results. Equipment can never be calibrated too often and is all too frequently neglected and then blamed for malfunction or error. In many cases error is attributed to either lack of or incorrect operator training. However a well-trained confident operator with well-maintained and calibrated equipment will in most cases come up with good traceable and accountable results, which are imperative for a successful water loss control program. After all how can water loss be calculated if the results of the flow and volumetric measurements are in doubt?

B.6   Pressure Measurement Equipment

In addition to good flow recording most water loss control programs require good pressure measurements. Pressure may also be level-measuring equipment because pressure is the driving force in any water system whether provided by pumps or gravity. Pressure dictates the nature frequency and volume of our leakage and physical losses and therefore must be taken very seriously.

B.6.1   Portable Loggers

Portable pressure or level data loggers are by far the easiest means of collecting field pressure data and transferring it to digital media. Data loggers come in all shapes and sizes with different configurations. Some have internal pressure sensors see Fig. B.34 and some have external sensors. However, one of the most important things to understand prior to using a pressure logger is the type of sensor it has and its limitations.

Selecting a Sensor

Pressure sensors are usually calibrated for a maximum pressure and will output either a 4- to- 20 mA signal or a frequency in relation to the pressure sensed. In loggers with internal sensors this is transparent to the user, as the interface is automatic; however, in external sensors this is vitally important see section below discussing portable pressure sensors.

> **P**ressure sensors are expensive, and can be easily damaged by excess pressure or transients.

It is however important to know the pressure limitations of the sensor to be used and ensure that it is not subjected to pressure higher than this value otherwise damage, in many cases irreparable, will occur.

Pressure sensors are expensive so be careful! In addition to selecting the right pressure rating for the job at hand it is also important to select a sensor, which will give the required resolution of measurement. The resolution of the sensor dictates the minimum

Figure B.34 Data logger with internal pressure sensor. (*Source*: Reinhard Sturm.)

pressure step that the logger can record. For example, a sensor may have a maximum rating of 100 m with a resolution of 1%. That means that the logger would record up to 100 m of pressure in steps of ±1 m. In some cases such as C-factor testing or level recording this may not be sufficient so a sensor with a higher resolution may be used, for example ,maximum pressure 100 m, resolution 0.1 m would give steps of ±10 cm. In the case of level recording specifically, usually the pressures to be measured are much lower and the requirement for resolution much greater as 10 cm in a large reservoir could relate to a huge volume of water. In this case an example might be of maximum pressure 10 m, resolution 0.1 m, which would give pressure steps of 10 mm, or resolution 0.01 m, which would give steps of 1 mm.

Obviously care must be taken not to use these lower pressure sensors in distribution situations where pressures would in many cases be much higher!

When selecting the correct pressure sensor or pressure logger for the job remember that the system pressure changes throughout the day, in many cases where head losses are great during peak demand pressure can be much higher at night even when a standard fixed outlet pressure reducing station is controlling the system. So if sizing equipment for use during the day remember to allow ample additional scale for higher nighttime pressures. If the operator desires to equalize pressures then he should read details of modulating pressure control in the pressure management section (see Chap. 18).

Testing Pressure and Level Sensors

In addition to selecting the right sensor it is also important that frequent calibration of the pressure sensors is undertaken as they do tend to drift with time and use and also sometimes with temperature change.

Testing is undertaken at two or more points, the first zero to ensure that the bottom end of the scale is secured and no offset is incurred, then, as a minimum a pressure close to the high end of the scale should be tested. It is advised to test several other points between the top and bottom of the scale too, however this is not always done when time constraints don't allow.

B.6.2   Dead Weight Testers

Pressures are usually induced with a dead weight tester. Dead weight testers come in two basic formats, with the mechanical version with weights, or the hydraulic version, which is usually digital. Either model provides a very efficient and accurate way of testing sensors.

If a dead weight tester is not available it is possible to test zero and a high-pressure against a calibrated column of water such as a water tower with a known and static (at the time) level. Alternatively a simple column can be built out of pipe at the workshop. Remember it is not the diameter which counts but the height so small diameter tube can be used.

Once the comparisons have been made the loggers and sensors are usually calibrated either electronically within the logger software or the data is downloaded in raw format into a spreadsheet and reformatted there (see Fig. B.35).

Sample Pressure Profile At Critical Node

c:\123r3\dist1e\D2P1.CSV

Water Mains Pressure Recording

Site No: 0002 390 Site Name: No 350 Centennial Pressure Range: 100.0 m

Macro Commands ALT F3
Note: Use select to pull in data
Note: Use view to check data and return
Note: Use CAL to regress data with calibration curve

Tabular Data

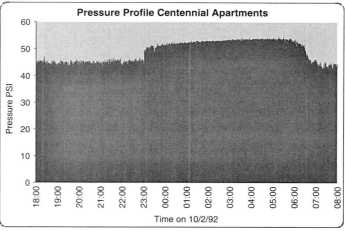

Day	Pressure
19 Oct 1992	
18:00	30.3
18:01	30.3
18:02	30.8
18:03	30.4
18:04	30.7
18:05	31.1
18:06	31
18:07	30.5
18:08	30.8
18:09	30.3
18:10	30.6
18:11	31.1
18:12	31
18:13	30.1
18:14	30.9
18:15	30.6
18:16	30.8
18:17	30.2
18:18	30.7
18:19	30.4
18:20	31
18:21	31.2
18:22	30.9
18:23	31.9
18:24	30.5
18:25	30.2
18:26	30.5

FIGURE B.35   Pressure data calibration sheet. (*Source*: Julian Thornton.)

B.6.3   Portable Charts

Portable charts are an older method of testing pressures and work perfectly well. The only drawback is that in most cases data points have to be manually taken from the chart and input into a spreadsheet for analysis. This can be time-consuming for large projects where many pressures will be measured and resolution requirements may be high.

When setting up a pressure chart care should be taken to select the right chart and clock mechanism for the job as with the flow charts mentioned above. Calibration of pressure charts is done in much the same way as the loggers and sensors mentioned above.

B.6.4   Portable Pressure Sensors

In some situations it is preferable to have portable pressure sensors, which can either be used with a number of different logger types for different applications or can be used directly with telemetry or SCADA applications.

Care must be taken to ensure that the output type is compatible with the logger input types and that the sensor can either be powered by an internal logger battery without causing flat batteries due to overly high draw down or that a portable battery pack or main power source is available.

Testing of pressure sensors is undertaken in exactly the same ways as stated above.

Checklist

- In the case of portable sensors ensure that the output matches the input of the device that will receive the signal.
- Ensure that sufficient and suitable power sources are available.
- In all cases check that the maximum pressure rating and resolution are suitable for the job.
- Ensure that all pressure-measuring devices are properly calibrated to zero and maximum pressure with other intermediate check points if possible.
- Handle sensor with care as they can tend to be fragile in some cases and should not be dropped or mishandled.
- Ensure that the sensor, logger, or chart to be used is waterproof if it will be used in a potentially underwater environment.
- Ensure that suitable test equipment is on hand or available locally for periodic checks.

B.6.5   Traditional Acoustic Leak-Detection Equipment

Successful leak detection very much depends on operator skills. Proper training is invaluable.

Acoustic leak-detection equipment has to be one of the simplest leak-detection devices to be used in the field, however the success of work undertaken with this type of equipment very much depends on the training of the operator.

It is very important that operators gain as much hands-on training with another skilled operator as possible prior to undertaking survey alone. Success very much depends on the operators decisions based on signals received and interpreted by the brain. While this section discusses some of the leak detection methodology related to the equipment more detail can be found in Chap. 16.

Mechanical and Electronic Listening Sticks

Listening sticks, or sounding rods as they are sometimes known as, work by making contact with a fitting which is within a distance of the leak where the vibration sound of the leak leaving the pipe can be heard. Listening sticks come in two basic formats: mechanical and electronic.

Mechanical listening sticks where probably one of the first types of leak pinpointing equipment to be manufactured and used widely. The very first ones were made out of solid wooden bars (dense wood) and can still be found in use. Later manufacturers started making stainless steel rods with diaphragms housed in sounding cavities to amplify the noise.

As operators demanded better performance, manufacturers started to improve technology by amplifying signals and providing filters to allow the operator to try to filter out ambient noise, which is one of the main problems when using this type of acoustic equipment. Ambient noise, which can interfere with a listening stick survey, could be from

- Traffic
- Demand
- Air traffic
- Gas and steam pipes

The use of any of the above listening sticks is simple. The operator identifies the area to be surveyed and then proceeds to make contact with pipe fittings, hydrants, and service connections until he finds a suspected leak sound. At this point it is a simple matter of identifying where the loudest sound can be heard.

In urban areas, listening sticks are often deployed at night, when traffic noise is lower. This allows the operator more freedom to identify leak sounds without interference from other sources. Prior to a listening stick survey the operator should identify the pipe route and material to best estimate the required distance between contact points. This is necessary as leak sound travels different distances in different pipe materials because of varying pipe attenuation. More details on leak survey can be found in Chap. 16.

Checklist

- Identify pipe route before survey.
- Identify pipe material and test distances before survey.
- If traffic is going to be a problem schedule the survey at night.
- Prepare documentation to show to residents identifying the operator and the reason that he is listening on their services.
- If using mechanical diaphragm type units check to ensure that the diaphragm is in good condition as listening sticks often get used as boring bars. This destroys the contact between the rod and the diaphragm.

Mechanical and Electronic Geophones

Geophones are used to listen for leaks from the surface and come in two basic forms, mechanical and electronic.

> **If Using Electronic Units Always Check the Batteries before Starting a Survey.**

Mechanical geophones have been in use for many years successfully and are still a very valid and robust equipment use by operators all over the world. The mechanical variety usually consists of brass disks with vibrating diaphragms inside. The two disks are connected by a hollow tube to a headset arrangement and are not unlike a doctor's stethoscope in appearance.

Electronic geophones were developed as the need for better performance pushed manufacturers of geophones into amplification and filtering techniques. More on geophones can be found in Chap. 16.

Geophones are used by listening above the pipe route for the sound of water escaping from the pipe. They are usually used in conjunction with the listening stick in a sonic survey. Care must be taken to listen directly above the pipe otherwise failure may result. (See Fig. B.36). There is a distinct difference in leak sound when listening on hard or soft surfaces. The operator must learn to identify the difference and use a sounding bar where ground surfaces do not transmit sound.

Checklist

- Identify pipe route prior to survey.
- Identify ground surfaces prior to survey.

FIGURE B.36   Listening for leaks using an electronic geophone. (*Source:* Health Consultants, Inc.)

- Be equipped with sounding bars where surfaces may be soft.
- For mechanical equipment check diaphragms and tubes frequently to ensure they are not damaged.
- For electronic equipment check batteries.

Tracer Gas Equipment

Tracer gas equipment consists of an injection manifold which connects the gas tank to the water main to be tested, usually with gauges to test bottle pressure and main pressure and a receiver unit which takes samples and compares them to air density.

Tracer gas survey is a very effective way of locating either very small leaks such as those found during hydrostatic testing or hard to find leaks, often on PVC or large-diameter pipes where the leak sound is not transmitted far, making the use of traditional acoustic equipment ineffective.

An easy way of testing this equipment to make sure it is functioning is to hold an unlit lighter (with gas escaping) in front of the sensor and ensure that there is a noticeable deflection on the needle. If there is no deflection, it may be that the pump or filter system is either clogged with dirt or has water inside.

B 6.6   New Technology Leak-Detection Equipment

Leak Noise Correlators

Leak noise correlators were first introduced commercially into the marketplace in the late 1970s, however technology has advanced in leaps and bounds over the last few years putting this type of equipment in the realm of new technology with changes every year. In essence a leak noise correlator consists of a receiver unit, two radio transmitters with sensors and/or hydrophones. The sensors pick up the leak sound from the water main or pipe being tested at two points, the idea being that the leak is bracketed between the two. Using the calculation $D = 2l + Td \times V$ (as shown in more detail in Chap. 16) the correlator identifies like signals, measures the time delay between one signal and the other and using a known or calculated velocity along with a measured distance between sensors, calculates the position of the leak.

Most modern correlators have a number of functions built-in to assist the operator in making a good and accurate leak location. Some of the functions available are

- Automatic filter selection
- Distance measurement
- Velocity calculation
- Multiple-pipe features
- Auto correlation for single sensor use at pipe ends
- Linear regression
- Leak location memory
- Printer
- Sensor assortment for different field situations

Leak noise correlators can be used over quite long lengths of pipe depending on the material and diameter of the pipe and the lack of ambient noise, which could interfere with leak sounds. Most correlators can manage in excess of 500 m and some of the newer digital versions up to 3000 m in ideal conditions.

Care must be taken when using correlators properly

- Measure the exact distance between sensors
- Properly identify the pipe material
- Properly identify the pipe diameter
- Keep the leak close to the center of the two sensors
- Measure the velocity in the pipe section (s)
- Identify peaks which are not leakage and eliminate

More detail on leak noise correlation can be found in Chap. 16.

Many operators record leak sounds on tape or digitally on computer. It is a good idea to have a few reference leak sounds recorded which could be used for testing the equipment on a periodic basis. Most faults, which occur with the equipment, are through misuse of sensor, which is quite sensitive to being dropped. Also cable connections and battery life need to be constantly checked and kept in good order.

Leak Noise Loggers

Leak noise loggers have been on the market since the 1980s in various formats but have recently taken a major leap in technology as they can now not only identify areas with potential leak sound by analyzing and recording noise at night when ambient noise is at a minimum, but also now correlate between the sensors.

Comparative Pairs of Meters

A common method of identifying leakage, in particular on transmission mains is by the use of volumetric comparison between a pair of meters, one fitted downstream of the other (see Fig. B.37).

If the meters in question are high-resolution permanent meters then it is often possible to undertake a volumetric balance. However, if the meters used for the test are temporary meters, then care should be taken to identify the potential error on each installation and identify a reasonable resolution for identification. If the latter is the case it may be preferable to undertake a night flow analysis comparison as opposed to a volumetric balance. This would depend on the nature of the piping and hydraulics of the system.

Obviously this method can only be used for identifying reasonably large leaks when used on large-diameter pipelines, but does in many cases offer an accountable solution to testing long lengths of transmission mains.

B.6.7   Meter Testing Equipment

Meter testing equipment comes in various forms, including permanent bench testing equipment and portable field testing equipment. As discussed earlier, meter testing can be undertaken in a number of ways, either comparatively or by means of a calibrated volume, the latter being preferable. Where possible, meters should be tested in the field to enable a complete test of the meter installation as well as unit accuracy.

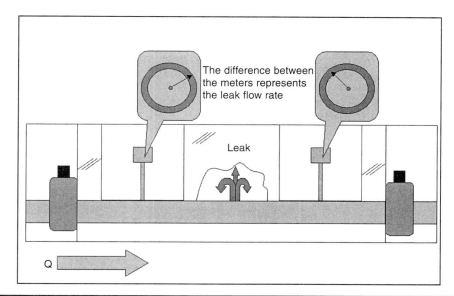

FIGURE B.37 *Volumetric comparison between a pair of meters can be used to find leakage.* (*Source:* Julian Thornton.)

Residential Meters

Residential meters are usually field-tested with a calibrated tank, which can be filled from an outside tap (while no other uses are occurring). Alternatively, a small portable tester can be fitted in line as per the kit shown in Fig. B.38.

ICI Meters

Industrial, commercial, and institutional (ICI) accounts usually utilize large-diameter meters. If the meter is to be removed for testing, it should be subjected to an approved volumetric test. If not, it may be tested using either a portable temporary meter such as an ultrasonic or insertion meter (watch for errors in the test meter) or by comparing volumes against a set of previously calibrated meters either on a large meter tester as shown in Fig. B.39 or on a test trailer.

Bulk Supply, Source, and Master Meters

Bulk supply, source, and master meters are also tested by one of the two methods discussed above for ICI accounts. Alternatively, insertion or ultrasonic meters can be used if care is taken with inaccuracy in the test equipment due to potentially incorrect location. With proper care, the latter method is valid.

When testing these meters it is important to identify not only the accuracy of the measuring unit itself but also any telemetry equipment which may be transmitting signals back to a base unit. More information on this topic can be found in Chap. 6.

B.7   Pressure Control Equipment

Pressure control equipment comes in various shapes, forms, and sizes, depending on the nature of the control required. In this section we will touch on some typical applications related to water loss control. Obviously, there are many other types of valves and

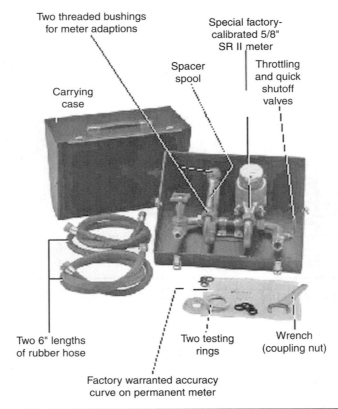

Two threaded bushings
for meter adaptions

Special factory-
calibrated 5/8"
SR II meter

Spacer
spool

Carrying
case

Throttling
and quick
shutoff
valves

Two 6" lengths
of rubber hose

Two testing
rings

Wrench
(coupling nut)

Factory warranted accuracy
curve on permanent meter

FIGURE B.38 Small portable meter tester. [*Source:* Invensys Metering Systems (formerly Sensus Technologies).]

FIGURE B.39 Large meter tester. [*Source:* Invensys Metering Systems (formerly Sensus Technologies).]

controllers, with other functions. More information on pressure management can be found in Chap. 18.

B.7.1   Types of Valves

The most common applications in pressure management schemes where water loss is the key factor are

- Pressure-reducing valves (see Fig. B.40)
- Pressure-sustaining valves (see Fig. B.41)
- Altitude valves (see Fig. B.42)
- Float control valves (see Fig. B.43)
- Flow control valves (see Fig. B.44)

More information on the use of these valves can be found in Chap. 18. Pressure valves are usually one of three types although others do exist:

- Diaphragm valves (see Fig. B.45)
- Piston valves (see Fig. B.46)
- Sleeve valves

Diaphragm valves usually come in two formats

1. Globe style
   - Straight-through
   - Angle
2. Y pattern

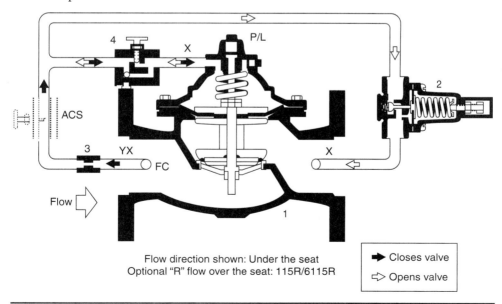

Flow direction shown: Under the seat
Optional "R" flow over the seat: 115R/6115R

➡ Closes valve
⇨ Opens valve

FIGURE B.40   Pressure-reducing valve. (*Source:* Watts ACV, Houston,Texas.)

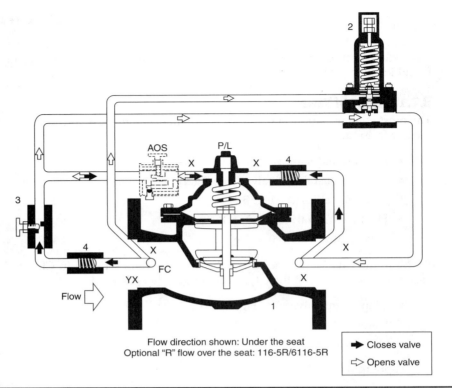

Flow direction shown: Under the seat
Optional "R" flow over the seat: 116-5R/6116-5R

➡ Closes valve
⇨ Opens valve

FIGURE B.41   Pressure-sustaining valve. (*Source:* Watts ACV, Houston, Texas.)

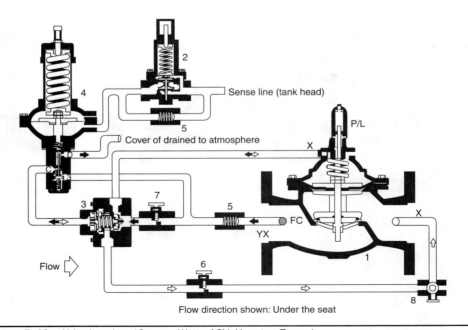

Flow direction shown: Under the seat

FIGURE B.42   Altitude valve. (*Source:* Watts ACV, Houston, Texas.)

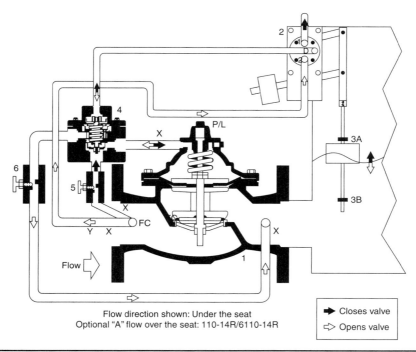

Flow direction shown: Under the seat
Optional "A" flow over the seat: 110-14R/6110-14R

➡ Closes valve
⇨ Opens valve

FIGURE B.43   Float control valve. (*Source:* Watts ACV, Houston, Texas.)

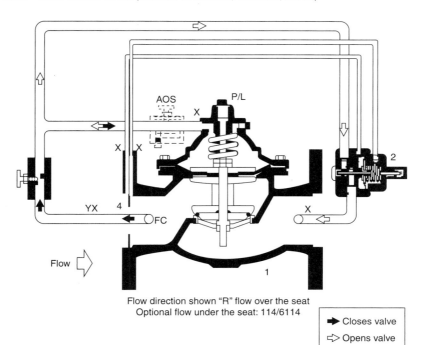

Flow direction shown "R" flow over the seat
Optional flow under the seat: 114/6114

➡ Closes valve
⇨ Opens valve

FIGURE B.44   Flow control valve. (*Source:* Watts ACV, Houston, Texas.)

– Hydraulically operated. diaphragm-actuated, automatic control valve
– Stem assembly is top and bottom guided
– QUAD ring seal/non-edged seat
– Can be serviced without removal from line

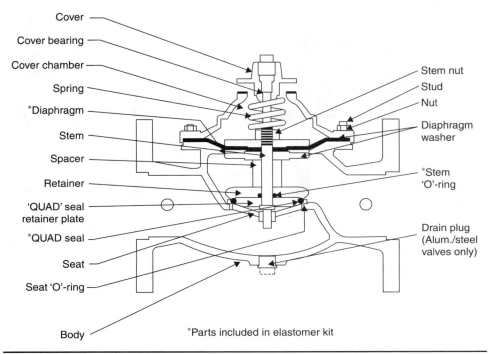

Labels (left)	Labels (right)
Cover	Stem nut
Cover bearing	Stud
Cover chamber	Nut
Spring	Diaphragm washer
*Diaphragm	*Stem 'O'-ring
Stem	Drain plug (Alum./steel valves only)
Spacer	
Retainer	
'QUAD' seal retainer plate	
*QUAD seal	
Seat	
Seat 'O'-ring	
Body	

*Parts included in elastomer kit

FIGURE B.45 Diaphragm type valve. (*Source*: Watts ACV, Houston,Texas.)

The make of valve or configuration of the valve assembly chosen for the job will depend on the nature of the installation and the availability of local support. Most manufacturers have excellent installation information, and most will provide start-up and operational support at very little extra cost.

All valves need regular maintenance to function properly over long periods of time. Most utilities like to settle on a particular make of valve and stock parts for those valves. This cuts down on having to stock identically sized parts for valves of various manufacturers if maintenance is to be done in-house.

B.7.2 Types of Controllers

Three basic types of controllers are available for specialized pressure leakage management. Other generic controllers may be adapted for use in this field. The three types in common uses are

- Time-based controllers
- Demand-based controllers
- Remote-node controllers

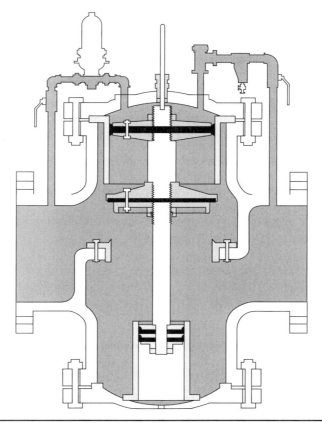

Figure B.46   Piston-type valve. (Source: Ross Valve Mfg. Co., Inc., Troy, N.Y.)

Time-Based Controllers

Time-based controllers work on an internal timer. The timer is set either to manipulate outlet pressure to various levels using an interface with a predetermined profile and pilot adaptor or to switch from one preset pilot to another by means of solenoid valves. Either of these scenarios works well but should be used in areas with fairly constant demand patterns, little seasonal and or weekend variation.

When using time-based controllers, care should be taken that the lowest pressures set can still meet emergency fire-fighting requirements.

Demand-Based Controllers

Demand-based controllers work by setting an outlet pressure by means of a pilot adaptor to a preset relationship between flow and pressure. Demand-based controllers combat head loss in water systems, ensuring that when demands are low pressure is at a minimum to reduce the effect of pressure on the leakage which is running in the system.

Demand-based controllers have the added benefit that they can be used to control pressure down below minimum fire-fighting requirements, as the controller will automatically adjust itself back to the required pressure when the hydrant is operated and the flow demand goes up. If a demand-based controller breaks at the lower pressure position, the controller is programmed to default the valve back to the higher set point, ensuring that water is available for emergency or peak demands.

Remote-Node Controllers

Remote-node controllers work by relaying a signal back to the valve and controller assembly from a remote node. The remote node selected is usually a critical node. The critical node may be a node which is at the highest elevation and therefore has the least pressure. Alternatively, it may be selected as an area with a special consumer or large consumer or an area with particularly high localized head losses.

The remote node is fitted with a pressure logger and the logger is programmed to communicate often by way of low-power radio or cell phone with the controller on a predetermined basis. The remote logger orders the controller to allow more or less pressure into the system by opening or closing the valve, in order to maintain a stable target pressure at the remote point.

This type of control, like demand-based control, is suitable for areas with changing profiles and a need for emergency response.

Checklist

- Make sure the valves to be used can be maintained and supported locally.
- Determine what type of area is to be controlled.
- Undertake a detailed demand analysis before installation.
- Select the right type of controller to meet the requirements of the area.
- If using radio or cell phone communications, be sure the equipment is operating on an authorized wavelength.
- If using demand-based control, be sure that the meter which is fitted to provide the pulse is suitably sized and the pulse generator is suitable for the controller.

B.6.8   Maintaining Equipment

In the last few sections we have talked in detail about the use of permanent and portable equipment for field measurements and testing. Without data from this type of equipment it is extremely hard to assess water system condition and improve water loss figures. However, this equipment needs to be properly maintained to ensure that it gives repeatable accurate results. Bad data in produces bad data out!

> **W**hen considering a budget for purchasing equipment, also consider the maintenance costs as without proper maintenance the equipment will become worthless.

Good Use and Practice

When installing permanent equipment or purchasing portable equipment for the first time, it is a good idea to identify a good practice list which should be adhered to by all operators using the equipment. Some things which might appear on the list are

- Regular maintenance
- Regular third-party testing

- Maintenance of the housing environment in the case of permanent equipment
- User log

Cables and Fittings

A weak point in most equipment is the cables and fittings. People misuse cables, using them to carry equipment and to lower equipment into holes. This puts unnecessary strain on the connections and can sometimes cause irreparable damage.

Fittings should be greased and cleaned regularly with a light suitable oil to ensure that they do not get dirt into them. Waterproof seals should be checked regularly and changed when in doubt. Water ingress into fittings is one of the most common reasons for failure.

Storage and Carriage

One of the areas of most likely damage to portable equipment is during shipping and transportation of the equipment, particularly if it is to be sent via a third party over long distances.

Equipment should be purchased from the manufacturer together with a suitable durable hard carry case. If the manufacturer cannot supply this type of case, then one should be procured from a third party. Cases can be expensive, but they will always pay off in the long run.

B.8   Summary

In this appendix we have discussed various types of portable and permanent field equipment, which are often required or encountered in a water loss control program. While it is impossible to show all of the equipment types and configurations in this book, operators are urged to familiarize themselves with their particular equipment prior to starting fieldwork. Manufacturers will be more than willing in most cases to supply the necessary engineering manuals and often on-site support to familiarize operators with their equipment.

> **M**ost equipment manufacturers will provide on-site training.

More details on the use of some of the equipment covered in this Appendix can be found in Chaps. 16 to 19.

APPENDIX C

Demand Profiling for Optimal Meter Sizing[*]

C.1 Introduction

Precise, customer-specific demand profiles are used to generate valuable types of water use data. A demand profile consists of rate-of-flow data describing water use versus time. Such data are typically gathered directly from a utility customer's existing meter installation using specialized flow recorders that attach to meters and log water usage per unit of time. See Figs. C.1 and C.2.

Demand profiles generated from existing meters provide data essential for making a variety of critical decisions. Data logged from water meters is more accurate than other measures because a water meter represents the most precise way to measure actual water use. Flow recorders accomplish their mission without interrupting the accurate registration of the water meter and, typically, without altering the existing meter configuration. In a small number of cases, adapters are required but are easily installed.

Applications for customer demand profiles may be grouped into three general categories: (1) meter sizing and maintenance, (2) water use audits, and (3) cost of service studies. While only the first application is discussed in detail here, it is worth remembering that the same data gathered for meter sizing purposes has other important applications and can benefit a variety of utility divisions, including distribution, metering, conservation, customer service, engineering, and finance. In the case of water use audits, demand profiles assist with conservation programs, leak detection, customer service, and hydraulic modeling. In the case of cost-of-service studies, demand profiles are used to obtain data regarding the variability of use by residential, commercial, industrial, and wholesale customer class groups. Because the same data can be used in support of all these applications, it is important when collecting the data to consider all of the potential applications for which the data may be of value presently or in the future. For example, if a cost-of-service study or hydraulic model requires only hourly demand data, you may still choose to store the data in 10 to 60-s increments so that the same data can be used for meter sizing and maintenance programs.

In general, a demand profile should accurately provide peak flow data and the percentage and volume of water used in critical flow ranges. Critical flow ranges include,

---

[*] This appendix was provided by permission of Brad Brainard of F. S. Brainard and Company and will eventually form part of the new AWWA M22.

589

FIGURE C.1   Data being captured from an existing meter. (*Source:* F. S. Brainard and Company.)

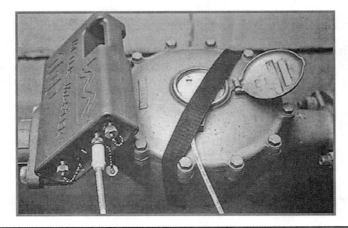

FIGURE C.2   Data being captured from an existing meter. (*Source:* F. S. Brainard and Company.)

as a minimum, flow below the specified accuracy range of a meter, flow at the cross-over range in a compound meter setting, and high flow. The objective is to size the meter properly for maximum accountability and revenue recovery without adversely effecting pressure levels or fire flow requirements. It is also important to consider meter maintenance costs. It may be that a 6-in turbine meter could better serve a customer with constant flows of 600 gpm than a 4-in turbine meter because, while both would accurately measure the flows, the 6-in turbine would experience less degradation from wear and tear. The most obvious direct benefit of proper meter sizing is the accurate measurement of water use; the more closely a meter is matched to a customer's usage pattern, the more water will be accounted for and billed. What is often not quite so obvious is the potential size of revenue gains associated with proper meter sizing.

Tim Edgar, in *The Large Water Meter Handbook,* illustrates this potential revenue gain by the case of a 100-unit apartment building with a 4-in turbine meter. The actual monthly consumption was 500,000 gal, but much of that volume was at low flow rates. Because the turbine meter was not accurate at flow rates less than 12 gpm, 15 percent of the volume went unrecorded and unbilled in both water and, as is very often the case, sewer charges. The result was a revenue loss of $1700.00 per year (at $3/1000 gal for combined water and sewer). As Edgar points out, if a utility has 100 such incorrectly sized meters, those 100 meters would cost a utility over $1 million in lost revenue over 6 years.[1]

As an example, the Boston Water and Sewer Commission began a downsizing program in 1990. John Sullivan, Boston's Director of Engineering, reported in presentations to the American Water Works Association that, between August 1990 and April 1992, the city had accounted for an additional 113,784 ft^3 of water per day (0.8 mcd). With just the meters downsized in the first year of the program, Boston anticipated the total increase in revenue over 5 years from combined water and sewer billings to be $6.8 million (1991 dollars). These savings would only be realized in systems with many oversized turbine meters.

While the most direct benefit of proper meter sizing is increased revenue and accountability, meters offer a distribution system much more valuable than just revenue enhancement. Any decision made by a utility related to water usage can only be as good as the consumption data collected from meters. In general, demand profiles provide valuable data to improve distribution system design, performance, and management. In addition to finding ways to increase accounted-for water levels and revenue, demand profiles help to identify service size requirements, clarify meter maintenance requirements, define water use characteristics for conservation programs, enhance customer satisfaction and awareness, improve hydraulic models, and establish equitable and justifiable rate structures. Additionally, with increased water scarcity and cost, conservation has become an important industry issue. For many utilities, conservation has become the most cost-effective means to improve water resource availability. All of these distribution system design, performance, and management objectives are dependent on the capability of a system's meters to account for usage as accurately as possible, which can only occur as a consequence of sizing meters properly for each and every application.

C.2   Recorder Design

C.2.1   Theory of Operation

Demand profiles are generated with electronic flow recorders. The portable flow recorders discussed here are also referred to as demand profilers, demand recorders, and data loggers. The devices pick up data from either the meter's internal drive magnets or the meter's pointer movement and store the data for later downloading into a desktop or handheld computer for analysis. These recorders can be moved from one meter site to the next with minimum effort and operate with standard meters, thereby eliminating the need for special registers. Typically, the magnetic or optical sensor is either strapped to the outside of a meter using Velcro or heavy-duty

Figure C.3   Sensor attached to meter with Velcro strap. (*Source:* F. S. Brainard and Company.)

tape or is integral to an adapter located between the meter body and the existing register. See Fig. C.3.

Because of potential adverse operating conditions (meter pits, temperature extremes, rough handling, public access), recorders should be submersible, durable, and securable. In order to provide extended data storage capability in remote locations, recorders should also offer substantial battery life. This section describes current technology for demand profiling. As new technologies evolve in this field, they should be evaluated in order to promote this area of knowledge and capability.

C.2.2   Recording Methods

Flow recorders using magnetic pickups sense the magnetic field generated by the magnetic coupling of a water meter's internal drive magnets and convert the magnetic flux change into a digital pulse that is logged into memory and later downloaded into a PC for analysis. Optical pickup devices sense the meter pointer passing beneath the sensor and also store the signal as digital pulses to be later downloaded. Each pulse is associated with a known volume of water. The principal advantage of a magnetic pickup is the higher resolution of data made possible by the rotation speed of a meter's magnets. In almost all cases, the drive magnets inside a meter rotate much faster than the sweep hand (pointer) on the register's dial face. In small meters, the number of magnet rotations per unit of time can be as high as approximately 30 per second at 20 gpm. At this rate, the magnets are rotating 900 times as fast as the sweep hand. In the case of turbine meters, the rotation speed of the magnets can vary greatly, from approximately 800 times the speed of the sweep hand to the same speed as the sweep hand. Available adapters can substantially increase the resolution of the data on many of the slower-magnet-speed meters by isolating an additional magnet with a higher rotation speed. Optical and mechanical adapters are available to enable compatibility with the older gear-driven meters, which preceded magnetic-drive meters.

C.2.3   Installing Magnetic Sensors

Because most meters have the magnetic coupling directly under the register, it is typically easy to pick up a reliable signal by placing the sensor on the side of the register.

Figure C.4 Sensor picks up pulses from register magnets. (*Source:* F. S. Brainard and Company.)

Almost without exception, the magnetic coupling is directly under the register in the case of all 2-in and smaller positive displacement and multijet meters. If the magnetic coupling is not directly under the register, it is typically in the center of the turbine rotor in the middle of the flow. In this case, the magnetic sensor must be placed on the side of the meter body in order to be as close to the drive magnets as possible. As discussed above, adapters are required for some meters, such as gear-driven meters. See Fig. C.4.

If the magnetic coupling is under the register but the register has shielding on the sides, the sensor may have to be located directly on top of the register in order to circumvent the shield. Because the recorder's magnetic sensor is essentially picking up the electromagnetic noise generated by a water meter, the sensor can be susceptible to picking up noise generated by other sources of electromagnetic noise such as motors, generators, and alarm systems. The recorder's sensing circuitry should be designed to consistently pick up the magnetic signal generated by a water meter's drive magnets, while minimizing the potential for picking up electromagnetic noise from other sources.

C.2.4 The Recorder's Data Storage Capacity

It is essential that a recorder have adequate data storage capacity in order to enable the recorder to store a substantial amount of data. As discussed in greater detail in Sec. C.3.3, flow data must be logged into memory in small time increments if accurate maximum and minimum flow rate data is to be ensured. The potential factor of difference in the observed maximum flow rate between a 10-second and a 60-second data storage interval monitoring the exact same flow is 6:1. The potential factor of difference in the observed maximum flow rate between a 10-second and a 300-second (5-min) data storage interval monitoring the exact same flow is 30:1. In other words, if a solitary flow usage of 200 gal occurred for just 10 second at a rate of 1200 gpm, whereas the 10-second data storage interval could detect this high flow rate of 1200 gpm, the 300-second data storage interval would observe a maximum flow rate of just 40 gpm because the 200 gal would be averaged over 5 minute rather than averaged over 10 second.

Obviously, this difference could have serious ramifications for a meter size selection. Frequently, users choose to store data for 1 week when assessing the size of a commercial/industrial user's meter, in order to ensure that a representative sample of flow data is gathered. If a user is to store 10-second data for 1 week, the recorder must be able to continuously store a minimum of 60,480 intervals of data. For other applications, such as cost-of-service studies and hydraulic modeling, a smaller data storage capacity is required than for meter sizing; however, if the data are to be used most efficiently, the storage capacity should provide for high-resolution data so that the data may be used effectively for the various applications.

C.3   Recording Data

C.3.1   Length of Record

As discussed above, many recorder users choose to store data from commercial/industrial sites for 1 week because certain high-rate water uses (e.g., a cleaning operation at a factory) may occur on only a particular day each week. It is important to discuss water usage with a customer prior to storing data, if possible, to ensure that the duration of the recording period is sufficient to get a representative sample of flow data. In the case of multitenant residential or hotels/motels, 24 hours of data may be sufficient as long as the data are collected during hot weather in the case of residential and high occupancy in the case of hotels/motels. Essentially, it is best to make some effort to understand a user's water use characteristics in order to select the optimum length of the data storage period. Experience with different types of users over time will also provide an indication of the optimum record length for different classes of users. The record length is critical and should be determined on a case-by-case basis.

C.3.2   Customer's Water Use Habits

Data should be recorded during a period in which the user experiences typical peak, average, and minimum flow rates and for duration sufficient to capture those rates. For example, it would not be appropriate to record data at a school or factory during a vacation period. Similarly, as mentioned above, you would want to record data for at least a week at an industrial site if there were evidence that the customer performed different operations on different days of the week. Seasonal cycles are as important to consider as weekly ones. Weather at different times of the year may substantially alter demand patterns. If a user uses a lot more water on a hot summer day, it is important to record data on such a day in order to capture peak flow data.

The personnel performing an analysis should anticipate potential changes in demand patterns. At a residential development, it would be important to consider the number of additional units currently under construction. It is also important to resurvey a user if the type of use changes. Commercial lease space can have a high rate of turnover. A warehousing or distribution company with substantially lower water usage could replace a bottling company. If the meter is not resized, the new user will be the beneficiary of a lot of free water.

C.3.3   The Recorder's Data Storage Interval

The data storage interval is the period of time over which a flow recorder counts pulses before that interval's pulse count is logged into memory. The interval determines the resolution of the raw data file from which all subsequent graphs and reports are generated: the shorter the interval, the greater the detail possible in subsequent graphs and reports. For example, a data storage interval of 10 second allows accurate data analysis for periods of 10 second or longer. The user selects the data storage interval before the recorder goes into the field. As long as the graph/report generating software allows for adjustment of the time interval over which maximum and minimum flow rates are calculated (see Sec. C.4.2), the data storage interval should be kept short, for example 10 second.

Keeping the data storage interval short is particularly important in order to provide sufficient data resolution to determine maximum flow rates accurately. In order to ensure the accurate identification of a maximum flow rate, the data storage interval cannot exceed 50% of the duration of a maximum flow event. For example, if an industrial customer has a particular operation which occurs just once each 30 minute, lasts 30 second, and uses 500 gal of water (i.e., a demand of 1000 gpm), identification of the 1000-gpm flow rate can only be assured if data are logged into memory at least once each 15 second. If the data storage interval is between 15 and 30 second, there is an increasing likelihood that the maximum flow rate will be understated due to the possibility that no data storage interval begins and ends within the 30-second event. If the data storage interval is more than 30 second, the likelihood becomes a certainty. In this particular example, a data storage interval of 15 second or less would show the 1000-gpm flow rate. On the other hand, if the data storage interval is 15 minute (900 second), the maximum flow rate would appear as only 33 gpm, because all that is known is that a total of 500 gal was used during a 15-minute period, and 500 gal divided by 15 minute is 33 gpm. If the data storage interval were 5 minute, a maximum flow rate of 100 gpm would be indicated. A lower maximum flow rate would be indicated if the 500-gal usage was divided between two 5-min data storage intervals. As can be seen, a serious meter sizing error can easily be made if the recorded data are not stored at a level of resolution sufficient to capture the actual maximum flow rate.

As another example, let's say that a small manufacturing company has an operation which periodically uses 250 gal of water for 10 second (which equates to a rate of 1500 gpm) in addition to its other uses. This scenario is simulated graphically in Figs. C.5 to C.7. In each case, the same data was used to create each graph; the only difference is the data storage interval, which, in this case, is also the interval used for maximum and minimum flow rate calculations. In the case shown in Fig. C.5, a data storage interval of 10 second is used and a true maximum flow rate of 1520 gpm is identified. In the case shown in Fig. C.6, the data

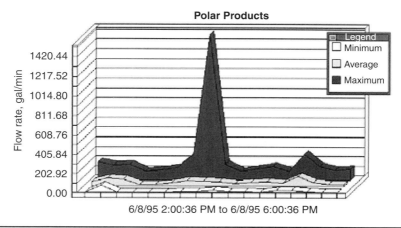

F‌IGURE C.5   Data storage interval of 10 second (Meter-Master Model 100 program). (*Source:* F. S. Brainard and Company.)

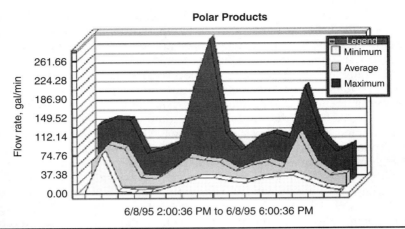

FIGURE C.6 Data storage interval of 60 second (Meter-Master Model 100 program). (*Source*: F. S. Brainard and Company.)

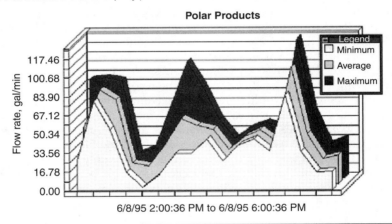

FIGURE C.7 Data storage interval of 300 second (5 minute) (Meter-Master Model 100 program). With this interval, the true maximum flow rate disappears into the rest of the data. (*Source:* F. S. Brainard and Company.)

storage interval is 60 second and the calculated maximum flow rate is reduced to 280 gpm. In the case shown in Fig. C.7, the data storage interval is 300 second (5 minute) and the true maximum flow rate disappears into the rest of the data.

Although the above examples exaggerate normal circumstances, they are intended to illustrate the potential for meter sizing errors if one ignores the importance of data resolution.

It should be noted that there are disadvantages to making the data storage interval too small. This interval defines the size of the downloaded data file and the length of time you can record before running out of memory. The same test recorded with a 5-second interval will take up to six times more memory than one stored with a 30-second interval. Furthermore, larger files take longer to download and to generate graphs and reports. Generally, a 10-second interval provides adequate detail and recording time for most applications. If you are making a long recording and a 10-second interval would use up all of the logger's memory before the recording is completed, lengthen the data storage interval. Another problem with too short an interval is discussed in Sec. C.3.4 and in Sec. C.4.2. Briefly, if too short an interval is used on a meter with slow-moving drive magnets (or sweep hand, in the case of optical sensors), skewing (exaggeration) of maximum and minimum flow rates can occur because there is too little data for accurate calculations. A recorder's operating instructions should identify such meters so that care is taken when selecting intervals for data presentation. Software design can improve the integrity of downloaded data by intelligently interpreting pulse data in order to minimize the potential for exaggerated maximum and minimum flow rates.

C.3.4   The Meter's Pulse Resolution

The meter's pulse resolution is defined as the number of pulses generated that equate to a unit of liquid measure. For magnetic pickups, the resolution is the number of meter magnet poles (as the magnets rotate) which equate to a unit of liquid measure. It is desirable that the internal magnets revolve as fast as possible without degrading the reliability of the meter; accordingly, the higher the number of magnet poles per unit of measure, the better. Faster magnets generate more pulses, which translates into greater data accuracy. For optical pickups, the same considerations apply to the speed of sweep-hand rotation. Therefore, it is important to have some knowledge concerning the speed at which a meter generates pulses. A flow recorder's operating instructions should provide guidance in this area.

The pulse resolution (or factor) is especially important when determining maximum and minimum flow rates. The issues are very similar to those discussed in Sec. C.3.3. Concerning maximum flow rates, if a magnet (or sweep hand) is rotating slowly, it is possible that a large, short-term usage could take place without any evidence of its occurrence. For example, if a 6-in turbine meter (meter "a") generates just one magnetic pulse for each 500 gal while another 6-in turbine (meter "b") generates one pulse for each 2 gal, the 250-gal usage at 1500 gpm described in the preceding section might not even be identified at all by a recorder attached to meter "a," while meter "b," with fast-moving magnets, would have provided 125 pulses to the recorder. Furthermore, if the recorder attached to the meter with the slow-moving magnets did detect one pulse within a 10-second interval, it might be erroneously assumed that 500 gal were used during that 10-second interval, which would equate to a flow rate of 3000 gpm because, if one pulse is logged in 10 second, this is the equivalent of 6 pulses per minute, and 6 pulses per minute multiplied by 500 gal per pulse equals 3000 gpm. Accordingly, a meter with fast-moving magnets can provide continuously accurate data throughout the flow ranges, whereas a meter with slow-moving magnets cannot. Likewise, using optical sensors, the faster the rotation of the sweep hand, the more accurate the resultant data will be. However, an optical sensor would have to detect numerous pulses per revolution of a sweep hand in order to approach the substantial level of accuracy achievable by a magnetic sensor.

Minimum flow rates identify leakage rates and affect the selection of turbine versus compound meters in larger applications. In order to ensure the accurate identification of minimum flow rates, as with maximum flow rates, a user must know which meters have slow-moving drive magnets. For example, if a meter's magnets are providing just one pulse for each 20 gal, and the current flow rate is a steady rate of just 5 gpm, only 1 pulse will be generated each 4 minute. If one observes the data in time increments smaller than once each 4 minute, the flow rate will appear to vary between zero and some amount greater than the actual flow rate of 5 gpm. As an illustration, if a 1-minute time interval is used for observing

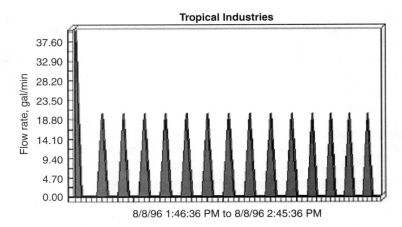

FIGURE C.8   Pulse resolution is extremely important (Meter-Master Model 100 program). (*Source:* F. S. Brainard and Company.)

the data, the flow rate will appear to equal zero for 3 of each 4 minute and 20 gpm for 1 of each 4 minute because each pulse, equaling 20 gal, will appear just once each 4 minute when a steady flow rate of 5 gpm is occurring. If a 4-minute time interval is used to observe the data, it will appear as if a steady flow rate of 5 gpm is occurring.

Figures C.8 and C.9 represent the scenario just described. Both graphs were generated from the exact same data, but the time increments used to view the data are 1 and 4 minute, respectively. Each pulse from the meter equals 20 gal, and they were spaced 4 minute apart (except during the initial interval shown). Software design can help by evaluating the data to determine the likelihood that raw pulse data should be averaged over longer periods of time because the pulse distribution indicates the presence of a constant flow rate.

Unless you are actually at the meter site watching the meter at the time of the event, it is not possible to know with certainty whether a periodic use of 20 gal is occurring or

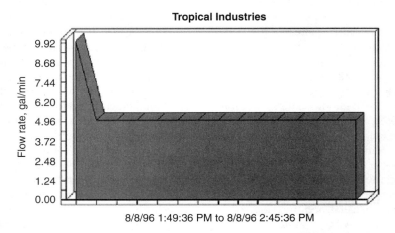

FIGURE C.9   Pulse resolution is extremely important (Meter-Master Model 100 program). (*Source:* F. S. Brainard and Company.)

a steady flow rate of 5 gpm is occurring. If each pulse from the meter equals a smaller amount of water, such as 1 gal, the true picture will be much clearer.

The key to getting accurate flow data is generating a sufficient number of pulses per time interval. In the case of magnetic pulses, all 2-in and smaller positive-displacement and multijet meters provide good pulse resolution, such that the data can reasonably be observed in time increments as small as 10 second. Because some turbine meters have magnets that rotate relatively slowly, the minimum time increment necessary for observing minimum (low) flow rate data, such as leakage rates, may be as long as 5, 10, or more minute, unless the software can interpret the data intelligently. Adapters which increase the magnetic pulse resolution are useful in determining accurate flow rate data because the flow data may be accurately viewed in smaller time increments, which minimizes the need to interpret the data using potentially inaccurate assumptions. The same considerations apply when using optical sensors. Meters with rapidly rotating sweep hands will provide more accurate flow data than meters with slow-moving sweep hands. Slow-moving sweep hands will not allow an optical sensor to achieve the resolution needed for accurate maximum and minimum flow rate calculations, unless the optical sensor generates numerous pulses per revolution of the sweephand, for example 50.

C.3.5 Meter Accuracy

When one uses a flow recorder, it is assumed that the meter to which it is attached is accurate. A flow recorder cannot determine meter accuracy, but it can determine the accurate meter configuration for a meter site. Because a flow recorder is only as accurate as the meter to which it is attached, routine meter testing is important when using recorders to determine the appropriate meter size. Because most meter inaccuracy involves underregistration of usage, a flow record on an underregistering meter can result in selection of an undersized meter.

Ideally, a meter should be tested for accuracy, and repaired/recalibrated if testing indicates that it is not accurate, prior to recording data for meter sizing purposes. As discussed in Sec. C.5.3, a demand profile performed in conjunction with a flow test may indicate that all of the flow is occurring in an accurate range of the meter, even though the meter is not accurate throughout the flow ranges. If this is the case, the meter does not need to be repaired/recalibrated because no accountability or revenue is currently being lost.

Flow recorders should be considered a valuable companion tool as part of a meter test program. As referred to in the previous paragraph, a flow recorder can identify the percentage of flow in low, medium, and high flow ranges. With this information, testing can be focused on the ranges in which most of the usage is occurring, and unnecessary and costly repairs can sometimes be avoided. If a flow record indicates that all of the flow at an oil refinery or brewery is occurring in a high flow range, it is not relevant whether or not the meter is accurate at low and medium flows.

C.4 Creating Reports/Graphs

C.4.1 Verifying Data Accuracy

One of the principal advantages of recording flow data directly from water meters rather than using alternative technologies, such as ultrasonic devices, is that the resultant flow data is based on and may be verified against the meter's registration. Graphs and reports generated from the data may be used with confidence because the accuracy is based on the premise that a water meter is the most accurate and reliable means to measure potable water use. However, if the accuracy of the data generated with a flow recorder is not verified by comparing the total volume observed by the flow recorder to the total volume registered by the water meter itself during the data storage period, this key advantage is lost.

Verification of data accuracy is critical and is accomplished (1) by requiring the user to enter the beginning and ending meter readings when downloading data and (2) by having an accurate meter magnetic pulse factor database so that the total volume registered by the meter may be compared to the total volume registered by the flow recorder. This procedure also requires the operator to take special care, when making a record of the meter readings, that the numbers are accurate and include digits down to the decimal. In order to read a meter down to the decimal, a digit for all rotating dials and painted on "zeros" must be read.

The sample software screen shown in Fig. C.10 requires the user to compare the meter's register volume to the flow recorder's observed volume. The numbers should either be extremely close or differ by an explicable margin. In this case, the electronically recorded total of 1291.774 gal compares favorably with the water meter's registered volume of 1295 gal during the same period. The software calculates register volume by subtracting the meter's beginning register reading from the ending register reading. This example is based on a magnetic pickup; it calculates the recorder volume by multiplying the total magnetic pulse count for the entire recording period times the magnetic pulse factor for that meter in the software's database. An explicable difference between the two total volumes would include differences due to change gears used in some meters for calibration purposes. Because the change gears are used to speed up and slow down the register to

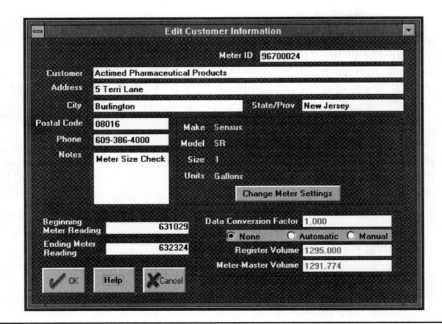

FIGURE C.10   Meter-Master Model 100 program. (Source: F. S. Brainard and Company.)

match the activity below in the meter's chamber, the recorder's volume could differ from the register's volume by as much as 15% even though both the meter and the recorder may have functioned 100% accurately. The software screen shown includes an automatic "data conversion factor" option so that the recorder's volume can automatically be calibrated to match the meter's volume 100% in such cases.

C.4.2 Data Resolution and the "Max–Min" Interval

Data resolution refers to the time intervals over which volume and maximum, average, and minimum flow rates are calculated. The sample software screen shown in Fig. C.11 displays volume, maximum, minimum, and average flow data in a grid format. In this case, the volume interval is the time interval represented by each line of data. A volume interval of 300 second will provide volume data as well as maximum, minimum, and average flow rates for each 5 minute of the survey. When creating a report or graph, the longer (larger) the volume interval, the shorter the report and the fewer the points plotted on a graph.

To compute a flow rate, the software calculates the number of pulses per unit of time selected. For example, if a 10-second Max–Min interval is selected, each volume interval is divided into 10-second increments, and the increments with the largest and the smallest pulse counts are converted to per-minute maximum and minimum flow rates, respectively. This represents one widely used method for calculating flow rate data. The considerations for selecting the Max–Min interval are similar to those related to the data storage interval; however, besides ensuring that the interval is sufficiently short for accurate flow rate calculations, one must ensure that the interval selected is not too short. In general, the maximum flow rate gets both larger and more accurate as the Max–Min interval gets smaller, until a

	Date/Time	Volume	Maximum	Minimum	Average
1	6/8/95 2:00:36 PM	153.226	22.177	0.806	10.215
2	6/8/95 2:15:36 PM	256.048	24.597	15.323	17.070
3	6/8/95 2:30:36 PM	228.629	24.194	0.806	15.242
4	6/8/95 2:45:36 PM	57.661	10.887	0.000	3.844
5	6/8/95 3:00:36 PM	50.000	12.500	0.403	3.333
6	6/8/95 3:15:36 PM	106.048	14.919	2.016	7.070
7	6/8/95 3:30:36 PM	177.419	37.903	4.032	11.828
8	6/8/95 3:45:36 PM	106.855	13.710	6.048	7.124
9	6/8/95 4:00:36 PM	147.177	18.548	6.048	9.812
10	6/8/95 4:15:36 PM	93.548	12.097	4.839	6.237
11	6/8/95 4:30:36 PM	128.226	16.935	4.032	8.548
12	6/8/95 4:45:36 PM	145.968	18.952	5.645	9.731
13	6/8/95 5:00:36 PM	116.532	16.935	6.452	7.769
14	6/8/95 5:15:36 PM	316.129	37.500	6.855	21.075
15	6/8/95 5:30:36 PM	134.677	18.145	4.435	8.978

Meter-Master Data Grid

Actimed Pharmaceutical Products — Data Storage Interval: 10 Seconds

To select a portion of data, click on row number in first column and drag cursor down or click on first row number, hold down SHIFT key, and then click on ending row number.

Edit Customer OK Help

Volume Interval (in seconds) 900
Max-Min Interval (in seconds) 60
Recalculate

Figure C.11 Meter-Master Model 100 program. (*Source:* F. S. Brainard and Company.)

point, after which the maximum flow rate exceeds reality. The point after which the maximum flow rate exceeds reality is a function of the meter's pulse resolution. The faster the magnet or sweep hand, the smaller the Max–Min interval can be without skewing the data. Similarly, the minimum flow rate gets both smaller and more accurate as the Max–Min interval gets smaller, until a point, after which the minimum flow rate becomes smaller than reality. Again, the rotation speed of the meter's magnet or sweep hand is the determining factor. It is important that a recorder's operating instructions provide guidance concerning this issue. If you familiarize yourself with those meters that generate few pulses per time interval and the volumetric equivalents of each pulse for such meters, the selection of appropriate volume and Max–Min intervals for viewing the data becomes more apparent. As mentioned previously, the software can also be designed to minimize the potential for exaggeration as the Max–Min interval is shortened.

Selection of the time intervals for viewing the data depends, in part, on the application. As discussed above it is important to consider the type of usage profile typically generated by each class of user. Usage at multifamily residential locations, for example, typically does not differ substantially in small time increments. Demand typically ramps steadily up and down in the morning and evening, which allows for longer time intervals when viewing the data. On the other hand, an industrial user may have high-volume wash cycles with short duration, requiring shorter time intervals for accurate maximum flow rate calculations.

C.4.3   Graph/Report Presentation Options

Software can present data in an endless variety of formats and styles. Generally, the software should provide options to view volume data, max/avg./min flow rate data, and rate-versus-volume data. The sample graphs shown in Figs. C.12 and C.13 display max/avg./min flow rate data and rate-versus-volume data. The max/avg./min graph is useful for identifying instantaneous maximum and minimum flow rates and the duration of events. The rate-versus-volume graph is useful for meter sizing and maintenance programs because it shows the percentage and volume of water being used in various flow ranges.

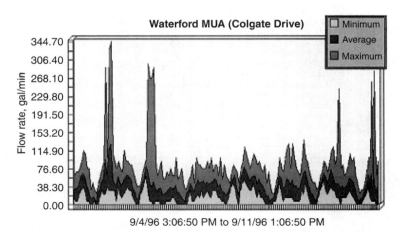

Figure C.12   Max./avg./min. graphs (Meter-Master Model 100 program). (*Source:* F. S. Brainard and Company.)

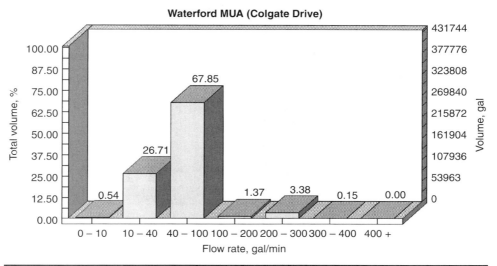

FIGURE C.13 Rate-versus-volume graph (Meter-Master Model 100 program). (*Source:* F. S. Brainard and Company.)

C.5 Using Demand Profiles to Size and Maintain Meters

C.5.1 Summary of Meter Sizing Benefits

The use of demand profiles for meter sizing applies to all users. Although relatively standard meter size and water use patterns characterize single-family residential customers, outdoor residential water use can differ substantially, requiring meter sizes larger than the norm. For users other than single-family residential, each customer generates a unique demand profile, and the meter should be sized accordingly. Although generic demand data can be developed for various customer class groups based on demographic and business type information, the cost of gathering customer-specific demand data is minimal when compared to the revenue and community relations benefits associated with maximizing meter accuracy and water use accountability.

Multifamily residential meters (e.g., apartment buildings) are the most consistently oversized meters because of both traditional fixtures count methods of sizing and the advent of efficient, low-volume fixtures. The graph displayed in Fig. C.14 is from an 8-in wholesale connection serving a small residential community. Although the specified accuracy range for an 8-in turbine meter is approximately 40 to 3500 gpm, the flow rate never exceeded 40 gpm at this site. Accordingly, the customer received a lot of free water. Replacement of the meter with one that is properly sized and configured will substantially increase both accounted-for water levels and revenue. The revenue gain is enhanced by higher sewer charges, which are typically a function of water charges. Although in some cases a smaller customer surcharge based on the meter size means less revenue to a utility in the short run, the overall water service cost to a community is reduced as a consequence of the lower capital costs associated with smaller meters.

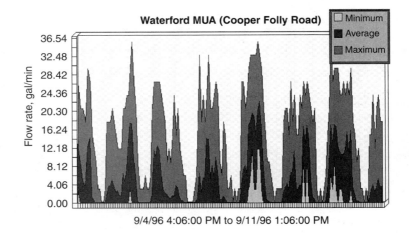

FIGURE C.14 The specified accuracy range for an 8-in turbine meter is approximately 40–3500 gpm (Meter-Master Model 100 program); the flow rate never exceeded 40 gpm at this site. (*Source*: F. S. Brainard and Company.)

Proper meter sizing has positive spillover effects with other programs. For example, a cost-of-service study in support of a rate structure design can only be fair and equitable if all of the sample sites have properly sized meters. Leak detection efforts are undermined if a meter is oversized, because low flows are needlessly undetectable and the meter's pulse resolution is less than it would be with a smaller meter. Similarly, hydraulic models, conservation efforts, and other programs all benefit from accurate use registration, which is dependent on proper meter sizing.

C.5.2 Compound versus Turbine Decisions

Many utilities experience shifting philosophies concerning the application of compound versus turbine meters. Compound meters are more expensive and have higher mainte-nance costs, but they register accurately through a broader range of flows. By compari-son, turbine meters are less expensive to purchase and maintain, but offer a smaller accuracy range. For each meter application there is an optimum solution, and a demand profile will enable you to make the correct decision in each instance. If a compound meter is installed when a turbine is more appropriate, excessive maintenance costs and prob-lems can be expected, and the utility will spend money unnecessarily. Conversely, if a turbine is installed when a compound is more appropriate, registration will be lost, and, once again, the utility will unnecessarily lose money and accountability.

A rate-versus-volume graph like that in Fig. C.13 enables a user to determine the amount of flow occurring in the crossover range of a compound meter setting that week. In this crossover range, there is a substantial drop in the level of accurate use registration by the meter setting because the turbine side of the compound setting is just starting to move, and, consequently, all flow through the turbine is below its accuracy range. If there is a meaningful amount of flow in the crossover range, an alternative compound meter size or a single meter setting should be considered.

C.5.3   Meter Maintenance Considerations

Another related use of demand profiles is meter maintenance programs, especially large meter maintenance programs. Some utilities consider demand profiles when making meter test, repair, and/or replace decisions because the demand data enables the utility to perform an accurate cost/benefit analysis of these three maintenance options on a case-by-case basis. For example, if a 10-in turbine meter tests 100% accurate in a high flow range, 90% accurate in a medium flow range, and 80% accurate in a low flow range, the conventional wisdom would average the three accuracies, which would equal 90%, and recommend repair. However, if a demand profile indicates that the flow rate never drops below 1000 gpm, the in-service meter accuracy for the subject application would equal 100% because all flow is occurring in a high flow range. With the advantage of a demand profile, costly and unnecessary service interruption and repair costs can be avoided and appropriate maintenance programs can be devised. Proper check valve operation in a compound meter setting can also be evaluated by ensuring that the turbine side does not move unless the small side exceeds a specified flow rate.

Water meters, like any piece of machinery, have optimum performance ranges, and projected test requirements can be related to a user's demand profile. If a 4-in meter is constantly being driven at a flow rate close to its high-end performance rating, more frequent repair requirements can be anticipated.

Reference

1.   Edgar, T. *Large Water Meter Handbook.* Dillsboro, N.C.: Flow Measurement Publishing, 1995, pp. 41–42.

Glossary

This section covers a wide range of water loss control terms used in this manual. Explanations are kept brief, details on methodology and further descriptions can be found in the relevant chapters of this manual. The glossary is organized in alphabetical order and is based on the full alphabetic list of water loss management definitions provided in the AWWARF publication "Leakage Management Technologies" (Ref1).

Active Leak Detection (ALD)   ALD is a pro-active policy a water utility implements if it decides to search for hidden leaks. ALD includes in its most basic form planned regular sounding with leak detection devices or instruments.

Annual Water Balance (AWB)   The result of a component analysis of inputs, exports, and uses of water within the distribution system. Every drop of water input to the system is allocated to one use component of the water balance, in accordance with the standard annual water balance definitions.

Apparent Losses   Includes all types of inaccuracies associated with customer metering as well as data handling errors (meter reading and billing), plus unauthorized consumption (theft or illegal use). It is important to note that reducing apparent losses will not reduce physical water losses but will recover lost revenue.

Authorized Consumption   The volume of metered and/or unmetered water taken by registered customers, the water supplier, and others who are implicitly or explicitly authorized to do so by the water supplier, for residential, commercial, and industrial purposes. It also includes water exported across operational boundaries.

Authorized consumption may include items such as fire fighting and training, flushing of mains and sewers, street cleaning, watering of municipal gardens, public fountains, frost protection, building water, and the like. These may be billed or unbilled, metered or unmetered.

Awareness Duration   Awareness Duration is the average time from the occurrence of a leak until the water utility becomes aware of its existence. The awareness time is influenced by the type of applied ALD policy.

Background Losses   Background Losses are individual events (small leaks and weeps) that will continue to flow, with flow rates too low to be detected by an active leak detection campaign unless either detected by chance or until they gradually worsen to the point that they can be detected.

Backlog of Leaks   Substantial number of hidden leaks accumulated over time because of the absence of an ALD program. The removal of the backlog of leaks is in most cases perfectly justified from an economic point of view and is often the beginning of regular active leak detection and repair.

Billed Authorized Consumption Those components of authorized consumption which are billed and produce revenue (also known as revenue water). Equal to billed metered consumption plus billed unmetered consumption.

Billed Metered Consumption All metered consumption which is also billed. This includes all groups of customers such as domestic, commercial, industrial, or institutional and also includes water transferred across operational boundaries (water exported) which is metered and billed.

Billed Unmetered Consumption All billed consumption which is calculated based on estimates or norms but is not metered. This might be a very small component in fully metered systems (e.g., billing based on estimates for the period a customer meter is out of order) but can be the key consumption component in systems without universal metering. This component might also include water transferred across operational boundaries (water exported) which is unmetered but billed.

Breaks Events with flow rates greater than those of background losses and therefore detectable by standard leak detection techniques. Breaks can be visible or hidden.

Breaks and Background Estimates (BABE) Concepts The breaks and background estimates (BABE) concepts were developed by the U.K. National Leakage Initiative between 1991 and 1993. The concepts were the first to model the components of physical leakage using a systematic approach. The concept recognizes that the annual volume of real losses consists of numerous leakage events, where each individual loss volume is influenced by flow rate and duration leak run time before it's repaired. The BABE concepts permit rational planning, management, and operational control of strategies for real loss reduction.

Component Analysis of Real Loss Determination and quantification of the components of real losses in order to calculate the expected level of real losses in a distribution system. The BABE concepts were the first component analysis model.

Component Analysis of Use Determination and quantification of the components of use (such as toilet flushing, dishwasher, shower use, and the like.) by building up the total use volume of a component from individual usage events, estimates of the volume used in each event, and the number of events in the period under consideration.

Current Annual Real Losses (CARL) The volume of water lost from all kind of leaks (breaks and background losses) during the reporting period. This includes water lost from (still) hidden breaks as well as from breaks which were found and repaired during the year. It also includes possible losses at the utility's storage tanks and is equal to the component real losses of the annual water balance.

Customer Metering Inaccuracies and Data Handling Errors Apparent water losses caused by customer meter inaccuracies and data handling errors in the meter reading and billing system.

Distribution System The totality of the network infrastructure, comprising service reservoirs, mains, service lines, valves, and fittings of all types used to transport and distribute water from the utility's treatment plants or points of delivery of imported treated water to the point of delivery to the customer. The distribution system includes the treated water transmission system.

District Metered Area (DMA) Hydraulically discreet part of the distribution network, ideally with one but sometimes with two or more inflow points equipped with bulk meters. District metering involves the permanent monitoring of minimum night flows into DMAs. It is a leakage management technique targeted at reducing the awareness duration for new leaks and to help prioritization of leak detection efforts.

Economic Level of Leakage (ELL)   The economic level of leakage is found by determining the level of losses where the sum of the cost of the real loss reduction and the cost of water lost is at a minimum.

Economic Meter Change out Frequency   The economic meter change out frequency for a utility is determined through an analysis of the ELAL and is usually expressed either in terms of the number of years the meter has been in service, or the volume that has passed through the meter, or a combination of the two criteria.

Exceptional Night Use   Individual night uses by commercial, industrial, or agricultural users where the flow rates used at night are a significant proportion of the minimum night flow recorded during a minimum night flow measurement. Exceptional night users are identified in advance of the measurement through local operational knowledge so that the use by these customers during the period of the measurement can be recorded and taken into account.

Fixed and Variable Area Discharge Path (FAVAD)   Losses from fixed area leakage paths vary according to the square root of the system pressure, while discharges from variable area paths vary according to pressure to the power of 1.5. As there will be a mixture of fixed and variable area leaks in any distribution system, loss rates vary with pressure to a power that normally lies between the limits of 0.5 and 1.5. The simplest versions of the FAVAD concept, suitable for most practical predictions, are

Leakage rate L (volume/unit time) varies with pressure N1 or $L1/L0 = (P1/P0)^{N1}$

The higher the N1 value, the more sensitive existing leakage flow rates will be to changes in pressures. The FAVAD concepts have for the first time allowed accurate forecasting of the increase or decrease of real losses due to a change in pressure.

Hidden Losses   The volume of hidden losses represents the quantity of water lost by leaks that are not currently being detected and repaired.

Infrastructure Condition Factor (ICF)   The infrastructure condition factor is the ratio between the actual level of background leakage in a zone and the calculated unavoidable background leakage of a well-maintained system.

Infrastructure Leakage Index (ILI)   The ILI is a performance indicator of how well a distribution network is managed (maintained, repaired, rehabilitated) for the control of real losses, at the current operating pressure. It is the ratio of current annual volume of real losses (CARL) to unavoidable annual real losses (UARL).

$$ILI = CARL/UARL$$

Being a ratio, the ILI has no units and thus it facilitates comparisons between countries that use different measurement units (metric, U.S., or imperial).

Leak Duration   The length of time for which a break runs is apportioned, in the BABE concepts, into three separate time components—awareness, location, and repair—the duration of each of which is separately estimated and modeled; leak duration equals awareness plus location plus repair time.

Leakage Management   Leakage management can be classified into two groups:

- Reactive leak detection
- Active leak detection (ALD)

Leakage on Service Connections up to point of Customer Metering   Water lost from leaks and breaks of service connections from (and including) the tapping point until the point of customer use. In metered systems this is the customer meter, in unmetered situations this is the first point of use (stop tap/tap) within the property. Leakage on service connections might be reported breaks but will predominately be small leaks which do not surface and which run for long periods (often years).

Leakage on Transmission and/or Distribution Mains   Water lost from leaks and breaks on transmission and distribution pipelines. These might either be small leaks which are still unreported (e.g., leaking joints) or large breaks which were reported and repaired but did leak for a certain period before that.

Location Duration   For reported leaks and breaks, this is the time it takes for the water utility to investigate the report of a leak or break and to correctly pinpoint its position so that a repair can be carried out. For unreported leaks and breaks, depending on the ALD method used, the location duration may be zero since the leak or break is detected during the leak detection survey and therefore awareness and location occur simultaneously.

Losses at Utility's Storage Facilities   Losses from leaking treated water storage facilities caused by, for example, operational or technical problems. These losses include leakage through the tank structure, overflows, evaporation, and the like.

N1 Factor   The N1 factor is used in the FAVAD concepts to calculate pressure/leakage relationships.

Leakage rate L (volume/unit time) varies with pressure N1 or $L1/L0 = (P1/P0)^{N1}$

The higher the N1 value, the more sensitive existing leakage flow rates will be to changes in pressures. N1 factors range between 0.5 (corrosion holes only in metallic systems) and 1.5 with occasional values of up to 2.5. In distribution systems with a mix of pipe materials, N1 values might be in the order of 1 to 1.15. Therefore, a linear relationship can be assumed initially until N1 step tests are carried out to derive better data.

N1 Step Test   The N1 step test is used to determine the N1 value for areas of the distribution network, and thereby determine from the N1 value how real losses in the zone are split between breaks and background losses. During the test supply pressure into the area is reduced in a series of steps. The reduction in flow into the zone and the change in pressure at the average zone point are recorded. The data is then analyzed to determine the "effective area" of leakage in the zone and to compare this with the change in effective area caused by the change in pressure. From this comparison, it is possible to determine the N1 value and to determine the ratio of fixed size holes (breaks) and variable sized holes (background leakage). Caution should be exercised when interpreting the results of this test in systems containing plastic pipe materials because breaks in plastic pipes can have N1 values of 1.5 and in certain circumstances up to 2.5.

Non-Revenue Water   Those components of system input which are not billed and do not produce revenue. Equal to unbilled authorized consumption plus real and apparent water losses.

Passive Leak Detection   Same as reactive leak detection.

Pressure Management   Pressure management is one of the fundamental elements of a well-designed leakage management strategy. Pressure management is best undertaken in conjunction with district metering. Pressure management seeks to optimize system pressures to minimize losses, while maintaining adequate levels of service.

Pressure Reducing Valve (PRV) Pressure reducing valves are traditionally understood as devices to be used in case of excessively high pressures, for example, in systems with widely varying altitudes. In the case of pressure management, PRVs are to be understood as control devices used to reduce, regulate, and manage operating pressures.

Reactive Leak Detection Reactive leak detection (also known as passive leak detection) is practiced in many water utilities—whether economically justified or not. Reactive leak detection is reacting to reported breaks or pressure drops, usually reported by customers or noted by the utility's own staff while carrying out other duties. Under normal circumstances, the overall level of leakage will continue to rise under reactive leakage control.

Real Losses These are the physical water losses from the pressurized system and the utility's storage tanks, up to the point of customer use. In metered systems, this is the customer meter. In unmetered situations, this is the first point of use within the property.

The annual volume lost through all types of leaks, breaks, and overflows depends on frequencies, flow rates, and average duration of individual leaks, breaks, and overflows.

Note: Although physical losses, after the point of customer use, are excluded from the assessment of real losses, this does not necessarily mean that they are not significant or worthy of attention for demand management purpose.

Recoverable Leakage Equivalent to hidden or excess losses.

Repair Time The time it takes a water utility to organize and affect shutoff the flow from the leak once it has been pinpointed.

Reported Breaks Reported beaks are those events that are brought to the attention of the water utility by the general public or the water utility's own personnel. A break or a leak that manifests itself at the surface will normally be reported to the water utility whether or not it causes nuisance such as flooding.

Revenue Water Those components of authorized consumption which are billed and produce revenue (also known as billed authorized consumption). Equal to billed metered consumption plus billed unmetered consumption.

Service Connections A service connection is defined as "the pipe connecting the main to the measurement point or the customer curb stop, as applicable." Where several registered customers or individually occupied premises share a physical connection, for example, apartment buildings, this will still be regarded as the one connection, irrespective of the configuration and number of customers or premises. The "number of service connections" variable is required for the calculation of several performance indicators. The N variable is also used to calculate the unavoidable annual real losses (UARL) in a system, by taking into consideration the unavoidable leakage expected to occur on service connections between the main and the curb stop or property line. It is then added to the other components of UARL (on mains, and on pipes between the curbstop/property line and the customer meter) to calculate the total UARL.

System Input Volume The volume of treated water input to that part of the water supply system to which the water balance calculation relates. Equal to own sources plus water imported

- *Own Sources*: The volume of (treated) water input to a distribution system from the water supplier's own sources allowing for known errors (e.g., source meter inaccuracies). The quantity should be measured after the utility's treatment plant(s). If there are no meters installed after the treatment plant, the output has to be estimated based on raw water input and treatment losses.

- It is important to note that Water Losses at raw water transmission pipelines and losses during the treatment process are not part of the annual water balance calculations.

- *Water Imported*: The volume of bulk supplies imported across operational boundaries. Water imported can be either
  - Measured at the boundary meter (if already treated)
  - Measured at the outflow of the treatment plant (if raw water is imported and there is a separate treatment plant)
  - In either case, corrected for known errors (e.g., transfer meter inaccuracies)

- *Mix of raw water*: If raw waters imported are mixed with own source raw water in the treatment plant, there is no need for a differentiation and the total production (output) of this one or more plant(s) is used as the basis for the system input. As always, corrections have to be made for known errors. As with the "own sources," it is important to note that water losses at raw water transmission systems and losses during the treatment process are not part of the annual water balance calculations. In case the utility has no distribution input meters, or they are not used and the key meters are the raw water input meters, because these are the meters that they buy the raw water on, the system input has to be based on the raw water meters and treatment plant use/loss has to be taken into account.

Top-Down Audit   The water audit is the process of identification and validation of the volumes which go into the water balance. It is called top-down since all components of the water balance are assessed and validated starting at the top with the system input volumes down to the consumption volumes. Finally the end result of the water balance is the volume of real losses.

Unauthorized Consumption   Any unauthorized use of water. This may include illegal water withdrawal from hydrants (e.g., for construction purposes), illegal connections, bypasses to customer meters, or meter tampering.

Unavoidable Annual Real Losses (UARL)   Real losses cannot be totally eliminated. The estimated volume of unavoidable annual real losses (UARL) represents the lowest technically achievable annual real losses for a well-maintained and well-managed system. Equations for calculating UARL for individual systems were developed and tested by the IWA Water Loss Task Force, allowing for

- *Background Leakage*: small leaks with flow rates too low for sonic detection if nonvisible

- *Reported leaks and breaks*: based on frequencies, typical flow rates, target average durations

- *Unreported leaks and breaks*: based on frequencies, typical flow rates, target average durations

- *Pressure/leakage rate relationships* (a linear relationship being assumed for most large systems)

The UARL equation recommended requires data on four key system-specific factors

- Length of mains
- Number of service connections
- Location of customer meter on service connection (relative to property line, or curbstop in North America)
- Average operating pressure

Unbilled Authorized Consumption   Those components of authorized consumption which are legitimate but not billed and therefore do not produce revenue. Equal to unbilled metered consumption plus unbilled unmetered consumption.

Unbilled Metered Consumption   Metered consumption which is for any reason unbilled. This might for example include metered consumption by the utility itself or water provided to institutions free of charge, including water transferred across operational boundaries (water exported) which is metered but unbilled.

Unbilled Unmetered Consumption   Any kind of authorized consumption which is neither billed nor metered. This component typically includes items such as fire fighting, flushing of mains and sewers, street cleaning, frost protection, and the like. It is a small component which is very often substantially overestimated. Theoretically this might also include water transferred across operational boundaries (water exported) which is unmetered and unbilled—although this is an unlikely case.

Unreported Breaks   Unreported breaks are those that are found by leak detectors undertaking an active leak detection program. These breaks go undetected without some form of active leak detection.

Water Audit   Equal to top-down audit—The water audit is the process of identification and validation of the volumes which go into the water balance.

Water Balance   Equal to annual water balance—represents the results of the water audit in form of a standardized water balance.

Water Losses   The difference between system input and authorized consumption. Water losses can be considered as a total volume for the whole system, or for partial systems such as transmission or distribution systems, or individual zones. Water losses consist of real losses and apparent losses.

References

1. Fanner, V. P., R. Sturm, J. Thornton, et al. *Leakage Management Technologies*. Denver, Colo.: AwwaRF and AWWA, 2007.

Index